AF796063

BD 021217801 6

P 551.465/NIH

Hydrodynamics of Estuaries and Fjords

FURTHER TITLES IN THIS SERIES

1 J.L. MERO
THE MINERAL RESOURCES OF THE SEA
2 L.M. FOMIN
THE DYNAMIC METHOD IN OCEANOGRAPHY
3 E.J.F. WOOD
MICROBIOLOGY OF OCEANS AND ESTUARIES
4 G. NEUMANN
OCEAN CURRENTS
5 N.G. JERLOV
OPTICAL OCEANOGRAPHY
6 V. VACQUIER
GEOMAGNETISM IN MARINE GEOLOGY
7 W.J. WALLACE
THE DEVELOPMENT OF THE CHLORINITY/SALINITY CONCEPT IN OCEANOGRAPHY
8 E. LISITZIN
SEA-LEVEL CHANGES
9 R.H. PARKER
THE STUDY OF BENTHIC COMMUNITIES
10 J.C.J. NIHOUL
MODELLING OF MARINE SYSTEMS
11 O.I. MAMAYEV
TEMPERATURE—SALINITY ANALYSIS OF WORLD OCEAN WATERS
12 E.J. FERGUSON WOOD and R.E. JOHANNES
TROPICAL MARINE POLLUTION
13 E. STEEMANN NIELSEN
MARINE PHOTOSYNTHESIS
14 N.G. JERLOV
MARINE OPTICS
15 G.P. GLASBY
MARINE MANGANESE DEPOSITS
16 V.M. KAMENKOVICH
FUNDAMENTALS OF OCEAN DYNAMICS
17 R.A. GEYER
SUBMERSIBLES AND THEIR USE IN OCEANOGRAPHY AND OCEAN ENGINEERING
18 J.W. CARUTHERS
FUNDAMENTALS OF MARINE ACOUSTICS
19 J.C.J. NIHOUL
BOTTOM TURBULENCE
20 P.H. LEBLOND and L.A. MYSAK
WAVES IN THE OCEAN
21 C.C. VON DER BORCH (Editor)
SYNTHESIS OF DEEP-SEA DRILLING RESULTS IN THE INDIAN OCEAN
22 P. DEHLINGER
MARINE GRAVITY

Elsevier Oceanography Series, 23

Hydrodynamics of Estuaries and Fjords

PROCEEDINGS OF THE 9th INTERNATIONAL LIEGE COLLOQUIUM ON OCEAN HYDRODYNAMICS

Edited by

JACQUES C.J. NIHOUL

Professor of Ocean Hydrodynamics,
University of Liège,
Liège, Belgium

ELSEVIER SCIENTIFIC PUBLISHING COMPANY
Amsterdam — Oxford — New York 1978

ELSEVIER SCIENTIFIC PUBLISHING COMPANY
335 Jan van Galenstraat
P.O. Box 211, Amsterdam, The Netherlands

Distributors for the United States and Canada:

ELSEVIER NORTH-HOLLAND INC.
52, Vanderbilt Avenue
New York, N.Y. 10017

Library of Congress Cataloging in Publication Data

Liège Colloquium on Ocean Hydrodynamics, 9th, 1977.
Hydrodynamics of estuaries and fjords.

(Elsevier oceanography series ; 23)
Bibliography: p.
Includes index.
1. Estuarine oceanography--Congresses.
2. Fjords--Congresses. 3. Hydrodynamics--Congresses. I. Nihoul, Jacques C. J.
II. Title.
GC96.5.L53 1977 551.4'609 78-1405
ISBN 0-444-41682-X

ISBN 0-444-41682-x (Vol. 23)
ISBN 0-444-41623-4 (Series)

© Elsevier Scientific Publishing Company, 1978
All rights reserved. No part of this publication may be reproduced, stored in a retrieval system or transmitted in any form or by any means, electronic, mechanical, photocopying, recording or otherwise, without the prior written permission of the publisher, Elsevier Scientific Publishing Company, P.O. Box 330, Amsterdam, The Netherlands

Printed in The Netherlands

UNIVERSITY OF BRADFORD
18 DEC 1978
212178 P551·465 NIH
LOCATION J. B. Priestley Library

Foreword

The International Liège Colloquia on Ocean Hydrodynamics are organized annually. Their topics differ from one year to another and try to address, as much as possible, recent problems and incentive new subjects in physical oceanography.

Assembling a group of active and eminent scientists from different countries and often different disciplines, they provide a forum for discussion and foster a mutually beneficial exchange of information opening on to a survey of major recent discoveries, essential mechanisms, impelling question-marks and valuable suggestions for future research.

Estuaries and Fjords have been extensively studied in the past and the choice of this subject for the 1977 Colloquium may appear a little out of line with the tradition.

Estuaries and Fjords however play an essential role in man's activities. Connecting the oceans and the inland rivers, they are natural transportation channels. They provide rational sites for harbors and industrial developments and simultaneously natural nursing grounds, rich in nutrients, for marine plants and animals.

Estuaries and Fjords, on the other hand, have become increasingly vulnerable, receiving the impact of modern expansion : continuously growing population, production and use of power, manufacture of new and more diversified materials, intensification of transportation and fishing effort.

Diversions of rivers, land reclamation, excessive siltation, dredging, dumping of chemical and biological wastes, while creating severe threats on the estuarine environment , have produced continuous, often drastic, modifications calling for further, more extensive, more elaborate and more interdisciplinary research.

The perfecting of new equipment, the constitution of more exhaustive data banks, coinciding with the development of mathematical modelling techniques and the intensive use of modern computers, have provided the means of a better understanding of estuarine dynamics.

The Scientific Organizing Committeé of the Ninth International Liège Colloquium on Ocean Hydrodynamics saw the desirability of bringing together, on the important and pressing subject of Estuaries and Fjords, specialists from different fields, experimentalists and modellers, hydrodynamicists, chemists and biologists.

The present book which may be regarded as the outcome of the colloquium comprises the proceedings of the meeting and specially commissioned contributions on observations, parameterization and modelling of estuarine dynamics.

Jacques C.J. NIHOUL.

The Scientific Organizing Committee of the Ninth International Liège Colloquium on Ocean Hydrodynamics and all the participants wish to express their gratitude to the Belgian Minister of Education, the National Science Foundation of Belgium, the University of Liège and the Office of Naval Research for their most valuable support.

LIST OF PARTICIPANTS

Ir. Y. ADAM, Secrétariat d'Etat à l'Environnement, Belgium.

Dr. G. ALLEN, Centre Océanologique de Bretagne, Brest, France.

M. A. BAH, Université de Liège, Belgium.

Dr. W. BAYENS, Secrétariat d'Etat à l'Environnement, Belgium.

Ir. G. BELHOMME, Université de Liège, Belgium.

M. A. BERQUIN, Mission d'Aménagement Basse Normandie, Caen, France.

M. G. BILLEN, Vrije Universiteit Brussel, Belgium.

Dr. R. BONNEFILLE, Electricité de France, Chatou, France.

Prof. M.J. BOWMAN, State University of New York at Stony Brook, U.S.A.

Dr. G.A. CANNON, PMEL/NOAA, Seattle, U.S.A.

Prof. G. CHABERT D'HIERES, Institut de Mécanique, Grenoble, France.

Dr. P.C. CHATWIN, University of Liverpool, U.K.

Dr. P.B. CREAN, University of British Columbia, Vancouver, Canada.

M. D.K. DEMPSTER, Water Research Center, Herts, U.K.

Prof. A. DISTECHE, Université de Liège, Belgium.

Dr. J. DRONKERS, Rijkswaterstaat, Rijswijk, The Netherlands.

Dr. K.R. DYER, Institute of Oceanographic Sciences, Taunton, U.K.

Dr. A.J. ELLIOTT, NATO ASW Research Centre, La Spezia, Italy.

Dr. D.M. FARMER, Institute of Ocean Sciences, Victoria, Canada.

Prof. H.G. GADE, Geophysical Institute, Bergen, Norway.

M. Y. GALLARDO, Ant. ORSTOM Centre Océanologique de Bretagne, Brest, France.

Dr. R.F. GRAMENDE, ELSEVIER Scientific Publishing Company, Amsterdam, The Netherlands.

M. P. HECQ, Université de Liège, Belgium.

M. H.B. HELLE, University of Bergen, Norway.

Dr. D.O. HODGINS, River and Harbour Lab., Trondheim, Norway.

Dr. J.L. HYACINTHE, CNEXO, Paris, France.

Prof. R.G. INGRAM, Mc Gill University, Montreal, Canada.

Ir. B. JAMART, University of Washington, Seattle, U.S.A.

Dr. M. KARELSE, Delft Hydraulics Laboratory, The Netherlands.

Ir. A. LANGERAK, Delft Hydraulics Laboratory, The Netherlands.

Dr. G. LEBON, Université de Liège, Belgium.

Dr. J.J. LEENDERTSE, RAND Corp., Santa Monica, U.S.A.

Dr. C. LE PROVOST, Institut de Mécanique, Grenoble, France.

Dr. D. LIU, RAND Corp., Santa Monica, U.S.A.

Ir. A. LOFFET, Université de Liège, Belgium.

M. J.P. MATHISEN, River and Harbour Lab., Trondheim, Norway.

M. D. MICHEL, Université Libre de Bruxelles, Belgium.

M. L.R. MUIR, Ocean & Aquatic Sciences, Burlington, Canada.

Prof. J.C.J. NIHOUL, Université de Liège, Belgium.

Dr. J.P. O'KANE, University College, Dublin, Ireland.

Prof. F.B. PEDERSEN, Technical University of Denmark, Lyngby-Copenhagen, D.K.

Dr. P.A. PERRELS, Delft Hydraulics Lab., The Netherlands.

Dr. J.J. PETERS, Waterbouwkundig Laboratorium, Borgerhout, Belgium.

Ir. G. PICHOT, Secrétariat d'Etat à l'Environnement, Belgium.

Dr. D. PRANDLE, Institute of Oceanographic Sciences, Birkenhead, U.K.

Dr. H.G. RAMMING, University of Hamburg, Germany.

M. R.P. REICHARD, University of New Hampshire, Durham, U.S.A.

Dr. F.C. RONDAY, Université de Liège, Belgium.

M. Y. RUNFOLA, Université de Liège, Belgium.

Dr. J.C. SALOMON, Lab. d'Océanographie Physique, Brest, France.

Dr. H.M. van SCHIEVEEN, Rijkswaterstaat, Rijswijk, The Netherlands.

Prof. J.D. SMITH, University of Washington, Seattle, U.S.A.

Prof. N.P. SMITH, University of Texas, Port Aransas, U.S.A.

Dr. R. SMITH, University of Cambridge, DAMPT, U.K.

Ir. J. SMITZ, Université de Liège, Belgium.

M. J. STRONACH, University of British Columbia, Vancouver, Canada.

Dr. P.J. SULLIVAN, University of Western Ontario, Canada.

M. H. SVENDSEN, University of Bergen, Norway.

M. M.J. TUCKER, Institute of Oceanographic Sciences, Taunton, U.K.

Prof. R.E. ULANOWICZ, University of Maryland, Solomons, U.S.A.

Dr. R.J. UNCLES, Inst. for Marine Environmental Research, Plymouth, U.K.

M. J. VOOGT, Rijkswaterstaat, Rijswijk, The Netherlands.

Prof. D.F. WINTER, University of Washington, Seattle, U.S.A.

Dr. J.T.F. ZIMMERMAN, NIOZ, Texel, The Netherlands.

CONTENTS

A SYNOPTIC VIEW OF A COASTAL PLAIN ESTUARY*

ROBERT E. ULANOWICZ AND DAVID A. FLEMER‡
University of Maryland, Center for Environmental and Estuarine Studies, Chesapeake Biological Laboratory, Solomons, Maryland 20688

James T. Allison, Donald R. Heinle, Curtis D. Mobley, James P. Brown, John Klepper, Kent Mountford, Michael A. Champ, Donald W. Lear, John W. Pierce, Robert Cory, Charles E. Lewis, James L. Raper, Susan K. Smith

ABSTRACT

During October, 1972 the Patuxent River Estuary was monitored intensively and synoptically over two tidal cycles to determine the spatial and temporal patterns of various hydrodynamic, chemical and biological features. Forty-one depths at eleven stations along nine transects were sampled simultaneously at hourly intervals for salinity, temperature, dissolved oxygen, chlorohyll a, particulate nitrogen, nitrate, nitrite, total kjeldahl nitrogen, ammonia, particulate carbohydrate, dissolved organic carbon, total hydrolizable phosphorous, dissolved inorganic phosphorous, suspended sediment, particle size distribution, and zooplankton. Tidal velocity was continuously monitored at each depth by recording current meters. Riverine input and meteorological conditions were relatively stable for two weeks preceeding the deployment.

This communication describes the calculation of the intrinsic rates of change of the observed variables from their measured distributions in the Estuary. The steady-state, one-dimensional equation of species continuity is employed to separate the advection and tidal dispersion of a hydrodynamically passive substance from its intrinsic rate of change at point. A new spatial transform is introduced for the purpose of interpolation and extrapolation of data.

*Contribution No. 766 , University of Maryland, Center for Environmental and Estuarine Studies.

‡Present Address: Division of Biological Services, U.S. Fish and Wildlife Service, Washington, DC 20240.

The intrinsic rate of change profiles reveal a region of heavy bloom activity in the upper estuary and a secondary bloom near the point in the River that most of the suspended material settles out. The changes in ammonia and nitrates are highly correlated to the productivity patterns. Phosphorous rates are less closely correlated to productivity. The perturbations that the Chalk Point steam electric power plant have on the heat and oxygen balances are easily discernible.

INTRODUCTION

Practically every ecologist who has planned a field study has had to grapple with the limitations finite manpower and equipment impose upon his ability to adequately sample his system over its spatial and temporal domains. Marine and estuarine ecologists are particularly limited by the size and accessibility of their study areas from viewing the manifold physical, chemical and biological processes synoptically. While the developing technology of remote sensing is beginning to alleviate this difficulty, there is still no substitute for <u>in situ</u> sampling through the water column and over its areal extent.

In the study described below the investigators have amassed a set of data on key physical, chemical and biological variables taken simultaneously over a net of stations along the Patuxent River Estuary, a tributary estuary of the Chesapeake Bay. The objectives behind such a data acquisition are threefold:

1. To serve as a data set for the purpose of calibrating a combined physical - chemical - biological model yet to be developed.

2. To enable the authors to estimate the magnitude of the rates of various processes as they occur along the Estuary.

3. To provide a reference set of data that investigators without recourse to synoptic data collection may use to evaluate their own hypothesis about estuarine ecosystem dynamics.

An opportunity to embark upon such an ambitious task occurred in the fall of 1972 during the acquisition of prototype data for the U.S. Army Corps of Engineer's Chesapeake Bay Study. The Chesapeake Biological Laboratory and the Chesapeake Bay Institute of the Johns Hopkins University were under contract to the Corps to deploy current meters and research vessels to measure tidal velocities and salinities in the mid-portion of the Bay.

To monitor the stations prescribed by the Corps in the Bay stem and major tributaries usually required several deployments of the available manpower. The Patuxent Estuary, however, was small enough to cover in a single deployment, yet large enough to serve as a replica of most estuarine processes.

The study called for the deployment of thirty-four meters at eight stations on six "transects" along the axis of the Estuary. The current meters (Braincon #1301 Histogram Recording Current Meters) recorded tidal speed and direction automatically every ten minutes. The salinity beside each meter was to be measured with an induction salinometer lowered from a shipboard at hourly intervals for thirteen hours of three consecutive daylight periods.

With all of the vessels and men deployed for this study it appeared to the authors that, for relatively little extra effort, a host of chemical and biological variables could be measured simultaneously with the currents and salinities. The result would be a "snapshot" of the Estuary giving detailed information about a complex of phenomena for a short period of time.

As extra manpower and equipment would be needed for such a survey beyond that of the two participating organizations, assistance was solicited from neighboring research groups in the Bay. The response was overwhelming. Nine research institutions volunteered boats, equipment and manpower to the effort. With the consent of the Corps the program was expanded to cover forty depths of eleven stations on nine transects along the Estuary.

Some of the details concerning the study area, sampling location, variables measured and data reduction are given in the following sections. Later, the authors present the analysis of the process rate profiles and attempt to relate these magnitudes to mechanisms occurring at various reaches of the Estuary.

STUDY AREA

The Patuxent River is a significant tributary of the Chesapeake Bay some 160 km in length and draining some 2494 km^2, all within the State of Maryland. The River rises some 48 km west of the city of Baltimore and flows southeast and south through the Piedmont Plateau to the fall line 90 km above the mouth. While the upper 32 km of this river is protected as a source of drinking water for the Washington Metropolitan Area, approximately 200 million liters per day of treated sewage enter the next 56 km. The region from

90 km to 48 km above the mouth is tidal, fresh-brackish water and is characterized by a narrow channel meandering through broad, marshy flats. The lower 40 km of the Estuary is a drowned river valley characterized by partially-mixed, two-layer flow, except near the mouth where occasional three-layer phenomena have been reported.

The study area is confined to the lower 72 km of the Estuary ending at a point where the Western Branch sub-tributary joins with the main stream. Eleven stations were established at nine distances along the river as shown on Fig. I. The coordinates of each station are listed in Table I. along with the depth in meters at which each current meter was suspended. The vertical spacing between the sensors was nominally 3 meters.

The lower four stations were sampled only for tidal current, salinity and temperature, whereas stations P-03-01 through P-07-01 were sampled for the full set of physical, chemical and biological data as described below.

The study period was from 0600 on 17 October through 0700 on 18 October, 1972 with samples taken at hourly intervals over the two tidal cycles and one diurnal period.

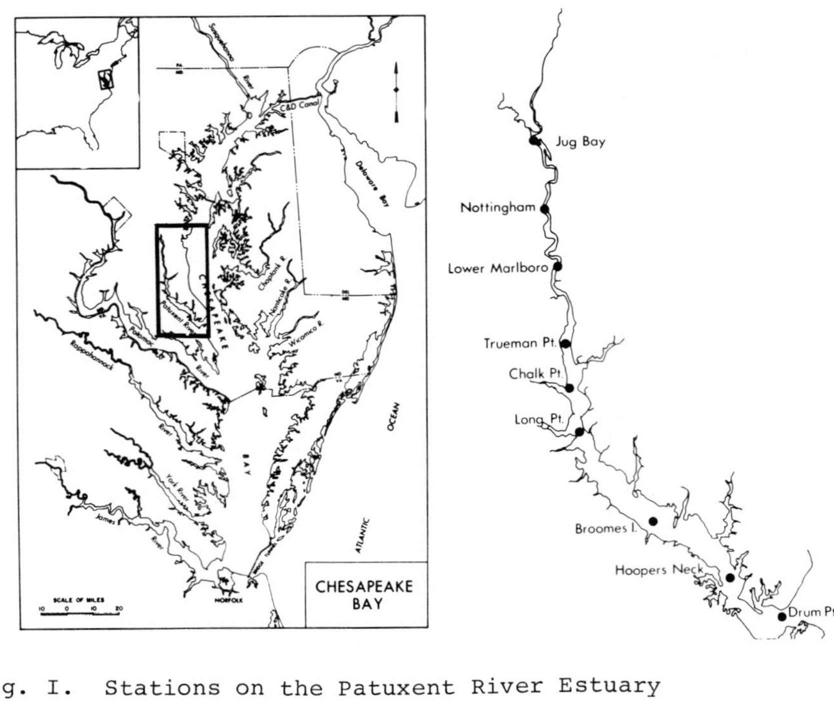

Fig. I. Stations on the Patuxent River Estuary

TABLE 1

Station location and depths

Station Designation	KM	Longitude	Latitude (N)	Depths (M)
P-01-01	0.0	76° 25' 17"	38° 18' 43"	0.6, 3.7, 6.7, 9.8, 12.2
P-01-02	0.0	76 25 17	38 18 55	0.6, 3.7, 6.7, 9.8, 12.8, 15.9
P-02-01	10.0	76 29 33	38 20 50	0.6, 3.7, 6.7
P-02-02	10.0	76 29 08	38 21 00	0.6, 3.7, 6.7, 9.8, 12.8, 15.9, 18.9, 21.5
P-03-01	22.6	76 35 07	38 24 42	1.2, 3.7, 6.7
P-04-01	33.4	76 39 55	38 29 38	1.2, 3.7, 6.7, 9.8
P-04-02	39.3	76 40 32	38 32 30	1.2, 3.7
P-05-01	43.6	76 40 44	38 34 46	1.2, 3.7
P-05-02	53.3	76 41 03	38 39 23	1.2, 3.7, 6.7
P-06-01	61.3	76 42 02	38 42 33	1.2, 3.7, 6.7
P-07-01	71.8	76 42 53	38 46 40	1.2

SAMPLING PROCEDURE

At each hour beginning on the hour the following sequence of sampling procedures was carried out at each of the seven stations for every depth at which a current meter was moored:

1. Conductivity and temperature were measured *in situ* by lowering an induction coil and thermocouple apparatus (Inter-Ocean 503A CST or Beckman RS-5 salinometer).

2. Dissolved oxygen was measured *in situ* at three stations equipped with the Inter-Ocean CST-DO units and from the remaining stations by immersing a YSI dissolved oxygen cell into freshly pumped water from the proper depth.

3. Approximately eight liters of water was pumped from the required depth and immediately processed as described below.

4. Zooplankton were filtered from 30 liters of water pumped from the prescribed depth through a 72 micron plankton net.

5. During the daylight hours Secchi disc extinction depths were read.

Aliquots of the water collected in step 3 above were immediately filtered and processed as follows:

Chlorophyll - Mg CO_3 was added to a 100-200 ml (exact amount recorded) aliquot and filtered through a GF/C glass fiber filter. The filtered material was immediately frozen for subsequent analysis in the laboratory.

Particulate Nitrogen - 100-200 ml of water was filtered on a different Millipore system and the GF/C filter and material were dehydrated for later analysis.

Particulate Carbohydrate - 100-200 ml was filtered through a GF/C filter which had been fired to remove any carbon. The residual material was frozen for later analysis.

Dissolved Nitrogen and Phosphorous - Two Whirl-Pax bags were filled with 75-100 ml of filtrate from the two preceeding filtrations and frozen for later chemical analysis.

Total Phosphorous and Organic Carbon - Unfiltered samples of 50-75 ml volume were frozen to be analyzed later.

Suspended Sediment - About 50-100 ml was filtered on a "tared" 47 mm Millipore filter to be dehydrated and weighed in the laboratory.

ANCILLARY COOPERATIVE STUDIES

In addition to the baseline measurements outlined above the schedules of three other Patuxent studies being conducted by cooperating institutions were altered to be cotemporaneous with the Patuxent synoptic survey.

Heinle and Flemer (1976) were directing monthly observations of mass transfer between a section of marsh and the Patuxent River channel. The subject marsh was within the synoptic survey area and the sampling protocol was very similar to that described above. Therefore, the 24-hour marsh study took place simultaneous with the synoptic survey.

The Philadelphia Academy of Natural Sciences, likewise, was conducting monthly cruises to measure the gross and net photosynthesis along the River by the oxygen-bottle method. Relative numbers of phytoplankton and bacterial taxa were also determined alongside the various stations of the synoptic survey while the study was underway (Mountford et. al. 1972).

The National Aeronautics and Space Administration facilities at Wallops Island and Langley, Virginia realized an opportunity to acquire ground-truth data from the synoptic operations and arranged to fly two C-147 and one C-130 missions to take black and white, color IR photographs and multi-spectral scans of the Patuxent during the daylight hours of the deployment (Ohlhorst, 1976).

The U.S. Coast and Geodetic Survey was also maintaining four automatic recording (six-minute interval) tidal height gauges along the Patuxent as part of the Corps' Chesapeake Bay Study.

CHEMICAL ANALYSIS OF SAMPLES

Immediately upon the termination of the deployment the samples were sorted and sent to the laboratories of five of the cooperating institutions. The Chesapeake Biological Laboratory performed the analysis for chlorophyll a, particulate carbohydrate and particulate nitrogen; the Department of Biology of the American University analyzed the samples for particulate and dissolved carbon; the Maryland State Water Resources Administration determined the values of total and dissolved phosphorous; and the Annapolis Field Office of the U.S. Environmental Protection Agency effected the measurement of ammonia, kjeldahl nitrogen, nitrate and nitrite. The sedimentology division of the Smithsonian Institution's Museum of Natural History weighed the sediment samples.

Active chlorophyll a was determined fluorometrically with a Turner Model 111 fluorometer (Yentsch and Menzel, 1963 and Holm-Hansen et. al., 1965). A specific absorption coefficient of 12.8 was used in the primary spectrophotometric calibration.

The Dumas method of high temperature oxidation was used to determine particulate nitrogen. Analysis were carried out on a Coleman Model 29A Nitrogen Analyzer equipped with a Model 29 combustion tube and syringe.

Particulate carbohydrate was determined by the anthrone reaction as described in Strickland and Parsons (1972).

Particulate and dissolved fractions of organic carbon were measured according to the methods described by Menzel and Vacarro (1964).

The remaining fractions of phosphorous and nitrogen were measured on Technicon Auto Analyzers according to Methods for Chemical Analysis of Water and Wastes published by the U.S. Environmental Protection Agency (1974).

The single reagent ascorbic acid reduction method (pp. 249-255) was used to obtain dissolved orthophosphorous, while the total hydrolyzable phosphorous values were the results of the colorometric ascorbic acid reduction method (pp. 256-263).

Total kjeldahl nitrogen values resulted from the automated phenate method (pp. 182-186); ammonia from the automated colorometric phenate method (pp. 168-172); and both nitrite and nitrate from the automated cadmium reduction method (pp. 207-212).

In summary, the tidal speed and direction were recorded at each depth at ten-minute intervals. Other variables measured each hour at the forty depths include salinity, temperature, dissolved oxygen, suspended sediment, chlorophyll a, particulate nitrogen, particulate carbohydrate, nitrate, nitrite, ammonia, total kjeldahl nitrogen, total hydrolyzable phosphorous, dissolved orthophosphorous, particulate organic carbon, dissolved organic carbon and zooplankton density.

Other variables observed on an opportunistic basis include gross and net photosynthesis, phytoplankton taxa and relative numbers, insolation, coliform counts, river flow and tidal height. Meteorological data from the Patuxent River Naval Air Station near the mouth of the Estuary are probably available, but have not been assembled to date.

All processed data is available to the public through the National Oceanographic Data Center*.

ESTIMATION OF PROCESS RATES

The primary objective of the Patuxent Synoptic Study as cited in the introduction is to enable the development of a combined physical - chemical - biological model of a coastal plain estuary. Ideally, if one is to set about modeling a system of such complexity, it is useful to develop a preliminary model based on fragmentary empirical data and other *a priori* estimates. Such an initial model is often a substantial aid in prescribing a data acquisition scheme.

Unfortunately, the opportunistic and *ad hoc* nature of this study did not allow for such preliminaries, and the authors must begin the modeling process after the data collection. The model structure (especially the chemical and biological sub-models) will thus be guided by the results of the initial data manipulations. The entire modeling procedure will then take on much of the nature of a *posterior* modeling as described elsewhere (Ulanowicz et. al., 1975 and 1978).

Under this approach the structure of the reaction kinetics results from comparing the rates at which species (inorganic, organic, and living) are appearing and disappearing with the amounts present. The data acquisition scheme described above will result in information on the stocks of the species. The rates at which they are intrinsically changing, however, is confounded by the association advection and dispersion in the Estuary. The remainder of this presentation will be devoted to the estimation of the process rates and the qualitative behavior evinced by the results.

*NOAA Master Reel #9008, Environmental Data Services, National Oceanographic Data Center, Washington, D.C. 20235, U.S.A.

The separation of the intrinsic rates from the hydrological transport requires a statement of mass balance. Since data was acquired from a string of single stations along the estuary, it is natural to begin with a one-dimensional mass balance, i.e., all variables are averaged over a cross-section of the estuary.

Since concentrations and velocities are available at frequent intervals, it is possible to state the equation of species conservation at various times during the tidal cycle. To do so, however, would yield results with little statistical significance. Therefore, a one-dimensional, tidally-averaged equation of species continuity is chosen to begin with:

$$A \frac{\partial C}{\partial t} = \frac{\partial}{\partial x} (CQ) - \frac{\partial}{\partial x} (KA \frac{\partial C}{\partial x}) + R \qquad (1)$$

where C is the tidally-averaged concentration
Q is the cumulative freshwater input up to point x
A is the local cross-sectional area
K is the longitudinal dispersion coefficient
R is the rate of appearance or disappearance of C
x is the distance upstream
t is the time

Now the middle of October, 1972 was a propitious time to perform the synoptic study, since the U.S. Geological Survey records indicate that riverine input to the lower estuary was virtually constant for the two weeks preceeding the observations. Hence, the River was, most likely, as close to tidally-averaged steady-state conditions as one could hope to achieve. During the measurement period a meteorological high pressure front did pass through the area causing a net loss of water from the Estuary, but the effect of this short-term phenomenon upon the steady-state gradients was probably small. Henceforth, the Estuary will be assumed at steady-state, and equation (1) can thus be solved for the "reactions term" as:

$$R = K \frac{d}{dx} (A \frac{dC}{dx}) + A \frac{dC}{dx} \frac{dK}{dx} - C \frac{dQ}{dx} - Q \frac{dC}{dx} \qquad (2)$$

Each term on the right-hand-side of equation (2) can be reasonably estimated - the concentration profiles are known from the measurements, the freshwater input profile can be evaluated with minor assumptions from USGS data, the areas are available from bathymetric charts, and the dispersion coefficient profile can be calculated from the observed salinities. There are, however, a number of numerical details associated with these estimates which should be discussed.

To begin with the values for the concentrations at a station are averaged over the station depths. In this averaging each point reading is weighted according to the fraction of the cross-sectional area associated with the particular depth. The resultant station values are subsequently averaged over the two tidal cycles (and one diurnal period) of the study. Each variable then has one "steady-state" value associated with each station at which measurements were taken. The calculated steady-state values are listed in Table II.

The longitudinal distances between stations are greater than is desirable, with distances of over 10 km separating the biological stations. Furthermore, the lower 22 kilometers were not covered by the chemical and biological sampling program. A rational method of interpolation and extrapolation of the variables and their derivatives is therefore, in order. Reasoning heuristically that longitudinal mixing becomes greater (in the absolute sense) as the Estuary cross-section increases, it would follow that longitudinal gradients are dampened as the Estuary widens. The cross-sectional area thus becomes a weighting factor for the existing gradients, and it is convenient to define a new independent variable, λ, characterizing longitudinal distance as:

$$d\lambda = \frac{dx}{A(x)} \tag{3}$$

or equivalently:

$$\lambda = \int_0^x \frac{dx}{A(x)} \tag{3a}$$

This transformation of the independent variable has the advantage that the transformed descriptions of advection and dispersion become independent of estuary geometry, i.e., equation (2) becomes:

$$AR = K\frac{d^2C}{d\lambda^2} + \frac{dC}{d\lambda}\frac{dK}{d\lambda} - C\frac{dQ}{d\lambda} - Q\frac{dC}{d\lambda} \tag{4}$$

Straightforward linear extrapolation of C (λ) and its derivatives into the downstream range gave more plausible results than similar efforts using several different non-linear regression schemes on C (x). The areas used in this transform are graphed in Figure 2.

Encouraged by the utility of this transform, the author proceeded to estimate C(λ) and its two derivatives by the simplest means possible. Concentrations at any longitudinal distance were

TABLE II

Averaged concentrations

Species (Units)	22.6 (KM)	33.4	39.3	43.6	53.3	61.3	71.8
Salinity (MG/L)	11810	10430	7640	6340	1410	300	203
Chlorophyll -A (G/L)	10.25	7.464	16.85	22.46	40.20	68.50	8.715
Ammonia	0.081	0.130	0.100	0.153	0.114	0.075	0.759
Nitrate-Nitrite (MGA/L)	0.166	0.174	0.101	0.118	0.224	0.717	2.150
Kjeldahl Nitrogen (MGA/L)	0.567	0.502	0.629	0.507	0.416	0.609	1.387
Dissolved Ortho Phosphate (MG/L)	0.026	0.052	0.055	0.019	0.058	0.071	0.767
Total Phosphorous (MG/L)	0.052	0.134	0.114	0.100	0.070	0.262	1.187
Dissolved Organic Carbon (MG/L)	3.192	3.483	3.196	3.190	4.278	4.962	3.562
Particulate Carbon (MG/L)	2.210	2.088	1.881	2.438	3.527	4.015	2.290
Suspended Material (MG/L)	18.651	-----	35.507	68.486	52.380	44.034	30.00
Dissolved Oxygen (MG/L)	9.42	9.34	8.98	8.62	10.80	12.11	7.13
Heat Content (KCAL/LITER)	16.83	16.38	15.69	16.02	15.12	14.62	13.89

approximated by linear interpolation of the two nearest stations. The derivatives at the mid-point between two stations were estimated by the difference quotient of the concentration change and the interval of λ. Derivatives at other points were acquired by linear inter polation and extrapolation. Second derivatives were calculated by a repeat application of the derivative scheme.

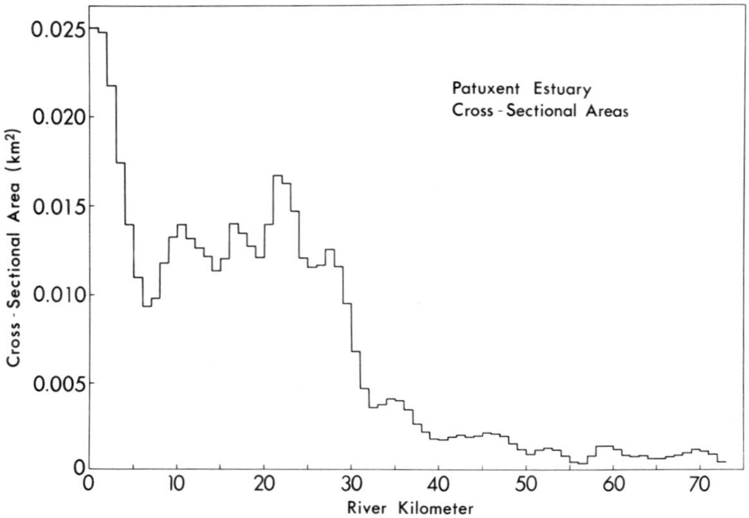

Fig. 2. Patuxent estuary cross-sectional areas.

Over 40% of the area of the Patuxent watershed lies adjacent to the study area, making it impractical to consider that all of the freshwater input occurs at the head of the model. By pro-rating the input according to area, one estimates that Q increases from about 4.75 M^3/sec at the head of the Estuary to around 8.10 M^3/sec at the mouth. Now observation of the watershed reveals that most of the feeder streams run perpendicular to the longitudinal axis of the Estuary and their mouths are evenly dispersed along both banks. An appropriate assumption, therefore, is that the rate of accumulation of freshwater input, dQ/dx, is nearly continuous along the main River axis and proportional to the width of the watershed at that point. Figure 3a shows the schematic representation of the watershed adjacent to the study area. Figure 3b below it illustrates the cumulative riverine input at any point in the Estuary. Under the above assumptions Q varies almost linearly along the region of interest.

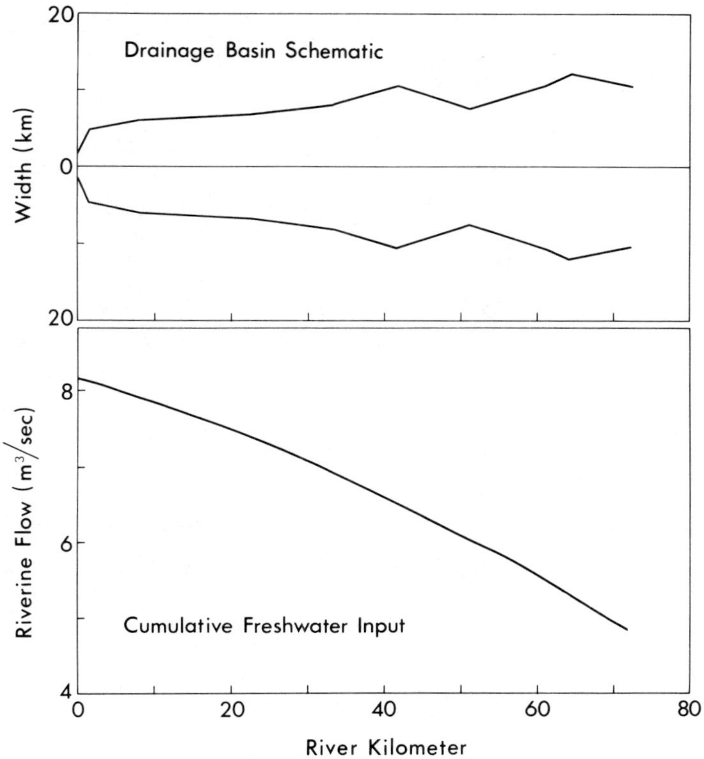

Fig. 3. (a) Drainage basin schematic showing width as a function of distance upstream (b) Cumulative freshwater input as calculated along the estuary

The only remaining terms from equation (4) to be estimated are K and $dK/d\lambda$. They may be calculated from the observed salinity profile. Salt, being a conservative substance, should have a zero intrinsic rate of change. There is a source term for salt, however, which arises from the input of residual salinity (C_r) associated with the freshwater input. Equation (4) is then written as:

$$\frac{d}{d\lambda}\left(K\frac{dC}{d\lambda} - CQ\right) = -C_r\frac{dQ}{d\lambda} \tag{5}$$

or $$\frac{d}{d\lambda}\left[K\frac{dC}{d\lambda} - (C-C_r)Q\right] = 0 \tag{5a}$$

Assuming that advection balances dispersion at steady-state allows one to estimate K as:

$$K = (C - C_r)Q\left(\frac{dC}{d\lambda}\right)^{-1} \qquad (6)$$

and subsequently calculate:

$$\frac{dK}{d\lambda} = \left[\frac{d}{d\lambda}(CQ) - C_r\frac{dQ}{d\lambda} - K\frac{d^2C}{d\lambda^2}\right]\left(\frac{dC}{d\lambda}\right)^{-1} \qquad (7)$$

To avoid the possibility of a negative value of K resulting from noise in the derivatives of the salinity, the salinity was approximated by the implicit function:

$$(S - .17)^{1.04776}\ [\lambda + 814.04(S-.17) - 10794] = 3255.7$$

where S is the salt content in parts per thousand and λ is measured in reciprocal kilometers. Figure 4 shows the measured and approximated salinities.

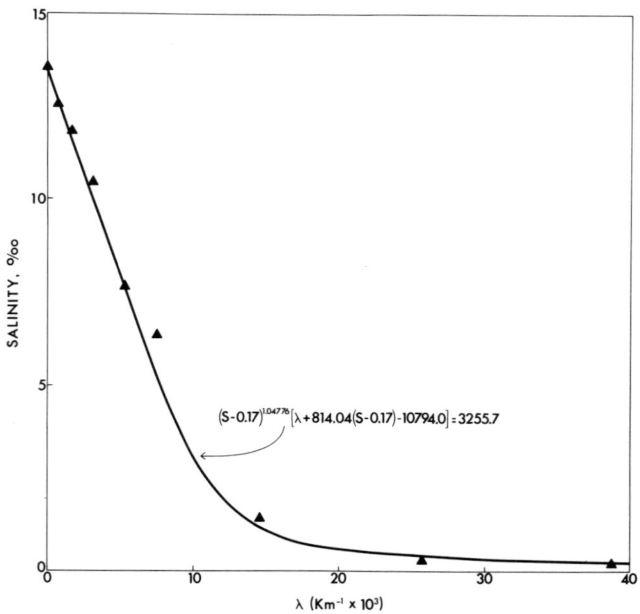

Fig. 4. Salinity as a function of reduced longitudinal coordinate (river month = 0.0).

The longitudinal variation of the dispersion coefficient is depicted in Figure 5. Qualitatively, the variation is similar to that obtained from the Escaut Estuary by Wollast (1973) and discussed by Ronday (1975). The dispersion coefficient declines upstream to a minimum near the point at which the Estuary narrows and rises thereafter to values higher than those found in the lower estuary.

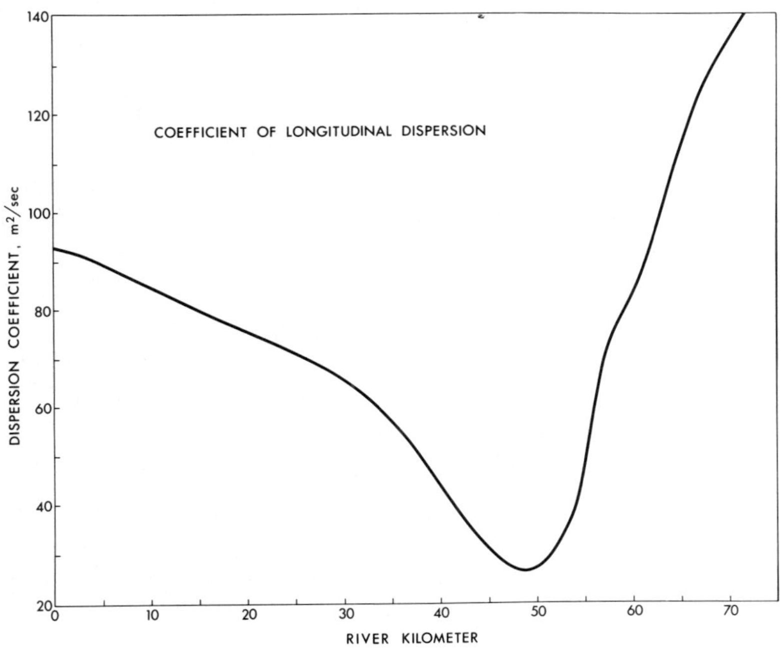

Fig. 5. Calculated coefficient of longitudinal dispersion.

Now Q, K and their derivatives have been estimated independently of the outlined interpolation scheme. A useful test of how compatible the interpolation estimates are with the assumptions used on Q and K would be to calculate the reaction rates of the salt as if it were a reactive substance. Performing such a balance yields a total gain of 0.41 metric tons of salt per day for the entire Estuary. This is an inconsequential fraction of the 5.7 million metric tons of salt present in the Estuary.

DISCUSSION OF THE INTRINSIC RATES OF CHANGE

The calculated profiles for the rates of change are depicted in Figure 6. A positive value for the rate of change indicates a source of the given material and a negative value denotes a sink. The reader will notice that the term "reaction" has been avoided where possible so as not to infer a priori the mechanism contributing to a given source or sink. Other mechanisms besides chemical or biological reactions which might contribute to the intrinsic rates of change include inputs associated with freshwater inflows and adsorption onto sedimenting material.

Chlorophyll a is often used as an indicator of primary productivity of an aquatic ecosystem. The appearance of chlorophyll a is then, indicative of an algal "bloom". A very significant bloom is observed in the upper estuary (60-72 km), and a secondary bloom is observed along the range from 39 to 45 km (Figure 6a). A sewage treatment plant introduces nutrients into the Western Branch which enters the mainstream of the River about two kilometers above the study area. It is reasonable to assume that the observed bloom is in response to this nutrient addition. The secondary bloom is coincident with the initial disappearance of suspended material and is possibly the result of light no longer being limiting to productivity. Chlorophyll a is lost in the remainder of the Estuary presumably due to herbivorous uptake. On balance the Estuary as a whole is a source of approximately 0.04 metric tons of chlorophyll per day.

It is of primary interest to follow the behavior of the nutrient species to see how they relate to the observed patterns of phytoplankton growth and death.

The most striking correlation to the productivity is exhibited by ammonia, Figure 6b. Its rate of change is practically inversely proportional to that of chlorophyll a. With the exception of a small reach of the Estuary (53-57 km), the appearance of one microgram of chlorophyll a is accompanied by the disappearance of approximately ten microgram atoms of ammonia (and vice-versa).

Nitrate and nitrite also exhibit close correlation to the primary production patterns (Figure 6c). The loss of these species is slightly heavier than that of ammonia in the upper estuary (>45 km) and the inverse correlation with primary production breaks down more drastically in the stretch from 47-57 km. In the lower estuary nitrates appear on almost a mole-for-mole basis with ammonia.

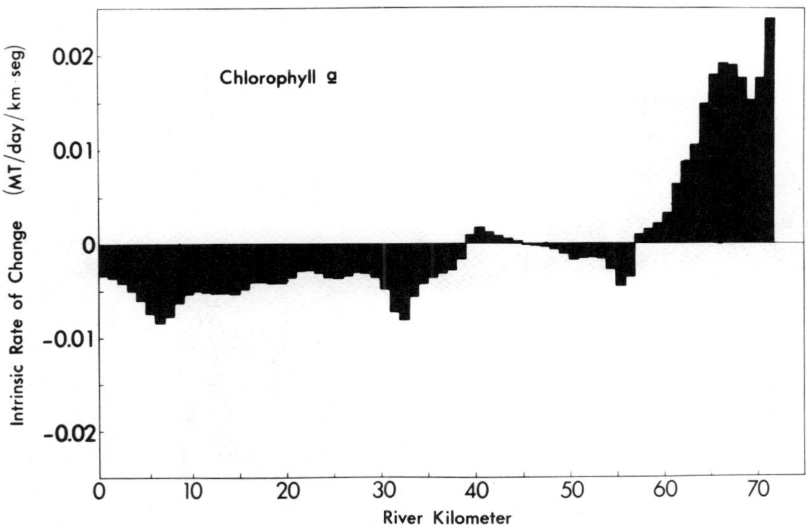

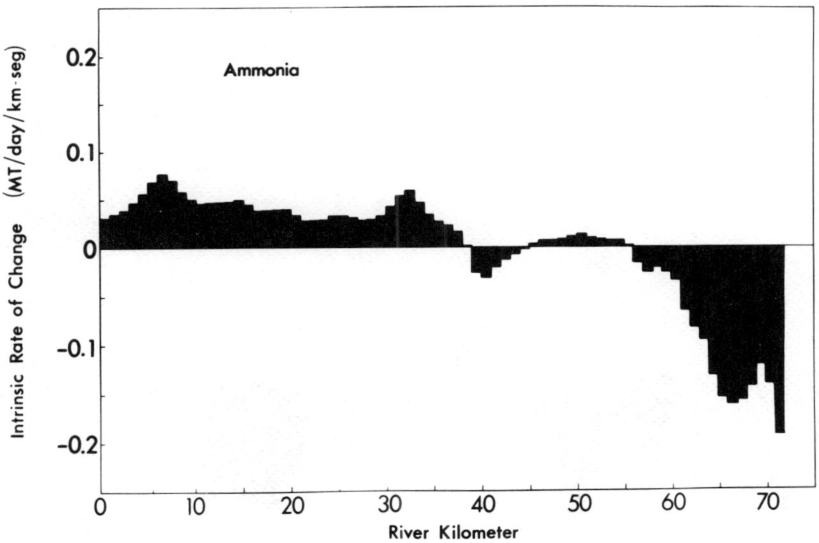

Fig. 6. Daily rates of appearance (+) or disappearance (-) of various substances for kilometer sequents of the Patuxent River Estuary.

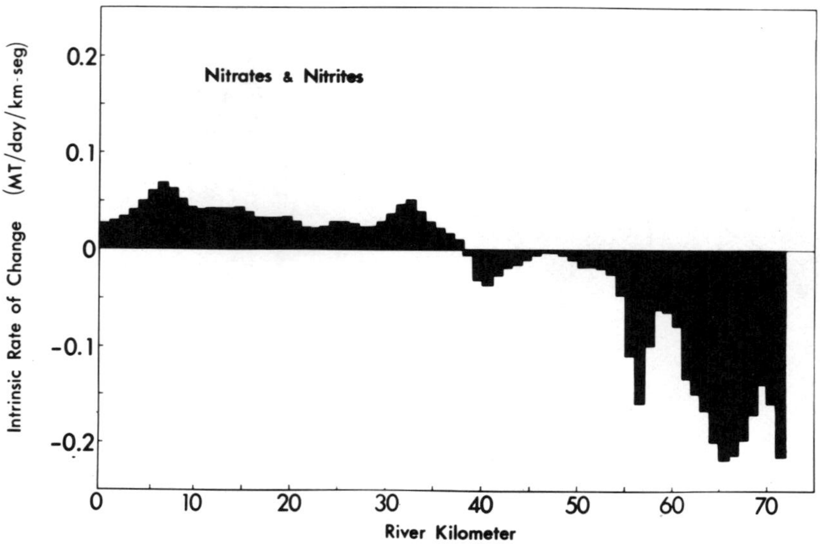
Nitrates & Nitrites
Intrinsic Rate of Change (MT/day/km·seg)
0.2
0.1
0
-0.1
-0.2
0
10
20
30
40
50
60
70
River Kilometer

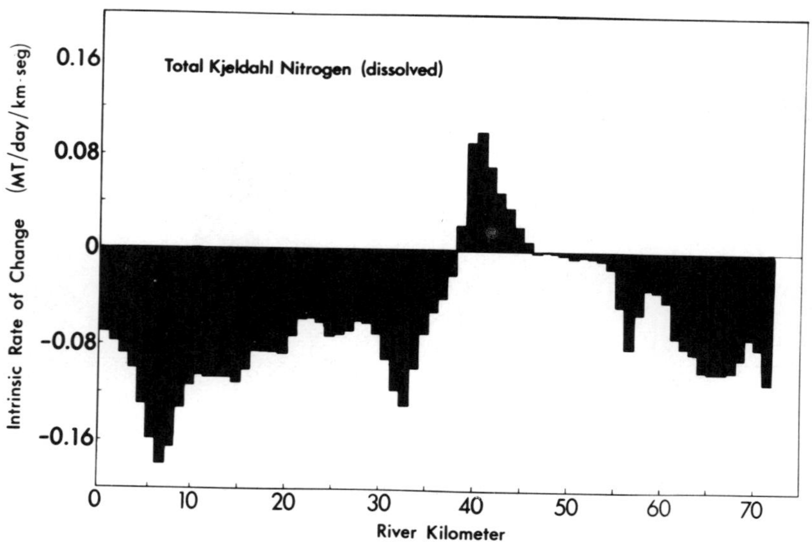
Total Kjeldahl Nitrogen (dissolved)
Intrinsic Rate of Change (MT/day/km·seg)
0.16
0.08
0
-0.08
-0.16
0
10
20
30
40
50
60
70
River Kilometer

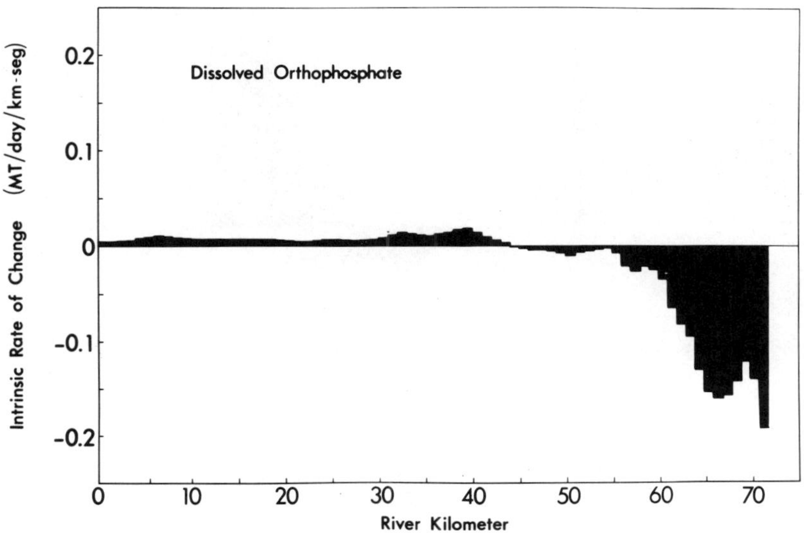
Dissolved Orthophosphate
Intrinsic Rate of Change (MT/day/km-seg)
0.2
0.1
0
-0.1
-0.2
0
10
20
30
40
50
60
70
River Kilometer

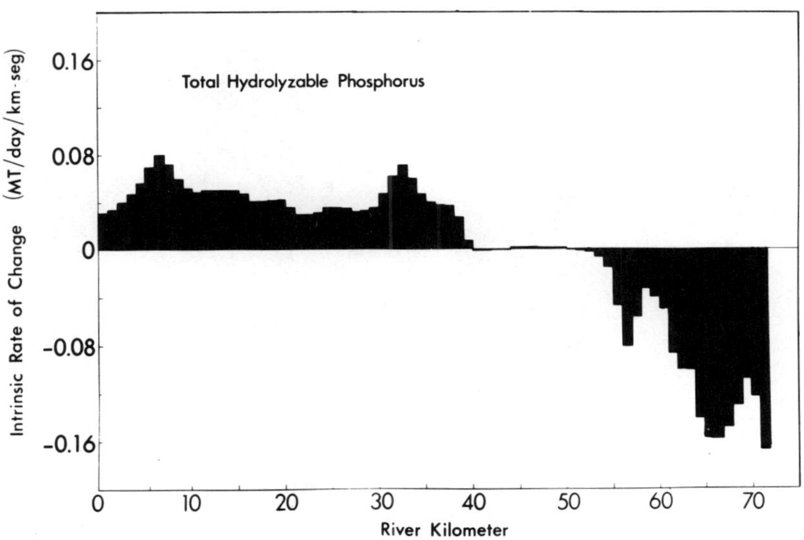
Total Hydrolyzable Phosphorus
Intrinsic Rate of Change (MT/day/km-seg)
0.16
0.08
0
-0.08
-0.16
0
10
20
30
40
50
60
70
River Kilometer

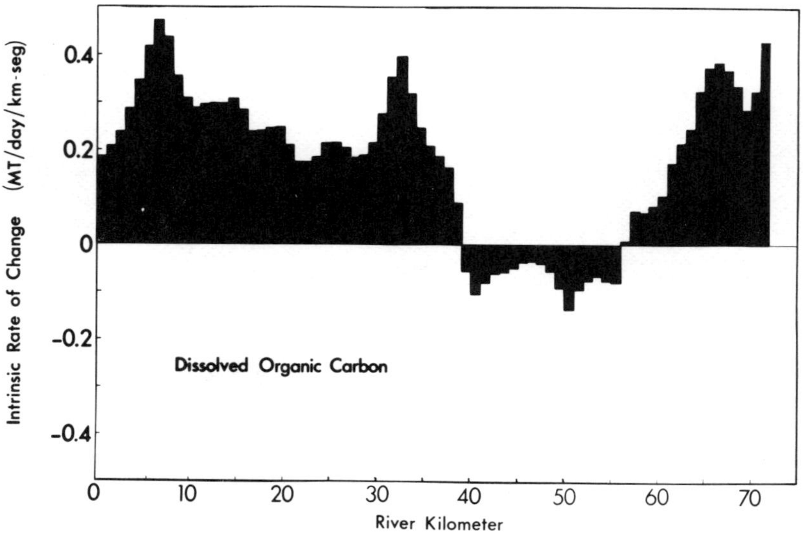
0.4
0.2
0
-0.2
-0.4
Intrinsic Rate of Change (MT/day/km-seg)
Dissolved Organic Carbon
0
10
20
30
40
50
60
70
River Kilometer

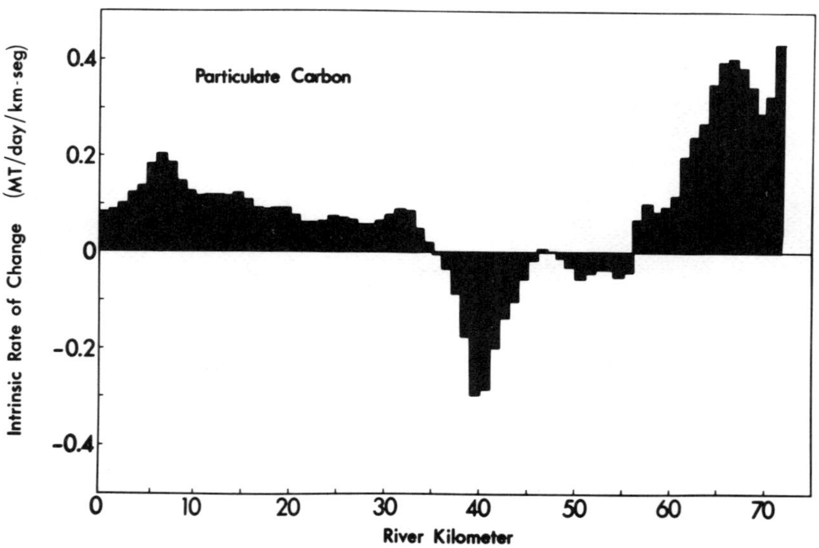
0.4
0.2
0
-0.2
-0.4
Intrinsic Rate of Change (MT/day/km-seg)
Particulate Carbon
0
10
20
30
40
50
60
70
River Kilometer

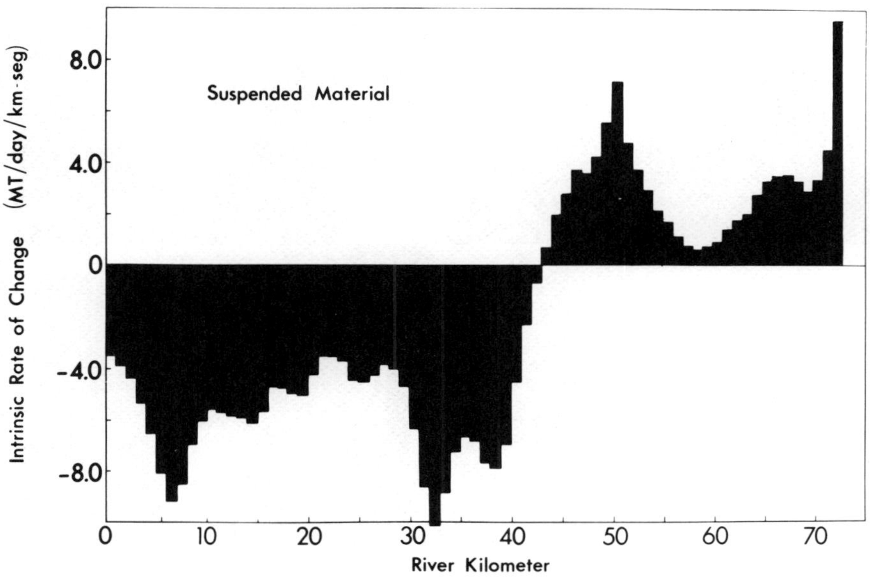

Suspended Material
8.0
4.0
0
-4.0
-8.0
Intrinsic Rate of Change (MT/day/km-seg)
0
10
20
30
40
50
60
70
River Kilometer

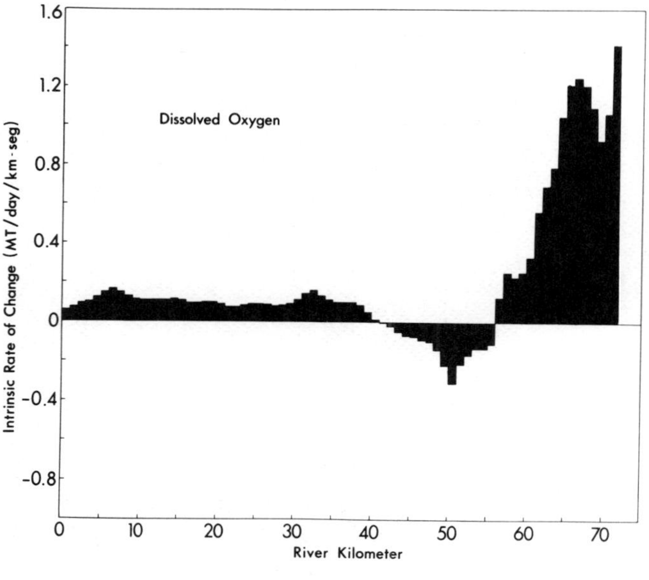

Dissolved Oxygen
1.6
1.2
0.8
0.4
0
-0.4
-0.8
Intrinsic Rate of Change (MT/day/km-seg)
0
10
20
30
40
50
60
70
River Kilometer

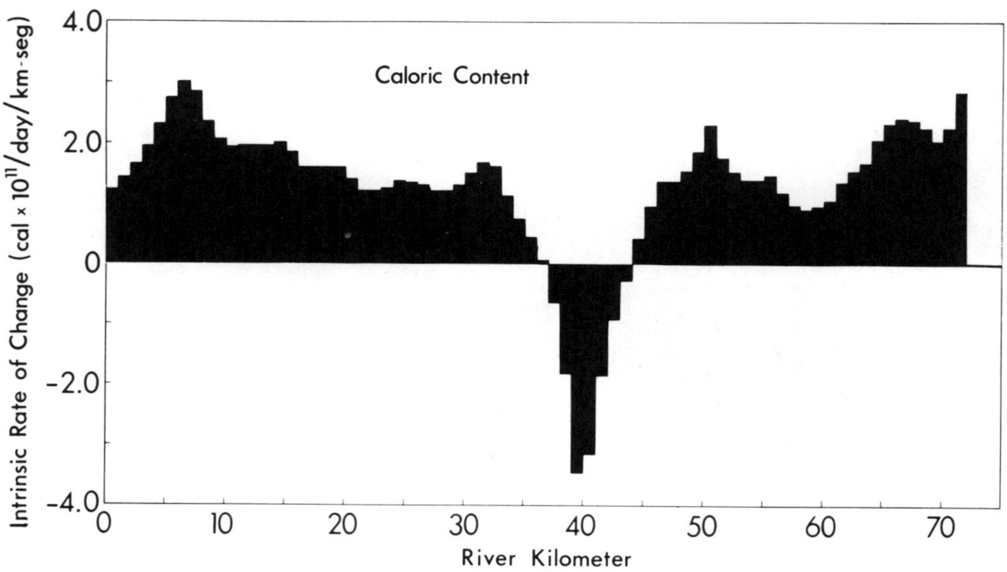
Caloric Content
Intrinsic Rate of Change (cal × 10¹¹/day/km-seg)
4.0
2.0
0
-2.0
-4.0
0
10
20
30
40
50
60
70
River Kilometer

Total (kjeldahl) nitrogen is lost throughout the entire length of the Estuary (see Figure 6d) with the exception of the reach from 38-45 km. The gain in total nitrogen coincides with the secondary bloom of phytoplankton just below the sediment trap.

There is loss of all species of nitrogen from the Estuary as a whole. Cumulative loss of total nitrogen amounts to about 4.7 metric ton atoms per day with 1.9 metric ton atoms of nitrate-nitrite and about 0.5 metric ton atoms of ammonia disappearing from the study area each day.

Phosphorous appears to be less correlated to productivity than was the case with the nitrogen species. Dissolved orthophosphate (see Figure 6e) was lost from the upper Estuary (>45 km) with heavy disappearance above 60 km. The lower Estuary hosted a small gain in the same species. Apparently, the dissolved phosphorous does not remain long in the water column after its addition from the Western Branch.

Total phosphorous (Figure 6f) behaves similarly, except that there is significantly more phosphorous gained in the lower Estuary (presumably in the particulate form). Total phosphorous is almost conserved over the whole range with a loss of only 0.4 metric ton atoms occurring per day. Dissolved phosphorous, in apposition, is lost at the rate of 1.7 metric ton atoms per day.

There are several hypotheses which might explain the observed patterns of phosphorous behavior. The phosphorous lost in the upper Estuary is likely due to adsorption to the suspended sediments. There does not appear to be any uptake of dissolved phosphorous in the region of the secondary bloom. The source of phosphorous in the lower Estuary is in question. It could originate in the main stem of the Bay, or it could conceivably be regenerated from the sediments.

Dissolved and particulate carbon (see Figures 6g, h) follow similar patterns. Both are accreted in the upper regions (>55 km) and the lower Estuary (<35 km), but the forms are lost in the transitional region. The bloom and detrital contributions from the marsh are likely sources of carbon in the upper reach. Metabolic products could possibly explain the source of carbon in the lower Estuary. The disappearance of chlorophyll a in the lower Estuary does not imply the absence of carbon fixation in these regions. It simply states that losses (e.g. consumption by grazers) exceeds production by growth. The productivity of the lower Estuary is revealed by the carbon figures. Over 20 metric ton atoms of carbon are produced each day by the study area with 13.5 ton atoms appearing in the dissolved phase and 6.7 ton atoms in the particulate form.

The calculations reveal (Figure 6i) that 150 metric tons of suspended material are lost to the system each day with most of that figure probably going to the sediments. The upper region where suspended material is accreted is well demarcated from the lower region (<43 km) where sedimentation dominates.

The bloom appears to be a net producer of oxygen, even when averaged over the diurnal period. The marked contribution of oxygen from the bloom dominates Figure 6j. Immediately downstream of the bloom area is a region (41-56 km) of oxygen depletion. The region of depletion is probably a combination of the Streeter - Phelps oxygen sag and the lowered oxygen solubility associated with the warmer water in the vicinity of the power plant out fall (km 46). The lower Estuary exhibits a gradual recovery from the sag.

The assumption of steady-state is probably weakest when made upon the thermal structure of the Estuary. During the Autumn the upper River is dominated by the cooler runoff water, whereas the lower Estuary is kept warmer by exchange with the heat reservoir of the Bay. Traveling with a parcel of river water, one would observe a warming trend as one proceeded downriver. Figure 6k shows that this is indeed what is observed throughout most of the Estuary, with the notable exception of a region downstream of the power plant effluent (37-43 km) where the excess heat from the power plant is being dissipated.

FUTURE WORK

The logical next step in the modeling process is the correlation of the calculated rates of change with the observed concentrations (or prescribed functions thereof) to culminate in a kinetic scheme for the chemical and biological species. This would result in a complete steady-state, tidally averaged model of the Patuxent kinetics. Using the kinetic scheme thus derived it would be interesting to release the assumption of steady-state and see how the derived model would respond to changes in riverine flow and nutrient input. With any degree of success in this endeavor, the next progression would be to a one-dimensional dynamic model which could be calibrated to the hourly data. The parameters derived from the tidally-averaged model should provide a convenient starting point for the calibration of the dynamic model.

No matter how complete the Patuxent model may become, its applicability will always be limited to the fall season unless it can be calibrated against data from other seasons. Hence, there is a desire on the part of the authors to repeat such synoptic surveys during other seasons of the year.

ACKNOWLEDGEMENTS

Besides those contributors listed on the title page, the authors would like to thank a host data collection assistants at the following institutions:

American University (Dept. of Biology)
Johns Hopkins University (Chesapeake Bay Institute)
The Chesapeake Biological Laboratory
Environmental Protection Agency (Region III, Annapolis Field Office)
National Aeronautics and Space Administration (Langley and Wallops, Virginia)
Philadelphia Academy of Sciences (Benedict Field Station)
Smithsonian Institute (Museum of Natural History and Chesapeake Bay Center)
U.S. Geological Survey
Maryland Water Resources Administration

The U.S. Army Corps of Engineers, Baltimore District, was most cooperative in allowing us to coordinate our activities with those prescribed under contract DACW31-70-C-0077.

Finally, Professor Nihoul and the University of Liege were most generous in assisting one of the authors (Ulanowicz) with travel expenses incurred in attending the Ninth Colloquium on Ocean Hydrodynamics.

REFERENCES

Environmental Protection Agency. 1974. Methods for Chemical Analysis of Water and Wastes. EPA-625-16-74-003 Office of Technology Transfer, Washington, D.C.

Heinle, D. R. and D. A. Flemer. 1976. Flows of Materials between Poorly Flooded Tidal Marshes and an Estuary. Mar. Biol. 35:359-373.

Holm-Hansen, O., C. J. Lorenzen, R. W. Holmes, and J. D. H. Strickland. 1965. Fluorometric determination of chlorophyll. J. Cons. Perm. Int. Explor. Mer. 30:3-15.

Menzel, D. W. and R. F. Vacarro. 1964. The measurement of dissolved organic and particulate carbon in seawater. Limnol. Oceanogr. 9:138-142.

Mountford, K., R. S. Mullen, and R. S. Shippen. 1972. Patuxent River Primary Productivity Profiles and Surface Phytoplankton Data. Benedict Estuarine Laboratory, Benedict, Maryland. Unpublished Manuscript.

Ohlhorst, C. W. 1976. Analysis of Six Broad Band Optical Filters for Measurement of Chlorophyll-a and Suspended Solids. NASATMX-339 National Aeronautics and Space Administration, Langley, Virginia.

Ronday, F. C. 1975. Etude de l'envasement et de la variation longitudinale du coefficient de dispersion dans des estuaires partiellement stratifiés. Annales des Travaux Publics de Belgique, No. 4.

Strickland, J. D. H. and T. R. Parsons. 1972. A Practical Handbook of Seawater Analysis. Fish. Res. Bd. Can., Ottawa.

Ulanowicz, R. E., D. A. Flemer, D. R. Heinle, and C. D. Mobley. 1975. The A-Posteriori Aspects of Estuarine Modeling. pp. 602-616 in Vol. 1, L. E. Cronin (ed.), Estuarine Research, Academic Press, New York.

Ulanowicz, R. E., D. A. Flemer, D. R. Heinle and R. T. Huff. 1978. The Empirical Modeling of an Ecosystem. Ecol. Modelling, 4 (in press).

Wollast, R. 1973. Origine et mécanismes de l'envasement de l'estuaire de l'Escaut. Service d'Océanographie chimique. Insitut de Chimie Industrielle. Université Libre de Bruxelles et Laboratorie de Recherches Hydrauliques - Bogerhout. Ministere des Travaux Publics de Belgique.

HYDRODYNAMICS OF THE SCHELDT ESTUARY

Jacques C.J. NIHOUL and François C. RONDAY
Université de Liège, Belgium

Jean J. PETERS and André STERLING
Ministère des Travaux Publics, Belgium

I. HYDRODYNAMIC CHARACTERISTICS OF THE SCHELDT ESTUARY

I. 1. Introduction

The Scheldt Estuary is the Southern branch of the Rhine - Meuse - Scheldt delta. The natural evolution of the delta has been, to a large extent, influenced by man's activities : embanking, closing of arms and creation of artificial fresh water lakes.

Since the closing in 1867 of the connection with the Western Scheldt, the Scheur and the Oostgat are the only openings of the Eastern Scheldt to the sea. The mouth of the Rhine is situated 80 km to the North.

The drainage basin of the Scheldt and its tributaries covers some 21.580 km^2 in the North-West of France, the West of Belgium and the South-West of the Netherlands.

The flow of the Scheldt River is generally small while tidal motions in the estuary are large, producing a fairly good mixing of fresh and sea waters.

A map of the Scheldt Estuary (fig. 1) shows, downstream of Wálsoorden, a complicated system of channels often referred to as "flood" and "ebb" channels according as the main water motion occurs during flood-tide or ebb-tide, respectively. (Such a classification is, in many cases oversimplified, especially for the Middelgat channel and the Gat Van Ossenisse channel between Terneuzen and Hansweert).

Upstream of Walsoorden, up to Gentbrugge, the river is characterized by a main channel, well defined, with, occasionally embryos of secondary channels upstream.

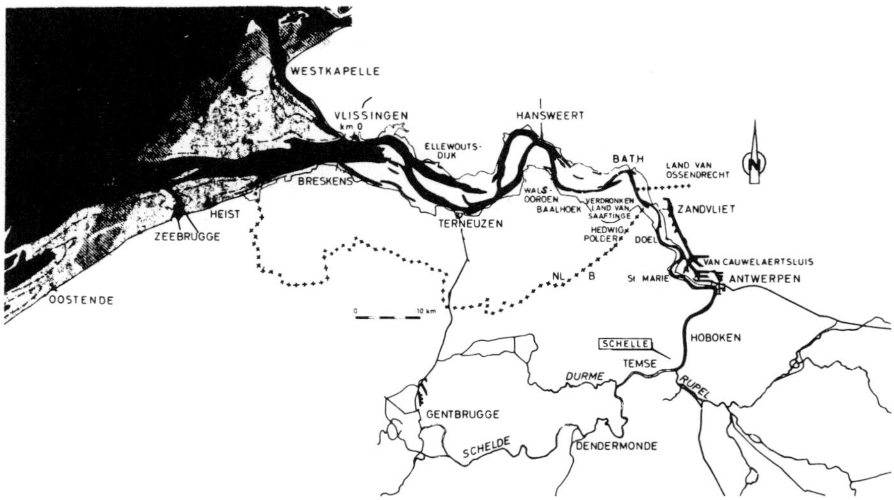

Fig. 1. Map of the Scheldt Estuary.

I. 2. Hydraulic and geometrical parameters of the Scheldt Estuary

Using the "hydraulic parameter" and the "geometrical parameter" introduced by Pritchard (Pritchard 1967) one can distinguish different zones in the Scheldt Estuary, characterized by different water and salt circulations.

The hydraulic parameter is the ratio of the volume of water flowing upstream the estuary through a given section during the flood tide (mentioned as "flood volume" in the figures) to the volume of fresh water flowing into the estuary upstream the section during a complete tidal cycle (mentioned as "fresh water volume" in the figures).

At Schelle, immediately downstream of the mouth of the Rupel and generally the upstream limit of the brackish water zone, the mean fresh water flow is estimated at some 100 m^3/s, corresponding roughly to a volume of $4{,}5.10^6 m^3$ evacuated over a tidal period. Monthly averaged fresh water flows less than 40 m^3/s or higher than 350 m^3/s at Schelle may be regarded as exceptional (Coen 1974).

The tide in the Scheldt estuary is essentially the lunar semi-diurnal tide M_2 with a period of 12 hours, 25 minutes. The tidal amplitude increases from 4 m in the open sea to 5 m at Hermiksen, upstream of Antwerpen and decreases then to values of the order of 2 m at Gentbrugge (Theuns and Coen 1973).

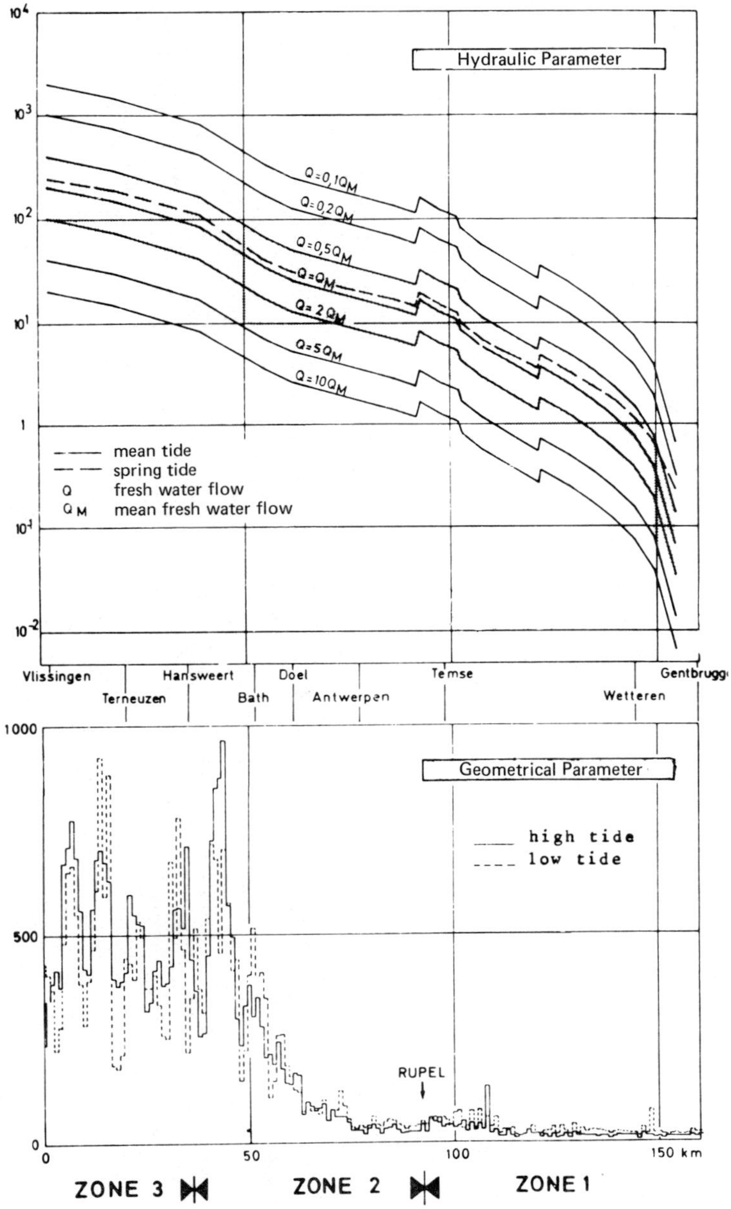

Fig. 2.

The hydraulic parameter varies considerably as a function of fresh water flow and position along the estuary (fig. 2). At a given section, for the same fresh water flow, it varies weakly with the tidal amplitude. For mean values of the fresh water flow and the tidal amplitude, the hydraulic ratio varies from 200 at Vlissingen to 100 near Hansweert (close to Walsoorden), 10 near the mouth of the Rupel and 1 or less at Gentbrugge.

The geometrical parameter is more difficult to define. In a rectangular channel the ratio of width to depth is meaningfull. In an irregular section of a shallow and wide estuary with many channels, sand-banks and tidal flats, the ratio of width to average cross-sectional depth varies with position along the estuary and with the tidal levels, sometimes very abruptly. Therefore the cross-sections were simplified and the variation of the width to mean ratio was considered between extreme water levels (fig. 2).

In the region of multiple channels, downstream of Walsoorden, this ratio oscillates between 100 and 1000 with an approximate mean value about 450. The irregular oscillations observed in the first 50 km from Vlissingen can be attributed to the existence of numerous sand-banks and shallow areas with a fairly complicated flow pattern around and over the banks and the possible formation of zones of "dead water". Upstream of Walsoorden, the width-depth ratio decreases rapidly and remains regularly small, of the order of 10 % of the mean downstreams values. In the region upstream of the Rupel tributary, the width to depth ratio loses much of its meaning. The water there is practically fresh water. The cross section is very small at low tide and the ratio may become artificially large where the depth becomes very small.

According to Pritchard's classification (Pritchard 1967), the Scheldt estuary can thus be divided in three zones.

(i) zone 1, from Gentbrugge to the mouth of the Rupel characterized by low values of both geometrical and hydraulic parameters.

(ii) zone 2, from the mouth of the Rupel to Walsoorden, characterized by intermediate values of the parameters indicating a partial mixing of the water.

(iii) zone 3, from Walsoorden to the sea, characterized by high values of the two parameters indicating a good mixing of the water (except perhaps during exceptional river flood periods in the winter when higher stratification occurs).

The three zones are indicated in fig. 3 which shows the variation of the wet section A(x) of the Scheldt at half-tide as a function of the distance x from Vlissingen.

I. 3. <u>Distribution of salinity in the Scheldt Estuary</u>.

Schematically, the salinity of the Scheldt Estuary varies from low values, characteristic of fresh water, near Gentbrugge to large values, characteristic of sea water, near Vlissingen.

The longitudinal salinity profile is of course a function of the fresh water flow which limits the intrusion of salt, but it also varies over a tidal period, being shifted upstream at flood-tide and downstream at ebb-tide. (For extremely small fresh water flows, the intrusion of salt may extend farther than the mouth of the Rupel; for extremely large fresh water flows, the salinity at Vlissingen may be appreciably smaller than the sea water value, indicating that a considerable amount of mixing still takes place beyond Vlissingen).

This situation is illustrated in fig. 4 : the shaded regions indicate the envelopes of the expected displacements of the longitudinal salinity profile over half a tidal period at respectively a typically large (left) and a typically small (right) flow of fresh water.

Lateral salinity gradients are generally negligible upstream of Walsoorden. The maximum salinity differences between the two banks of the river are observed between Walsoorden (∿ km 40) and the Dutch-Belgian border (∿ km 60) where occasionally, differences up to 4 ‰ are found. In the mean, however, over a tidal period, the differences remain very small.

Fig. 5 shows in illustration the distribution of chlorinity observed on March 31st 1971 at slack water in a cross section of the Scheldt Estuary situated near the Dutch-Belgian border. One can see that there is a small lateral gradient.

Downstream of Walsoorden, the existence of flood and ebb channels separated by large banks and the associated complicated circulation pattern may produce lateral variations of salinity and, in particular, occasional salinity differences from one channel to the other.

Vertical salinity profiles show, in general, an increasing salinity from the surface to the bottom with strong irregular variations over a tidal cycle (fig. 6). The difference between mean surface and mean bottom salinity is generally less than 1 ‰ with maximum values up to 2 °/$_{\circ\circ}$ near the Dutch-Belgian border (De Pauw and Peters 1973).

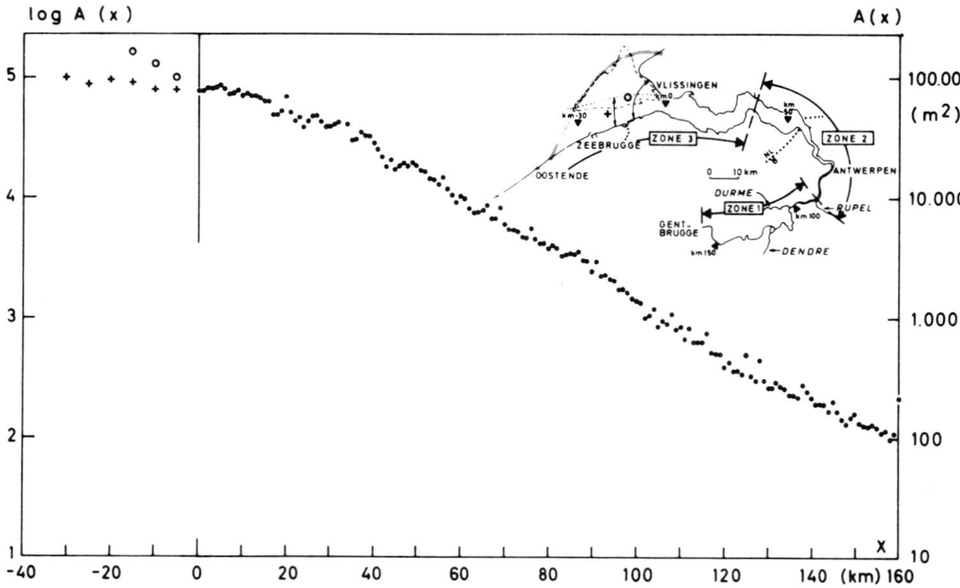

Fig. 3. : Longitudinal profile of the wet section of the Scheldt Estuary.

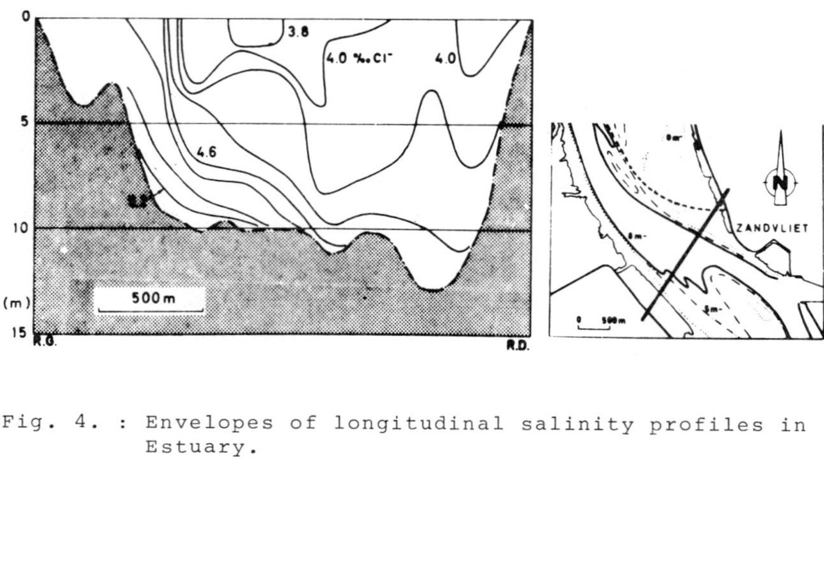

Fig. 4. : Envelopes of longitudinal salinity profiles in the Scheldt Estuary.

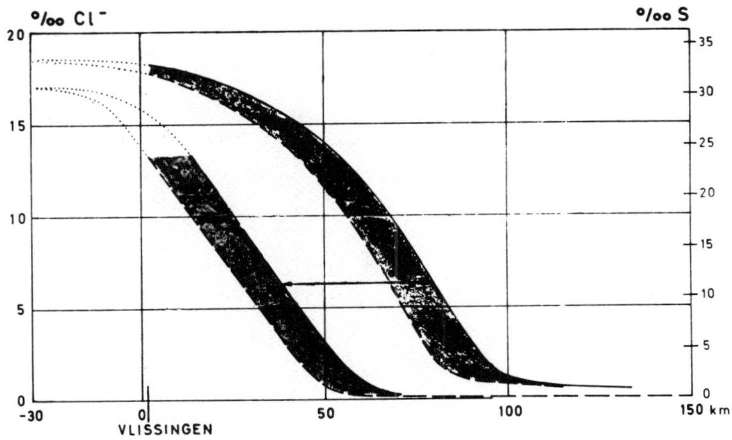

Fig. 5. : Distribution of chlorinity observed on March 31st 1971 at slack water in a cross section of the Scheldt Estuary near the Dutch-Belgian Border.

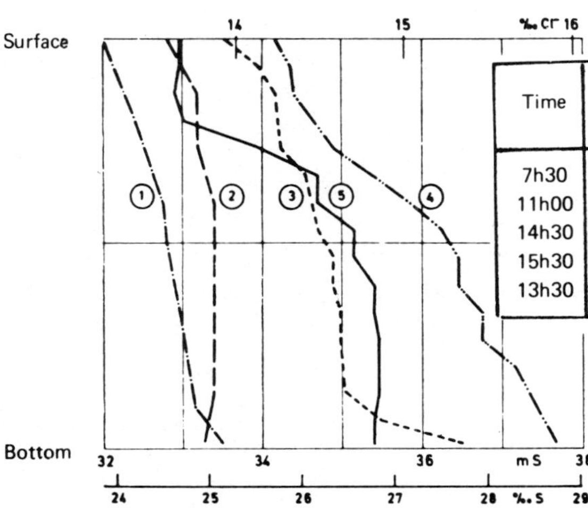

Time	Tide	Date	Position Longitude	Position Latitude
7h30	Flood	31.1.73	51°24'30" N	3°54'40"E
11h00	Flood	1.2.73	"	"
14h30	Ebb	31.1.73	"	"
15h30	Ebb	1.2.73	51°24'50" N	3°55'30"E
13h30	Ebb	30.1.73	51°24'30" N	3°54'40"E

Fig. 6. : Vertical salinity profiles in a section of the Scheldt Estuary situated in the Middelgat Channel (km 30).

The observations of the distribution of salinity in the Scheldt Estuary seem to confirm the existence of three distinct zones as suggested in section 2. :

(i) zone 1, upstream of the Rupel, is typically a region of fresh water of the Scheldt and its tributaries.

(ii) zone 2, between the Rupel and Walsoorden, is representative of a partially stratified estuary. The longitudinal and vertical salinity gradients are maximum in that region and more particularly near the Dutch-Belgian border (∿ km 60) where the maximum variations of salinity over a tidal period are also observed.

(iii) zone 3, downstream of Walsoorden, is fairly well-mixed. Variations of salinity are small except occasionally in secondary channels and around sand banks.

I. 4. Tidal and residual currents in the Scheldt Estuary

The water circulation in the Scheldt Estuary is dominated by the tides. Tidal currents can reach values of 1 m/sec or more (fig. 7) while the order of magnitude of the residual currents (defined as the mean currents over one or several tidal periods) varies typically from 1 cm/sec at the mouth of the estuary to 10 cm/sec in the region of the mouth of the Rupel. Vertical variations may be important (Peters 1974).

The residual currents are due to the fresh water inflow upstream, the non-linear residue of tidal oscillations and the presence of salinity gradients. They will be discussed in more details in part III.

II. HYDRODYNAMIC MODELS APPLICABLE TO THE SCHELDT ESTUARY

II.1. Introduction

The dynamics of the Scheldt Estuary is determined by the presence, on the one hand, of strong tidal currents with subsequent turbulence and mixing, and by the existence on the other hand of salinity gradients and subsequent stratification and buoyancy effects.

As mentioned in part I, the estuary is only partially mixed, at least in zone 2, and the tri-dimensional hydrodynamic equations describing the estuary are, to begin with, more complicated that the equations on which mathematical models of well-mixed shallow continental seas are based.

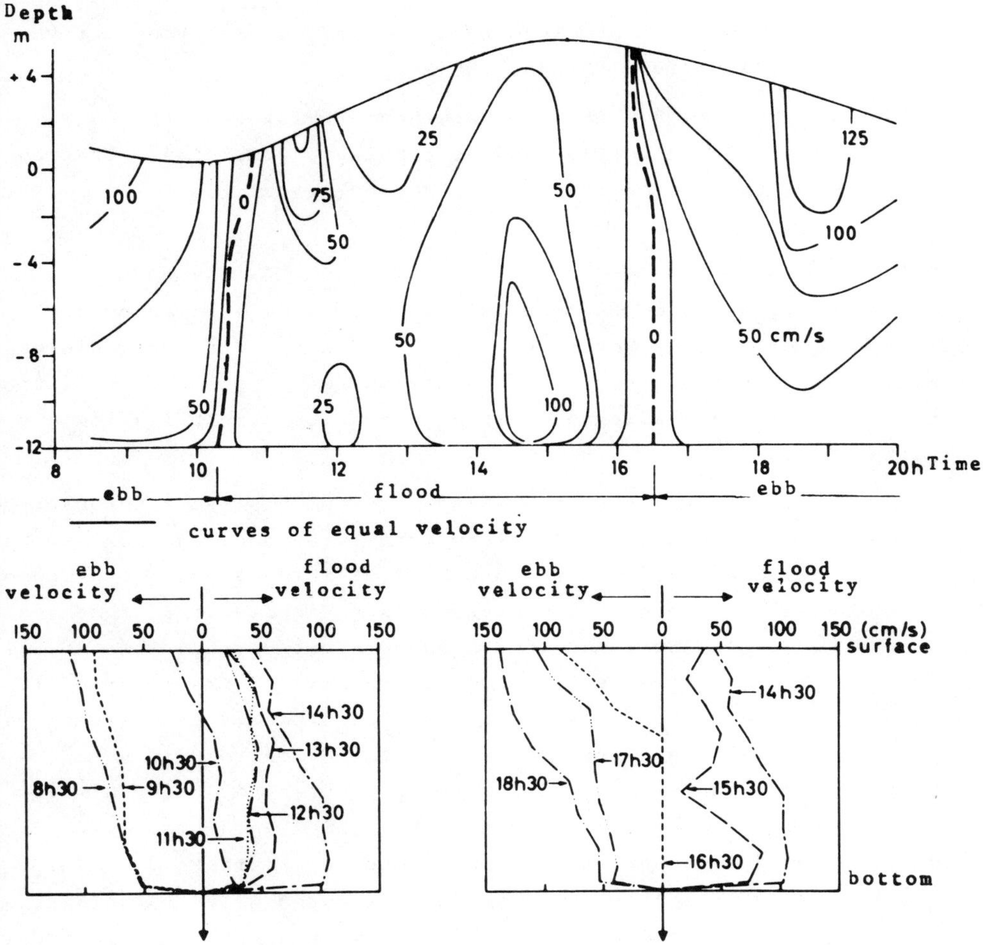

Fig. 7. : Current velocity profiles measured at a fixed station near the Dutch-Belgian border, Oct. 15th, 1970.

However, the morphology of the estuary (limited in width and depth and extending essentially streamwise) suggests different simplifications and it is often justified to work with two-dimensional and one-dimensional models integrated over the width or (and) the depth. Such models may be very different from one estuary to another ; particular characteristics of the river determining the important effects and those which can be neglected.

In the following, one discusses the hydrodynamic models applicable to zone 2 and zone 3 of the Scheldt estuary (the problem of zone 1 is less typical and can be treated by classical river mechanics). Simplifications will be made ; based on the examination of the orders of magnitude provided by the existing data bank for the Scheldt (e.g. Jäger 1973, Ronday 1975).

Clearly, they are amenable to permanent revision as the data bank is completed and refined and as simulation exercises, confronting models with observations, reveal other priorities.

II.2. Boussinesq equations

The Hydrodynamics of the Scheldt Estuary is determined by the coexistence of fresh water flow, strong tidal and wind currents generating turbulent fluctuations and mixing, and small but still significant salinity gradients,sources of density stratification and gravity effects.

It is obviously not possible - and not even desired - to describe the dynamics of the estuary in every detail reproducing micro-turbulent fluctuations, the time and length scales of which are small as compared to the main hydrodynamic processes (say, smaller than time scales of the order of 1 minute and length scales smaller than 1 meter while characteristic time scales of tidal motions are of the order of 1 hour or more and characteristic time scales of residual currents several orders of magnitude larger). The mathematical model is basically concerned with mean values which, from a theoretical point of view, may be defined as stochastic averaged but which, practically, are time averaged over a period of time (a couple of minutes, say, for the Scheldt) sufficiently large to eliminate the micro-turbulent fluctuations and sufficiently small to leave the main processes unaffected (Nihoul 1975).

The micro-turbulent fluctuations are not, however, completely eliminated from the equations by the averaging. They subsist in the non-linear terms and their effect - which is essentially an enhanced dispersion, similar to the molecular diffusion but many times more efficient - must be parameterized with the help of the so-called "eddy diffusivities".

Although the water density of the Scheldt is not constant, its variations, associated essentially with the salinity gradients, are small and Boussinesq approximation can be applied.

Thus if $\underset{\sim}{u}$ is the mean velocity, p the mean pressure and a the mean (buoyancy being defined as the deviation of water density from a constant reference value, divided by that constant value and multiplie by the acceleration of gravity g) the Boussinesq equations can be written, the vertical axis $\underset{\sim}{e}_3$ pointing upwards (e.g. Nihoul and Ronday 1976)

$$\nabla \cdot \underset{\sim}{u} = 0 \tag{1}$$

$$\frac{\partial a}{\partial t} + \nabla \cdot \underset{\sim}{u}\, a = \frac{\partial}{\partial x_3} (\lambda \frac{\partial a}{\partial x_3}) \tag{2}$$

$$\frac{\partial \underset{\sim}{u}}{\partial t} + \nabla \cdot \underset{\sim}{u}\underset{\sim}{u} + 2\, \underset{\sim}{\Omega} \wedge \underset{\sim}{u} = - \nabla q - a\, \underset{\sim}{e}_3 + \frac{\partial}{\partial x_3} \left(\nu \frac{\partial \underset{\sim}{u}}{\partial x_3} \right) \tag{3}$$

where λ is the eddy diffusivity, $\underset{\sim}{\Omega}$ the earth rotation vector, ν the eddy viscosity and where, p and ρ_m denoting respectively the pressure and the constant reference density,

$$q = \frac{p}{\rho_m} + g\, x_3$$

In eqs. (2) and (3) horizontal turbulent dispersion has been neglected because characteristic length scales of horizontal variations are always much larger than vertical length scales and this effect is negligible as compared to vertical turbulent dispersion. (One emphasizes here that, with an averaging time of the order of a minute, the turbulent dispersion discussed here is that of micro-turbulent fairly homogeneous and isotropic eddies and that the corresponding eddy diffusivities are comparable in the horizontal and the vertical directions. There is, of course, an important horizontal dispersion associated with variable and irregular currents but this effect is, in eqs. (2) and (3) included in the advection terms and will appear explicitly later).

II.3. Three-dimensional hydrodynamic models.

Eqs. (1), (2) and (3) form a system of five scalar equations for the five unknowns u_1, u_2, u_3, q and a.

In rectangular coordinates, appropriate to a rectilinear channel, they can be written :

$$\frac{\partial u_1}{\partial x_1} + \frac{\partial u_2}{\partial x_2} + \frac{\partial u_3}{\partial x_3} = 0 \tag{4}$$

$$\frac{\partial a}{\partial t} + \frac{\partial}{\partial x_1}(u_1\, a) + \frac{\partial}{\partial x_2}(u_2\, a) + \frac{\partial}{\partial x_3}(u_3\, a) = \frac{\partial}{\partial x_3}\left(\lambda \frac{\partial a}{\partial x_3}\right) \tag{5}$$

$$\frac{\partial u_1}{\partial t} + \frac{\partial}{\partial x_1}(u_1\, u_1) + \frac{\partial}{\partial x_2}(u_1\, u_2) + \frac{\partial}{\partial x_3}(u_1\, u_3) + 2\,(\Omega_2\, u_3 - \Omega_3\, u_2) = - \frac{\partial q}{\partial x_1} + \frac{\partial}{\partial x_3}\left(\nu \frac{\partial u_1}{\partial x_3}\right) \tag{6}$$

$$\frac{\partial u_2}{\partial t} + \frac{\partial}{\partial x_1}(u_1\, u_2) + \frac{\partial}{\partial x_2}(u_2\, u_2) + \frac{\partial}{\partial x_3}(u_3\, u_2) + 2\,(\Omega_3\, u_1 - \Omega_1\, u_3) = -\frac{\partial q}{\partial x_2} + \frac{\partial}{\partial x_3}\left(\nu\,\frac{\partial u_2}{\partial x_3}\right) \quad (7)$$

$$\frac{\partial u_3}{\partial t} + \frac{\partial}{\partial x_1}(u_1\, u_3) + \frac{\partial}{\partial x_2}(u_2\, u_3) + \frac{\partial}{\partial x_3}(u_3\, u_3) + 2\,(\Omega_1\, u_2 - \Omega_2\, u_1) = -\frac{\partial q}{\partial x_3} - a + \frac{\partial}{\partial x_3}\left(\nu\,\frac{\partial u_3}{\partial x_3}\right) \quad (8)$$

For the Scheldt Estuary, where the radius of curvature R of the bends is always much larger than the characteristic length scale l_2 of transverse variations, it is readily shown (Nihoul et Ronday 1976) that these equations remain valid if the x_1-axis is taken along the central streamline of the river, the x_2-axis perpendicular to it in the transversal direction and the x_3-axis vertical (oriented upwards with the origin at the mean river level) provided the vertical component of the earth rotation vector is corrected to include the effect of curvature.

With a good approximation, it is sufficient to replace, in the equations, $2\Omega_3$ by

$$f = 2\,\Omega_3 - \frac{v_c}{R} \quad (9)$$

where v_c is the circumferential velocity (Nihoul and Ronday 1976).

For a typical velocity v_c of 1m/sec and a radius of curvature of some 10 km, the two terms in the right-hand side of (8) are of the order 10^{-4} sec^{-1} and thus

$$2\,\Omega_1 \sim 2\,\Omega_2 \sim 2\,\Omega_3 \sim f \sim 10^{-4}\ \text{sec}^{-1}$$

In rectilinear sections, R is infinite and the correction vanishes.

One can estimate the orders of magnitude of the different terms of eqs. (4) to (8) for different values of the variables within the range of observations in the Scheldt. One finds that several terms are always much smaller than the others and can be neglected.

This can be shown simply by considering a typical case.

Let, in MKS units, (Jäger 1973) :

$u_1 \sim 1$; $u_2 \sim 3\cdot10^{-2}$; $u_3 \sim 3\cdot10^{-4}$; $a \sim 10^{-1}$

be characteristic values of the velocity components and the buoyancy and

$l_1 \sim 3 \cdot 10^4;\ l_2 \sim 10^3;\ l_3 \sim 10;\ t_c \sim 10^4 \sim f^{-1}$

characteristic length scales and time scales.

The Richardson number (measuring the importance of the vertical stratification) is found to be

$$Ri = \frac{\dfrac{\partial a}{\partial x_3}}{\left\| \dfrac{\partial \underset{\sim}{u}}{\partial x_3} \right\|^2} \sim 1 \tag{10}$$

The separation between the mean values $\underset{\sim}{u}$, a and the micro-turbulent fluctuations is effected by integration over a period of time τ of the order of one or two minutes ($\tau \sim 10^2$).

According to Kolmogorov theory, the corresponding eddy viscosity, in a non stratified medium, is given by

$$\nu_o \sim \varepsilon\, \tau^2 \tag{11}$$

where ε is the rate of turbulent energy transfer from the larger eddies to the smaller ones where dissipation takes place.

In stratified waters, the eddy viscosity ν and the eddy diffusivity λ depend on the Richardson number. One can write (Munk and Anderson 1948)

$$\nu = \frac{\nu_o}{(1 + 10\ Ri)^{1/2}} \tag{12}$$

$$\lambda = \frac{\nu_o}{(1 + 3{,}33\ Ri)^{3/2}} \tag{13}$$

Taking $\varepsilon \sim 10^{-5}$ as a typical value for the Scheldt Estuary (Nihoul et Ronday 1976), one finds, using (10)

$\nu \sim 3 \cdot 10^{-2} m^2/sec$
$\lambda \sim 10^{-2}\ m^2/sec$

These values are in good agreement with observations (e.g. Bowden 1965, Fisher 1972).

Comparing orders of magnitude, one finds (Nihoul and Ronday 1976) :

(i) that all terms in the left-hand side of eq.(8) and the vertical turbulent diffusion in the right-hand side are completely negligible as compared to buoyancy which can only be balanced by the vertical gradient of q , i.e.

$$a = -\frac{\partial q}{\partial x_3} \qquad (14)$$

(ii) that the transverse gradient of q is essentially balanced by Coriolis and eventual curvature effects and, to a smaller extent, by turbulent diffusion and time variations of the transverse velocity, yielding

$$\frac{\partial q}{\partial x_2} \lesssim 10^{-4} \qquad \text{(MKS)} \qquad (15)$$

(iii) that, in eq.(6), to the contrary, earth rotation and curvature effects can be neglected as compared to turbulent diffusion.

To estimate the longitudinal gradient of q, it is convenient to write

$$q = q_l + q_v + q_t \qquad (16)$$

where q_1 is the cross-section average of q and depends on x_1 and t only ; q_v is the width-average of $q - q_1$ and depends on x_1 , x_3 and t ; q_t is equal to $q - q_1 - q_v$ and depends on the four variables x_1, x_2 , x_3 and t.

Substituting in (14) and (15), one gets

$$\frac{\partial q_t}{\partial x_2} \lesssim 10^{-4}; \; q_t \lesssim 10^{-1}$$

$$a = -\frac{\partial q_v}{\partial x_3} - \frac{\partial q_t}{\partial x_3} \sim -\frac{\partial q_v}{\partial x_3}$$

$$q_v \sim 0\,(l_3\, a) \sim 1$$

An estimate of q_1 can be based on the value of q at the free surface

$$q = \frac{p_a}{\rho_m} + g\,\zeta \qquad (17)$$

where ζ is the surface elevation.

For elevations of the order of 1 m,

$$q_1 \sim 10; \frac{\partial q}{\partial x_1} \sim \frac{\partial q_1}{\partial x_1} \sim 3 \cdot 10^{-4}$$

To sum up, the equations of a three-dimensional hydrodynamic model of the Scheldt Estuary can be written

$$\nabla \cdot \underset{\sim}{u} = 0 \qquad (18)$$

$$a = -\frac{\partial q}{\partial x_3} \qquad (19)$$

$$\frac{\partial a}{\partial t} + \nabla \cdot \underset{\sim}{u} a = \frac{\partial}{\partial x_3}\left(\lambda \frac{\partial a}{\partial x_3}\right) \qquad (20)$$

$$\frac{\partial u_1}{\partial t} + \nabla \cdot \underset{\sim}{u} u_1 = -\frac{\partial q}{\partial x_1} + \frac{\partial}{\partial x_3}\left(\nu \frac{\partial u_1}{\partial x_3}\right) \qquad (21)$$

$$\frac{\partial u_2}{\partial t} + f u_1 = -\frac{\partial q}{\partial x_2} + \frac{\partial}{\partial x_3}\left(\nu \frac{\partial u_2}{\partial x_3}\right) \qquad (22)$$

Eqs. (18) to (22) form a complete system of five scalar equations for the five unknowns u_1, u_2, u_3, a, q.

II.4. Two-dimensional models assuming transversal homogeneity

Early models of estuaries traditionally assumed a sufficiently good transversal homogeneity to neglect, in the equations, all terms containing derivatives with respect to x_2 (e.g. Pritchard 1967).

These terms, in eqs. (18), (20) and (21) also contain the transverse velocity u_2 and the hypothesis amounts to considering that, in section II.3., either the velocity u_2 has been overestimated or the characteristic length scale l_2 has been underestimated.

If this is the case, eq. (18) reduces to

$$\frac{\partial u_1}{\partial x_1} + \frac{\partial u_3}{\partial x_3} = 0 \qquad (23)$$

and one can define a stream function ψ such that

$$u_1 = \frac{\partial \psi}{\partial x_3} \qquad (24)$$

$$u_3 = -\frac{\partial \psi}{\partial x_1} \qquad (25)$$

The hydrodynamic equations can then be reduced to 3, i.e.

$$a = -\frac{\partial q}{\partial x_3} \qquad (26)$$

$$\frac{\partial a}{\partial t} + \frac{\partial \psi}{\partial x_3}\frac{\partial a}{\partial x_1} - \frac{\partial \psi}{\partial x_1}\frac{\partial a}{\partial x_3} = \frac{\partial}{\partial x_3}\left(\lambda \frac{\partial a}{\partial x_3}\right) \tag{27}$$

$$\frac{\partial^2 \psi}{\partial t\,\partial x_3} + \frac{\partial \psi}{\partial x_3}\frac{\partial^2 \psi}{\partial x_3\,\partial x_1} - \frac{\partial \psi}{\partial x_1}\frac{\partial^2 \psi}{\partial x_3^{\,2}} = -\frac{\partial q}{\partial x_1} + \frac{\partial}{\partial x_3}\left(\nu \frac{\partial^2 \psi}{\partial x_3^{\,2}}\right) \tag{28}$$

for the three variables ψ, a, q ; the latest being also regarded as a function of t, x_1 and x_3 only.

Eq. (22) is ignored in this type of model as it provides a useless relation between uninteresting variables u_2 and q_t. However, a closer examination of that equation shows that one must take such models with some reservation.

Indeed fu_1 reaches typical values of order 10^{-4} and must be balanced by another term at least. If l_2 is larger than estimated (say, 10^4), either q_t is larger ($q_t \sim 1$ instead of 10^{-1}) or u_2 is larger. In the first case, q_t is a non negligible contribution to q in equations like (21), and one may not assume that q does not depend on x_2. In the second case, an increase of u_2 compensating the decrease of l_2 , it is likely that terms containing $\frac{\partial u_2}{\partial x_2}$ or $u_2 \frac{\partial}{\partial x_2}$ cannot be neglected.

The model would remain valid if u_2 was smaller than estimated ($u_2 << 3\ 10^{-2}$) but observations indicate that, in the Scheldt Estuary, this condition is far from being realized in many places.

II.5. Width-integrated models

If one is not primarily interested in transverse variations, it is advantageous to integrate the evolution equations from one bank to the other and to study the longitudinal and vertical variations of mean or integrated values over the width of the estuary.

Let

$$x_2 = D\ (x_1, x_3) \tag{29}$$

and

$$x_2 = -\Delta\ (x_1\ x_3) \tag{30}$$

be the equations of the banks.

The banks must be stream surfaces. Thus

$$u_1 \frac{\partial D}{\partial x_1} + u_3 \frac{\partial D}{\partial x_3} = u_2 \qquad \text{at} \qquad x_2 = D \tag{31}$$

$$u_1 \frac{\partial \Delta}{\partial x_1} + u_3 \frac{\partial \Delta}{\partial x_3} = - u_2 \qquad \text{at} \qquad x_2 = - \Delta \tag{32}$$

Let

$$V_1 = L\, \bar{u}_1 = \int_{-\Delta}^{D} u_1 \, dx_2 \tag{33}$$

$$V_3 = L\, \bar{u}_3 = \int_{-\Delta}^{D} u_3 \, dx_2 \tag{34}$$

$$A = L\, \bar{a} = \int_{-\Delta}^{D} a \, dx_2 \tag{35}$$

where $\overline{y}$ is the width-averaged of y($y = u_1$, u_2, a) and where

$$L = \Delta + D \tag{36}$$

Setting $\tilde{y} = y - \overline{y}$ ($y = u_1$, u_2, a), one writes

$$u_1 = \overline{u_1} + \tilde{u}_1 \tag{37}$$

$$u_3 = \overline{u_3} + \tilde{u}_3 \tag{38}$$

$$a = \bar{a} + \tilde{a} \tag{39}$$

Integrating eqs. (18) and (20) from $- \Delta$ to D, one obtains

$$\begin{aligned} \int_{-\Delta}^{D} \left(\frac{\partial u_1}{\partial x_1} + \frac{\partial u_3}{\partial x_3} \right) dx_2 + (u_2)_D - (u_2)_\Delta &= \frac{\partial V_1}{\partial x_1} - (u_1)_D \frac{\partial D}{\partial x_1} - (u_1)_\Delta \frac{\partial \Delta}{\partial x_1} + \frac{\partial V_3}{\partial x_3} \\ - (u_3)_D \frac{\partial D}{\partial x_3} - (u_3)_\Delta \frac{\partial \Delta}{\partial x_3} + (u_2)_D - (u_2)_\Delta &= \frac{\partial V_1}{\partial x_1} + \frac{\partial V_3}{\partial x_3} = 0 \end{aligned} \tag{40}$$

$$\begin{aligned} &\int_{-\Delta}^{D} \left\{ \frac{\partial a}{\partial t} + \frac{\partial}{\partial x_1}(u_1\, a) + \frac{\partial}{\partial x_3}(u_3\, a) \right\} dx_2 + (u_2\, a)_D - (u_2\, a)_\Delta - \int_{-\Delta}^{D} \left(\frac{\partial}{\partial x_3} \lambda \frac{\partial a}{\partial x_3} \right) dx_2 \\ &= \frac{\partial A}{\partial t} + \frac{\partial}{\partial x_1} \int_{-\Delta}^{D} u_1\, a\, dx_2 - (u_1\, a)_D \frac{\partial D}{\partial x_1} - (u_1\, a)_\Delta \frac{\partial \Delta}{\partial x_1} + \frac{\partial}{\partial x_3} \int_{-\Delta}^{D} u_3\, a\, dx_2 \\ &- (u_3\, a)_D \frac{\partial D}{\partial x_3} - (u_3\, a)_\Delta \frac{\partial \Delta}{\partial x_3} + (u_2\, a)_D - (u_2\, a)_\Delta - \frac{\partial}{\partial x_3}\left(\lambda \frac{\partial A}{\partial x_3} \right) + \frac{\partial}{\partial x_3} \left[\lambda \left(a_D \frac{\partial D}{\partial x_3} + a_\Delta \frac{\partial \Delta}{\partial x_3} \right) \right] \\ &+ \left(\lambda \frac{\partial a}{\partial x_3} \right)_D \frac{\partial D}{\partial x_3} + \left(\lambda \frac{\partial a}{\partial x_3} \right)_\Delta \frac{\partial \Delta}{\partial x_3} = \frac{\partial A}{\partial t} + \frac{\partial}{\partial x_1} (L^{-1} V_1 A) + \frac{\partial}{\partial x_1} \int_{-\Delta}^{D} \tilde{u}_1\, \tilde{a}\, dx_2 \\ &+ \frac{\partial}{\partial x_3} (L^{-1} V_3 A) + \frac{\partial}{\partial x_3} \int_{-\Delta}^{D} \tilde{u}_3\, \tilde{a}\, dx_3 - \frac{\partial}{\partial x_3} \left(\lambda \frac{\partial A}{\partial x_3} \right) = 0 \end{aligned} \tag{41}$$

underlined terms cancelling as a result of (31) and (32) and the last four terms of (41) being negligible as compared to $\frac{\partial}{\partial x_3}(\lambda \frac{\partial A}{\partial x_3})$. Indeed :

$$\frac{\partial}{\partial x_3}\left(\lambda \frac{\partial A}{\partial x_3}\right) \sim 0\left(\frac{\lambda a L}{l_3^{\,2}}\right) \sim 10^{-2}$$

While, for instance

$$\left(\lambda \frac{\partial a}{\partial x_3}\right)_D \frac{\partial D}{\partial x_3} \sim 0\left(\frac{\lambda a}{l_3}\,\frac{\partial D}{\partial x_3}\right) \sim 10^{-4}$$

assuming a maximum of 10 meters of width variation over depth.

Similarly, integrating eq. (21), one obtains

$$\frac{\partial V_1}{\partial t} + \frac{\partial}{\partial x_1}(L^{-1}\, V_1\, V_1) + \frac{\partial}{\partial x_3}(L^{-1}\, V_1\, V_3) + \frac{\partial}{\partial x_1}\int_{-\Delta}^{D} \tilde{u}_1\, \tilde{u}_1\, dx_2 + \frac{\partial}{\partial x_3}\int_{-\Delta}^{D} \tilde{u}_1\, \tilde{u}_3\, dx_2 = -\, L\,\frac{\partial \bar{q}}{\partial x_1} + \frac{\partial}{\partial x_3}\left(\nu\,\frac{\partial V_1}{\partial x_3}\right) \tag{42}$$

q_v being negligible, the approximation $q \sim \bar{q}$ has been made. This simplification results from eq. (22) which plays no other role in the integrated model.

Integrating eq. (19), one finds, with the same approximation

$$A = -\, L\,\frac{\partial \bar{q}}{\partial x_3} \tag{43}$$

One can see that in eqs (41) and (42), integrals of products of deviations around the mean appear. Observations reveal that these terms (the structure of which is reminiscent of the Reynolds stresses) are responsible of longitudinal and vertical dispersion similar to turbulent dispersion but considerably more efficient.

On the model of turbulent diffusion, one sets

$$\frac{\partial}{\partial x_1}\int_{-\Delta}^{D} (-\,\tilde{u}_1\,\tilde{a})\, dx_2 = \frac{\partial}{\partial x_1}\left(\Lambda_1\,\frac{\partial A}{\partial x_1}\right) \tag{44}$$

$$\frac{\partial}{\partial x_3}\int_{-\Delta}^{D} (-\,\tilde{u}_3\,\tilde{a})\, dx_2 + \frac{\partial}{\partial x_3}\left(\lambda\,\frac{\partial A}{\partial x_3}\right) = \frac{\partial}{\partial x_3}\left(\Lambda_3\,\frac{\partial A}{\partial x_3}\right) \tag{45}$$

$$\frac{\partial}{\partial x_1}\int_{-\Delta}^{D} (-\,\tilde{u}_1\,\tilde{u}_1)\, dx_2 = \frac{\partial}{\partial x_1}\left(N_1\,\frac{\partial V_1}{\partial x_1}\right) \tag{46}$$

$$\frac{\partial}{\partial x_3}\int_{-\Delta}^{D} (-\,\tilde{u}_1\,\tilde{u}_3)\, dx_2 + \frac{\partial}{\partial x_3}\left(\nu\,\frac{\partial V_1}{\partial x_3}\right) = \frac{\partial}{\partial x_3}\left(N_3\,\frac{\partial V_1}{\partial x_3}\right) \tag{47}$$

where N_1, N_3, Λ_1 , Λ_3 are new diffusivity coefficients.

According to observations, (e.g. Ronday 1975)[1] :

$N_1 \sim 10^2$ à 10^3 m^2/s; $\Lambda_1 \sim 10^2$ m^2/s

$N_3 \sim 10^{-2}$ à 10^{-3} m^2/s; $\Lambda_3 \sim 10^{-2}$ m^2/s

Eq. (40) suggests the definition of a stream function such that

$$V_1 = \frac{\partial \psi}{\partial x_3} \tag{48}$$

$$V_3 = - \frac{\partial \psi}{\partial x_1} \tag{49}$$

Eq. (40) is then identically verified. Eqs. (41), (42) and (43) can be written

$$\frac{\partial A}{\partial t} + \frac{\partial}{\partial x_1}\left(L^{-1} \frac{\partial \psi}{\partial x_3} A\right) - \frac{\partial}{\partial x_3}\left(L^{-1} \frac{\partial \psi}{\partial x_1} A\right) = \frac{\partial}{\partial x_1}\left(\Lambda_1 \frac{\partial A}{\partial x_1}\right) + \frac{\partial}{\partial x_3}\left(\Lambda_3 \frac{\partial A}{\partial x_3}\right) \tag{50}$$

$$\frac{\partial^2 \psi}{\partial t \, \partial x_3} + \frac{\partial}{\partial x_1}\left(L^{-1} \frac{\partial \psi}{\partial x_3} \frac{\partial \psi}{\partial x_3}\right) - \frac{\partial}{\partial x_3}\left(L^{-1} \frac{\partial \psi}{\partial x_3} \frac{\partial \psi}{\partial x_1}\right) = - L \frac{\partial \bar{q}}{\partial x_1} + \frac{\partial}{\partial x_1}\left(N_1 \frac{\partial^2 \psi}{\partial x_1 \, \partial x_3}\right) + \frac{\partial}{\partial x_3}\left(N_3 \frac{\partial^2 \psi}{\partial x_3^{\,2}}\right) \tag{51}$$

$$A = - L \frac{\partial \bar{q}}{\partial x_3} \tag{52}$$

They form a complete system for the three variables ψ , A and $\bar{q}$.

II.6. Depth-integrated models

Depth-integrated models have been extensively used in shallow continental seas hydrodynamics (e.g. Nihoul 1975). In the case of a partially stratified estuary like the Scheldt Estuary (in zone 2, at least), their utility is less evident. On the one hand, they tend to mask the effects of vertical stratification which, in the gravity field, often play an important role. On the other hand, they are often more difficult to handle than width-integrated models. For instance, although both types of models are two-dimensional as a result of the integration over one space coordinate, in depth-integrated models, one cannot introduce a stream function and subsequently reduce the number of variables if the system is not in a steady state.

---

[1] N_1, N_3, Λ_1 and Λ_3 are in general not constant along the estuary. They must be regarded as control parameters to be determined by theoretical reflexion and simulation exercises based on the data bank.

Indeed, if $x_3 = \zeta$ and $x_3 = -h$ are the equations of the surface and the bottom, respectively, one must have

$$\frac{\partial \zeta}{\partial t} + u_1 \frac{\partial \zeta}{\partial x_1} + u_2 \frac{\partial \zeta}{\partial x_2} = u_3 \quad \text{in} \quad x_3 = \zeta \tag{53}$$

$$u_1 \frac{\partial h}{\partial x_1} + u_2 \frac{\partial h}{\partial x_2} = -u_3 \quad \text{in} \quad x_3 = -h \tag{54}$$

Defining

$$U_1 = \int_{-h}^{\zeta} u_1 \, dx_3 \tag{55}$$

$$U_2 = \int_{-h}^{\zeta} u_2 \, dx_3 \tag{56}$$

and integrating (18) from $-h$ to ζ, one obtains

$$\int_{-h}^{\zeta} \left(\frac{\partial u_1}{\partial x_1} + \frac{\partial u_2}{\partial x_2} \right) dx_3 + (u_3)_\zeta - (u_3)_{-h} = \frac{\partial U_1}{\partial x_1} - (u_1)_\zeta \frac{\partial \zeta}{\partial x_1} - (u_1)_{-h} \frac{\partial h}{\partial x_1} + (u_3)_\zeta$$

$$+ \frac{\partial U_2}{\partial x_2} - (u_2)_\zeta \frac{\partial \zeta}{\partial x_2} - (u_2)_{-h} \frac{\partial h}{\partial x_2} - (u_3)_{-h} = \frac{\partial \zeta}{\partial t} + \frac{\partial U_1}{\partial x_1} + \frac{\partial U_2}{\partial x_2} = 0 \tag{57}$$

using (53) and (54).

In wide, well-mixed estuaries with flood - and ebb - channels, like zone 3 of the Scheldt, it may be interesting to combine, to the width-integrated model, a depth-integrated model to have additional information on the lateral distribution of velocity.

In partially mixed estuaries, depth-integration is only advantageous if the vertical distribution of buoyancy is known from observations. Otherwise, the integration over depth, transforming partial differential equations into integral equations*, does not lead to any real simplification, even if it appears to reduce the number of variables.

II.7. Cross-section integrated models

In a first approach, when one is mainly interested in longitudinal variations, one may use a cross-section integrated model.

The operation of integration over the cross-section is similar, mathematically, to the integration over the width and the details will not be reproduced here.

---

* because

$$q = \frac{p_a}{\rho_m} + g\zeta + \int_{x_3}^{\zeta} a \, dx_3$$

The final one-dimensional equations appropriate to the Scheldt Estuary are given in (Nihoul and Ronday 1976).

III. RESIDUAL CIRCULATION AND SILT DEPOSITION IN THE SCHELDT ESTUARY

III. 1. Flocculation and variations of the suspended load in the Scheldt Estuary

Critical salinity values of 1 to 5 °/₀₀ are typically found in the central reach of the estuary (fig. 4). At such values of the salinity, intense flocculation and sedimentation take place and the suspended load decreases abruptly. (There is about one order of magnitude difference between the values of turbidity and COD upstream and downstream of the critical region).

Variations of the river flow rate Q and the subsequent displacement of the region of critical salinity reflect on the sedimentation and tend to spread the deposition of suspended matter over most of zone 2 (fig. 8).

It is found however that bottom sediments tend to accumulate in a narrower region than one might expect and this is explained in the following with the help of a two-dimensional width-integrated residual circulation model.

III. 2. Model of residual circulation

The time-dependent width-integrated hydrodynamic equations have been derived in section II.5. If V_1, V_2 and A represent respectively the integrals of u_1, u_2 and a over the width, these equations can be written

$$\frac{\partial V_1}{\partial x_1} + \frac{\partial V_3}{\partial x_3} = 0 \tag{58}$$

$$\frac{\partial A}{\partial t} + \frac{\partial}{\partial x_1}(L^{-1} V_1 A) + \frac{\partial}{\partial x_3}(L^{-1} V_3 A) = \frac{\partial}{\partial x_1}\left(\Lambda_1 \frac{\partial A}{\partial x_1}\right) + \frac{\partial}{\partial x_3}\left(\Lambda_3 \frac{\partial A}{\partial x_3}\right) \tag{59}$$

$$A = - L \frac{\partial \bar{q}}{\partial x_3} \tag{60}$$

$$\frac{\partial V_1}{\partial t} + \frac{\partial}{\partial x_1}(L^{-1} V_1 V_1) + \frac{\partial}{\partial x_3}(L^{-1} V_1 V_3) = - L \frac{\partial \bar{q}}{\partial x_1} + \frac{\partial}{\partial x_1}\left(N_1 \frac{\partial V_1}{\partial x_1}\right) + \frac{\partial}{\partial x_3}\left(N_3 \frac{\partial V_1}{\partial x_3}\right) \tag{61}$$

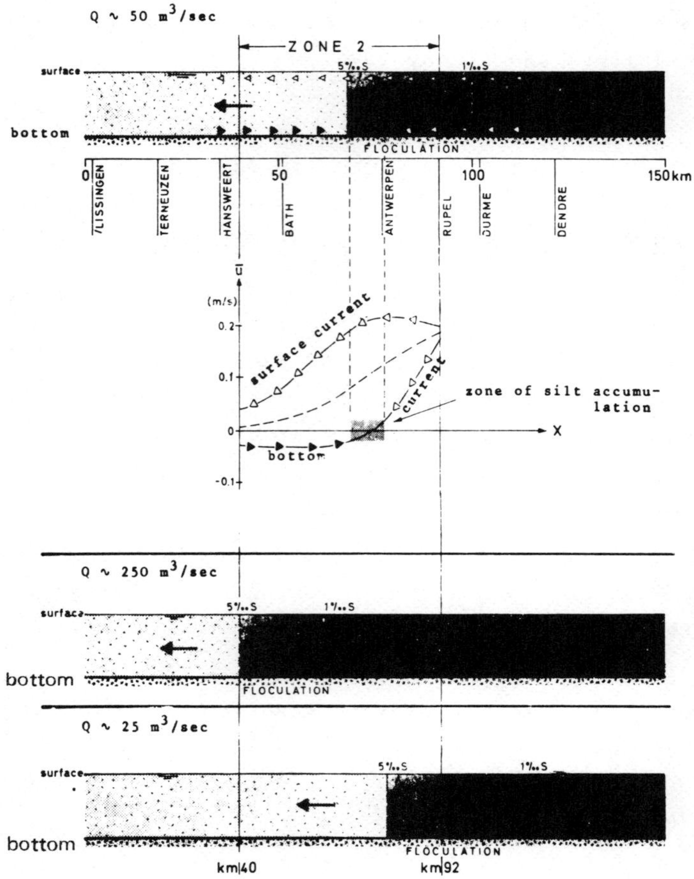

Fig. 8. : Flocculation in the Scheldt Estuary for different river flows.

The residual circulation can be defined as the mean circulation over a period of time sufficiently large to eliminate tidal oscillations and transitory wind induced currents.

The residual circulation is essentially related to the river flow rate and to gravity currents associated with the stratification. A width-integrated model is thus appropriate. The residual width-integrated equations are derived from eqs. (58) to (61) by further averaging over time. In this process, time derivatives are replaced by finite differences which can be made as small as desired by increasing the period of time over which the averages are taken. Thus if the residual currents are defined over a sufficiently long period of time, they can be described by steady state equations.

If < >indicates a time average and if

$$U_i = \langle V_i \rangle \qquad i = 1,2 \tag{62}$$

$$V_i = U_i + W_i \qquad i = 1,2 \tag{63}$$

$$B = \langle A \rangle \tag{64}$$

$$A = B + C \tag{65}$$

$$\pi = \langle \bar{q} \rangle \tag{66}$$

one gets, integrating eq. (58) to (61) over time :

$$\frac{\partial U_1}{\partial x_1} + \frac{\partial U_3}{\partial x_3} = 0 \tag{67}$$

$$\begin{aligned} \frac{\partial}{\partial x_1}(L^{-1} U_1 B) + \frac{\partial}{\partial x_3}(L^{-1} U_3 B) &= \frac{\partial}{\partial x_1}\left\{ \Lambda_1 \frac{\partial B}{\partial x_1} + L^{-1} \langle - W_1 C \rangle \right\} \\ &+ \frac{\partial}{\partial x_3}\left\{ \Lambda_3 \frac{\partial B}{\partial x_3} + L^{-1} \langle - W_3 C \rangle \right\} \end{aligned} \tag{68}$$

$$\begin{aligned} \frac{\partial}{\partial x_1}(L^{-1} U_1 U_1) + \frac{\partial}{\partial x_3}(L^{-1} U_1 U_3) &= - L \frac{\partial \pi}{\partial x_1} \\ &+ \frac{\partial}{\partial x_1}\left\{ N_1 \frac{\partial U_1}{\partial x_1} + L^{-1} \langle - W_1 W_1 \rangle \right\} \\ &+ \frac{\partial}{\partial x_3}\left\{ N_3 \frac{\partial U_1}{\partial x_3} + L^{-1} \langle - W_3 W_1 \rangle \right\} \end{aligned} \tag{69}$$

$$B = - L \frac{\partial \pi}{\partial x_3} \tag{70}$$

One can see that the average of the quadratic terms gives again two contributions ; the first of which containing the product of the means and the second the mean product of the fluctuations.

The latter contributes to the dispersion and combines to turbulent mixing and shear effect. In a first approach, one can parameterize the general effect of the dispersion processes, on the model of turbulent dispersion, introducing new dispersion coefficients K_1, K_3, M_1, M_3 such that

$$\Lambda_1 \frac{\partial B}{\partial x_1} + L^{-1} \langle - W_1 C \rangle = K_1 \frac{\partial B}{\partial x_1} \tag{71}$$

$$\Lambda_3 \frac{\partial B}{\partial x_3} + L^{-1} \langle - W_3 C \rangle = K_3 \frac{\partial B}{\partial x_3} \tag{72}$$

$$N_1 \frac{\partial U_1}{\partial x_1} + L^{-1} \langle - W_1 W_1 \rangle = M_1 \frac{\partial U_1}{\partial x_1} \tag{73}$$

$$N_3 \frac{\partial U_1}{\partial x_3} + L^{-1} \langle - W_3 W_1 \rangle = M_3 \frac{\partial U_1}{\partial x_3} \tag{74}$$

with, according to observations (MKS) (e.g.Ronday 1975) :

$K_1 \sim 10^2$; $K_3 \sim 10^{-2}$; $M_1 \sim 10^3$; $M_3 \sim 10^{-1}$

K_1, K_3, M_1 and M_3 must be regarded as control parameters to be determined by inspection of the data bank. In general, they are functions of x_1 even if, in first approximation, they can be assumed independent of x_3.

If l_1 and l_3 are characteristic length scales of respectively horizontal and vertical variations of the velocity field, eq. (67) gives

$$\frac{U_1}{l_1} \sim \frac{U_3}{l_3}$$

In eq. (69), the terms in the left-hand side are of the order

$$\frac{L^{-1} U_1 U_1}{l_1}$$

The second term of the right-hand side is of the order

$$\frac{M_1 U_1}{l_1^2}$$

while the last term of the right hand side is of the order

$$\frac{M_3 U_1}{l_3^2}$$

Taking (Ronday 1975),

$l_3 \sim 10$; $l_1 \sim 10^4$: $L^{-1} U_1 \sim 10^{-2}$

one can see that vertical dispersion is several orders of magnitude larger than advection and horizontal dispersion and must accordingly be balanced by $- L \frac{\partial \pi}{\partial x_1}$.

A similar simplification is not, in general, possible for eq. (68) which does not contain a term analogous to the pressure gradient and where vertical dispersion is comparatively less important ; vertical salinity gradients being smaller than vertical velocity gradients in many cases.

Eq. (67) suggests the introduction of a residual stream function ψ_o such that :

$$U_1 = \frac{\partial \psi_o}{\partial x_3} \tag{75}$$

$$U_3 = - \frac{\partial \psi_o}{\partial x_1} \tag{76}$$

Eliminating U_1, U_3 and B, one obtains

$$\frac{\partial \psi_o}{\partial x_3} \frac{\partial^2 \pi}{\partial x_1 \partial x_3} - \frac{\partial \psi_o}{\partial x_1} \frac{\partial^2 \pi}{\partial x_3{}^2} = \frac{\partial}{\partial x_1} \left\{ K_1 \frac{\partial}{\partial x_1} \left(L \frac{\partial \pi}{\partial x_3} \right) \right\} + \frac{\partial}{\partial x_3} \left\{ K_3 \frac{\partial}{\partial x_3} \left(L \frac{\partial \pi}{\partial x_3} \right) \right\} \tag{77}$$

$$\frac{\partial}{\partial x_3} \left(M_3 \frac{\partial^2 \psi_o}{\partial x_3{}^2} \right) = L \frac{\partial \pi}{\partial x_1} \tag{78}$$

Eqs. (77) and (78) form a complete system for the two variables ψ_o and π.

III. 3. Application to the deposition of silt in the Scheldt Estuary

The residual circulation model has been applied to a section of the Scheldt Estuary from the Rupel to the sea (zones 2 and 3), taking into account the geometry of the basin (for instance, the function $L(x_1,x_3)$) and the appropriate boundary conditions.

To determine the dispersion coefficients, observed salinity and SiO_2* distributions were used. The model was run in preliminary hindcasting exercises until by successive improvement of K_1, K_3 and M_3 a satisfactory agreement was found between predictions and observations.

* The buoyancy equation (68) is easily adapted to describe the distribution of SiO_2. In the right-hand side two additional terms appear ; the first one representing the consumption by diatoms, the second one the lateral inputs.

Fig. (9) represents the lines of equal width-averaged residual horizontal velocity calculated by the model, in its final, calibrated form, for a flow rate equal to four times its low water value.

Currents are expressed in m/sec. The curve in dot and dashes (—.—.—.) noted n=4 indicates the line of zero residual current. It is readily seen on the salinity scale printed on the top of the diagram (and corresponding to the same flow rate) that the residual current goes to zero, at the bottom, at the very place where critical salinity values and flocculation are observed.

Mud sedimentation and accumulation thus occurs in a region where the average sediment transport on and near the bottom is negligible.

For reference, two curves in dashes (— — —) noted n=1 and n=10 respectively, indicate the lines of zero residual current for flow rates equal to the low water value and 10 times that value. The salinity scale is, of course, shifted accordingly and the zone of precipitation appears to coincide in most cases with the region of zero bottom current.

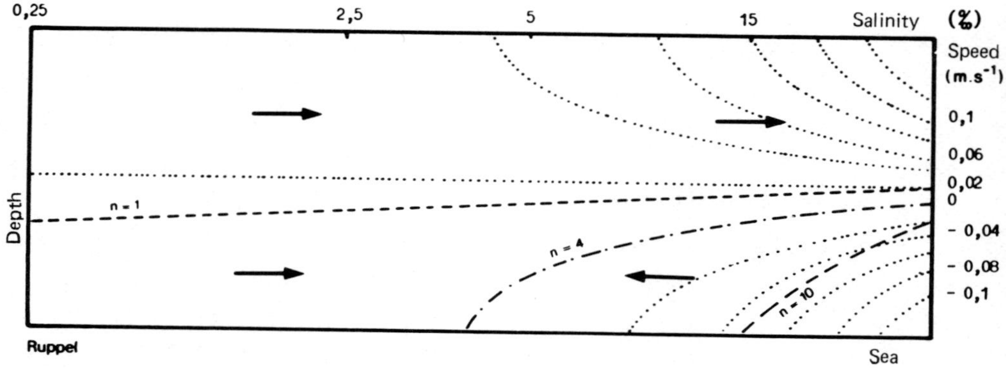

Fig. 9. : Lines of equal width-averaged residual horizontal velocity calculated by the model.

However, when the flow rate increases or decreases and comes back to typical values (n $\sim$ 4), a net motion near the bottom appears in the regions where it was zero for lower and higher flow rates and directions are such that, as indicated by the arrows in fig. 9, they tend to carry back the freshly deposited sediments to the median zone.

IV. ACKNOWLEDGEMENT

This work was conducted in the scope of the Belgian National Environment Program, sponsored by the Science Policy Administration, Office of the Prime Minister.

V. REFERENCES

Bowden K.F. (1965), J. Fluid Mech. 21, 83.
Coen I. (1974), *Debieten van het Scheldebekken*, Antwerpse Zeediensten, Ministerie van Openbare Werken, Belgium.
De Pauw N and Peters J.J. (1973), *Contribution to the Study of the Salinity Distribution and Circulation in the Western Scheldt Estuary*, Belgian National Environment Program, Sea Project, Ministry for Science Policy, Belgium.
Fisher H.B. (1972), J. Fluid Mech., 53, 671.
Jäger Ph. (1973), *Dynamical parameters of the Scheldt Estuary*, Belgian National Environment Program, Sea Project (compiled from "Stormvloeden of de Schelde", 1966, Ministerie van Openbare Werken, vol. II, IV, V), Ministry for Science Policy, Belgium.
Munk W.H. and Anderson E.R. (1948), J. Mar. Res. 7, 276.
Peters J.J. (1974), *Model voor de studie van de verontreiniging van het Scheldeestuarium*, Symposium The Golden Delta, 2, Pudoc.
Nihoul J.C.J. (1975), *Modelling of Marine Systems*, Elsevier Publ., Amsterdam.
Nihoul J.C.J. and Ronday F.C. (1976), *Modèles d'un estuaire partiellement stratifié*, in Recherche et Technique au Service de l'Environnement, edited by Le Conseil Scientifique de l'Environnement, Liège, 315-338.
Pritchard D.W. (1967), *Estuaries*, AAAS Publ., USA.
Ronday F.C. (1975), Annales des Travaux Publics de Belgique, 4, 1.
Theuns J. and Coen I. (1973), Annales des Travaux Publics de Belgique, 3, 139.

MATHEMATICAL MODEL OF WATER QUALITY IN A HIGHLY POLLUTED ESTUARY

G. BILLEN
Laboratorium voor Ekologie en Systematiek, Vrije Universiteit Brussel, Belgium.

J. SMITZ
Mécanique des Fluides Géophysiques. Environnement. Institut de Mathématique, Université de Liège, Belgium.

INTRODUCTION

This paper describes a general model of water quality in high polluted rivers and estuaries, showing the relations between the oxido-reduction balance, bacterial activity and hydrodynamic processes. Systematical application is made to the Escaut river estuary.

The Scheldt estuary (120 km long) is heavily polluted above Antwerp (km 80) by important amounts of organic matter. The deterioration of chemical and biological properties of the water by this domestic pollution is not the direct consequence of the presence of organic load, but the result of the intense heterotrophic activity which degrades this charge : oxygen is rapidly entirely depleted ; other oxidants (MnO_2 , NO_3^- , Fe(OH), ...) are used by anaerobic metabolisms, and the related reduced forms (Mn^{++}, NH_4^{++}, Fe^{++},...) are produced (km 120 - km 70).

Near km 60, owing to increasing salinity, flocculation and sedimentation of the suspended organic matter occur (Wollast 1973). Under the conjugated increase of salinity and disparition of organic matter, the bacterial activity falls down, and a phase of recuperation begins (km 30 - km 60), acceleration by mixing with sea-water, saturated in oxygen. The different oxidants are successively regenerated ; the last step is the reapparition of oxygen. Typical longitudinal profiles of oxido-reduction forms are shown in Fig. 1.a.b.

In this case, the classical models of river pollution cannot be used :

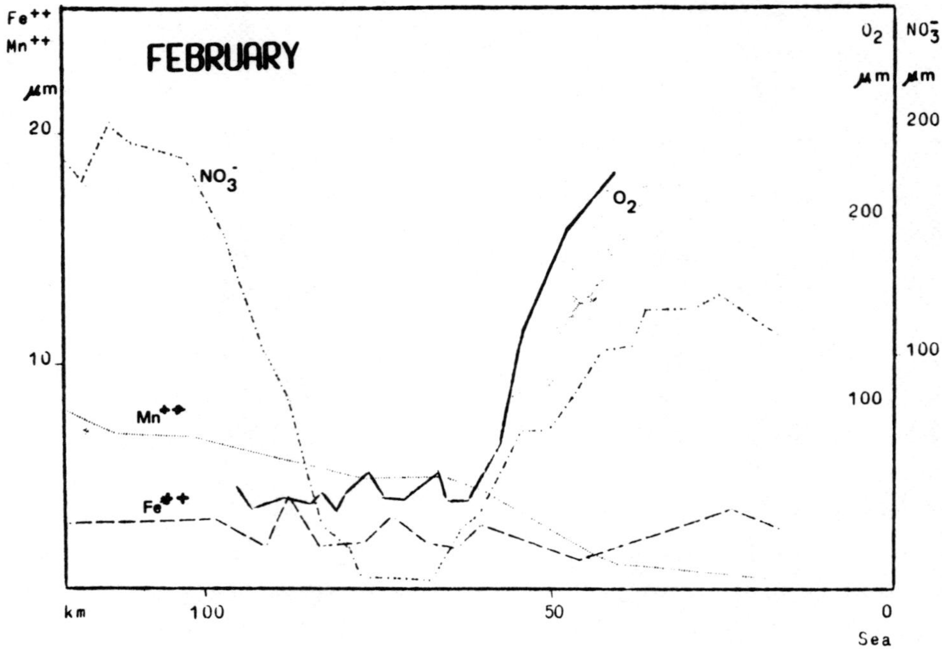
Fe++
Mn++
μm
FEBRUARY
O2
NO3-
μm
μm
20
200
NO3-
O2
200
10
100
Mn++
100
Fe++
km
100
50
0
Sea

Fig. 1.a.

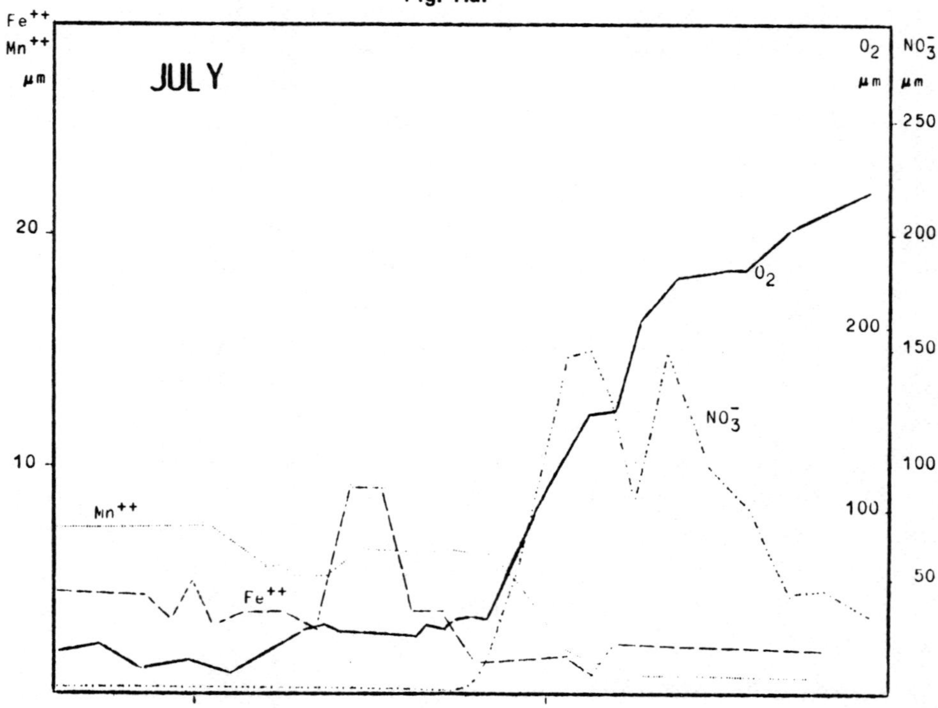
Fe++
Mn++
μm
JULY
O2
NO3-
μm
μm
250
20
200
O2
200
150
NO3-
10
100
Mn++
100
50
Fe++

Fig. 1.b.

i) the organic load is very important, and is not a limiting factor of the heterotrophic activity ;

ii) other oxidants than oxygen are used, and must be considered as state variables.

It is however possible to describe correctly the evolution of the chemical composition of the water by a complete oxido-reduction balance, using a few assumption about internal thermodynamic equilibrium.

PRINCIPLE OF THE MODEL

The bacterial activity, which controls the oxido-reduction processes, can be considered as an electron flux H(s,t) imposed to the system, according to the equation :

$$CH_2O + H_2O \xrightarrow{bact} CO_2 + 4H^+ + 4\ e^- \qquad (1)$$

The electron flux induces one or several of the following reactions :

a) $4e^- + \underline{O_2} + 4H^+ \rightleftarrows 2H_2O \qquad X_1$

b) $8e^- + \underline{NO_3^-} + 10H^+ \rightleftarrows NH_4^+ + 3H_2O \qquad X_2 \rightleftarrows Y_2$

c) $2e^- + \underline{MnO_2} + 4H^+ \rightleftarrows Mn^{++} + 2H_2O \qquad X_3 \rightleftarrows Y_3$

d) $1e^- + \underline{Fe(OH)_3} + 3H^+ \rightleftarrows Fe^{++} + 3H_2O \qquad X_4 \rightleftarrows Y_4$

d') $\underline{Fe^{++} + HCO_3^-} \rightleftarrows FeCO_3 + H^+ \qquad X'_4 \rightleftarrows Y'_4$

e) $14e^- + \underline{2SO_4^{--}} + Fe^{++} + 16H^+ \rightleftarrows FeS_2 + 8H_2O \qquad X_5 \rightleftarrows Y_5$

Some of these reactions are biologically mediated, and the presence of the responsible organisms has been demonstrated in the Scheldt Estuary (oxygen consumption, nitrification and denitrification,...).

Instantaneous bacterial activity has been measured experimentally (using a technique of dark ^{14}C bicarbonate incorporation). In the model, it is thus possible to introduce this activity, translated in electron flux, as a control parameter.

Then, the total electron flux H imposed to the system must be divided between the different chemical and biological pathways, i.e. between reactions a) to e).

In a first step, it is assumed that an internal thermodynamic equilibrium is reached for the reactions a) to e) in every section of the estuary, so that the following Nernst relations hold :

$$E_h = E_h^o + b \log \frac{X_i}{Y_i} \qquad (2)$$

where E_h is the local redox potential, X_i and Y_i the concentrations of oxidized and reduced forms. This assumption is equivalent to consider that the bacteria only use the thermodynamically favorable half reaction in their energy yielding metabolism. The validity of this assumption has been discussed by Stumm (1966), Thorstenson (1970) and verified by Billen (1974) for the special case of nitrification in the Scheldt Estuary.

In a second step, to take into account the population dynamics of the nitrifying bacteria, and to obtain a better correlation between calculated and measured concentrations, it is possible to introduce a kinetic limitation of the nitrate production term (the couple $NH_4^+ - NO_3^-$ being outside the thermodynamic equilibrium).

MATHEMATICAL FORMULATION

If X(s,t) is a mean cross-section concentration, we can write the conservation equation as :

$$\frac{\partial}{\partial t} X + u \frac{\partial}{\partial x} X = \frac{1}{A} \frac{\partial}{\partial s} (A\ K \frac{\partial}{\partial x} X) + P - C \qquad (3)$$

where

t is the time,

s is the longitudinal coordinate along estuary axis,

u is the velocity,

A is the section,

K is the diffusion coefficient,

P and C are respectively the production and consumption terms.

The values of u(s,t), A(s,t), K(s,t) are derived from hydrodynamic models of the estuary (Wollast, 1973, Ronday 1975, Nihoul and Ronday 1976).

Starting from a few boundary conditions (concentrations upstream - km 120 - and downstream - km 0 -), and from the values of the bacterial activity, the model describes the evolution of the water composition along a longitudinal profile. It is not important at this stage to calculate the organic load profile, because :

i) the organic load is so high that it does not constitute the limiting factor of bacterial activity ;

ii) the fate of organic matter is determined mainly by flocculation and sedimentation processes, and not by bacterial degradation.

If $\nabla\Delta$ represents the "hydrodynamic operator"

$$\nabla\Delta \equiv \frac{\partial}{\partial t} + u \frac{\partial}{\partial s} - \frac{1}{A} \frac{\partial}{\partial s} (A\ K \frac{\partial}{\partial s})$$

we can write for every oxidants X_i and every reduced form Y_i :

$$\nabla\Delta\ X_i(s,t) = P_i(s,t) - C_i(s,t) \qquad (i = 1,5) \qquad (4)$$

$$\nabla\Delta\ Y_i(s,t) = - P_i(s,t) + C_i(s,t) \qquad (i = 1,5) \qquad (5)$$

where the P_i's (s,t) (production term) and the C_i's (s,t) (consumption term) are unknown functions.

The problem can be solved by the introduction of associated variables Z_i and F :

i) $Z_i \equiv X_i + Y_i \qquad (i = 2,5) \qquad (6)$

As the operator $\nabla\Delta$ is linear, the variables $Z_i(s,t)$ are "conservative" :

$$\nabla\Delta\ Z_i = 0 \qquad (i = 2,5) \qquad (7)$$

The solutions of the four equations (7) provide directly the local maximum value for the X_i's and the Y_i's, and the value of the sum

$X_i(s,t) + Y_i(s,t)$

ii) $F(s,t) = \sum_i \nu_i\ X_i(s,t) \qquad (8)$

where the ν_i are stoechiometric coefficients for the electron fluxes.
As $\nabla\Delta$ is linear :

$$\nabla\Delta\ F(s,t) = \Sigma\ \nu_i\ P_i - \Sigma\ \nu_i\ C_i \qquad (9)$$

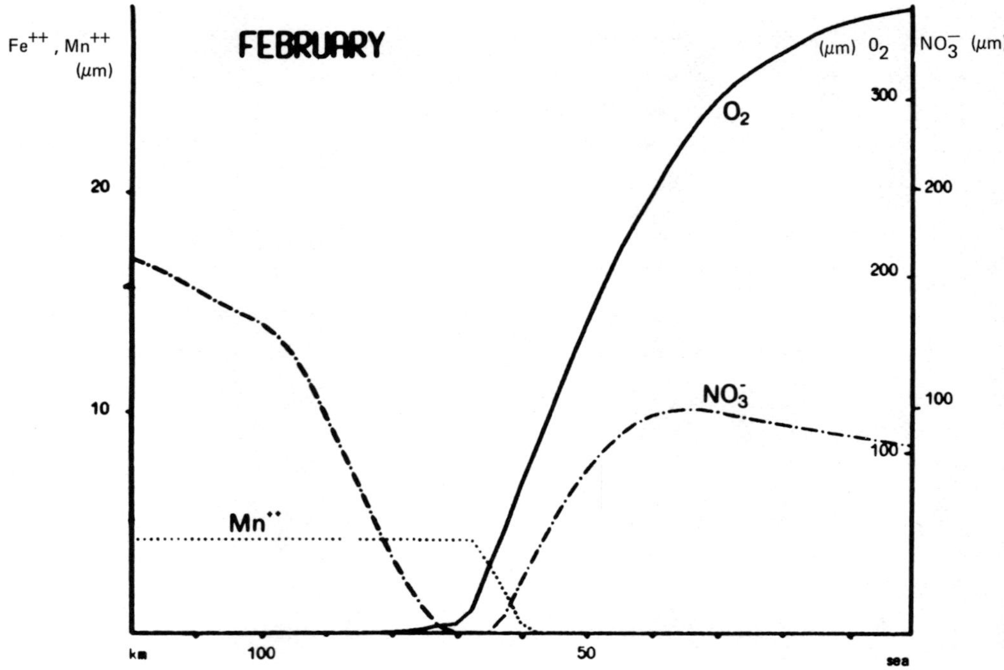

FEBRUARY
Fe++ , Mn++
(μm)
(μm) O2
NO3- (μm)
20
10
300
200
200
100
100
O2
NO3-
Mn++
km
100
50
sea

Fig. 2.a.

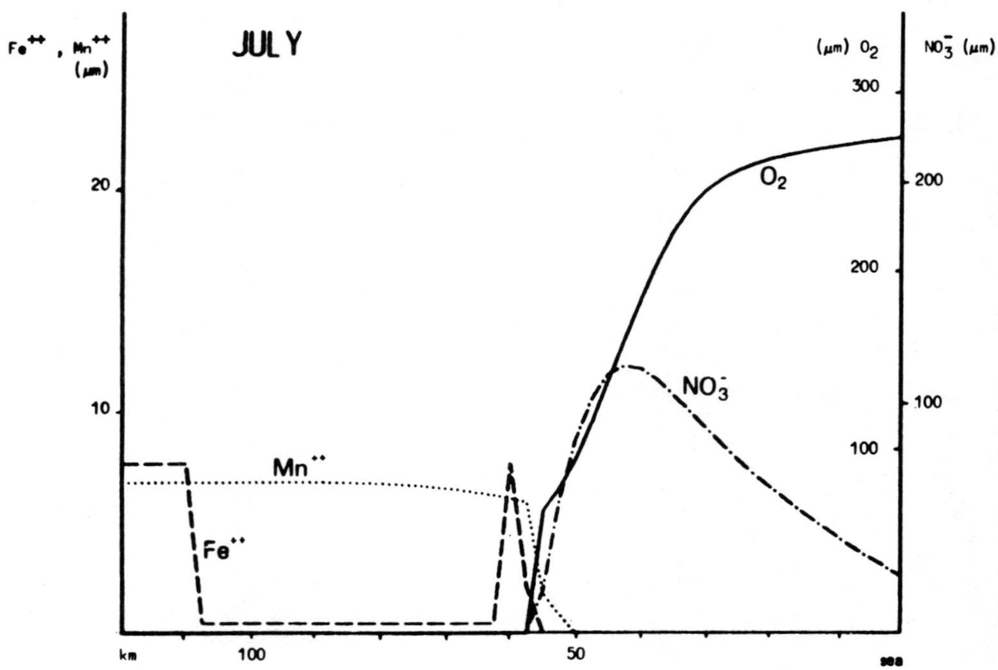

JULY
Fe++ , Mn++
(μm)
(μm) O2
NO3- (μm)
300
20
200
200
10
100
100
O2
NO3-
Mn++
Fe++
km
100
50
sea

Fig. 2.b.

In (9) the sum of the consumption terms $\Sigma \nu_i C_i$ is the total electron flux H(s,t) due to the bacterial activity ; the sum of the production terms $\Sigma \nu_i P_i$ reduces to the reaeration, calculated by a relation

$$\kappa (X_1^{sat} - X_1) \tag{10}$$

where κ is a known constant.

The equation (9) - which represents a global oxido-reduction budget - can be written

$$\nabla\Delta F(s,t) = \nu_1 \kappa (X_1^{sat} - X_1) - H(s,t) \tag{11}$$

The set of equations (2) - (11), with the corresponding boundary conditions, allows a complete determination of the X_i's (s,t) and the Y_i's (s,t).

The numerical discretization of the equations provides tridiagonal matricial equations, which are calculated by recurrent algorithms (Adam and Runfola 1971, Adam 1975). As the equation (11) contains the oxygen concentration X_1 explicitly, the term

$$(X_1^{sat} - X_1)$$

is calculated with the value of X_1 at the preceeding time step (if X_1 varies slowly), or with an iterative computation scheme.

During this numerical iterative process, a kinetic limitation of the local nitrate production term is introduced, to reproduce the activity rate of nitrifying bacteria in the downstream part of the estuary.

The results of the calculations are shown at fig. 2.a.b. (winter situation : february - summer situation : july).

ACKNOWLEDGEMENT

This work was conducted in the scope of the Belgian National Environment Program, sponsored by the Science Policy Administration, Office of the Prime Minister.

REFERENCES

Adam, Y. and Runfola Y., 1971. Numerical Resolution of diffusion equation, Rapport N.9, Progr. Nat. Environnement Physique et Biologique, Projet Mer.

Adam, Y., 1975. A Hermitian finite difference method for the solution of parabolic equations, to be published.

Billen, G., 1975. Nitrification in the Scheldt Estuary (Belgium and the Netherlands). *Estuarine and Coastal Marine Science*, 3, 79-89.

Billen G. and Smitz J., 1975. A mathematical model of oxido-reduction processes in the Scheldt Estuary, Math Modelsea - ICES Hydrography Committee C.M. 1975, C:21.

Billen, G., Smitz, J., Somville, M. and Wollast, R., 1976. Degradation de la matière organique et processus d'oxido-reduction dans l'Estuaire de l'Escaut. Prog. Nat. R-D Environnement - Projet Mer - Rapport Final - vol X, p. 102 - 152.

Breck, W.G., 1974. Redox potentials, in The Sea, Goldberg ed., vol. 5 , Wiley, New York.

Nihoul, J.C.J., 1975. Modelling of Marine Systems, Elsevier Publ., Amsterdam.

Nihoul, J.C.J., Ronday, F.C., Smitz, J. and Billen, G., 1977. Hydrodynamic and water quality model of the Scheldt Estuary. Marsh-Estuarine Simulation Symposia. Georgetown, South Carolina, Janvier 6-8, 1977, in press.

Overbeck, J. and Daley, R.J., 1973. Some precautionary comments on the Romanenko technique for estimating heterotrophic bacterial production. *Bull. Ecol. Res. Comm.* (Stockholm), 17, 342-344.

Romanenko, V.I., 1964. Heterotrophic CO_2 assimilation by bacterial flora of water. *Mikrobiol.*, 33, 779-683.

Ronday, F.C., 1975. Etude de l'envasement et de la variation longitudinale du coefficient de dispersion dans les estuaires partiellement stratifiés. *Annales des Travaux Publics*, 4, 1975.

Thorstenson, D.C., 1970. Equilibrium distribution of small organic molecules in natural waters. *Geochim. Cosmochim. Acta*, 34, 745-770.

Wollast, R., 1973. Circulation, accumulation et bilan de masse dans l'estuaire de l'Escaut, Rapport de synthèse 1972, Commission Interministérielle de la Politique Scientifique (Belgium).

MODELLING OF BIOLOGICAL AND CHEMICAL PROCESSES IN THE SCHELDT ESTUARY.

R. WOLLAST
Laboratory of Oceanography, University of Brussels (Belgium).

INTRODUCTION

Research chemists and biologists involved in a dynamical description of the behaviour of chemical species or living organisms in estuarine systems are faced with the difficult problem of evaluating the in situ rates of transformation of the species considered, and distinguishing concentration changes due to the mixing of water masses from changes related to biological or chemical processes.

Generally, they do not have the opportunity to use elaborated hydrodynamical models, for two reasons, either because these models are inexistant, or too complicated in order to include in a tractable manner, the kinetic terms describing the evolution of the chemical and biological parameters.

Hereafter, we intend to show by means of a few examples, that the use of simplified unidimensional and stationnary models of the estuarine system constitute a first approach allowing a better understanding of the chemical and the biological processes occuring in the system, when considering long term evolutions of the parameters.

The basic principle of these models is to use the distribution of salinity, which is a conservative parameter, in order to evaluate the mixing processes of fresh water and sea-water.

The longitudinal distribution is then simply reduced to :

$$\frac{d}{dx}(u.S) = \frac{1}{A}\frac{d}{dx}\left(A\ K\ \frac{dS}{dx}\right) \qquad (1)$$

The longitudinal mixing coefficient K deduced from the salinity profile then includes effects related to the complicated hydrodynamical circulation of estuarine system and, to a certain extent, non-stationnary effects due to changes of the fresh water discharge and the tidal amplitude.

For a non-conservative substance, the longitudinal profile will be described by :

$$\frac{d}{dx}(uC) = \frac{1}{A}\frac{d}{dx}\left(A\ K\ \frac{dC}{dx}\right) + P - C \qquad (2)$$

if P and C are respectively the production and the consumption terms affecting the considered constituant.

The principal aim of the model is however for the biologist or the chemist to evaluate their importance and their dependence on other environmental parameters.

The utilisation of the same longitudinal mixing coefficient in order to describe the hydrodynamical behaviour of such substances, and particularly those introduced in the estuary by the fresh water flow, must be considered with great care. It is for example easy to understand that such a simplification is out of the question when the estuary is vertically stratified or when the residence time of the fresh water is short in relation to the fluctuations of the water discharge. Thus, the model must be tested with other conservative parameters, ideally characteristic of the fresh water flow.

It will be shown in the following paragraphs that the simplified unidimensional model may be applied in the case of the vertically well mixed Scheldt estuary, and that it resulted in the identification of several important mechanisms and rate constants about the behaviour of various chemical species in the estuarine system.

THE BEHAVIOUR OF DISSOLVED SILICA

The estuaries constitute an important source of dissolved silica for the marine environment since river water has a mean content of 15 mgr SiO_2/l compared to that of 6 mgr SiO_2/l for the mean ocean and values as low as 0,1 mgr SiO_2/l for surface sea-water. It is an essential nutrient for diatoms, which are the dominant phytoplancton species in many areas.

The behaviour of dissolved silica in the estuarine systems is at the present time a subject of controversy. The consumption of dissolved silica which varies from 10, 20 % in many estuaries (Burton and Liss, 1973) to 80, 90 % in estuaries like the Scheldt (Wollast and De Broeu, 1971) and the Rhine (Van Bennekom, 1974) is explained either by chemical reactions with the suspended matter or by an intense activity of brackish diatoms.

The conservative or non conservative character of chemical species is well illustrated by plotting the evolution of its concentration in the estuary as a function of the salinity (Fig. 1). In the case of dissolved silica, for the month of February, one obtains a fairly good linear relation, corresponding to a simple mixing law and a strong deviation from the mixing line for September indicating an intense consumption of dissolved silica in the estuarine zone. We will use the particular behaviour of this compound for testing the possibility of using the simplified unidimensional model to describe the longitudinal profile of a conservative substance, such as dissolved Si during the winter, as well as to evaluate the rate and consumption mechanism of the same compound during the period from spring to fall.

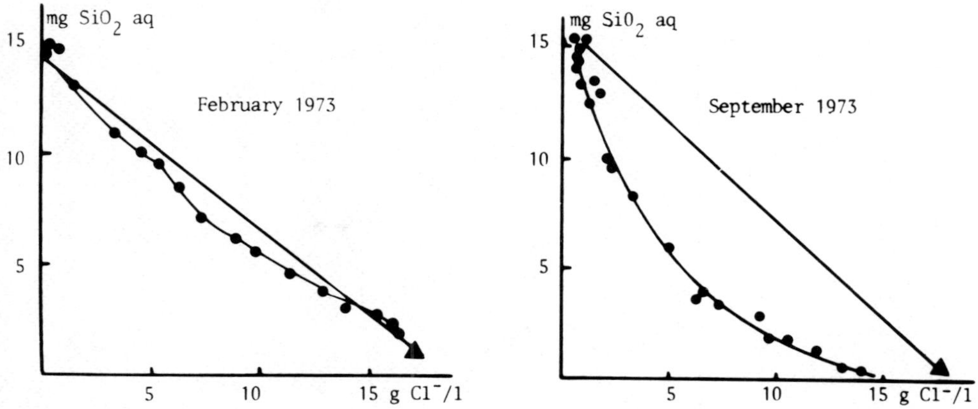

Fig. 1. Evolution of dissolved silica as a function of salinity during February 1973 and September 1973 compared to the mixing line. The points ▲ correspond to the composition of the coastal sea-water.

Distribution of silica for conservative conditions

The evolution of the longitudinal mixing coefficient K along the estuary calculated from the salinity profile during the month of February is given in figure 2. It should be noted that the longitudinal concentration profiles are always measured by following the low water slag starting from the mouth of the river.

The calculated distribution of dissolved silica, considered as conservative and submitted to the same mixing processes as salinity (same K) is compared to the measured distribution in figure 3. The agreement is very satisfactory and justifies the use of the simplified model in the case of the Scheldt.

Distribution of silica for non-conservative conditions

The montly evolution of the longitudinal profile of dissolved silica over one year shows that, if this compound behaves as a fairly conservative substance during the winter, it undergoes an uptake in the estuary beginning in May, increasing during the summer and decreasing rapidly after September. This already suggests that the removal is related to biological processes.

In order to estimate more quantitatively the rate of removal and to localize the area of silica uptake, the net consumption term (P – C) of equation 2 was estimated for successive 5 km long sections along sections along the estuary, using again the salinity profile as a tracer of the mixing properties of the water masses.

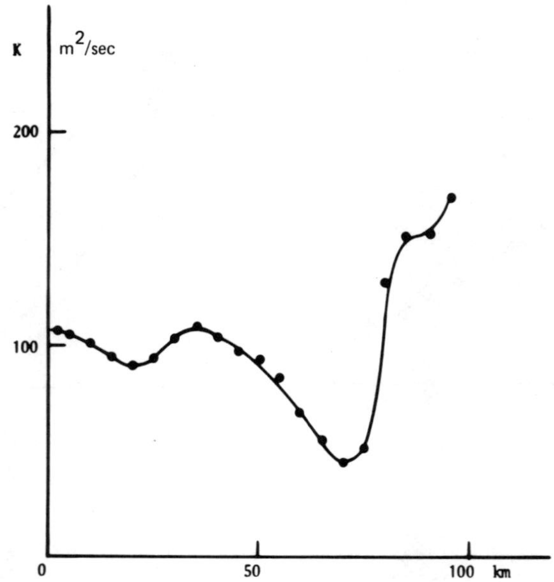

Fig. 2. Longitudinal mixing coefficient calculated from the salinity profile for the river Scheldt in February 1973.

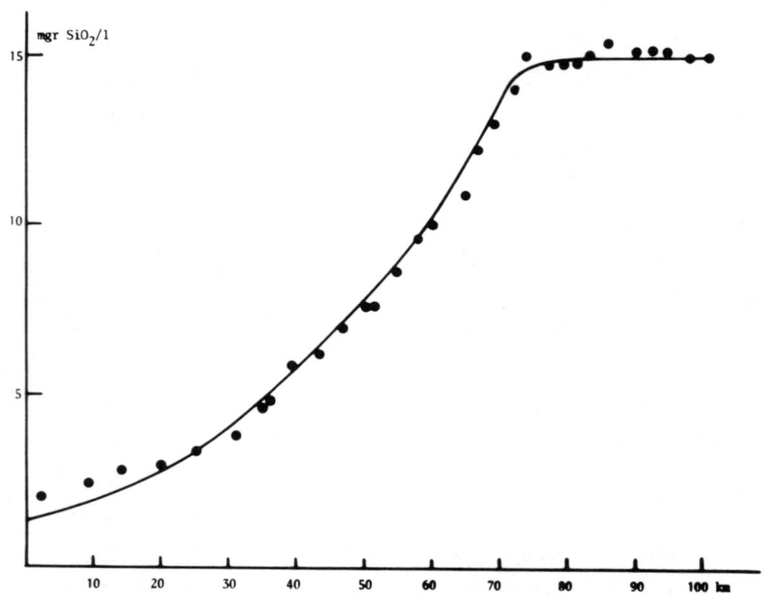

Fig. 3. Calculated and measured profile of dissolved silica for the river Scheldt in February 1973 assuming a conservative behaviour for silica.

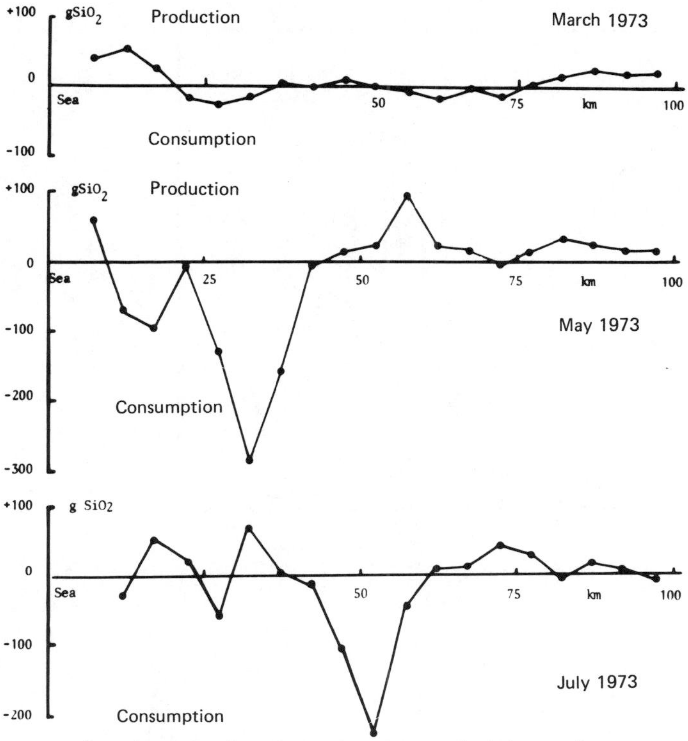

Fig. 4. Calculated production and consumption terms for dissolved silica as a function of the distance to the sea.

The results of these computations for three typical months are represented in figure 4. These profiles indicate that the uptake of silica is restricted in well defined zones of the lower part of the estuary. This conclusion corresponds to well known facts about the biological activity in the Scheldt estuary.

The high turbidity (Fig. 5.) of the upper part of the estuary (above km 60) strongly inhibits the activity of the phytoplancton and explains the conservative behaviour of dissolved silica in that region. On the other hand, the activity of the diatoms is characterized by successive blooms of restricted expance usually starting early in the year near the mouth and progressing upwards during summer and fall. This is well demontrated in figure 6 which shows the evolution of the number of diatom cells recorded by De Pauw (1975) at two stations of the Scheldt during three successive years.

Another direct proof of the role of the phytoplancton in the silica uptake observed in the Scheldt, as well as the validity of the consumption rate evaluated with the help of the model, was obtained by comparing the calculated uptake of silica to the primary productivity measured in the zone of uptake by $H^{14}CO_3$ incorporation.

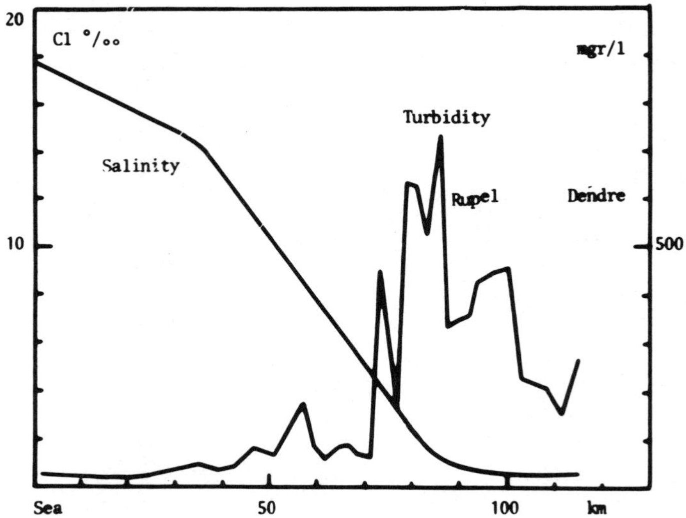

Fig. 5. Longitudinal turbidity profile in the Scheldt estuary

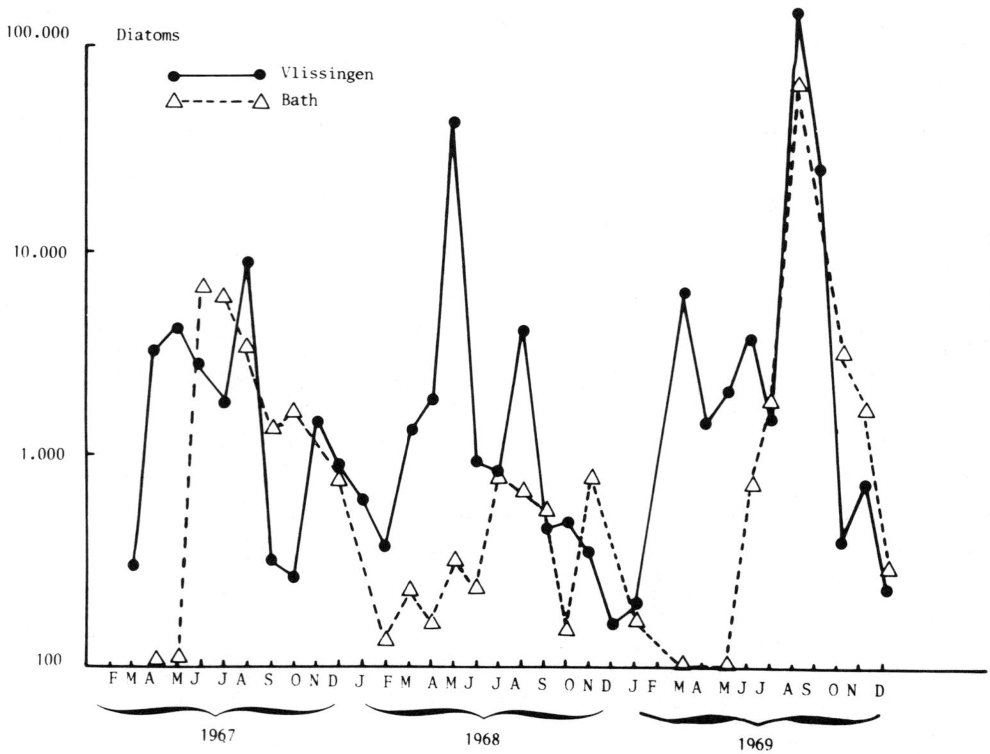

Fig. 6. Seasonnal evolution of the diatoms at Vlissingen (km 5) and Bath (km 40) (after De Pauw, 1975)

During the month of September 1973 the zone of highest primary productivity was situated between km 35 and km 65 from the mouth and reached 50 mg C/m^2.day. The mean rate of uptake of silica deduced from the model for the same region was estimated at 138 mg SiO_2/m^2.day. The weight ratio of silica to C uptake is thus 2.76 in agreement with a weight ratio of SiO_2/C equal to 2.3 based on the mean composition of marine diatoms (Lizitsine, 1972).

Taking these facts into account, we have used the unidimensional model in order to predict the longitudinal distribution of silica in the estuary, where the biological uptake of SiO_2 is simply proportionnal to the concentration of diatom cells (C_D) as measured by De Pauw (1975).

The equation is then simply :

$$- K \frac{d^2C}{dX^2} + \frac{1}{A}\frac{dC}{dX} \left(-K \frac{d}{dX} - V\right) + \frac{1}{A}\frac{dV}{dX} (C - C') + k\, C_D \qquad (3)$$

where V is the fresh water discharge and C' the silica concentration in the various small tributaries.

The results of the calculation are summarized in figure 7. The upper curve corresponds to a very low activity of the diatoms achieved during the winter and the lower curve to a maximum of the activity with a pronounced bloom of diatoms between km 50 and km 40 reached during the summer.

As we can see, this model is very useful in order to evaluate the activity of the diatoms and allows one to relate this activity with various environmental factors. The model was also used in order to evaluate the amount of dissolved silica discharge by the estuary into the North sea (figure 8).

In fact, the amount of dissolved silica delivered to the North sea is considerably reduced from May until September. The same situation occurs in the river Rhine (Van Bennekom, 1974). These unusual situations must be related to the eutrophication of these rivers due to their high concentration levels of dissolved nitrogen and phosphorus.

This low input of dissolved silica has a considerable effect on the phytoplancton composition of the North sea where silica may become limitant (Van Bennekom et al., 1975).

NITRIFICATION IN THE SCHELDT ESTUARY

Nitrification means the oxydation of ammonia into nitrite and nitrate caused by the activity of autotrophic bacteria. This process is very important because it modifies the speciation of the inorganic nitrogen and effects its assimilation rate by the phytoplancton. Nitrification also consumes large quantities of dissolved oxygen and may affect the quality of the estuarine water.

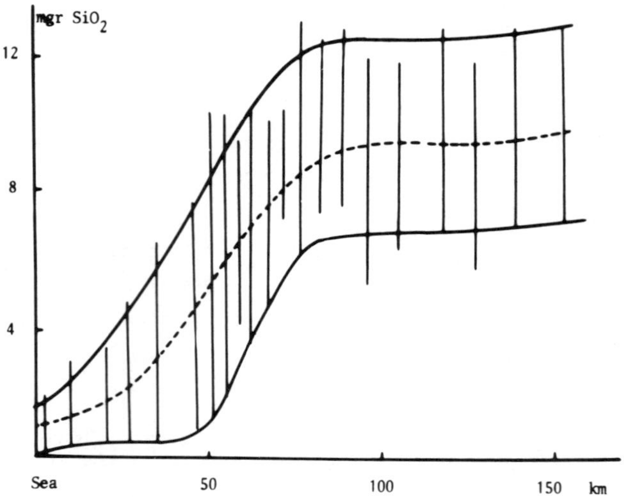

Fig. 7. Computed longitudinal profils of dissolved silica for winter (upper curve), summer (lower curve) and a medium situation. The vertical dashes correspond to the observed evolution of dissolved silica over one year (from De Pauw, 1975)

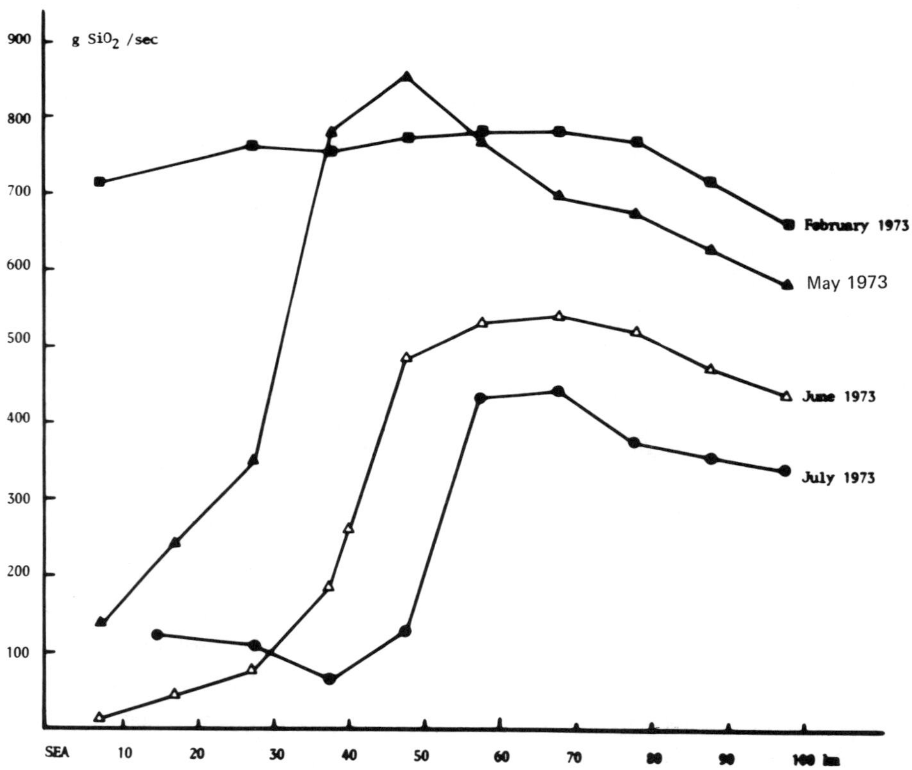

Fig. 8. Evolution of the flux of dissolved silica in the Scheldt.

It is generally admitted that the nitrification process occurs only when the heterotrophic degradation of organic matter is completed, but there is no physiological argument to support this hypothesis.

We have thus tried to develop a model of the nitrification in the Scheldt estuary which occurs intensively in the lower part of the estuary where the organic load has severely decreased and oxydative conditions are restored. However, this model takes into account known physiological properties of bacteria and is based furthermore, on several in-situ or laboratory experiments (Billen, 1975, Somville, 1975).

First of all it is important to underline that the oxydation of ammonia into nitrite and nitrate constitutes the sole source of energy in the metabolism of the nitrifying bacteria. Thus nitrification can only occur in the area of the estuary where this oxydation process is exoenergetic. This condition may be expressed thermodynamically by introducing the value of the oxydo-reduction potential Eh, above which nitrate and nitrite become more stable than ammonia. In the case of the Scheldt the zone where this condition is fulfilled is restricted to the lower part of the estuary and its extent depends upon the fresh water discharge, the pollution load, the temperature, etc...

Even when the thermodynamical conditions are favourable in the estuary, the nitrification takes place rather slowly and nitrification is rarely complete. Some in-situ observations clarify this particular behaviour.

Comparative counts of nitrifying bacteria grown on fresh water or sea-water mediums show that the nitrifying bacterial populations of the Scheldt estuary are essentially of continental origin. Even in the coastal zone near the estuary no halophile population is developing.

The nitrifying organisms show no activity in the upper zone of the estuary because of the unfavourable redox conditions and nitrate appears only when the critical value of Eh is reached (figure 9). On the other hand, the growth of the fresh water populations is rapidly inhibed with the increase of salinity and the bacterial populations are more rapidely diluted by sea-water than they can expand by reproduction (figure 10a).

The in-situ measured activity of the nitrifying bacteria (figure 10b) reflects both the inhibition due to unfavourable redox condition in the upper part of the estuary and the rapid dilution of the fresh water population by the sea-water in the lower part of the estuary.

Modelling of the nitrification

A bacteriological model of nitrification must necessarily first of all define the evolution of the biomass B of the nitrifying organisms, which may be decomposed into a growth term and a mortality term.

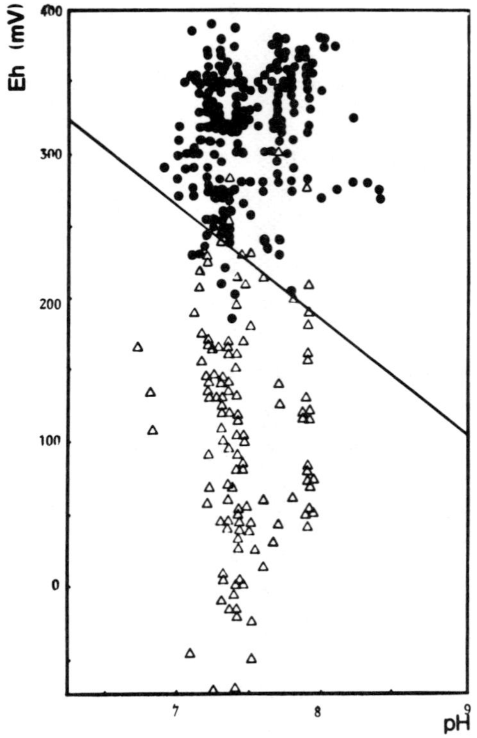

Fig. 9. Relation between nitrification and the oxydo-reduction conditions in the Scheldt

● nitrification
△ no nitrification

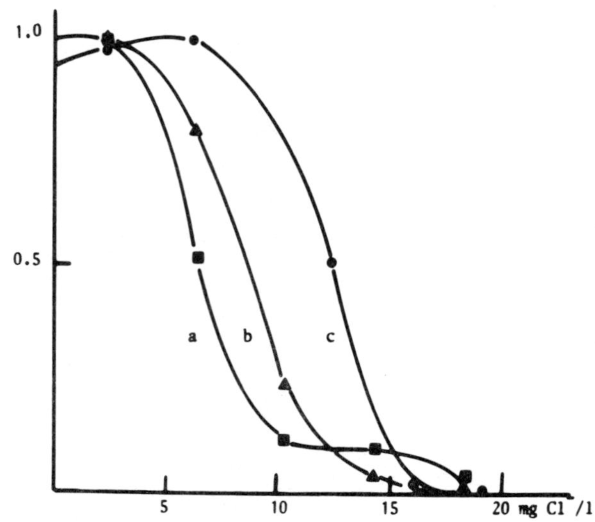

Fig. 10a. Activity index of nitrifying bacteria as a function of salinity. Curves a, b,c correspond to the activity of cultures adapted progressively to higher salinities.

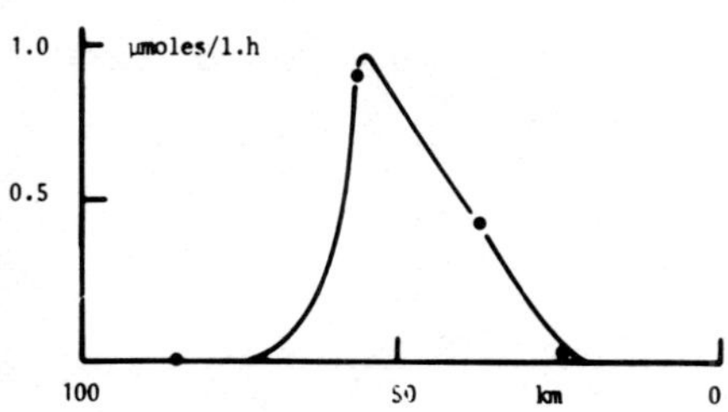

Fig. 10b. In-situ measurements of the nitrification activity during october 1975.

If $\nabla\Delta$ represents the "hydrodynamic" operator :

$$\nabla\Delta = \frac{\partial}{\partial t} + u\frac{\partial}{\partial X} - \frac{1}{A}\frac{\partial}{\partial X}\left(A\ K\frac{\partial}{\partial X}\right) \qquad (4)$$

the evolution of the bacterial biomass may be written :

$$\nabla\Delta\ B = KB - MB$$

where K and M are respectively the growth and mortality coeffecients. The growth coefficient is a function of temperature, salinity, concentration of ammonia and Eh.

If k is defined as the optimal growth coefficient, the influence of these various parameters may be conveniently described by the following relation :

$$K = k\ .\ f_1\ (T)\ .\ f_2\ (S)\ .\ f_3\ (NH_4)\ .\ (\ Eh) \qquad (5)$$

f_1, f_2 and f_3 being equal to 1 for the optimal values of temperature, salinity and ammonia concentration. The function (Eh) expresses simply that nitrification is only possible above a given redox potential where the oxydation of ammonia becomes exoenergetic :

$$(Eh) = \begin{cases} 0 & Eh < E_N \\ 1 & Eh > E_N \end{cases}$$

The value of k and f_1, f_2 and f_3 were based on data from literature and laboratory experiments.

According to Painter (1970), k varies between 2 10^{-5} and 5 10^{-6} sec^{-1}. Carlucci and Strickland (1968) have determined the influence of the temperature which may be described by

$$f_1 = 10^{\frac{T-28}{15}} \qquad (0 < T < 28°\ C)$$

f_2 was adjusted from our laboratory experiments and approximated at :

$$f_2 = \exp\ (-\ \{Cl^-\}/8)$$

$\{Cl^-\}$ being expressed in gr/liter.

The dependence of the activity with respect to the concentration of ammonia was described by the classical Michaëlis-Menten relation

$$f_3 = \frac{(NH_4)}{K_m + (NH_4)}$$

Painter (1970) reports values for K_m of between 70 to 700 μmole/l with a reasonable mean situated around 250 μmole/l.

The mortality of the fresh water populations of nitrifying bacteria is also a function of salinity. This influence is again described empirically by an exponential equation based on experimental results :

$$M = m \frac{(Cl^-) - 8}{3}$$

The nitrification activity is then simply related to the rate of growth of the bacterial population where the increase of nitrate + βKB is exactly compensated by a decrease of ammonia - βKB. β is an efficiency factor characteristic of the bacteria of between 10^{-7} and 5 10^{-6} μmoles/bact. according to Carlucci and Strickland (1968) and Watson (1965).

This model was applied to the description of the profile of nitrate and ammonia in the Scheldt during the winter period*.

The comparisons between the calculated and measured concentrations are shown in figure 11. The relation is very satisfactory. The however model shows to an interesting contradiction. If we use the admitted values for β of between 10^{-7} and 5 10^{-6} μmoles/bact., the number of the nitrifying bacteria predicted by the model is three orders of magnitude higher than the number observed in the water column. Inversely, the adjustement in the model of the number of nitrifying bacteria consistent with observed values also lead to a factor of β three order of magnitude higher, which in our opinion, is unrealistic.

Any attempt to improve the bacterial counts in the water column failed and this lead us to the suggestion that the benthic activity of the nitrifying bacteria is the predominant contribution in the oxydation of ammonia into nitrate. The importance of this process has been suggested previously in the case of the English rivers (Curtis et al., 1975) and in the coastal region of the North Sea (Billen , in press).

In fact, the concentration of the nitrogen species in the estuary, besides the winter, is further complicated by the uptake of these nutrients by the phytoplancton.

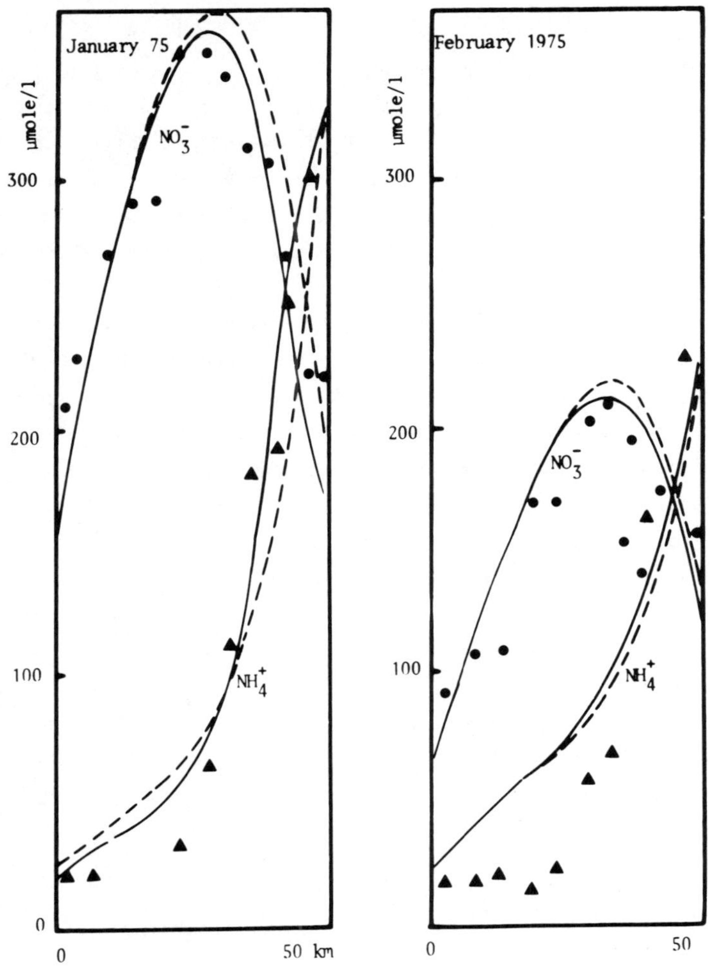

Fig. 11. Calculated curves and measured points of the longitudinal profiles of nitrate and ammonia. The continuous lines correspond to a computation based on the observed number of bacteria and an unusual high efficiency value (β = 2 and 3 10^{-3} µmoles/l). The dotted curves correspond to a reasonable value of β taken from the litterature (β = 5 10^{-6} µmoles/l) but a number of bacteria 1.000 times the measured values.

CONCLUSIONS

The chemical and biological processes occuring in natural environments are known to be very complicated according to the intricated interactions of numerous factors. In the case of the estuarine systems, the mixing of fresh water and sea water induces furthermore large changes of ionic strenght, chemical composition, speciation and distribution of organisms along the estuary, which renders the situation even more complicated than in most more homogeneous environments.

These processes are thus described generally by rather approximate semi-empirical relations as shown for example in this paper by the behaviour of dissolved silica and nitrate in the Scheldt estuary. These approximations do not justify in many cases the use of elaborated hydrodynamical models and to our opinion a stationnary one dimensional model constitutes a sufficient first approach.

Management decisions are often requiered urgently in order to prevent the increasing pollution of many estuaries. In this respect, the use of these simple models constitutes an usefull tool realising a compromise between fiability, cost and time requirements.

AKNOWLEDGEMENTS

This research was supported by the Belgian National Program of Research and Development on the Physical and Biological Environment, spondered and conducted by the Department of Science Policy, Prime Minister's Office (1970-1976).

REFERENCES

Billen, G., 1975. Nitrification in the Scheldt estuary (Belgium and the Netherlands). Estuarine and Coastal Marine Science, 3,279-289.

Billen, G., in press. A budjet of nitrogen recycling in the North sea sediments off Belgian coast. Submitted to Estuarine and Coastal Marine Science.

Burton, J.D. and Liss, P.S., 1973. Processes of supply and removal of dissolved silicon in the oceans. Geochim. Cosmochim. Acta, 37, 1761-1773.

Carlucci, A.F. and Strickland, J.D.H., 1968. The isolation, purification and some kinetic studies of marine nitrifying bacteria. Journal of experimental marine Biology and Ecology, 2, 156-166.

Curtis, E.J.C., Durrant, K. and Harman, H.M.I., 1975. Nitrification in rivers in the Trent basin. Water Res., 9, 255-268.

De Pauw, C., 1975. Bijdrage tot de kennis van milieu en plancton in het Westerschelde estuarium, Thèse de doctorat, Rijksuniversiteit Gent.

Lisitzin, A.P., 1972. Sedimentation in the world ocean, Soc. of Econ. Paleon. and Mineralog., Special Public., 17.

Painter, H.A., 1970. A review of literature on inorganic nitrogen metabolism in micro-organisme, Water Res., 4, 393-450.

Somville, M., 1975. Nitrification dans l'estuaire de l'Escaut, Mémoire de licence en sciences chimiques, Université de Bruxelles.

Van Bennekom, A.J., Krijgsman-Van Hartingsveld,E., Van der Veer, G.C.M. and Van Voorst, H.F.J., 1974. The seasonal cycles of reactive silicate and suspended diatoms in the Dutch Wadden Sea, Neth. J. of Sea Res., 8 (2-3), 174-207.

Van Bennekom, A.J., Gieskes, W.W.C. and Thijssen, S.B., 1975. Eutrophication of Dutch coastal waters, Proc. R. Soc. London, B 189, 359-374.

Watson, S.W., 1963. Autotrophic nitrification in the ocean, in : Symposium on Marine Microbiology, Oppenheimer, C.H. (Editor), CC. Thomas publ., Springfield.

Wollast, R. and De Broeu, F., 1971. Study of the behaviour of dissolved silica in the estuary of the Scheldt, Geochim. and Cosmochim. Acta, 35, 613-620.

TURBULENT MIXING IN A SALT WEDGE ESTUARY

GEORGE B. GARDNER AND J. DUNGAN SMITH
Department of Oceanography, University of Washington

ABSTRACT

Experiments carried out by Partch and Smith (1977) at anchor stations in the Duwamish River in Seattle, Washington showed a period of intense vertical salt flux during the ebb tide. In order to examine the longitudinal structure of these mixing events, as well as to procure data relevant to the overall dynamics of the estuary, a system was developed to permit measurement of the velocity and density fields from a moving vessel. The new system was used in an experiment in March 1977. Data obtained during this experiment indicate that the intense mixing events may be triggered by a hydraulic jump that occurs at a sharp change in river depth. The relatively high salinity water formed at the jump advects downstream, but vertical mixing continues with a salt flux of at least 0.5 $gm/m^2 \cdot sec$. It is postulated that this continued mixing is related to an internal hydraulic instability. As it occurs during periods of minimum shear, shear instability is ruled out as a source of energy. In addition to the intense mixing events, an internal hydraulic jump that forms at the 16th Avenue bridge around the time of maximum ebb is described. While this apparently is not related to the intense mixing events, it is of interest in its own right, and indicates the importance of internal hydraulics to the dynamics of salt wedge estuaries.

INTRODUCTION

In order to determine the predominant mixing mechanisms in a typical salt wedge estuary, a project was instigated several years ago to investigate turbulent processes in the Duwamish River. This estuary enters Puget Sound in the southern part of Seattle, Washington. Although heavily industrialized along its banks, it also has a sizable salmon fishery and has been studied previously from several environmental points of view. The background hydraulics and oceanography are described by Dawson and Tilley (1972), Santos and Stoner (1972), and Stoner (1972).

This is contribution 1003 from the University of Washington. The work described herein was supported by NSF Grant DES-75-15154.

For river discharges in excess of 30 m^3/s the estuary is of the salt wedge type throughout its length; whereas for lower river flows, such as normally occur in summer, the lower 5.6 km portion grades into a type 2B of the Hansen and Rattray (1966) classification. The salt wedge nature of the estuary is due to high runoff velocities and occurs in spite of a large tidal range (up to 3.5 m). These circumstances could produce qualitative differences between the mixing processes found in the Duwamish and those in certain other salt wedge estuaries, such as the Mississippi, which have a much smaller tidal range. Nevertheless, the former is, at very least, typical of an important subclass of salt wedge estuaries.

The lower 10 km of the Duwamish River have been dredged and straightened by the U.S. Army Corps of Engineers, producing an estuary comprised of several straight reaches connected by short curves, and an estuary of relatively uniform depth. Figure 1 is a map of the lower portion of the Duwamish River. The segment of primary concern in this paper lies between the two lateral lines upstream and downstream of the 16th Avenue South bridge. Figure 2 is a sketch showing the depth profile and the general longitudinal salinity field under the flow conditions that existed during the experiment described in the present paper. The relatively simple topography and its proximity to the University of Washington make the Duwamish River an ideal laboratory for turbulent mixing studies.

The first experiments conducted during the turbulent mixing project focused on elucidating the temporal variation of temperature, salinity, mean velocity, turbulent kinetic energy, Reynolds stress and turbulent salt flux profiles procured from anchor stations maintained for periods of two to four days. Some results from this part of the project have been described by Partch and Smith (1977). Their data were obtained at the 8.5 km point marked in Figure 1. Partch and Smith found that 50% or more of the flux of salt across the interface occurred during intense mixing events lasting only a few hours. These events were made evident through increases in surface salinity and surface mixed layer thickness as well as through direct salt flux calculations, the computed value of $\overline{s'w'}$ varying from 0.25 $gm/m^2 \cdot s$ during quiet periods to 1.24 $gm/m^2 \cdot s$ during the events. The intense mixing events also were reflected in greatly increased turbulent kinetic energies. These ranged from 4 cm^2/s^2 during quiet periods to 20 cm^2/s^2 during intense mixing events. The mixing events occurred during periods of minimum shear so Kelvin-Helmholtz instabilities could not have been the main source of turbulent energy. Further, the increasing surface layer depth and surface salinity eliminated entrainment by bottom-induced turbulence as the primary source. However, Partch and Smith showed that the mixing events corresponded to periods of critical or near critical

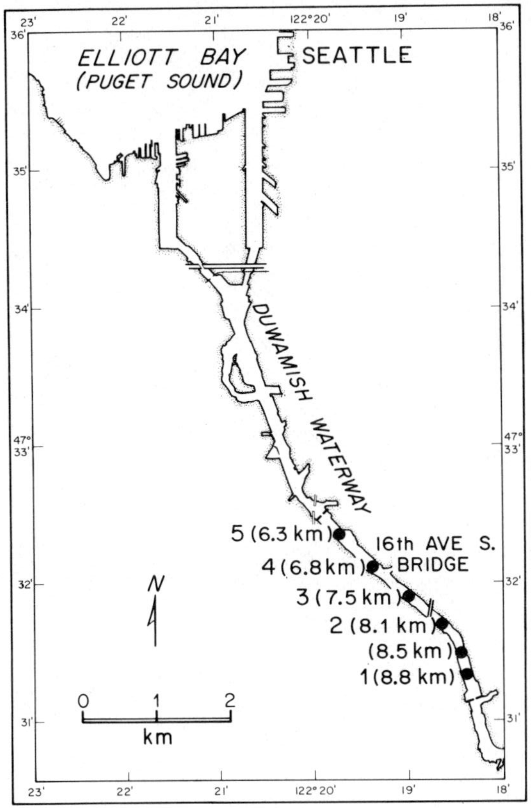

Figure 1. Map of the Duwamish Waterway. The dashed, lateral lines mark the limits of the study area. Salinity data from the numbered positions are described in the text of the paper. Kilometer 8.5 is the location at which the measurements reported by Partch and Smith (1977) were made.

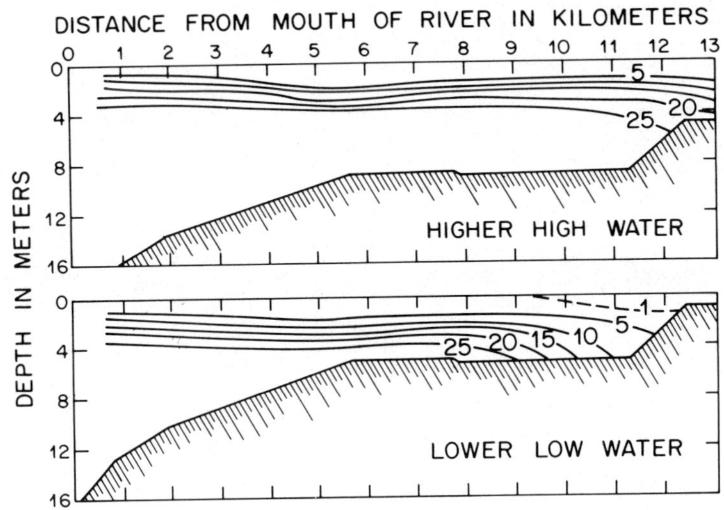

Figure 2. Schematic longitudinal sections of the salinity field at high and low water. This sketch is based on the results of Dawson and Tilley (1972), but has adjusted for the fresh water runoff conditions that existed during the March 1977 experiment.

internal Froude number and suggested that they were the result of breaking internal waves, an internal bore or an internal hydraulic jump.

If such intense mixing events occur over a significant segment of the estuary, and are common in other salt wedge estuaries, then they are of obvious importance in the understanding of circulation in such systems. Unfortunately procurement of data from one or several stations is not sufficient to characterize these features satisfactorily. Therefore, to investigate the longitudinal structure of the intense mixing events, and also to examine the more general aspects of estuarine circulation, a system of instrument deployment that permits measurements to be made from a moving ship was developed. With this system the 3 km reach marked on Figure 1 could be covered with a round trip time of one hour.

This underway measurement system was used for an experiment in the Duwamish between 11 and 14 March 1977. During this period the fresh water inflow varied from 33 m^3/s to 35 m^3/s, compared to 13.3 m^3/s to 14.0 m^3/s during the period studied by Partch and Smith. Under low runoff conditions such as examined by Partch and Smith, the density profile is of type A, shown on Figure 3, at the start of each ebb. The intense mixing event changes this to a type B, and sometimes a second event results in a type C profile. During the experiment described in this paper, the type A profile rarely, if ever occurred, the runoff being sufficient to maintain a surface mixed layer at all times. To confirm this statement, Figure 4 shows a sequence of salinity profiles covering one tidal day at intervals of approximately 2 hours. The intense mixing event is apparent as the increasing surface salinity between 1112 and 1444, 12 March 1977.

EXPERIMENTAL TECHNIQUE

Instrumentation

Two instrument systems were deployed during the March experiments. Both were designed to be used from a ship at speeds up to 4 kts, relative to the water. To provide accurate measurements of velocity and salinity at several, fixed, depths, a mast was suspended from an I-beam extending well to starboard from the bow of the research vessel. This mast supported 9 triplets of mechanical current meters as well as two pairs of temperature and conductivity cells. In contrast, vertical profiles of salinity were obtained from a CTD deployed from the stern of the research vessel. A two-axis electromagnetic current meter, oriented to measure longitudinal and vertical velocity components, was attached to the CTD. This current meter obviously cannot yield accurate measurements by itself due to unresolved motion of the instrument relative to the ship, but does provide a means of interpolating between the fixed depth velocity measurements from the mast. In order to allow the

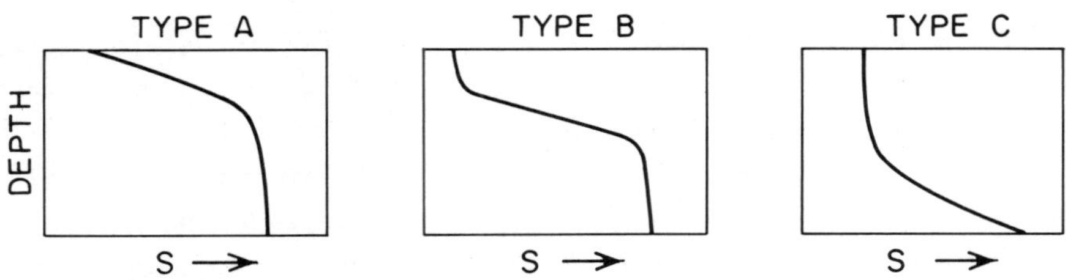

Figure 3. Schematic diagram of the three types of salinity profiles found in the Duwamish River by Partch and Smith.

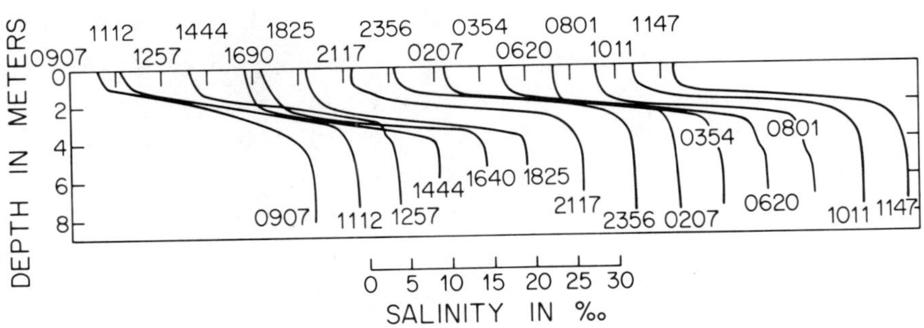

Figure 4. Sequence of alinity profiles from position 2 obtained between 0907, 12 March 1977 and 1147, 13 March 1977. The vertical lines at the top of the plot are origins for the curves, each of which is identified by the time at which it was taken.

profiling system to be lowered to within one meter of the bottom from the moving vessel, a depth sounder head was attached to the CTD, with the readout visible to the winch operator.

During the past year the ship-mounted mast system was deployed on three major cruises and on each it proved to be a very useful tool for measuring the spatial structure of near-surface velocity and salinity fields. The mast itself consists of a 13 m length of 2 1/2" OD thick walled steel tubing. This piece is strengthened and fared with a 3" x 1/4" steel plate welded to the afterside. The 1/4" plate also provides attachment points for the current meter junction boxes and for the cables connecting the underwater sensors with the ship's electronics laboratory.

The mast is attached to an I-beam by a clamp that holds the mast torsionally rigid yet allows it to pivot about an axis parallel to the supporting I-beam. Lateral support for the mast is provided by stays that attach to the I-beam about 1 m from the clamp. In operation, approximately 230 kgm of lead in the form of 23 kgm donut-shaped weights is attached to the bottom of the unit. The mast then is pulled back so that it makes an angle of 30° with the vertical using a wire attached near its lower end. This provides a stable attitude for the current meters at relative speeds up to the point where drag on the mast causes it to swing back at an angle greater than 30°. Depending on how far the mast is extended into the water, this maximum speed varies from three to four kts. Figure 5 shows the mast in operation in the Duwamish estuary. Note the typical lack of surface waves in the background.

The current meters attached to the mast are triplets of velocity component sensing mechanical devices, Smith (1974), oriented so that all meters sense a portion of the mean flow. The velocity is calculated using calibration data, then is rotated into a normal u,v,w, coordinate system. The techniques for accomplishing these tasks are described by Smith (1974).

Operation in shallow water with large tidally related or topographically induced variations in depth requires the depth to which the mast extends to be readily adjustable. This is accomplished by loosening the clamp, and raising or lowering the mast with the aid of a hand winch attached to the I-beam. The latter is within reach of the research vessel. Eyes are attached to the stays at several points so that their length can be changed rapidly.

With the mast pulled up to its shallowest setting (i.e. where the top triplet is just below the clamp) it can be rotated into the horizontal plane then swung inboard so that the sensors can be serviced. This relatively easy access is an important feature when using rather delicate current meters in an urban estuary at high speeds for periods of several days. It is possible to

Figure 5. Photograph of the ship mounted mast in operation in the Duwamish River.

bring the mast over the deck, change one or two triplets and clean the rest, then resume operation within 30 minutes. Figure 6 shows the system with the mast raised for servicing.

The I-beam supporting the mast is attached across the rails of the ship about 1.2 meters aft of the bow. The mast enters the water 1.5 meters aft of the point where the bow intersects still water, and is 2.3 meters, or about 1.7 times the vessel width, seaward of the hull at this point. Interaction of the bow wave produced by the mast with small surface waves makes measurement in the top 50 to 100 cm impractical. Below the surface, the ship hull appears even farther away due to its flair. The bow wave from the ship is considerably closer to the ship than to the mast, thus does not interfere with the current measurements. Tests in still water have confirmed that errors produced by the proximity of the ship's hull are small (less than 5%).

In order to make accurate velocity measurements from a moving ship, the ship motion must be accurately determined and subtracted from the measured relative velocity. This is accomplished with a microwave navigation system which measures distances from shore based transponders. This data like the

Figure 6. Photograph of the ship mounted mast raised for servicing the temperature, conductivity, and velocity sensors.

current, temperature and salinity information is recorded on magnetic tape through a NOVA 1200 based system. The same minicomputer is used for analysis of data.

Procedure

The system described above provides the ability to sample velocity and density fields accurately from a moving research vessel. In designing an experiment to use this capability, one must consider the trade-off between the length of estuary to be covered and the time available to cover it. In a tidal estuary such as the Duwamish, it is necessary to sample each position often enough to follow tidal fluctuations. While a 12 hour tidal cycle could be sampled reasonably at a 3 hour interval, the short-lived intense mixing events described by Smith and Partch (1977) required an interval of one hour or less. It was found that a 3 km reach of the estuary could be covered with this round trip time. In the March 1977 experiment, during which most of the data described in this paper were obtained, this 3 km reach was sampled continuously for 66 hours, with no gaps in excess of 2 hours. This unique set of data, which comprises the basis of the rest of the paper, greatly increases our understanding of the dynamics and mixing of the Duwamish estuary, and, by extension, similar salt wedge estuaries elsewhere.

LONGITUDINAL STRUCTURE OF INTENSE MIXING EVENTS

The March 1977 experiment obtained salinity, temperature and velocity measurements over the 3 km reach shown on Figure 1; however, to date, only the salinity data have been analyzed in detail. Not only are the intense mixing events clearly evident in the measured salinity fields, but also the longitudinal extent of this data set provides insight into the nature of these important features. Partch and Smith (1977) show that the intense mixing events are charactertized by an increase in both surface salinity and thickness of the surface mixed layer. In order to study the time and space relationships of these mixing events, the surface salinity s_o, the salinity at the bottom of the surface mixed layer s_1, and the thickness of the surface mixed layer h_1 were determined from the salinity profiles. Figure 7 indicates the technique for

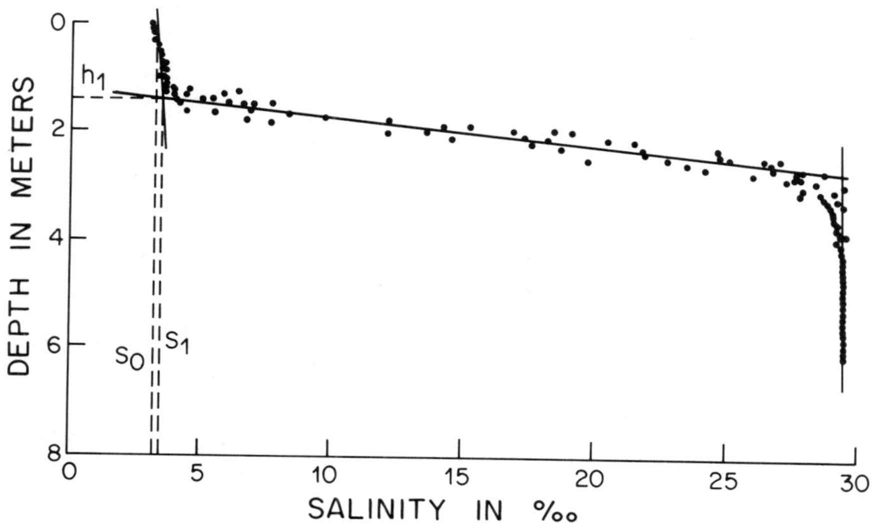

Figure 7. Sample salinity profile, showing the method of picking s_1 and h_1.

picking s_1 and h_1. Each profile was approximated by three straight line segments representing the surface mixed layer, the halocline and the bottom mixed layer, then s_1 and h_1 were found from the intersection of the upper two lines. This technique provides consistent, objective estimates of the surface layer salinity and thickness. The procedure was applied to profiles obtained each time the research vessel passed one of the five locations marked on Figure 1. This provided about 80 samples of s_o, s_1, and h_1 at each of the stations over the 66 hour experiment.

The samples obtained in this manner were at unequal time intervals, and each position was sampled at a different time; however, it was desirable for the intended analysis to have the data equally spaced in time with common intervals between stations, so it was necessary to interpolate between samples. The scheme used for this interpolation, described by Akima (1970), minimizes spurious fluctuations between sample points. The s_o, s_1, and h_1 records were interpolated to six minute intervals, beginning at 1000, 11 March 1977. To reduce noise in the records as much as possible, they were each separated into several time series beginning at the high water before the lower low water and compounded; that is the records were split at the high water before the stronger ebb of each tidal day. Curves of average s_o, s_1 and h_1 values then were plotted against time after high water. These are shown in Figure 8. Also marked on this figure are the average times and heights predicted for the two low waters and other high water. The experiment lasted slightly less than three tidal days, so the curves represent averages of two or three samples at each time. An estimate of the variability was obtained by calculating the standard deviation for each mean, and averaging these for each smoothed curve. The results are shown by the error bars adjacent to the curves.

The intense mixing events are evident in Figure 8 as the large peak in surface salinity during the strong ebb and the somewhat smaller but broader rise in salinity during the weak ebb. The salinity does not drop off to a low value after the weak ebb because the tidal excursion is insufficient to advect fresh water past the station after the intense mixing event ends. The secondary peaks at positions 4 and 5 during the strong flood probably are the result of a different process. A feature of the s_o curves, that may be related to this secondary peak, is a phase lag in the intense mixing event with downstream distance. The salinity peak produced by the intense mixing event will advect downstream until the surface current reverses; then it will advect back upstream. Thus, the high salinity patch can be carried back to the downstream stations. Calculations for the flow conditions on 11 March indicate that this could have been the case. As the maximum shear occurs on the flood, it also is possible that the secondary peaks are the result of shear instabilities. However, the results of Partch and Smith (1977) suggest that significant mixing by shear instabilities is unlikely. This question may be answered when the velocity data are analyzed and the tidal excursion for the surface laver can be calculated more accurately. In addition a gradient Richardson number field can be calculated and regions of subcritical Richardson number will be evident.

Figure 8 shows the same general features for s_1, although this time

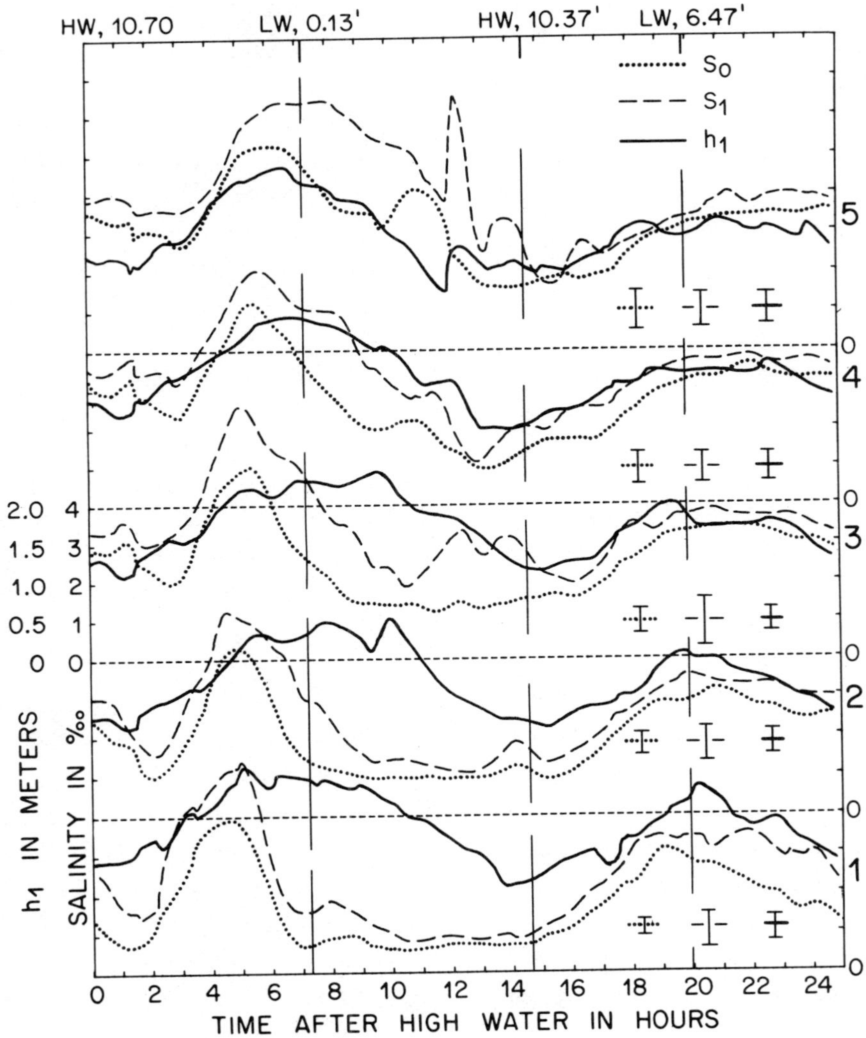

Figure 8. Surface salinity s_o, salinity at the bottom of the surface mixed layer s_1, and surface mixed layer depth h_1 as functions of time after high water. These curves are compounded from data over slightly less than three tidal days. Error bars indicate the average of the standard deviations for each curve.

series is noisier. The additional noise is partly related to uncertainty in defining the bottom of the surface layer and partly related to the effects of additional processes active at the base of this layer. The cause of the sharp spike during the strong flood at position 5 has been examined using the raw data and the spike appears to be spurious.

There are two main differences between the h_1 curves and those for s_o and s_1. First, there is a greater similarity between the weak ebb and strong ebb portions, and second, the phase shift noted for s_o and s_1 is not apparent; indeed, the peak seems to occur sooner downstream than upstream. It should be noted that while a peak in salinity during the ebb is opposite to what advection would produce, advection of the salt wedge should produce a peak in h_1 at low water, which is approximately the case in Figure 8.

The phase shift in the s_o and s_1 curves is particularly significant in understanding the nature of the intense mixing events. To get a somewhat more quantitative measure of this phase shift, the cross-correlation was calculated between position 1 and the other positions, using the interpolated, not the compounded data. Figure 9 shows the cross-correlation coefficients for lags between -25 and +25 hours. The difference between h_1 and s_o and s_1, is much more apparent in the correlations. The two salinities show almost no semi-diurnal peak, while h_1 has a relatively strong peak at this period. This difference, and the difference in the phase shifts, also apparent in the correlations, indicates that the h_1 curves may be dominated by advective effects, and changes in the slope of the pycnocline with the tidal phase, while the salinity curves are dominated by the intense mixing events.

The lags at maximum correlation in the s_o curves were used to calculate an effective downstream speed for the intense mixing events. The result is a spatially uniform value of 35 cm/sec. Based on velocity data from previous studies, this is consistent with an advective model for the mixing events. However, the maximum salinity increases downstream, so the data require the mixing to continue as the region of increased salinity advects down the estuary. The intense mixing event is well defined at the upstream-most position that was sampled. Therefore it must have had its origin upstream of the experimental section. The rise in s_o begins about 1 hour after high water, which indicates a maximum upstream position 1.26 km from position 1 if the intense mixing does not begin before high water and does not move downstream faster than 35 cm/sec. There is a sharp change in the estuary from a shallow, undredged river to a dredged channel at 1.1 km upstream of position 1. This suggests that an internal hydraulic jump forming at the transition from shallower to deeper water initiates the intense mixing. Such a jump would produce considerable turbulence and a large turbulent salt flux. Moreover, as the turbulent, higher salinity

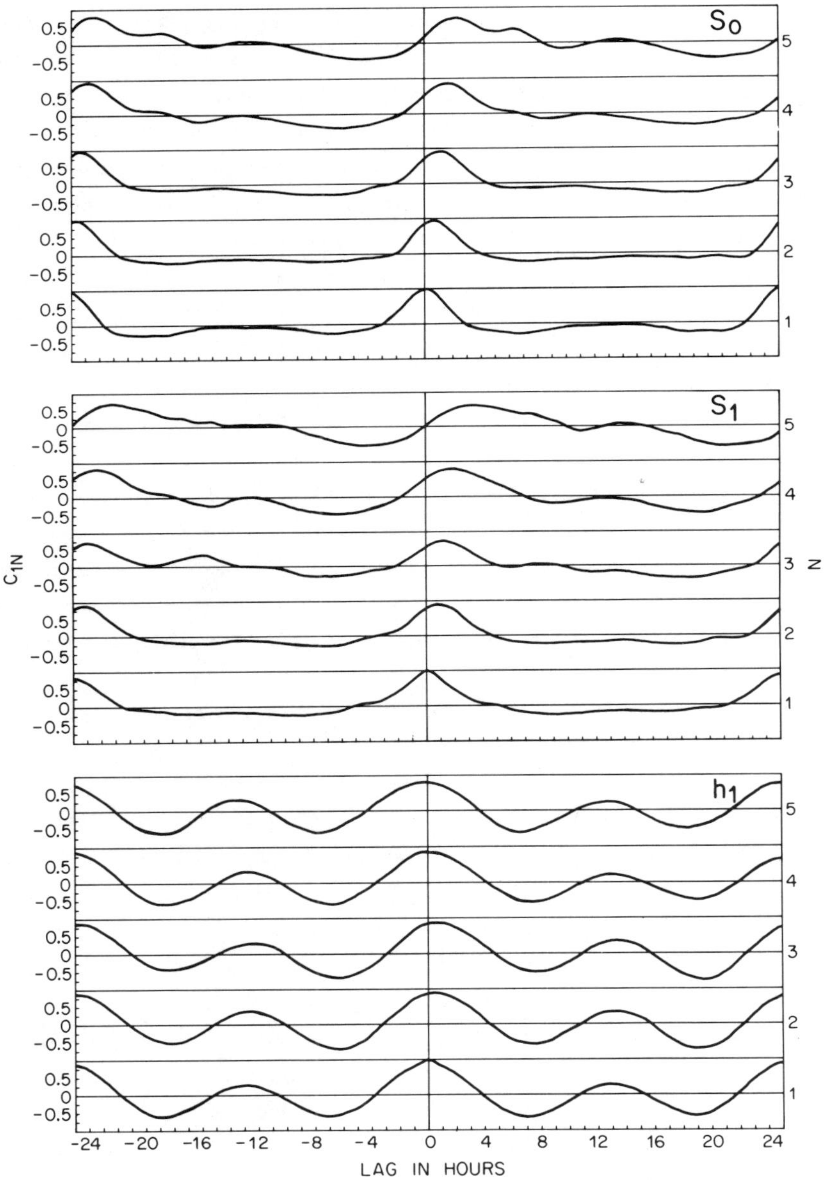

Figure 9. Cross correlation coefficients for s_o, s_1, and h_1 at positions 1 to 5 relative to s_o, s_1, and h_1 at position 1.

water is advected downstream from this point it would produce a peak in surface salinity.

It is reasonable to assume that the low salinity water, which appears at position 1 about 1/2 hour before low water, indicates that the jump had disappeared by the time that water passed the step in the bottom topography. If the salt wedge extended beyond the step for all stages of the tide then the jump should continue until low water as long as the wedge extends beyond the step. However, reference to Figure 2 indicates that it probably does not extend upstream of the step during the latter stages of the strong ebb. In addition the predicted tide curve indicates that there is less than 2m of water over the step during the last hour or two of this ebb and examination of h_1 in Figure 8 confirms that the salt wedge would not extend upstream of the step at this river surface elevation. The hour between high water and the first rise in salinity at position 1 also provides a reasonable estimate of the lag between the end of the jump and the arrival of minimum salinity water at this position. Therefore, the minimum salinity water should be seen about one hour before low water, which is consistent with Figure 8. On the weak ebb, the peak in s_o occurs at position 1 about 0.5 hours before low water, indicating that the peak intensity of the jump occurs about 1.5 hours before low water. At 35 cm/sec, the peak salinity would move 1.89 km in 1.5 hours. This is an upper limit for the actual distance traveled because 35 cm/sec probably is too high for the advective velocity at the end of the weak ebb, but it does suggest that the peak in s_o should reach below position 1, and probably to the region of position 2. This also is consistent with Figure 8.

This simple model provides a good explanation for several major features of the salinity data displayed in Figure 8. However, it still is necessary to account for the significant increase in the peak surface salinity as the patch of higher salinity water moves downstream from the jump. Turbulence generated at the jump would decay in much less than the available four hours so that an additional source of turbulent energy is required for the continued mixing. The near critical internal Froude number profile for the upper estuary, described by Partch and Smith (1977) suggests that an internal hydraulic instability is the most likely source. The salinity increase appears to begin at times consistent with the arrival of the patch of water generated by the jump suggesting that the instability must be triggered by that patch. One possibility is that the increased salinity in the surface layer reduces the internal wave speed sufficiently to change a near critical internal Froude number to a super critical one. For a two layer system without shear and with small density difference, the internal wave speed is given by:

$$c = \left(\frac{g \frac{\Delta\rho}{\rho} hh'}{h + h'} \right)$$

where h and h' are the thicknesses of the two layers, $\Delta\rho$ is the change in density across the interface and ρ is the mean density. Using the measured values of salinity, h_1 and total depth for position 1, both before the salinity increase begins and at the time of peak salinity, yields c = 51 cm/sec and c = 53 cm/sec respectively. The change in internal wave speed is too small to be significant, moreover, the change in h_1 dominates over the change in $\Delta\rho$, so the adjustment in wave speed is in the wrong direction to trigger an instability These values for phase speed will be modified slightly when shear is included in the calculation, however the times of interest are those of minimum velocity difference between levels so it seems unlikely that an increased surface salinity, alone, can account for the instability that results in continued mixing as the turbulent patch advects downstream. This, in turn, suggests that extraction of turbulent energy from the near critical flow depends in some way on the already existing turbulence field.

The magnitude of the vertical salt flux downstream of the jump can be estimated from the increase in salt content in the surface layer. To this end, the compounded time series of s_o, s_1, and h_1 were combined into a time series of $h_1 (s_o + s_1)/2$; that is they were combined into a time series of the product of mean surface layer salinity and surface layer depth. Graphs of this estimate of upper layer salt content per unit horizontal surface area are shown in Figure 10. The units in Figure 10 are noted as cm-gm/kgm; i.e., the product of surface layer thickness measured in cms, and salinity expressed as grams of salt per killogram of water. However, the fluid density is approximately 10^3 kgm/m^3 so to the accuracy of the analysis they also can be thought of as salt contents measured in tens of grams of salt per square meter of river surface area. Referring to Figure 10, we find that between position 1 and position 5, the peak surface layer salt content increases from 17770 gm/m^2 to 20530 gm/m^2 in 1.5 hours. The width of the river is approximately the same at these two positions, so the vertical salt flux is 0.51 gm/m^2·sec if the salinity anomaly curves are assumed to be the same width, similar in shape and scaled by the peak values. However, the assumptions are negated to some degree by longitudinal diffusion, so the estimate is a lower limit. A proper calculation using the time averaged value yields 0.63 gm/m^2·sec. In view of the difference in flow conditions, this estimate is in reasonable agreement with the value of 1.24 gm/m^2·sec measured by Partch and Smith (1977).

If, as postulated here, the mixing event is initiated by an internal hydraulic jump at the step in bottom topography, 1.1 km upstream of position 1,

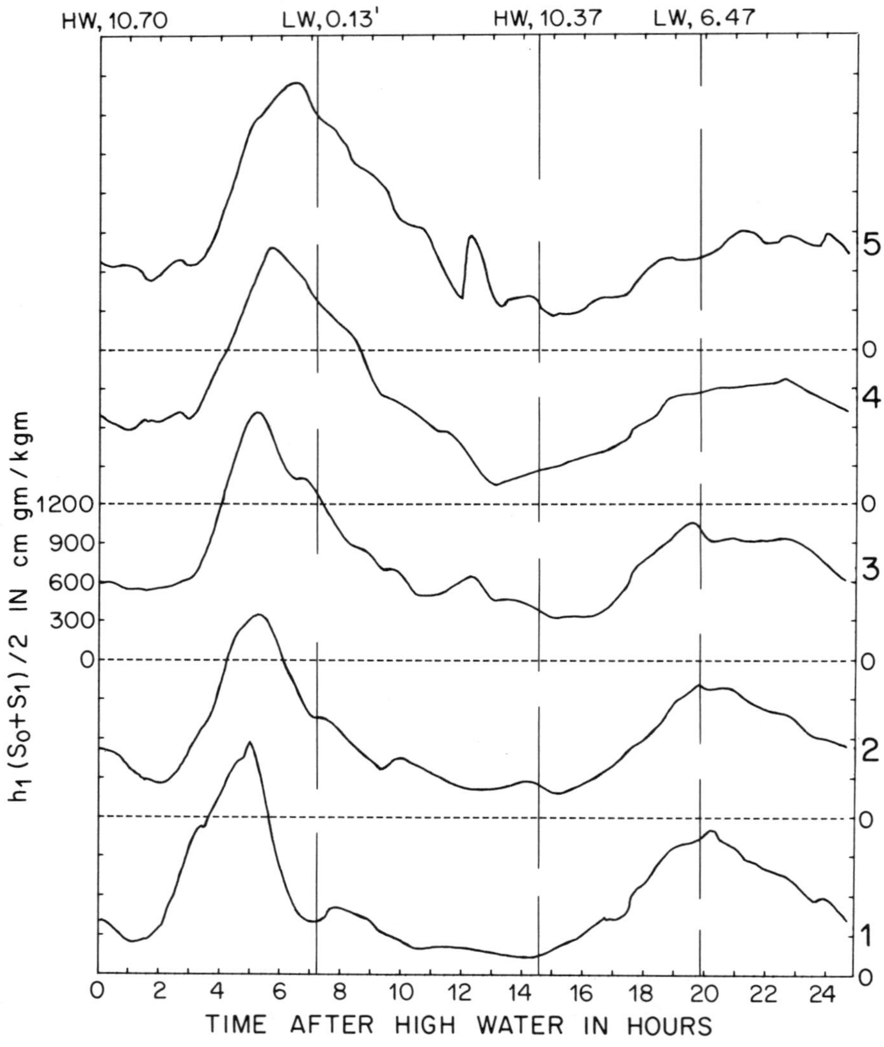

Figure 10. Salt content of surface mixed layer per unit horizontal area as function of time after high water.

an estimate can be made of the amount of salt added to the upper layer in the immediate vicinity of the jump. Figure 10 indicates that the increase in salt content at position 1 is about 14000 gm/m^2. Using the value of 0.63 $gm/m^2 \cdot sec$, calculated in the previous paragraph, as the salt flux downstream of the jump, one finds the amount of salt transported into the surface layer in the approximately 1 hour required to advect the turbulent patch from the jump to position 1 to be 2268 gm/m^2. This indicates that something like 11732 gm/m^2 of salt must be transported into the surface layer in the immediate vicinity of the step. Using 35 cm/sec as the surface layer velocity, the salt flux in the vicinity of the jump must be 4100/L in $gm/m^2 \cdot sec$ where L is the length of the mixing region at the jump measured in meters. As the extent of the jump is likely to be between 10 and 100 m it is clear that the intensity of mixing in the immediate vicinity of the step is several orders of magnitude greater than that further downstream.

This advective model for the intense mixing events is consistent with all observations obtained to date, and provides additional insight into the nature of these events. Analysis of velocity data from the ship-mounted mast system no doubt will provide refinements, and hopefully it will elucidate the mechanism by which turbulence production continues downstream of the step. With this increased understanding it should be possible to determine the general conditions necessary for the formation of intense mixing events, and to investigate the possibility of similar processes occurring in other salt wedge estuaries.

EFFECTS OF A BRIDGE

In addition to turbulent exchange caused in the upper estuary by the intense mixing events, constrictions associated with bridge piers were found to be important in triggering local mixing. The mechanism by which the latter is accomplished was not known prior to the March cruise and it is of interest to examine transects taken in the neighborhood of the 16th Avenue bridge for information about this process. By reference to the salinity profiles in Figure 4, it can be seen that the upper conductivity cell, mounted 1.33 m below the river surface, was in the surface layer throughout the tidal cycle; whereas the lower conductivity cell, mounted 2.20 m below the surface, was in the pycnocline throughout the high flow part of the ebb. Table 1 shows that the intense mixing events peaked at the bridge site about 1.5 hours after maximum flow suggesting that they might be distinguishable from any hydraulic effects produced under the bridge at high flow.

TABLE 1

Times of intense mixing events at stations near the bridge

Date	Position	Time of mixing events Start	Peak	End	Time of maximum ebb
11 March 1977	2	1130	1300	1430	1152
	3	1148	1324	1448	1152
12 March 1977	2	1212	1418	1512	1255
	3	1218	1448	1606	1255
13 March 1977	2	1300	1548	1706	1403
	3	1306	1554	1830	1403

River geometry in the neighborhood of the 16th Avenue bridge is shown in Figure 11. In addition, two structures that protect the bridge piers from

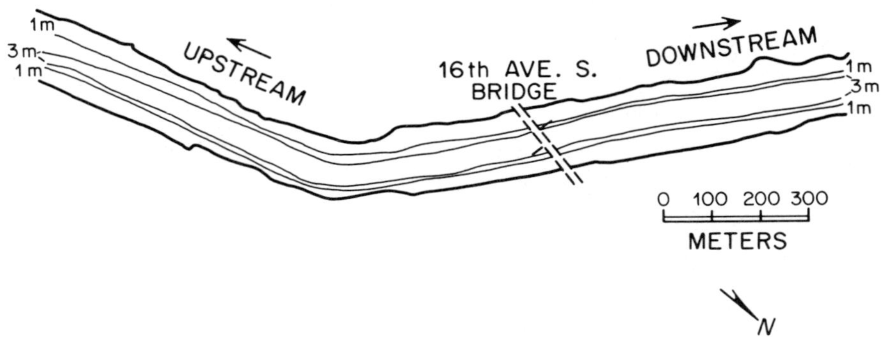

Figure 11. Plan view of the Duwamish River in the vicinity of the 16th Avenue S. bridge.

barge collisions are indicated as solid lines extending out from the upstream and downstream sides of the bridge toward the north and south banks, respectively. These bumpers are comprised of a tight network of pilings, thus serve to restrict the flow as it passes under the bridge. This effect is especially pronounced in the upper layers due to the protrusion of these structures into shallow water. It should be noted that the contours given on Figure 11 are for mean lower low water; thus, the actual water depth at the 1 meter contour during the maximum ebb is 2.5 meters. A tracing of an echo sounding profile made down the axis of the channel through this region is shown in Figure 12. The increased depth and rough topography upstream of the 16th Avenue bridge occurs because the area was dredged fairly recently with a dragline.

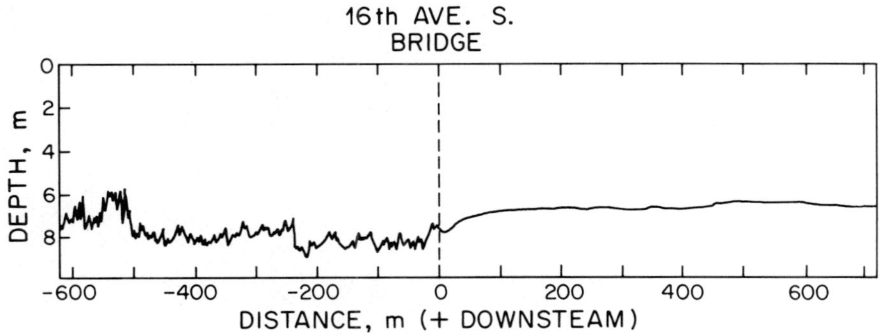

Figure 12. Longitudinal section showing bottom topography in the vicinity of the 16th Avenue S. bridge across the Duwamish River.

Good quality records of the salinity and velocity fields in the neighborhood of the 16th Avenue bridge are available for 11, 12 and 13 March 1977. These indicate that additional mixing occurred only during the high velocity ebbs, that is only in the early afternoon on these three days. A typical sequence of salinity profiles for the mid pycnocline near the bridge on 13 March is given in Figure 13. The bridge is taken as the origin of the distance axis and the times at which it was passed are given on the right hand side of the figure. At the left a reference salinity is noted to which the scale at the far right of the figure can be applied to obtain actual salinity values at any time and location. Profiles of the 2.2 m salinity, for the part of the tidal cycle not shown, are similar to those for 1558 and 1032 and vary significantly only in their spatially mean values. These follow the trend of s_o in Figure 8. Although there appears to be an increase in low frequency internal wave activity by 1124 and higher frequency internal wave activity by 1231, the main period of interest is between 1230 and 1430. During this interval, which also corresponds with the earliest part of the intense mixing event, there is a distinct rise in salinity upstream of the bridge, a rapid drop at the bridge, and a subsequent rise downstream of the structure. The peak of this event, around 1340, is shortly before the predicted time of maximum current under the bridge (1403). At this time the salinity 500 meters upstream of the bridge is about 18 $^o/oo$, whereas that 500 meters downstream is only about 10 $^o/oo$. Furthermore, there is substantial, moderate amplitude internal wave activity shown in the

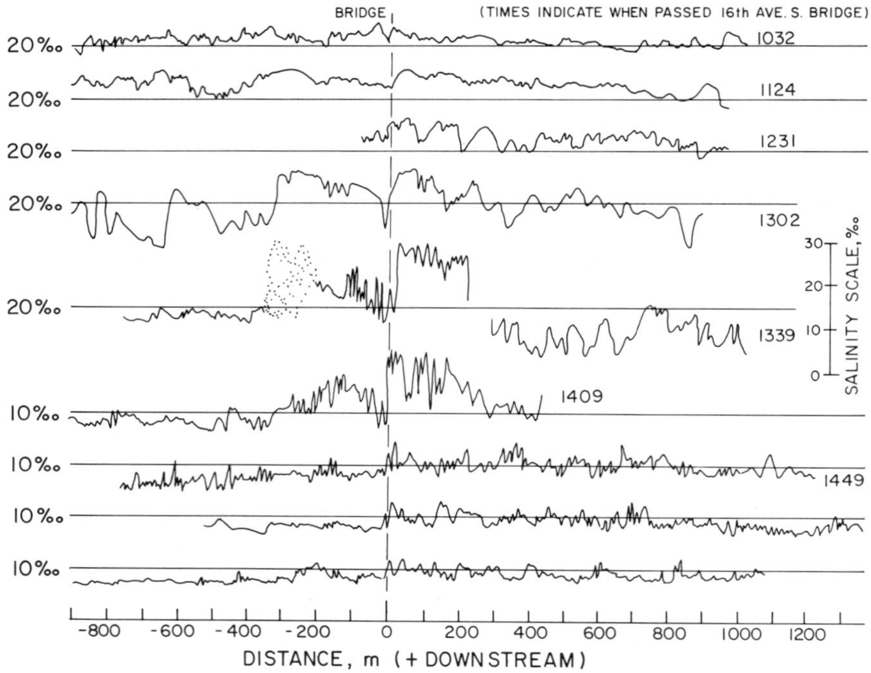

Figure 13. Salinity profiles in the neighborhood of the 16th Avenue S. bridge centered around the time of maximum ebb on 13 March 1977.

records for the downstream section at 1340. Here it should be noted that the pycnocline is very steep and that a depression of only 30 cm results in a 10 $^{o}/oo$ change in salinity. Also note that the ship-mounted mast moves with a speed about 5 times that of the longest internal waves, thus, the spatial structure of these low frequency internal waves is not displayed accurately in the traces.

Figure 14 is an analagous record showing the temporal structure of the salinity field between 1142 and 1343 on 12 March. The data for 11 March, in Figure 15, are similar. It is clear that the salient features of all three sets of data are associated with the presence of the bridge, and that the spatial structure displayed by the 1259 section on 12 March is to be expected near the peak each large ebb at least under runoff conditions such as those encountered in mid March 1977. The general features of the 1259 trace are those to be expected from a sensor in the pycnocline of a region in which there is a time dependent

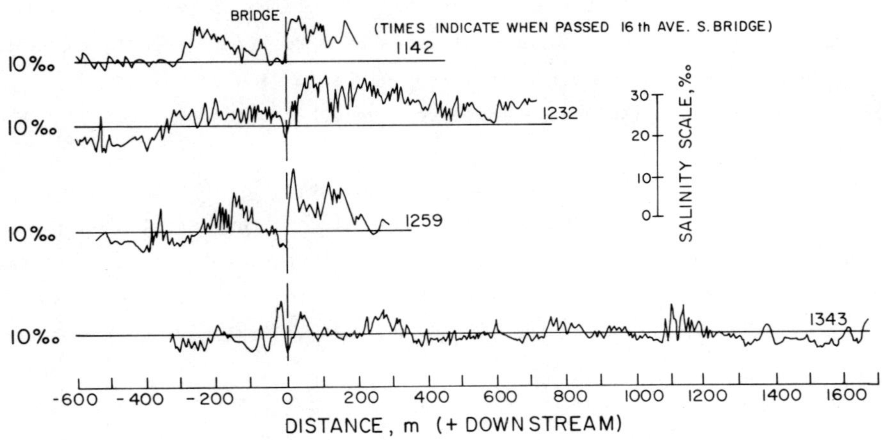

Figure 14. Salinity profiles in the neighborhood of the 16th Avenue S. bridge centered around the time of maximum ebb on 12 March 1977.

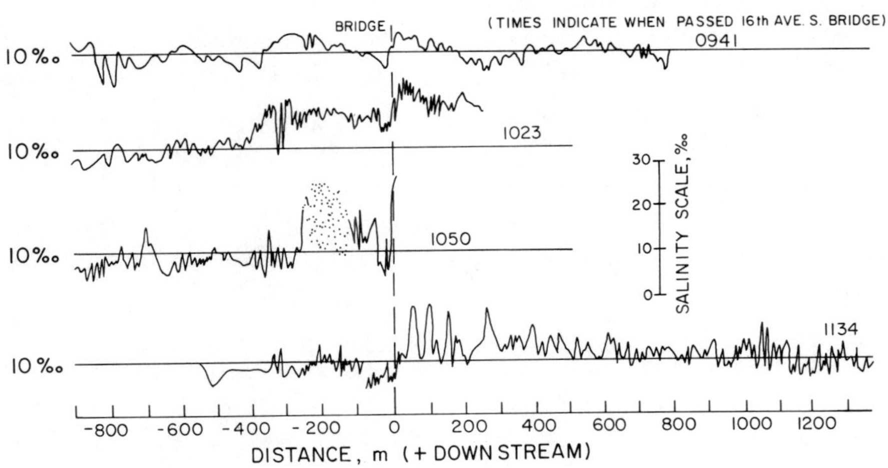

Figure 15. Salinity profiles in the neighborhood of the 16th Avenue S. bridge centered around the time of maximum ebb on 11 March 1977.

internal hydraulic jump. As a rise in salinity corresponds to a rise in interface level, the hump centered around -200 m can be seen to represent the upstream propagating long wave necessitated by conservation of mass and momentum in the unsteady case. Further, on the salinity traces for 1339 on 13 March and 1050 on 11 March the interface at the top of this hump breaks down into very high frequency, high-mode internal waves or more likely into turbulence. This is shown by the disconnected salinity values on Figures 13 and 15 respectively. The available data indicate a smooth progression from the rather low frequency oscillations, shown by the 0941 trace for 11 March or by the 1302 trace for 13 March, to the relatively high frequency internal motions shown in the vicinity of the bridge at 1023 on 11 March and at 1132 on 12 March, to turbulence as indicated by the disconnected points at 1050 on 11 March or 1339 on 13 March. Reference to Table 1 shows that in each case the climax event occurred somewhat before the maximum ebb and several hours before the intense mixing peak.

A series of downstream velocity component traces for various levels beneath the river surface is presented for 1259 on 12 March in Figure 16. In addition, two salinity records procured with the conductivity and temperature sensors mounted on the mast are included. The first of these provides a measure of the salinity in the mixed layer and clearly demonstrates a 2 $^{o}/oo$ jump between the upstream and downstream sides of the bridge. The second duplicates the 1259 trace on Figure 15 and is included to show the phase of the velocity structure with respect to interface morphology.

All velocity records have been corrected for ship motion. To do this the position data were block averaged for one record (0.4096 sec), edited to remove unreasonably large or small values, then averaged over a 30.72 sec interval. Ship speed was calculated by differentiating the 30.72 sec average position information. Finally these were smoothed using a fifth order polynomial. Although not our normal means of processing navigation data, this procedure allowed results to be procured rapidly and the quiet oceanographic conditions encountered in the Duwamish Waterway permitted it to yield an accurate ship velocity time series.

The downstream velocity component profiles of Figure 16 can be grouped into three categories: those for z = 43, 87, 130 and 173 cm represent the surface mixed layer; those for z = 217, 260 and 303 represent the pycnocline and those for 346 and 389 cm represent the salt wedge. From these traces it is clear that the constriction due to the bridge piers and their protective bumpers causes a substantial increase in flow speed at all depths, although this effect is strongest in the surface layer and in the upper pycnocline agreement with what would be expected from the geometry of the river channel

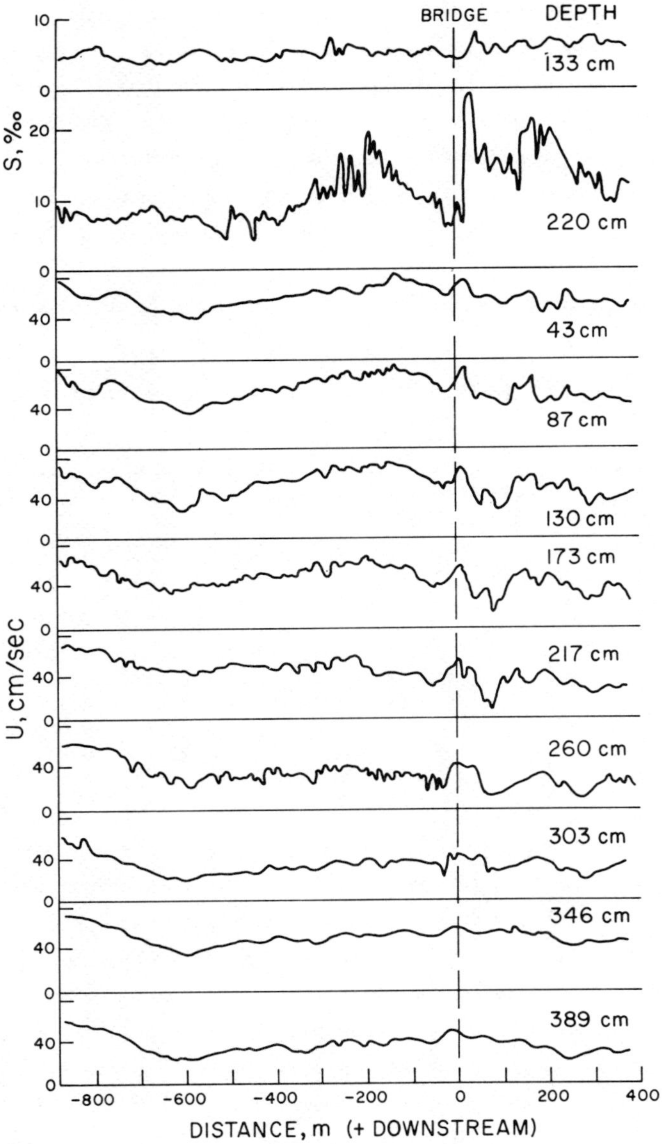

Figure 16. Downstream velocity component traces for 1259 on 12 March plus surface layer and mid pycnocline salinity profiles.

and the shape of the obstruction. Away from the bridge there is a distinct thinning of the surface layer and this is associated with a substantially increased downstream flow velocity, an effect that is especially clear over the large upstream hump centered near -200 m. However, these enhanced flow velocities are restricted to the pycnocline and surface mixed layer and primarily to the latter. For example, in the middle of the surface layer the flow speed is increased by nearly a factor of two over that in the mid pycnocline or below. This enhanced surface flow speed results in a drop in upper Pycnocline Richardson number relative to what otherwise would have been the case at this location. However, the decrease is small and the Richardson number is still of order 1 in the middle of the layer.

The Richardson number data of Table 2 indicate fully turbulent upper and

TABLE 2

Richardson numbers for various depths and horizontal positions in the neighborhood of the 16th Avenue bridge

Depth in meters	Downstream distance from bridge in meters: -900	-600	-130	0	40
.65	-	0.27	3.42	0.51	0.32
1.09	0.42	0.33	0.38	0.55	0.16
1.52	2.98	4.70	0.28	16.84	0.26
1.95	4.66	1.81	1.13	38.01	0.98
2.39	3.29	0.36	3.89	0.86	6.66
2.82	61.13	11.43	35.49	22.29	6.94
3.25	1.92	1.34	0.50	0.77	1.54
3.68	0.25	0.31	0.13	0.23	0.31

lower layers separated by a region of limited mixing. Near critical Richardson numbers associated with the upper layer penetrate into the upper pycnocline at two locations; the first is over the crest of the upstream hump and the second is in the rapidly decelerating region of the hydraulic jump. In both cases there is evidence from other measurements that turbulence is produced at least intermittently in these regions, and it is suggested that the additional shears are provided by high frequency internal waves that propagate through the sites. Unfortunately the analysis scheme used to reduce the March data makes it difficult to resolve such features.

Figure 17 shows all three velocity components at the 173 cm level, that is near the base of the surface layer. There appears to be a small positive vertical component of velocity associated with the upstream hump in the pycnocline as well as a flow towards the northeast bank in this region of the river. A

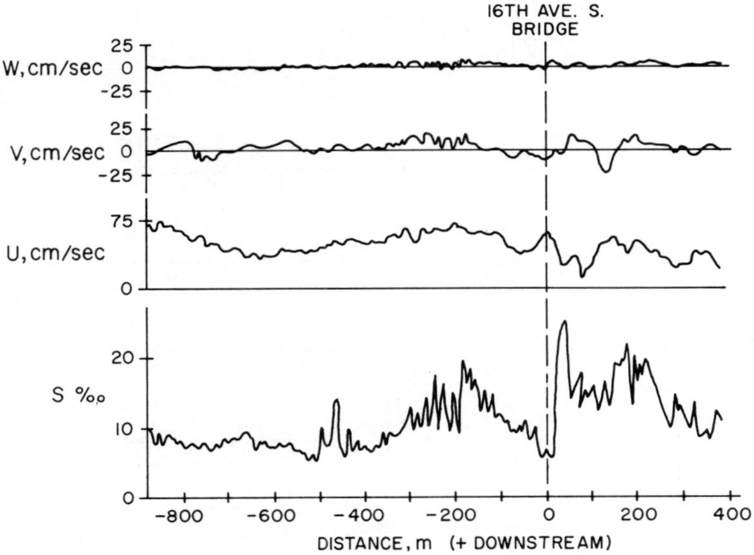

Figure 17. Velocity component profiles for a position near the base of the surface mixed layer centered around 1259 on 12 March. The horizontal velocity component u is positive in the downstream direction, the cross-stream velocity component v is positive toward the northeast bank and the vertical velocity component w is positive upward.

complicated internal wave field, associated with the upstream hump, is evident in both the vertical and cross-stream velocity components. Substantial vertical velocity components also are found in the region downstream of the jump. However, here they are associated with longer wavelength features. The cross-stream velocity component field can be best understood by reference to the series of profiles presented in Figure 18 and to the channel geometry displayed in Figure 11. The former shows that between -200 and -600 m the flow is toward the northeast bank in the surface layer and toward the southwest bank in the pycnocline and bottom layers. However, it is in this region that the channel curves necessitating a secondary flow. In the case at hand the maximum return flow in the secondary circulation occurs not in the bottom region as it does in an unstratified river but in the lower to central part of the pycnocline where the downstream velocity, hence the centrifugal force, is reduced. The maximum flow toward the outer part of the bend is in the mid to lower mixed layer with a near zero cross-stream component at the river surface.

Superimposed on this general secondary circulation, is a substantial flow toward the northeast bank centered just under the upstream hump. The amplitude of this disturbance is maximum in the center of the pycnocline and falls off rapidly with distance into the surface layer. Below the pycnocline

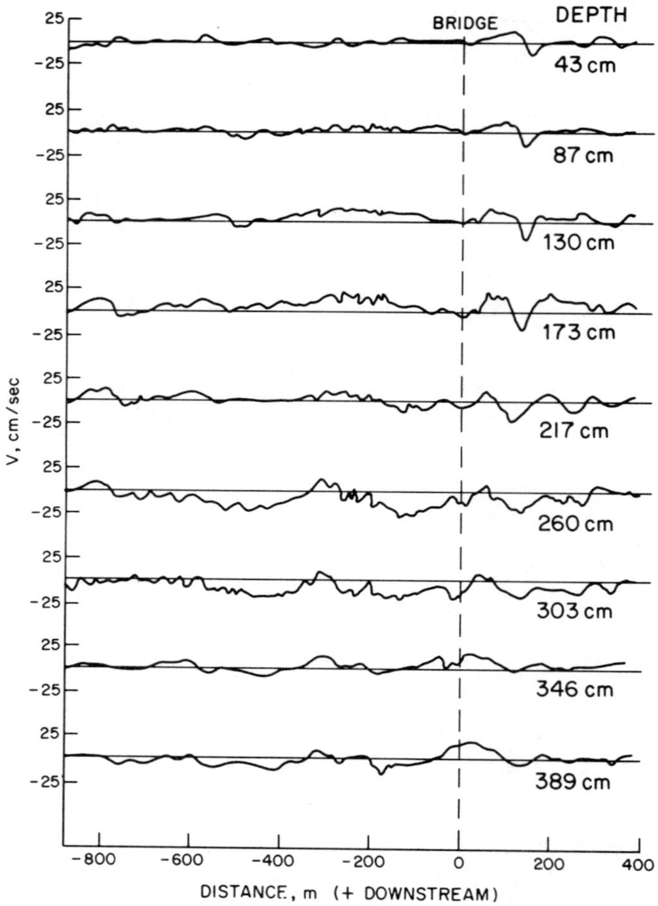

Figure 18. Cross-stream velocity components for the period centered around 1259 on 12 March. Positive values correspond to flow in the northeast direction, that is toward the outside of the bend located between -200 and -600 m. Note the distinct secondary circulation in this region.

the cross-stream velocity component at this location falls off more slowly. The cause of this lack of two-dimensionality in the upstream hump is not known and cannot be investigated in much detail as only transects down the center of the channel are available. Nevertheless, resoltuion of the velocity field is sufficient to guarantee that these non zero horizontal and vertical velocity components are associated with thinning of the mixed layer and that they are not spurious features. From Figure 11 it can be seen that the reach upstream of -600 m is straight and that in this region the spatially averaged vertical and cross-stream velocity components are zero as shown in Figures 17 and 18.

As in the case of the intense mixing events, the data on flow near the 16th Avenue bridge reported in this paper put additional constraints on the nature of the processes causing a salt flux from the lower to the upper layer, but do not permit it to be characterized fully. In the case of the disturbance near the bridge, we have shown (1) that the pycnocline is distorted just before maximum ebb, but only on the strongest ebb each day, in a manner not unlike an internal hydraulic jump, (2) that the surface layer salinity increases by 2 $^o/oo$ over a region several 10's of meters in horizontal extent just downstream of the bridge during this period, (3) that acceleration of the surface layer due to the upstream hump in the pycnocline causes increased shear and reduces the Richardson number field in this region, and (4) that shear associated with the downstream end of the hydraulic jump likewise results in reduced Richardson numbers. In addition, two records with chaotic salinity traces in the pycnocline region are suggestive of turbulent flow if not conclusive demonstrations thereof. These are associated with the upstream region of low Richardson number just like the obvious 2 $^o/oo$ jump in surface layer salinity is associated with the downstream end of the apparent hydraulic jump in which low Richardson numbers also are present. The 2 $^o/oo$ increase in salinity is equivalent to a horizontal salt flux of $2m^2 gm/kgm \cdot sec$ which yields a vertical salt flux of $2000/L\ gm/m^2 \cdot sec$ or $100\ gm/m^2 \cdot sec$ when distributed over L = 20m. The $2000/L\ gm/m^2 \cdot sec$ value is about half of the analagous salt flux at the upstream jump. Although the downstream mixing process also acts over a shorter period of time it provides a measurable contribution to the overall salt balance of the estuary.

REFERENCES

Akima, H., 1970. A new method of interpolation and smooth curve fitting based on local procedures. Journal of the Association of Computing Machinery, 12: 589-602

Dawson, W.A. and Tilley, L.J., 1972. Measurement of salt wedge excursion distance in the Duwamish River Estuary, Seattle, Washington, by means of the dissolved-oxygen gradient. U.S. Geological Survey Water Supply Paper, 1873-D, 27 pp.

Hansen, D.V. and Rattray, M., Jr., 1966, New dimensions in estuary classification. Limnology and Oceanography 11: 319-326.

Partch, E.N. and Smith, J.D., 1977. Time dependent mixing in a salt wedge estuary. Estuarine and Coastal Marine Science, 5:

Santos, J.F. and Stoner, J.D., 1972. Physical, chemical and biological aspects of the Duwamish River Estuary, King County, Washington, 1963-67. U.S. Geological Survey Water Supply Paper. 1873-C, 74 pp.

Smith, J.D., 1974. Turbulent structure of the surface boundary layer in an ice-covered ocean. Rapp. P. -v. Reun, cons. int. Explor. Mer, 1967: 53-65.

Stoner, J.D., 1972. Determination of mass balance and entrainment in the stratified Duwamish River Estuary, King County, Washington. U.S. Geological Survey Water Supply Paper. 1873-F, 17 pp.

A TWO-DIMENSIONAL NUMERICAL MODEL FOR SALT INTRUSION IN ESTUARIES

P.A.J. PERRELS and M. KARELSE
Research Engineers, Delft Hydraulics Laboratory, The Netherlands

ABSTRACT

A two-dimensional laterally integrated numerical model has been developed to represent the vertical velocity and salinity distribution along an estuary. The governing equations which express the conservation of mass, momentum and salt content, are solved by a finite difference method in combination with a splitting technique.

The model has been applied to the Delft tidal salinity flume, which may be considered as a two-dimensional tidal flow characteristic for estuaries such as the Rotterdam Waterway. By this application several assumptions about the effect of stratification on the vertical diffusion were tested. Preliminary results of the comparison of computed and measured data will be shown in this Paper.

This Paper is the result of a study which is incorporated in a basic research programma T.O.W. (working group "Stromen en transportverschijnselen") executed by Rijkswaterstaat (Public Works and Water Control Department), the Delft Hydraulics Laboratory and other research institutes.

1 INTRODUCTION

Estuaries are regions of water, which are connected to the sea or ocean at one end and fed by sources of fresh water (rivers) at the landward boundaries. In these regions saline sea-water and fresh river-water meet each other. Because of mixing of salt and fresh water the distribution of salinity in an estuary is a gradually varying function of space and time.

The major factors determining the salinity distribution are:

- the tidal motion within the estuary (as governed by the varying tidal elevation at the sea-entrance and the estuary geometry)
- the fresh river-water discharge
- the density difference between fresh water and saline sea-water
- the estuary geometry (tributaries, groins, harbours)
- wind influence
- Coriolis effects

Depending on the magnitude of these quantities varying degrees of stratification are possible.

Most of the mathematical models, which have been developed to calculate the salinity distribution in estuaries, are descriptive rather than predictive. Mathematical models having predictive capability need physical information on the spatial and temporal functions for turbulent eddy and mass diffusivities and physical relevant boundary conditions which are general applicable. The present knowledge on for instance the effect of stratification on the vertical eddy and mass diffusivities is limited (Delft Hydraulics Laboratory, 1974 and Fischer, 1976), consequently models are descriptive rather than predictive.

A general representation of salinity distribution in an estuary would require an unsteady, three-dimensional approach. For many situations however a simplified model may give satisfactory information in a much more economical way. For a lateral uniform situation, for instance, a vertical two-dimensional model is an obvious schematization.

Many mathematical models used for salinity intrusion problems in estuaries are one-dimensional (Harleman et. al., 1974), in that they use cross-sectional integrated forms of the equations of mass, momentum and salt content. These models require as input data information on the dispersion, i.e. the integrated effect of variation of velocity and concentration over the cross-section. This in itself, limits their predictive capability (Abraham et. al., 1975). For broad estuaries two-dimensional models with two horizontal dimensions have been developed (Leendertse, 1970). All these depth-integrated models are primarily applicable for mixed estuaries, in which the density differences between bottom and surface are small.

For stratified estuaries, in which there are two fluids of different densities separated by a distinct interface, two-layer models are available (Vreugdenhil, 1970): models without mixing, in which the upper-layer consists of fresh water and the under-layer of salt sea-water and models with vertical mixing, in which the vertical exchange of volume and mass between the two-layers results in a salinity that varies in longitudinal direction.

For partly mixed estuaries, which are characterized by gradually varying density in both horizontal and vertical directions, two-dimensional, laterally integrated models have been developed. Of these models known from literature (Hinwood et. al., 1975), that of Hamilton (1973 and 1975) is the most detailed. This model however still has some drawbacks, such as the treatment of the free surface and the bottom configuration, where extrapolation is needed because for the numerical model a fixed grid is used. For the bottom stress the quadratic friction law is applied in which, however, a constant distance to the bottom is used. At the downstream boundary a salt distribution is needed as function of place and time. At the free surface the first derivative of c is set equal to zero instead of using the kinematic condition. Furthermore, a space-staggered grid is used, which makes the implementation of boundary conditions containing first or higher derivatives less straight forward.

The more recent model of Boericke et. al. (1977) pays much attention to the exchange coefficients and also uses co-ordinate transformation. The computation of the velocities is rather simplified by neglecting the convective terms, although this yields inaccurate velocity profiles in tidal streams (Abbott, 1959). Boundary conditions for the velocity and the concentration at the bottom and at the free surface are not mentioned.

During the past two years a new two-dimensional laterally integrated model has been developed at the Delft Hydraulics Laboratory for predicting the unsteady velocity and salinity distribution in a partly mixed estuary, based on the shallow water approximation and a mixing length approach for the closure problem. Within those limits the model is kept as general as possible with an exact representation of the bottom configuration and the free surface and the possibility to improve several boundary conditions at the upstream and the downstream boundaries.

The free surface and the bottom configuration are reproduced exactly by means of a transformation. For the bottom stress a roughness coefficient is used, depending on the local circumstances. At the downstream boundary a function is needed, describing the transition of the salt concentration from an ebb to a flood situation.

The objectives of the design of the two-dimensional laterally integrated model, presented in this paper, are

- To judge the application of this model with respect to a partly mixed estuary, in which the vertical transport of momentum and salt is dependent on the Richardson-number (Delft Hydraulics Laboratory, 1974).
  A necessary first step is to find generally applicable relations for the Richardson-dependency of the vertical turbulent exchange of momentum and mass and for the transition function in the sea-ward boundary condition of the salt balance.
- To judge the application of this model with respect to a mixed estuary, in which the influence of the density differences on the vertical diffusion can be neglected; in this case the vertical exchange is known from literature (Delft Hydraulics Laboratory, 1973 and Fischer, 1973).
- To study the influence of density effects on the longitudinal dispersion coefficient of an one-dimensional model. Application of this knowledge will make the use of depth-averaged models for wide estuaries and of cross-sectional averaged models more justified.
- To get information for three-dimensional models about the ability to represent the vertical flow structure and salt distribution.

This Paper presents a description of the mathematical model and contains a comparison of the computed velocity and salinity distributions with experimental data from the Delft tidal salinity flume (Appendix I). This tidal flume may be considered as a nearly two-dimensional estuary in which it is possible to represent the characteristics (of the tidal motion and the salinity distribution) of estuaries like the Rotterdam Waterway. Tests in the tidal flume with bottom roughness have been used for the verification of the numerical model. Preliminary results of the comparison of computed and measured data will be shown in this Paper. The verification of the numerical model with flume data is still going on and will be published later when completed.

2 MATHEMATICAL MODEL

Formulation

After integration over the width and if the shallow water approximation is made, the equations for vertical two-dimensional density currents are: (Delft Hydraulics Laboratory, 1973)

The horizontal momentum equation:

$$\frac{\partial u}{\partial t} + \frac{1}{b}\frac{\partial}{\partial x}(bu^2) + \frac{\partial}{\partial z}(uw) - \frac{1}{b}\frac{\partial}{\partial x}(b\varepsilon_x \frac{\partial u}{\partial x}) - \frac{\partial}{\partial z}(\varepsilon_z \frac{\partial u}{\partial z}) = -\frac{1}{\rho}\frac{\partial p}{\partial x} \tag{1}$$

The hydrostatically distributed pressure:

$$p = p_s + \rho_o g\,(\xi - z) + g\int_z^{\xi} \sigma d\xi \tag{2}$$

The continuity equation:

$$\frac{1}{b}\frac{\partial}{\partial x}(bu) + \frac{\partial w}{\partial z} = 0 \tag{3}$$

The salt balance:

$$\frac{\partial c}{\partial t} + \frac{1}{b}\frac{\partial}{\partial x}(buc) + \frac{\partial}{\partial z}(wc) - \frac{1}{b}\frac{\partial}{\partial x}(bD_x \frac{\partial c}{\partial x}) - \frac{\partial}{\partial z}(D_z \frac{\partial c}{\partial z}) = 0 \tag{4}$$

The equation of state:

$$\rho = \rho_o + \sigma = \rho_o + \beta c \quad \text{in which } \beta \text{ is of the order } 0.75 \tag{5}$$

Integration of (3) over the height, and substitution of the kinematic boundary condition, yields:

$$\frac{\partial \xi}{\partial t} + \frac{1}{b}\frac{\partial}{\partial x} \{b \int_{z_b}^{\xi} u \partial z\} = 0 \tag{6}$$

The width b is supposed to be a function of x only, because strong dependency on z would give large errors by integration over the width (Vreugdenhil, 1974). To complete the model expressions are necessary for the turbulent viscosity and diffusivity coefficients.

For the present model a mixing length approach has been chosen which applies well to circumstances where the shallow water approximation can be made.

$$\varepsilon_z = l_m^2 \left|\frac{\partial u}{\partial z}\right| f\,(Ri) \tag{7}$$

$$D_z = l_m^2 \left|\frac{\partial u}{\partial z}\right| g\,(Ri) \tag{8}$$

in which the mixing length l_m is defined by:

$$l_m = \kappa\,(z+z_o) \qquad \text{if } 0 \leqslant z \leqslant 0.25\,H \tag{9a}$$

$$= \kappa\,(0.25\,H+z_o) \quad \text{if } 0.25\,H \leqslant z \leqslant H \tag{9b}$$

and the Richardson-number Ri by:

$$Ri = -\,g\,\frac{\partial \rho}{\partial z} / \left[\rho_o \left(\frac{\partial u}{\partial z}\right)^2\right]$$

$$z_o = 1/30\;k_s$$

in which k_s is the roughness length.
For the damping functions f (Ri) en g (Ri) many formulas are available (Delft Hydraulics Laboratory, 1974). A definite choice is one of the subjects of investigation.
The horizontal viscosity and diffusion terms:

$$\frac{1}{b}\frac{\partial}{\partial x}\left(b\varepsilon_x \frac{\partial u}{\partial z}\right) \text{ and } \frac{1}{b}\frac{\partial}{\partial x}\left(bD_x \frac{\partial c}{\partial x}\right)$$

are commonly neglected because their influence is small; however, for long periods the horizontal turbulent diffusion may give a significant contribution. In the model both terms are included for numerical reasons (Delft Hydraulics Laboratory, 1975).

Boundary conditions

The mathematical model requires boundary conditions and initial conditions for the tidal motion as well as for the salinity. The question of whether the model is descriptive or predictive depends on the formulation of the vertical momentum and mass exchange and on the way in which the boundary conditions are treated. If observed data are used which cannot be predicted, like the vertical tidal motion at sea can be predicted, the model is only descriptive.

For the sea-boundary ($x = 0$) the conditions are:

for the water-elevation: $\xi = \xi_o (t)$

for the velocity : $u = u_o (t,z)$

$$\text{or: } \frac{\partial^2 u}{\partial x^2} = 0$$

for the salinity : $c = c_{max}\, g_o (t,z)$ if $u > 0$ (flood tide)

$$\text{and: } \frac{\partial^2 c}{\partial x^2} = 0 \qquad \text{if } u \leqslant 0 \text{ (ebb tide)}$$

The form of $g_o (t,z)$ is one of the subjects of investigation, as a first approach a piece-wise linearized function of t, independent of z was used.

At the upstream boundary of the estuary $x = L'$ the following conditions are valid:

for the velocities : $u = u_{L'} (t,z)$

for the salinity : $c = 0$

Knowing the river-discharge and assuming a logarithmic velocity profile in the fresh water-region of the estuary, the function $u_{L'}$ has been determined (Appendix II) The only restrictive aspect of this boundary condition is that L' ought to be chosen large enough to be outside the zone of salinity intrusion

At the bottom $z = z_b$ the conditions are:

$$u = 0$$

$$w = 0$$

$$\frac{\partial c}{\partial z} = 0$$

At the water-surface the conditions are:

$$\frac{\partial u}{\partial n} = 0 \text{ (no wind)}$$

$$wc - D_z \frac{\partial c}{\partial z} = 0 \text{ (no salt through the water-surface)}$$

At the water-surface also the kinematic boundary conditions holds:

$$\frac{\partial \xi}{\partial t} = w - u \frac{\partial \xi}{\partial x}$$

a condition which is already used by the derivation of (6).

Transformation

In general the longitudinal area of interest will not be a rectangle, because of the variations in bottom and free surface. For a good representation of the flow a good description of the form of the free surface and the bottom is necessary. For a numerical approach with finite differences however a rectangular grid, which coincidences with the boundaries will be preferable. Therefore the area of interest is transformed by a simple transformation into a rectangle (Delft Hydraulics Laboratory, 1976 and Jamet, 1970).

If the position of the free surface is described by:

$$\xi = \xi\ (t,x) \tag{10}$$

and the position of the bottom by:

$$z_b = z_b\ (x) \tag{11}$$

then the transformation reads:

$$z' = (z-z_b)\ f_b\ (x,t) \tag{12}$$

where f_b is given by:

$$f_b = \frac{1}{\{\xi\ (x,t) - z_b\ (x)\}} \tag{13}$$

This transformation adds some terms to the equations, but simplifies the finite difference approach at the boundaries considerable.

Numerical approach

The mathematical model is solved with a finite difference method, so a splitting method can be applied (Richtmeyer, 1967 and Roache, 1972).
The momentum equation and the continuity equation for the salt are split into the spatial directions (Delft Hydraulics Laboratory, 1975).
The part in the x-direction of the momentum equation is solved explicitly, while the part in the z-direction is solved implicitly.
The implicit technique in the z-direction removes the most severe stability condition. The stability condition in the x-direction depends on the value of ε_x, which means that a suitable choice of ε_x permits a larger time step τ (Jamet, 1970 and Delft Hydraulics Laboratory, 1975).

$$\tau \leqslant \frac{2\varepsilon_x}{\left[\max\ |u| + \frac{2\varepsilon_x}{\Delta x}\right]^2} \tag{14}$$

This mixed approach was found to be most economical due to the fact that the ratio of the horizontal and the vertical dimensions of the problem area is usually large. If this ratio is closer to one, the same approach in the x- as in the z-direction would be more likely. For the continuity equation of the salt, the same difference technique is used. Special attention is paid to the treatment of the continuity equation and of the boundary conditions, particularly the boundary conditions involving a derivative. By the use of this mixed approach the difficulty of finding boundary conditions at the intermediate level is circumvented.

For the continuity equation a fourth order scheme in Δx and second order in Δz is used, and for the boundary conditions second order schemes are used. In this way inconvenient perturbations are avoided and a good accuracy is achieved. A special treatment is also applied near the bottom, where the gradients in the velocity profile are very large. So a very small step should be used, which in an uniform grid is not very attractive from an economical point of view. Therefore locally a special treatment is used.

Near the bottom the convective terms and the horizontal turbulent viscosity are neglected, so the momentum equation simplifies to:

$$\frac{\partial u}{\partial t} + \frac{\partial u}{\partial z}(\varepsilon_z \frac{\partial u}{\partial z}) = -\frac{1}{\rho}\frac{\partial p}{\partial x} \qquad (15)$$

which can be written as:

$$\frac{\partial}{\partial z}(\varepsilon_z \frac{\partial u}{\partial z}) = -\frac{1}{\rho}\frac{\partial p}{\partial x} - \frac{\partial u}{\partial t} \qquad (16)$$

Now the right-hand side of (16) is supposed to be independent of z, therefore (16) can be solved by integrating twice.

The constants that arise by the integration can be found by substituting the boundary conditions: $u = 0$ at $z = 0$

and: $u = u(\Delta z)$ at $z = \Delta z$

In this way an approximation for $\frac{\partial u}{\partial z}$ near the bottom can be achieved, which will then be substituted in the difference equation.

The difference equations are given in Appendix II.

3 VERIFICATION

As a first test for the numerical model a comparison with flume data of the Delft tidal salinity flume has been chosen mainly for the following reasons:

- In the tidal flume the tidal flow is nearly two-dimensional, in that there is no variation of density but some variation of velocity in transverse direction.

- The flow conditions are well adjustable.
- The system has been measured systematically.

For a description of the flume and the flume data see Appendix I.

The verification is split into two phases, in order to have a clear distinction of the different mechanisms that influence the tidal motion and concentration distribution. In the first phase, computations for homogeneous circumstances are performed and compared with the measurements from the homogeneous test T22. These computations were meant to calibrate the water-level variations and the velocity profiles. After some adaptions of the bottom roughness, a reasonable agreement was achieved for the water-level variation (Figure 1) and for the velocity profiles (Figure 2). The deviation in the vertical tidal motion is mainly caused by the difference in the adjustment of the river-boundary between the numerical model (boundary at $x = L$ is Q_L) and the tidal flume (boundary at $x = L'$ is $Q\ (L',t)$, see Figure 3). The velocity profiles are shown at two characteristic times at two distances from the downstream boundary. The standard deviation in the measurement is of the order of $\sigma_u \approx 0.015$ m/s.

The data for the computation were:

- Boundary conditions and flow parameters according to Table A2
- $z_o = 0.004$ m
- $\varepsilon_x = D_x = 0.37\ m^2/s$
- $\varepsilon_z = D_z$ according Equations (7), (8) and (9) with $f\ (Ri) = g\ (Ri) = 1$
- $\Delta x = 3.66$ m ($= L/49$)
- $\Delta z = H/12$
- $\tau\ = 2.79375$ s ($= 1/200\ T$)
- $Nt = 800$ ($= 4T$)
- CPU-time = 160 s (at a CDC 6600)

In the second phase of verifications the inhomogeneous test T180 was used. In this case the influence of the density differences on the flow ought to be represented. The major difficulty of these computations lies in correct representation of the Richardson-dependency of the coefficient of vertical, turbulent viscosity and diffusivity (Equations (7) and (8)). For the present computations the demping functions given by Van Rees (1975) turned out to give the best results. In Figure 4 the density distribution is shown at four characteristic times. The data for this computation were the same as in the homogeneous computation except:

$$\left.\begin{array}{l} f\ (Ri) = \exp\ (-\ 4Ri) \\ g\ (Ri) = \exp\ (-\ 15Ri) \end{array}\right\}$$ according Van Rees (1975), see also Delft Hydraulics Laboratory (1974)

and CPU-time = 180 s

Conclusions

Preliminary results are given of computed tidal motion and density distribution which show reasonable agreement with the experiments. Much work is still to be done primarily for determination of the physical coefficients. However, it may be concluded

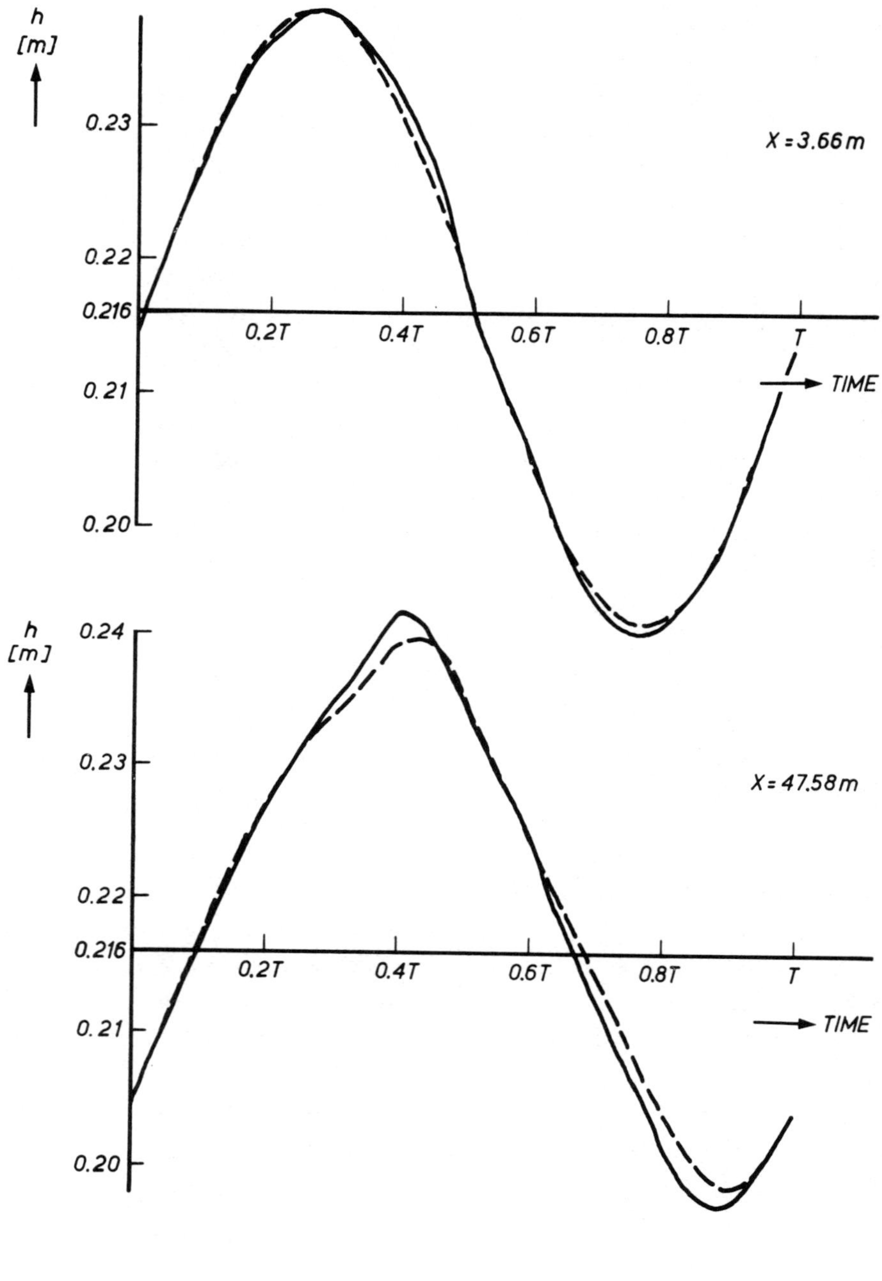

FIG. 1 COMPARISON COMPUTED AND MEASURED VERTICAL TIDAL MOTIONS
TEST T 22

——— MEASURED
– – – COMPUTED
T : TIDAL PERIOD

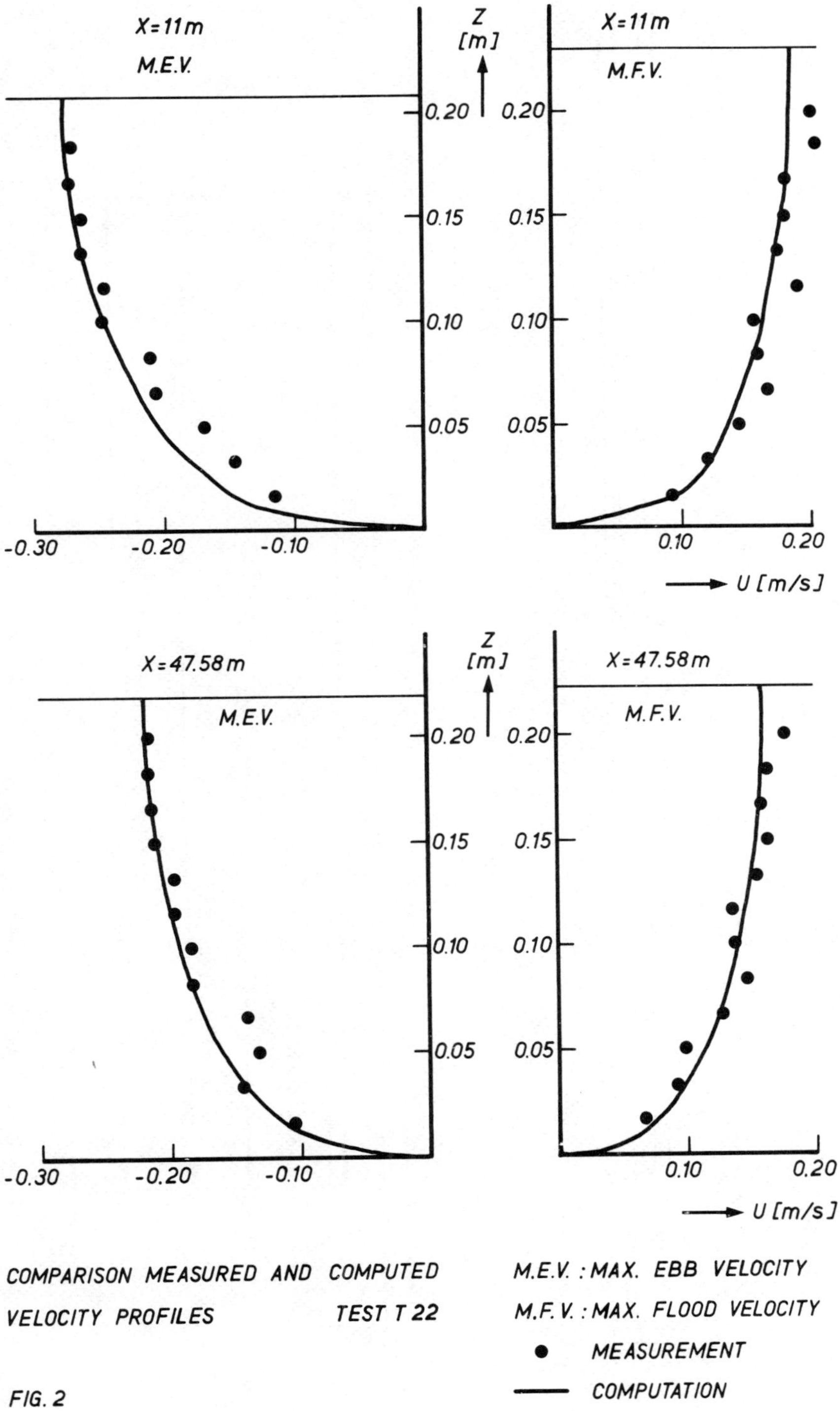

COMPARISON MEASURED AND COMPUTED VELOCITY PROFILES TEST T 22

FIG. 2

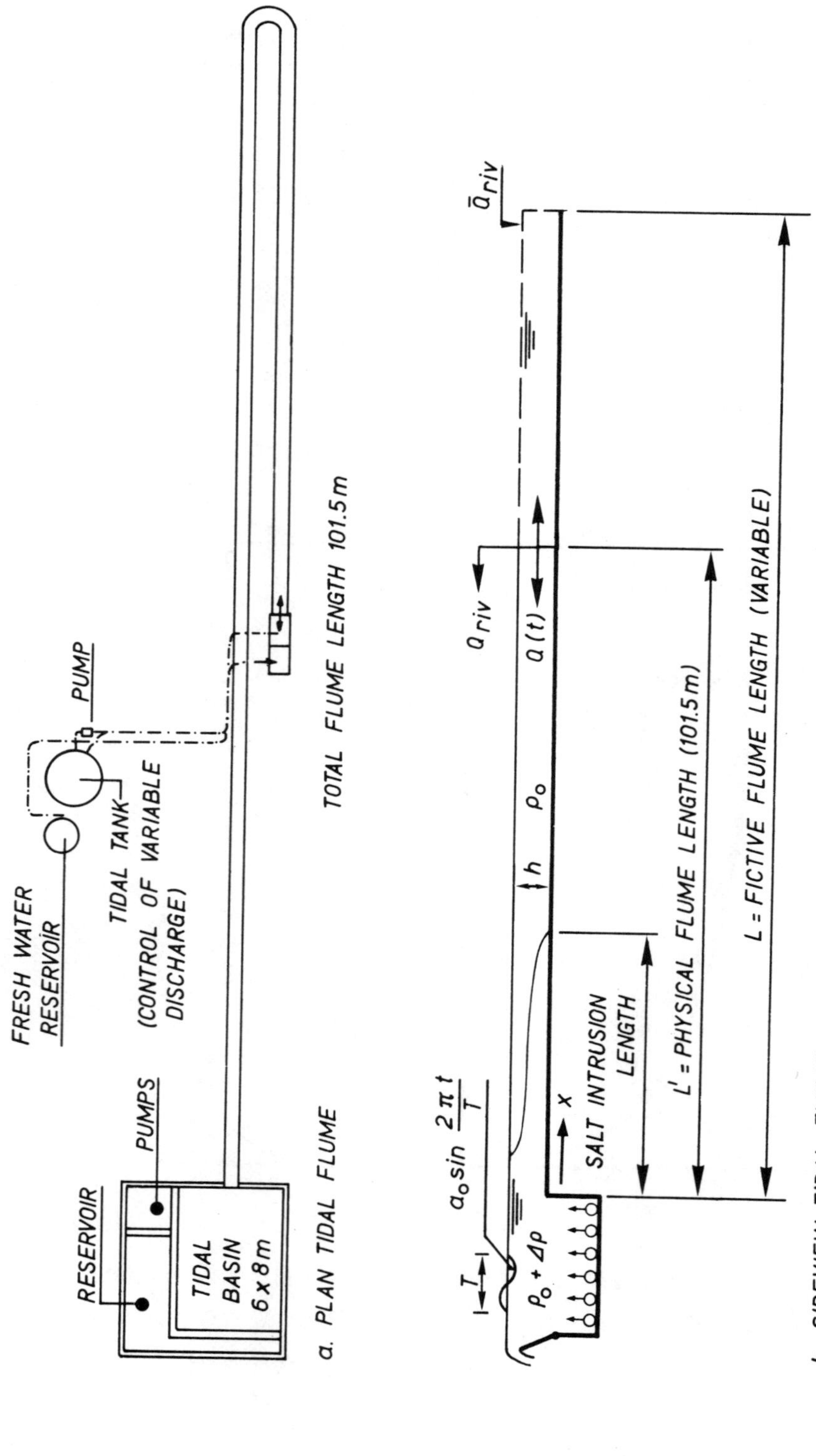

FIG. 3 - SETUP OF TIDAL SALINITY FLUME WITH NOTATIONS

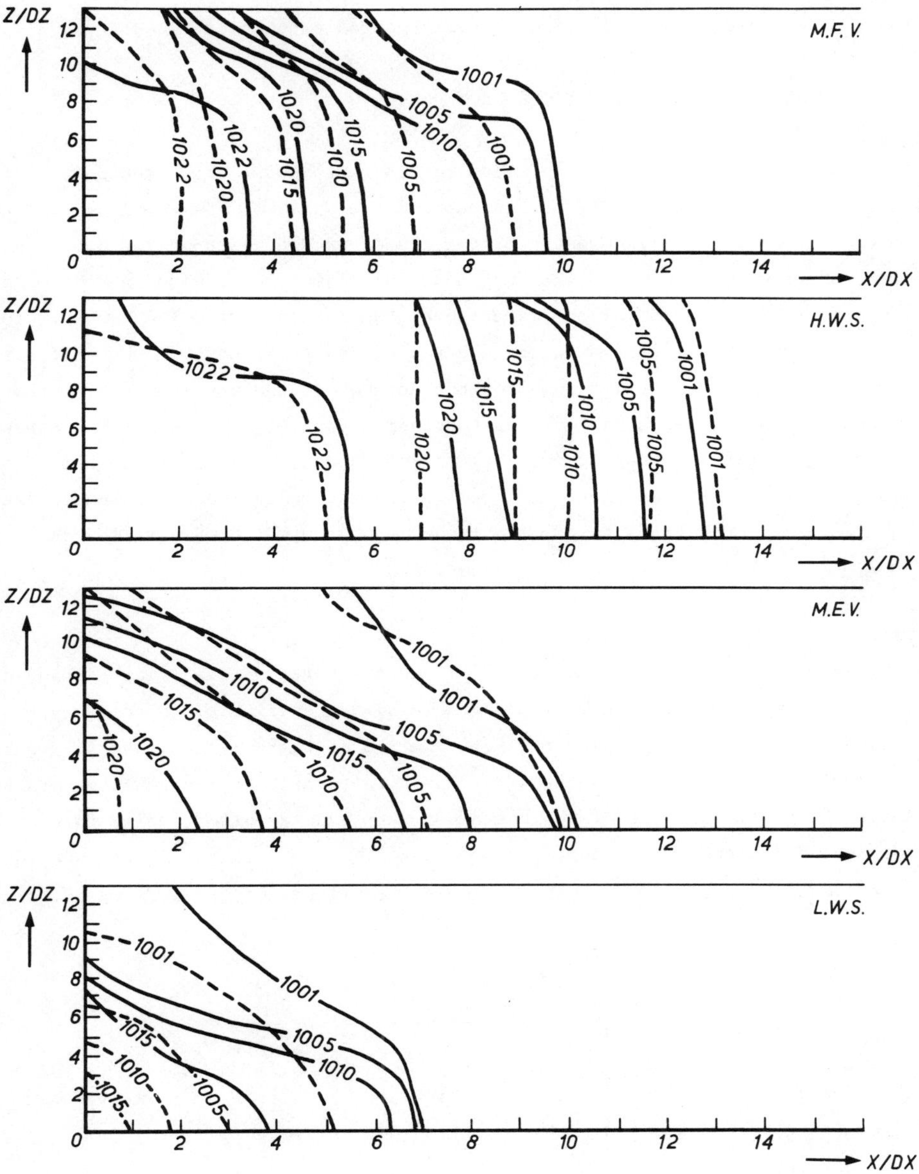

DENSITY DISTRIBUTION IN THE VERTICAL PLAIN

FIG. 4

MEASURED

COMPUTED

DX = 3.66 m

DZ = 1.66 cm

on these preliminary results that the present numerical model is able to represent the partly mixed tidal flow, regarding the tidal motion and salt distribution, in a descriptive way. More extensive comparison with the available flume tests is still going on and ought to prove the predictive character of this numerical model.

APPENDIX I: DATA FROM THE DELFT TIDAL FLUME

Description of the flume

The lucid flume for the experiments has a rectangular cross-section, 0.672 m wide and 0.50 m high. Two straight sections and the bend between them have a total length of 100 m (Figure 3). Downstreams the flume ends in a sea-basin, 8 m long, 6 m wide with a bottom, 1,1 m below the bottom of the flume. By means of a control valve any periodic tidal movement of the water-level can be generated. The density of the sea-water is kept constant by means of a circulation system which pumps salt water into the basin through perforated tubes on the bottom. At the upstream end of the flume is equipment to supply separately a constant and a variable discharge of fresh water. This makes it possible to reproduce tidal movements which can occur in flumes longer than 100 m. The variable discharge of fresh water is programmed according to one-dimensional tidal computations for flumes longer than 100 m. For a detailed description of the flume see Van Rees et. al. (1969) and Rigter (1973).

Experiments used for verification

Several tests, with plates (2 x 2 cm) on the bottom of the flume arranged in a diagonal pattern to obtain the desired roughness, are available. In this Paper two of these flume tests, which are used for verification of the numerical model, are represented. The tidal quantities in these two experiments are the same. However in test T22 there was no density difference between river- and sea-water, while in test T180 this density difference was about 22 kg/m^3 (see Table AI). So the difference in the tidal motion between these two tests is caused by effects of the density difference.

TABLE AI

Boundary Conditions and Flow Parameters

quality	symbol	test T22	test T180
depth (averaged over T)	h	0.216 m	0.216 m
fictive length of flume	L	179.34 m	179.34 m
Chézy-coefficient	C	19 m$^{\frac{1}{2}}$/s	19 m$^{\frac{1}{2}}$/s
tidal period	T	558.75 s	558.75 s
tidal amplitude at sea	A_o	0.025 m	0.025 m
fresh water-discharge	Q_L	0.0029 m^3/s	0.0029 m^3/s
density differences between river- and sea-water	$\Delta\rho$	0	22 kg/m^3

APPENDIX II

The finite difference equations for the conservation of momentum in the x-direction are:

$$\frac{u^{*}_{i,j} - u^{n}_{i,j}}{\tau} = -\frac{1}{b_i}\left\{\frac{b_{i+1}(u^{n}_{i+1,j})^2 - b_{i-1}(u^{n}_{i+1,j})^2}{2\,\Delta x}\right\} +$$

$$+ \varepsilon_x \left\{\frac{u^{n}_{i+1,j} - 2\,u^{n}_{i,j} + u^{n}_{i-1,j}}{\Delta x^2}\right\} - \frac{1}{\rho_{i,j}}\left\{\frac{p^{n}_{i+1,j} - p^{n}_{i-1,j}}{2\,\Delta x}\right\}$$

$$\frac{u^{n+1}_{i,j} - u^{*}_{i,j}}{\tau} = -\left\{\frac{u^{n+1}_{i,j+1} - u^{n+1}_{i,j-1}}{2\,\Delta z}\right\} TF^{n}_{1_{i,j}} +$$

$$-\left\{\frac{(u^{n+1}_{i,j+1}\ u^{n}_{i,j+1}) - (u^{n+1}_{i,j-1}\ u^{n}_{i,j-1})}{2\,\Delta z}\right\} TF^{n}_{2_{i,j}} +$$

$$-\left\{\frac{(u^{n+1}_{i,j+1}\ w^{n}_{i,j+1}) - (u^{n+1}_{i,j-1}\ w^{n}_{i,j-1})}{2\,\Delta z}\right\} TF^{n}_{3_{i,}} +$$

$$+\left\{\frac{\dfrac{(\varepsilon^{n}_{z_{i,j+1}} + \varepsilon^{n}_{z_{i,j}})}{2}\ \dfrac{(u^{n+1}_{i,j+1} - u^{n+1}_{i,j})}{\Delta z}}{\Delta z} + \right.$$

$$\left. - \frac{\dfrac{(\varepsilon^{n}_{z_{i,j}} + \varepsilon^{n}_{z_{i,j-1}})}{2}\ \dfrac{(u^{n+1}_{i,j} - u^{n+1}_{i,j-1})}{\Delta z}}{\Delta z}\right\} (TF^{n}_{3_i})^2$$

The finite difference equation for the conservation of mass is:

$$\frac{w^{n+1}_{i,j} - w^{n+1}_{i,j-1}}{\Delta z} = -\frac{(u^{n+1}_{i,j} - u^{n+1}_{i,j-1})}{\Delta z}\ \frac{(TF^{n}_{2_{i,j}} + TF^{n}_{2_{i,j-1}})}{2\ TF^{n}_{3_i}} +$$

$$-\frac{4}{3}\left[\frac{1}{2b_i}\left\{\frac{(b_{i+1}\ u^{n+1}_{i+1,j} - b_{i-1}\ u^{n+1}_{i-1,j})}{2\,\Delta x} + \right.\right.$$

$$\left.\left. + \frac{(b_{i+1}\ u^{n+1}_{i+1,j-1} - b_{i-1}\ u^{n+1}_{i-1,j-1})}{2\,\Delta x}\right\}\right] +$$

$$+\frac{1}{3}\left[\frac{1}{2b_i}\left\{\frac{(b_{i+2}\ u^{n+1}_{i+2,j} - b_{i-2}\ u^{n+1}_{i-2,j})}{4\,\Delta x} + \right.\right.$$

$$+ \frac{(b_{i+2}\, u^{n+1}_{i+2,j-1} - b_{i-2}\, u^{n+1}_{i-2,j-1})}{4\, \Delta x}\Big\} \Bigg]$$

The finite difference equation for the position of the free surface is:

$$\frac{\xi^{n+1}_i - \xi^n_i}{\tau} = -\frac{4}{3}\Big\{ \sum_{j=1}^{N_z} \frac{(u^{n+1}_{i+1,j} + u^{n+1}_{i+1,j-1})}{2} \Delta z\ TF^n_{3_{i+1}}\ b_{i+1} +$$

$$- \sum_{j=1}^{N_z} \frac{(u^{n+1}_{i-1,j} + u^{n+1}_{i-1,j-1})}{2} \Delta z\ TF^n_{3_{i-1}}\ b_{i-1}\Big\} \frac{1}{2\, \Delta x\, b_i} +$$

$$+ \frac{1}{3}\Big\{ \sum_{j=1}^{N_z} \frac{(u^{n+1}_{i+2,j} + u^{n+1}_{i+2,j-1})}{2} \Delta z\ TF^n_{3_{i+2}}\ b_{i+2} +$$

$$- \sum_{j=1}^{N_z} \frac{(u^{n+1}_{i-2,j} - u^{n+1}_{i-2,j-1})}{2} \Delta z\ TF^n_{3_{i-2}}\ b_{i-2}\Big\} \frac{1}{4\, \Delta x\, b_i}$$

The finite difference equations for the conservation of the salt content are:

$$\frac{c^{*}_{i,j} - c^n_{i,j}}{\tau} = -\frac{1}{b_i}\Big\{\frac{b_{i+1}\, u^{n+1}_{i+1,j}\, c^n_{i+1,j} - b_{i-1}\, u^{n+1}_{i-1,j}\, c^n_{i-1,j}}{2\, \Delta x}\Big\} +$$

$$+ D_x \Big\{\frac{c^n_{i+1,j} - 2\, c^n_{i,j} + c^n_{i-1,j}}{\Delta x^2}\Big\}$$

$$\frac{c^{n+1}_{i,j} - c^{*}_{i,j}}{\tau} = -\Big\{\frac{c^{n+1}_{i,j+1} - c^{n+1}_{i,j-1}}{2\, \Delta z}\Big\}\ TF^n_{1_{i,j}} +$$

$$- \Big\{\frac{u^{n+1}_{i,j+1}\, c^{n+1}_{i,j+1} - u^{n+1}_{i,j-1}\, c^{n+1}_{i,j-1}}{2\, \Delta z}\Big\}\ TF^n_{2_{i,j}} +$$

$$- \Big\{\frac{w^{n+1}_{i,j+1}\, c^{n+1}_{i,j+1} - w^{n+1}_{i,j-1}\, c^{n+1}_{i,j-1}}{2\, \Delta z}\Big\}\ TF^n_{3_i} +$$

$$+ \Bigg[\Big\{\frac{\frac{(D^n_{z_{i,j+1}} + D^n_{z_{i,j}})}{2} \frac{(c^{n+1}_{i,j+1} - c^{n+1}_{i,j})}{\Delta z}}{\Delta z}\Big\} +$$

$$- \Big\{\frac{\frac{(D^n_{z_{i,j}} + D^n_{z_{i,j-1}})}{2} \frac{(c^{n+1}_{i,j} - c^{n+1}_{i,j-1})}{\Delta z}}{\Delta z}\Big\} \Bigg] (TF^n_{3_i})^2$$

LIST OF SYMBOLS

a_o	tidal amplitude at x = 0
b	width
c	concentration
$c^n_{i,j}$	discretised c
D_x, D_z	turbulent diffusion coefficient for salt in the x- and z-direction respectively
f_b	transfer function
f (Ri)	ration between the mixing lengths for momentum in neutral and stratified conditions, depending on Ri
g	acceleration due to gravity
g_o	density profile at x = 0
g (Ri)	ration between the mixing lengths for salt in neutral and stratified conditions, depending on Ri
h	water-depth
H	transformed water-depth
u_o	velocity profile at x = 0
$u_{L'}$	velocity profile at x = L'
k_s	roughness length
L	fictive length of flume (Figure 3)
L	length of flume (Figure 3)
l_m	mixing length in neutral conditions
n	normal direction
p	pressure
p_s	pressure at the free surface
Q_L	river-discharge
Ri	Richardson-number
t	time
T	tidal period
TF_1, TF_2, TF_3	transfer coefficients
u	velocity component in x-direction
$u^n_{i,j}$	discretised velocity component u
w	velocity component in z-direction
$w^n_{i,j}$	discretised velocity component w
x	longitudinal direction
z	vertical direction
z_b	position of the bottom
z_o	coefficient of bottom roughness
$\Delta x, \Delta z$	spatial steps in x- and z-direction respectively
$\varepsilon_x, \varepsilon_z$	turbulent diffusion coefficient for momentum in the x- and z-direction respectively

κ	Von Karman-coefficient
ζ	integration variable
ρ	density
ρ_o	density of fresh water
σ	density differences
τ	time step
ξ	position of the free surface
ξ_o	position of the free surface at x = 0

REFERENCES

1 Abbott, M.R., 1960. Boundary layer effects in estuaries. Journ. Marine Research 18 no. 2: 82-100.

2 Abraham, G., Karelse, M. and Lases, W.B.P.M., 1975. Data requirement for one-dimensional mathematical modelling of salinity intrusion in estuaries. XVIth IAHR Congress, paper C32.

3 Boericke, R.R. and Hogan, J.M., 1977. An X-Z hydraulic/thermal model for estuaries Journ. ASCE, HY1: 19-37.

4 Delft Hydraulics Laboratory, 1973. Computational methods for the vertical distribution of flow in shallow water. Report on literature study, project W 152, Delft.

5 Delft Hydraulics Laboratory, 1974. Momentum and mass transfer in stratified flows. Report on literature study, project R 880, Delft.

6 Delft Hydraulics Laboratory, 1975. Berekening dichtheidsstroom, keuze van de differentiemethode. Report R 897-I, Delft.

7 Delft Hydraulics Laboratory, 1976. Berekening van stroming in een getijrivier, het homogene deel. Report R 897-III, Delft.

8 Fischer, H.B., 1973. Longitudinal dispersion and turbulent mixing in open channel flow. In: Ann. Review Fluid Mechn. 5: 59-78.

9 Fischer, H.B., 1976. Mixing and dispersion in estuaries. In: Ann. Review Fluid Mechn. 8: 107-133.

10 Hamilton, P., 1973. A numerical model of the vertical circulation of tidal estuaries and its application to the Rotterdam Waterway. Univ. of Washington, Dept. of Oceanography, Seattle.

11 Hamilton, P., 1975. A numerical model of the vertical circulation of tidal estuaries and its application to the Rotterdam Waterway. Geoph. Journ. Royal Astron. Soc. 40: 1-21.

12 Harleman, D.R.F. and Thatcher, M.L., 1974. Longitudinal dispersion and unsteady salinity intrusion in estuaries. La Houille Blanche no. 1/2: 25-33.

13 Hinwood, J.B. and Wallis, I.G., 1975a. Classification of model of tidal waters. Journ. ASCE 101, HY10: 1315-1332.

14 Hinwood, J.B. and Wallis, I.G., 1975b. Review of models of tidal waters. Journ. ASCE 101, HY11: 1405-1422.

15 Jamet, P., Lascaux, P. and Raviart, D.A., 1970. Une méthode de résolution numérique des équations de Navier-Stokes. Num. Math. 16. Springer-Verlag.

16 Kutler, P., Lomax, H. and Warming, R.F., 1972. Computation of space shuttle flow fields using non-centered finite difference schemes. AIAA-paper no. 72-193.

17 Leendertse, J.J., 1970. A water-quality simulation model for well-mixed estuaries and coastal seas; Vol. 1, Principle of computation. Memor. Rand Corp., RM-6230 RC, Santa Monica.

18 Van Rees, A.J. and Rigter, B.P., 1969. Flume study on salinity intrusion in estuaries. XIIIth IAHR Congress, paper C33.

19 Van Rees, A.J., 1975. Experimental results on exchange coefficients for non-homogeneous flow, XVIth IAHR Congress, paper C36.

20 Richtmeyer, R.D. and Morton, K.W., 1967. Difference methods for initial value problems. New York, Interscience Publishers.

21 Rigter, B.P., 1973. Minimum length of salt intrusion in estuaries. Proc. ASCE 99, HY9: 1475-1496.

22 Roache, P.J., 1972. Computational fluid dynamics. Albuquerque, Hermosa Publishers.
23 Vreugdenhil, C.B., 1970. Computation of gravity currents in estuaries. Delft Hydraulics Laboratory, publ. no. 86, Delft.
24 Vreugdenhil, C.B., 1974. Approximations in mathematical models for stratified flow. Delft Hydraulics Laboratory, Report S 114-IV, Delft.

THE EFFECT OF METEOROLOGICAL FORCING ON THE CHESAPEAKE BAY: THE COUPLING BETWEEN AN ESTUARINE SYSTEM AND ITS ADJACENT COASTAL WATERS

ALAN J. ELLIOTT[1] and DONG-PING WANG[2]

[1]SACLANT ASW Research Centre, Viale San Bartolomeo 400, 19026 La Spezia, Italy

[2]Chesapeake Bay Institute, The Johns Hopkins University, Baltimore, Md., 21218

ABSTRACT

Surface elevation and current meter records from the Potomac estuary are combined with elevation and wind stress data over the Chesapeake Bay to investigate the coupling between the Potomac, the Bay and the coastal ocean. The dominant sea level fluctuations in the Chesapeake Bay were found to be generated at the mouth of the Bay by the action of Ekman dynamics. Winds which blew parallel to the coast caused fluctuations in the mean sea level at the Bay mouth, these fluctuations then travelled northward up the Bay. Other fluctuations were due to an Ekman effect within the Bay itself and also due to longitudinal seiche motions.

The oscillations within the Potomac were due, in part, to the local forcing but were also due to co-oscillation with the Bay. Low frequency sea level fluctuations within the Potomac were the result of the Ekman effects in the coastal water; these disturbances, which had originated at the mouth of the Bay, appeared to influence the entire estuarine system.

The results suggest that future estuarine studies should include the coupling with the coastal ocean, both in modelling and in observational invesgations. The important time scales for the non-tidal fluctuations extend at least to monthly and seasonal periods. Therefore long term (several months to several years) monitoring of the wind, sea level, density and currents is required before the forcing and response mechanisms will be fully understood.

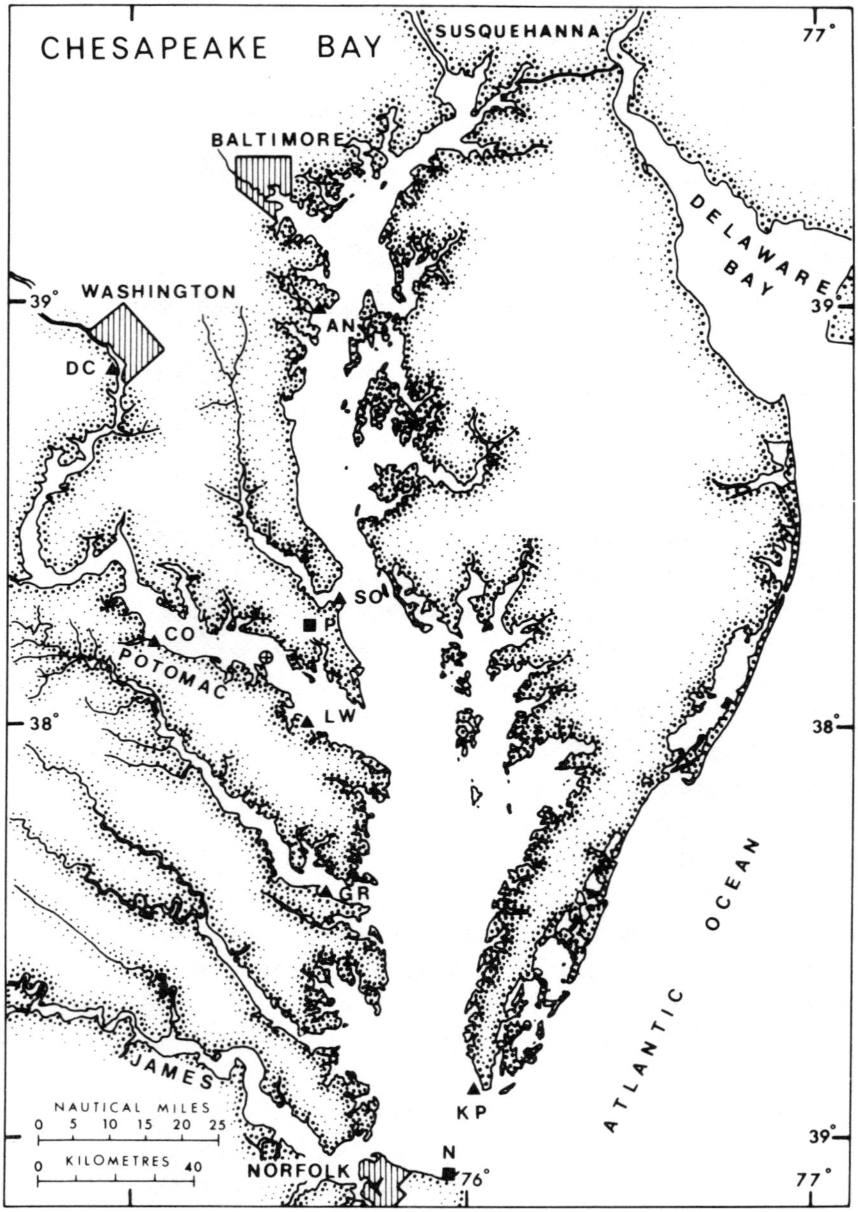

Fig. 1 The Chesapeake Bay estuarine system.

▲ Tide stations: AN, Annapolis; SO, Solomons; DC, Washington; CO, Colonial Beach; LW, Lewisetta; GR, Grey Point; KP, Kiptopeake Beach

■ Wind towers : P, Patuxent River; N, Norfolk

The position of the long-term current meter mooring in the Potomac is shown circled.

INTRODUCTION

The Chesapeake Bay and its tributaries form one of the largest estuarine systems in the world. The Bay itself extends a total distance of about 170 nautical miles (315 km) from its mouth near Norfolk, Virginia, to its head where it meets the Susquehanna River (Fig. 1). The tidal wave (semidiurnal tide) takes slightly longer than 13 hours to travel from the mouth of the Bay to its head; as a result the next tidal wave enters the Bay before the head has been reached and so two high tides can simultaneously be contained within the Bay. The Bay is large enough for rotational effects to be important: in the lower and mid portions of the Bay the tide advances as a Kelvin wave and tidal range and velocities are significantly greater along the eastern shore. The Kelvin wave is absent from the region near the head of the Bay where friction and reflection cause the tide to resemble a standing wave. By assuming that the tide is a purely progressive wave along the entire length of the Bay, the estimated phase speed for long waves is 480 km/day. It is also possible to estimate that the period of a longitudinal seiche within the Bay will have a fundamental value of around 2-2.5 days.

Previous studies within the Chesapeake Bay system have been directed towards an understanding of the density driven internal circulation (Pritchard; 1954, 1956) and have consisted of short duration studies during calm weather. It is only in recent years that self-recording instruments have made it possible to collect fairly long records during varying weather conditions. The aim of this paper is to review our understanding of the response of the Chesapeake system to meteorological forcing in the light of some recent observations (Elliott and Hendrix, 1976; Elliott, 1977; Wang and Elliott, 1977) and to present an overall picture of the mechanisms involved.

YEAR-LONG OBSERVATIONS IN THE POTOMAC ESTUARY

For a one-year period, from July 1974 through July 1975, current measurements were made at three depths in the Potomac estuary (Elliott, 1977). Meteorological data were also collected and the complete data set was filtered and then further averaged with 24 hour non-overlapping blocks to produce a sequence of daily mean values. (A 25 hour box-car filter was used to remove the tidal signals; while not ideal this filter was considered to be sufficient in view of the large nontidal components and it had the further advantage of not causing an excessive loss of data at the ends of the

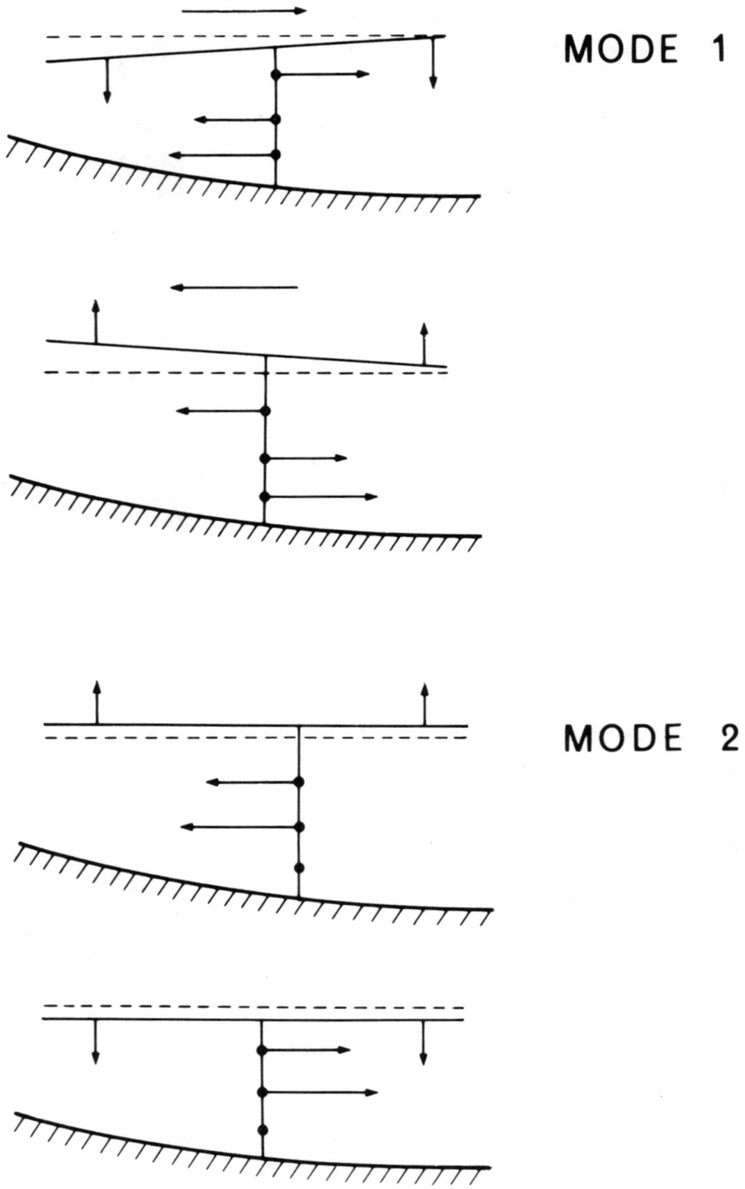

Fig. 2 Schematic representation of the two dominant modes observed in the response of the Potomac.

records.) The resulting low-pass signals showed marked current fluctuations with periods of 2-5 days and longer. The fluctuations had r.m.s. values of around 4-6 $cm \cdot s^{-1}$, exceeding the long term mean values by a factor of about 2. When averaged over the year-long period the mean flow was directed seaward in the surface layers and landward at mid-depth and near the bottom, i.e. consistent with the circulation usually associated with a partially-mixed estuary. This 'classical' circulation was observed for 43% of the time and had a mean duration of 2.5 days for each occurrence. However, a landward directed flow at all three levels (storage) and the reverse of the classical estuarine circulation (landward flow at the surface, outflow near the bottom) were the second most common flow patterns, each occurring for about 20% of the time and having mean durations of around 1.5 days.

The data were first analyzed by methods which did not consider the frequency dependence of the response. In particular, a multiple regression analysis was made but it was found that only 50%-60% of the fluctuations in each of the current records could be related to local meteorological forcing. The poor results from the regression were in fact caused by the presence of two distinct modes of response, one of which was due to non-local forcing. The two modes were separated using EOF analysis in the time domain (Wallace and Dickinson, 1972) and the results are presented schematically in Fig. 2.

The first mode was associated with the local wind forcing and contained 47% of the total velocity fluctuations. A downstream wind blew water out of the estuary causing a reduction in the mean water level and setting up a surface slope towards the mouth. The surface flow was directed seaward, while a return flow took place at mid-depth and near the bottom. This mode was reversible, i.e. an upstream wind stress could cause an increase in the mean water level with the mean surface sloping upwards toward the head. The surface flow would then be directed landward while the deeper water would flow seawards, i.e. the reverse of the circulation usually associated with a partially-mixed estuary.

The second mode, which accounted for approximately 30% of the total velocity fluctuations, was not related to the local wind (less than 1% of the wind fluctuations could be related to Mode 2, in contrast 80% of the wind fluctuations were connected with Mode 1). In addition, Mode 2 was not related to the local surface slope but was characterized by the rise and fall of the mean surface and by current fluctuations at mid-depth and near the surface. Landward net flow was associated with a rise in the mean elevation and, conversely, seaward net flow was accompanied by a reduction in the mean water level. The near-bottom flow was not influenced by this response, but for the upper portion of the water column the second mode was as significant as the local forcing (Mode 1).

The results showed that, for the Potomac, a knowledge of the local wind field is not sufficient for predicting the net currents since only about 50% of the fluctuations can be explained in terms of local forcing. This is in contrast to the results obtained by Weisberg (1976) in Narragansett Bay, and suggests that the second mode (far-field response), which contaminated the effects of local forcing, was due to the interaction between the Potomac and the Chesapeake Bay.

The above modal analysis was independent of frequency, i.e. the two modes were extracted from the observations but no time scales were assigned to them. Fig. 3 shows current, elevation and wind stress during the period of April 15 to May 18, 1975. This particular month-long period contained two distinct regimes. During April 15 to 26, the local wind was relatively strong, having a peak value of around 1 $dyn \cdot cm^{-2}$. The bottom current (U_{40}) and the surface elevation (E_3) appeared to be coherent with the downstream wind stress (τ_1); yet, while the current fluctuations seemed to be coherent at all three depths, there was the suggestion of an upward phase propagation. Visual inspection does not therefore suggest a simple Mode 1 type response although the surface current responded in part (and was in phase with) local wind. The dominant time scale appeared to be around 4-5 days.

The second part of the record, May 4 to 18, was characterized by significantly weaker local winds. The currents and surface fluctuations, however, were comparable in magnitude to those observed earlier. The coherence between bottom current (U_{40}) and elevation (E_3) was fair, and there appeared to be no time lag in the response at the three depths. The mid-depth and bottom currents were in phase during this period while the surface flow appeared to be 180° out of phase. The fluctuations had a time scale of 2-3 days and part of the response seemed to be due to the local wind; however, the relatively large amplitudes suggest that the non-local contribution was significant at these higher frequencies.

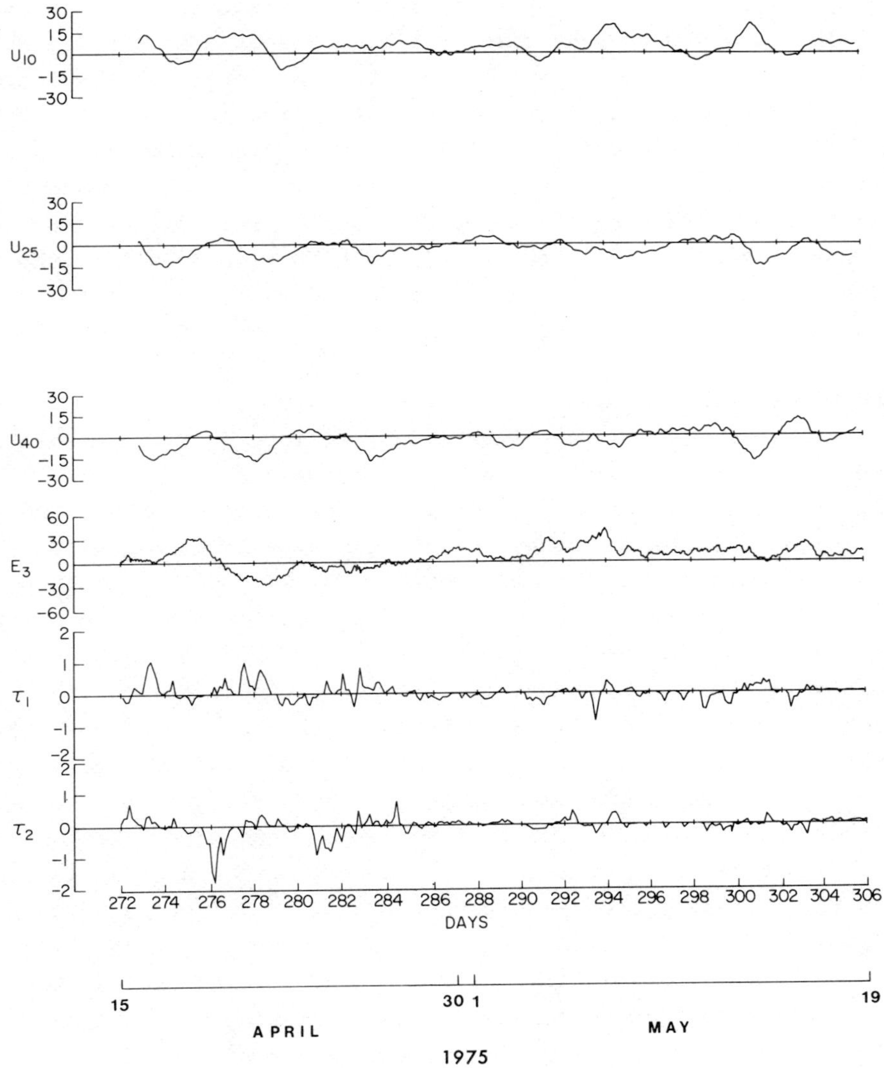

Fig. 3 Example of the Potomac data (April 15 to May 18, 1975); non-tidal velocities are in $cm.s^{-1}$, E_3 is the non-tidal elevation at Lewisetta in cm, τ_1 and τ_2 are the down-stream and cross-stream components of wind stress in $dyn.cm^{-2}$.

FREQUENCY DEPENDENT ANALYSIS

To investigate the frequency dependence of the response and to study the coupling between the Bay and the Potomac, Wang and Elliott (1977) used spectral techniques and also EOF analysis in the frequency domain. Data collected during the first two months of the year-long study were used (mid-July to mid-September, 1974), which includes wind, surface elevation and bottom current in the Potomac, plus a series of 3-day intensive current measurements (Elliott and Hendrix, 1976). In addition, the following discussion also includes surface elevation, slope and wind stress over the Chesapeake Bay (Fig. 1). For this analysis the data series were filtered using a Lanczos filter to remove the tidal and higher frequencies; the filter had a half-amplitude point of 34 hours and passed 95% of the energy at 50 hours.

Wind Stress over the Bay

As part of the year-long study, Elliott (1977) compared the wind stress at two locations near the Potomac estuary. One wind station (Patuxent River Naval Station) was located near the mouth of the Potomac, while the other (Quantico Marine Base) was situated about 100 km upstream from the mouth. The wind data from these two stations were found to be coherent, although the magnitude of the wind was higher at the more exposed Patuxent site. In contrast to the similarity between these two stations, wind data near the Atlantic coast at Norfolk (Fig. 1) was coherent with the wind at Patuxent River only for periods longer than five days. The poor coherence between the wind fluctuations at these two locations for shorter periods was probably a result of sea-breeze and other local effects which would be confined to the coastal region. Fig. 4 shows the spectra of the N-S and E-W components of wind stress at Norfolk and Patuxent River. At Norfolk the principal wind direction was N-S and the wind stress was significantly stronger than at the Patuxent site. The spectra for both locations displayed distinct peaks at periods of 20, 5 and 2.5 days. While the principal axis of the wind at Norfolk was in the N-S direction, the wind at Patuxent was oriented more in a NE-SW direction. However, at the 2.5 day period the wind energy was concentrated in the N-S direction at both locations.

Response of the Chesapeake Bay to Winds

Non-tidal sea surface fluctuations can be the result of both atmospheric pressure variations and also the response to surface wind stress. In the present study, however, it was found that the coherence between atmospheric pressure and mean sea level was low, suggesting that the wind fluctuations were the major driving force. The sea level spectra for four locations in the Chesapeake Bay are shown in Fig. 5. The marked similarity to the shape of the wind spectra (Fig. 4) confirms that the surface fluctuations were induced by the wind field.

The 20 day fluctuations at the Bay mouth were the result of an Ekman flux in the adjacent coastal water, i.e., sea level increased at the Bay mouth when the coastal wind was blowing to the south, and decreased when the coastal winds blew to the north. These fluctuations were then damped in amplitude as they propagated up the Bay away from the mouth (Fig. 5).

The 5-day fluctuations near the Bay mouth were driven by both the N-S and the E-W winds. The amplitude at this frequency increased slightly in the upper Bay (Fig. 5), which suggests local forcing. An increase of sea level in the upper Bay was associated with winds blowing to the west, a decrease with winds blowing to the east. This suggests that Ekman effects within the Bay itself were influencing the elevations in the upper Bay.

The 2.5 day sea level fluctuations had higher amplitudes near the head of the Bay and they decreased seawards. The fluctuations were generated within the Bay by the local N-S winds. The presence of seiche oscillations is suggested by the increase in amplitude towards the head of the Bay. In addition, the period of oscillation, 2.5 days, is of the correct order for a longitudinal seiche within the main portion of the Bay.

Interaction between the Potomac and the Chesapeake Bay

This section now returns to the discussion of the response of the Potomac and considers the frequency dependence of the response in the light of the results presented in the previous sections. Fig. 6 shows the spectra and coherence functions calculated from the 2-month long records of surface elevation, surface slope and near-bottom current in the Potomac estuary. All three spectra had distinct peaks at periods of 20, 5 and 2.5 days. However, whereas the slope and elevation spectra were similar in shape to those obtained for wind and elevation within the Bay, the bottom current showed a relatively higher amplitude at the 2.5 day period. Sea level and surface slopes were significantly coherent at 20 and 2.5 days and, to a lesser extent, at 5 days. In contrast sea level and bottom current were coherent at 5 and 2.5 days — but not at 20 days.

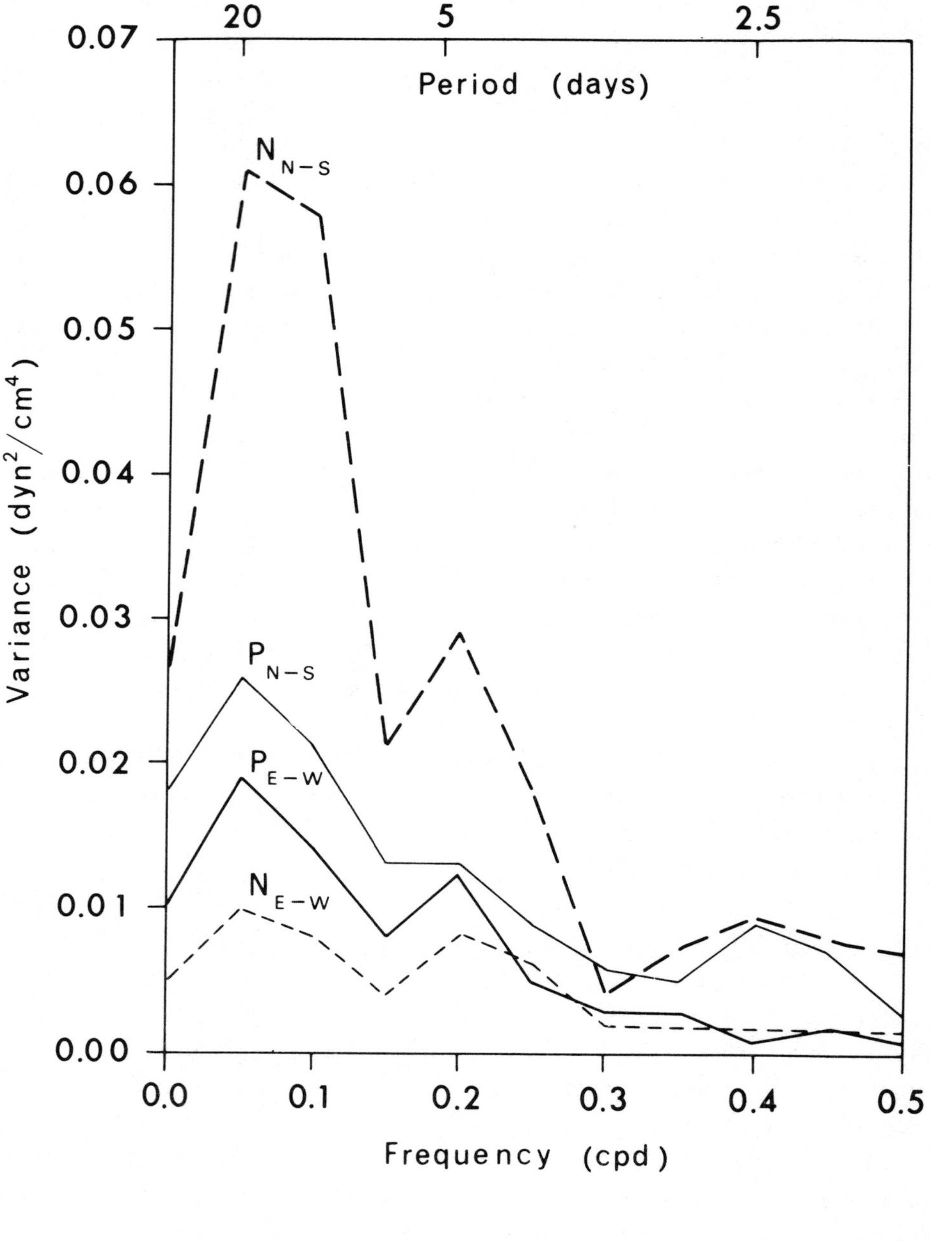

Fig. 4 Spectra of the N-S and E-W components of wind stress at Patuxent River and Norfolk.

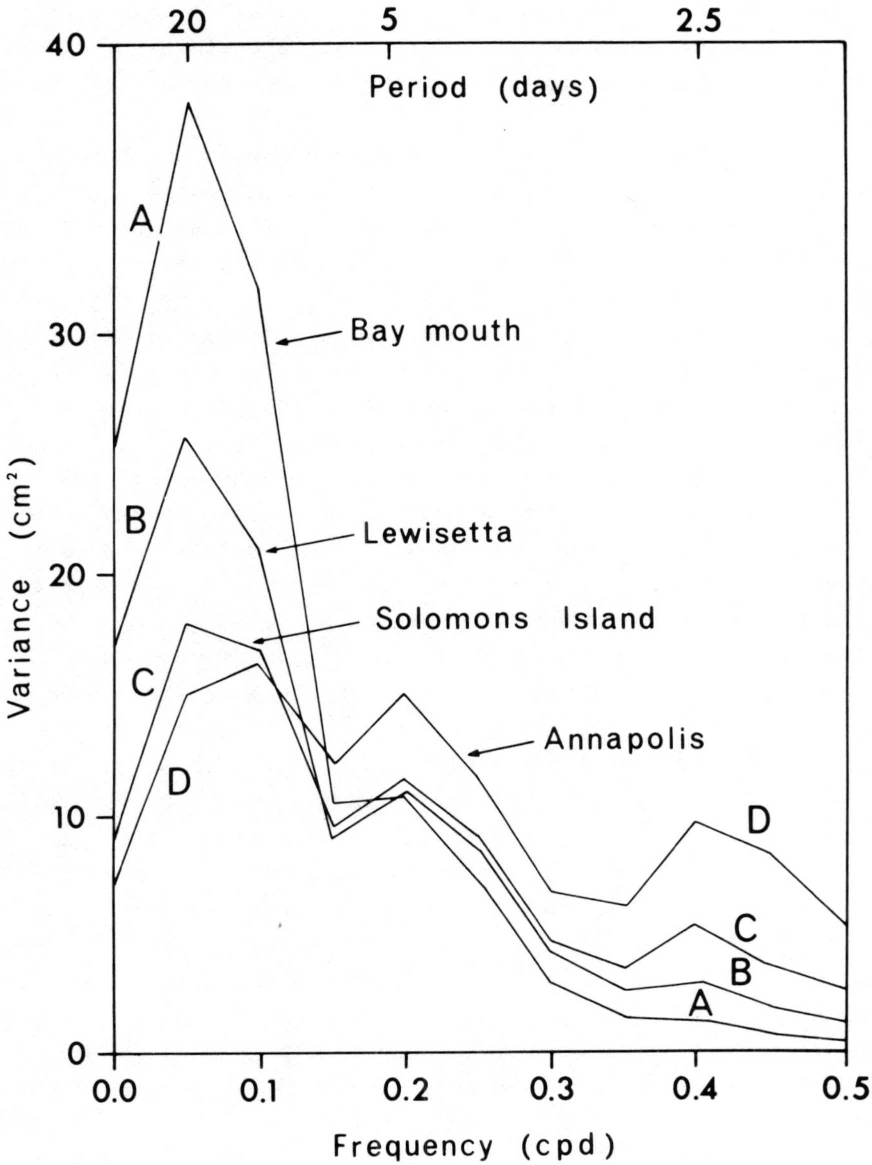

Fig. 5 Sea level spectra at four locations along the longitudinal axis of Chesapeake Bay.

The analysis of the year-long records had shown that the surface elevation was mainly associated with a Mode 2 response (exchange of surface water between the Potomac and Bay). In addition, we know that the 20 and 5 day sea level fluctuations were mainly generated at the mouth of Chesapeake Bay. Therefore the fluctuations which were observed in the Potomac appear to have been generated in the following manner:

1. The 20 day fluctuations were caused by the exchange of surface water with the Bay, i.e., they were basically a Mode 2 response. These fluctuations were the direct result of sea level variations at the ocean boundary caused by Ekman effects in the coastal ocean. The fluctuations were damped as they progressed up the Chesapeake Bay but they were still of sufficient amplitude when they reached the mouth of the Potomac to cause sea-level and surface current fluctuations within the estuary. They did not appear to influence the Potomac near-bottom currents.

2. The 5 day fluctuations in the Potomac are also believed to have originated, in part, at the mouth of the Bay and to have been caused by co-oscillation with the Bay (i.e., Mode 2). However, the increase of sea level in the upper Bay, and the significant correlation between bottom current and surface slope at this period (Fig. 6) suggest that local forcing was also important (i.e. Mode 1). Thus, the 5 day fluctuations consisted of both the Mode 1 and Mode 2 response.

3. The 2.5 day fluctuations appeared to be due to the seiche oscillations within the upper Bay. At this frequency the bottom current was coherent with the local surface slope, which suggests local forcing. On the other hand, the bottom current was also coherent with the surface elevation, which suggests that exchang between the two estuaries occurred over the whole water column. Thus, the 2.5 day motion in the Potomac had characteristics different from the Mode 1 or Mode 2 response. The time-domain EOF analysis fails to reveal the 2.5 day type response, presumably because the wind and sea level variations were dominated by the longer period fluctuations (Figs. 4 and 5).

INTERNAL ADJUSTMENT: A FURTHER DRIVING MECHANISM

During September of 1974 an attempt was made to repeat a James River type of analysis (Pritchard; 1954, 1956) by using modern equipment to collect data in the Potomac estuary. Two cross-sections of the Potomac were occupied continually during two 5 day periods and

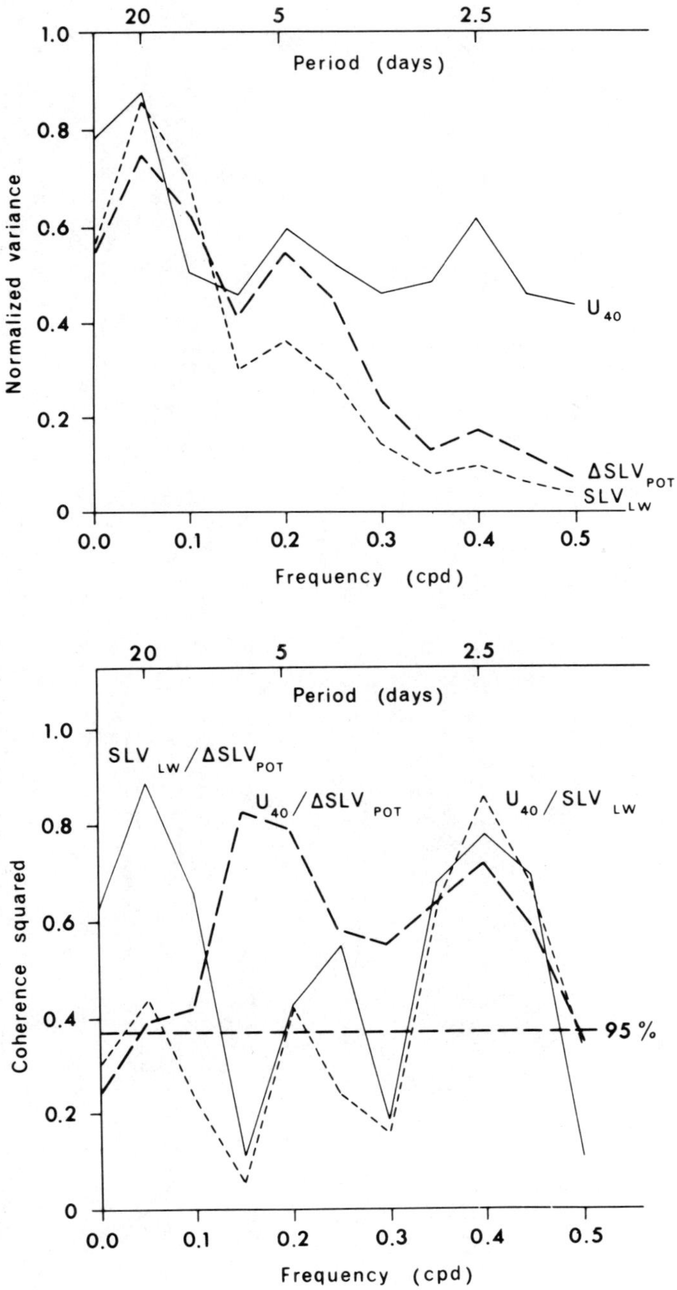

Fig. 6 Spectra and coherence functions for near-bottom current, sea level and surface slope within the Potomac.

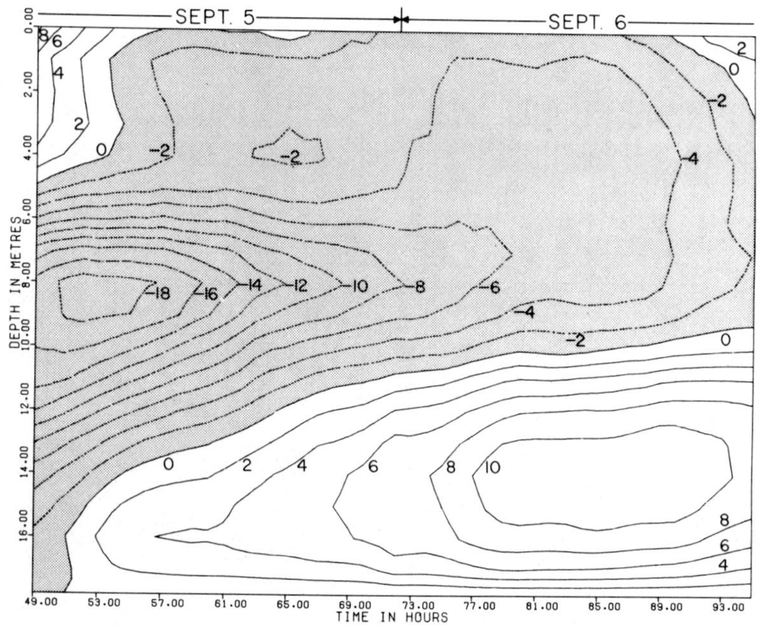

Fig. 7 Depth-time contour plot of the non-tidal current ($cm.s^{-1}$) in the Potomac estuary, September 5-6, 1974. (The time axis has its origin at 0000 hrs on September 3.)

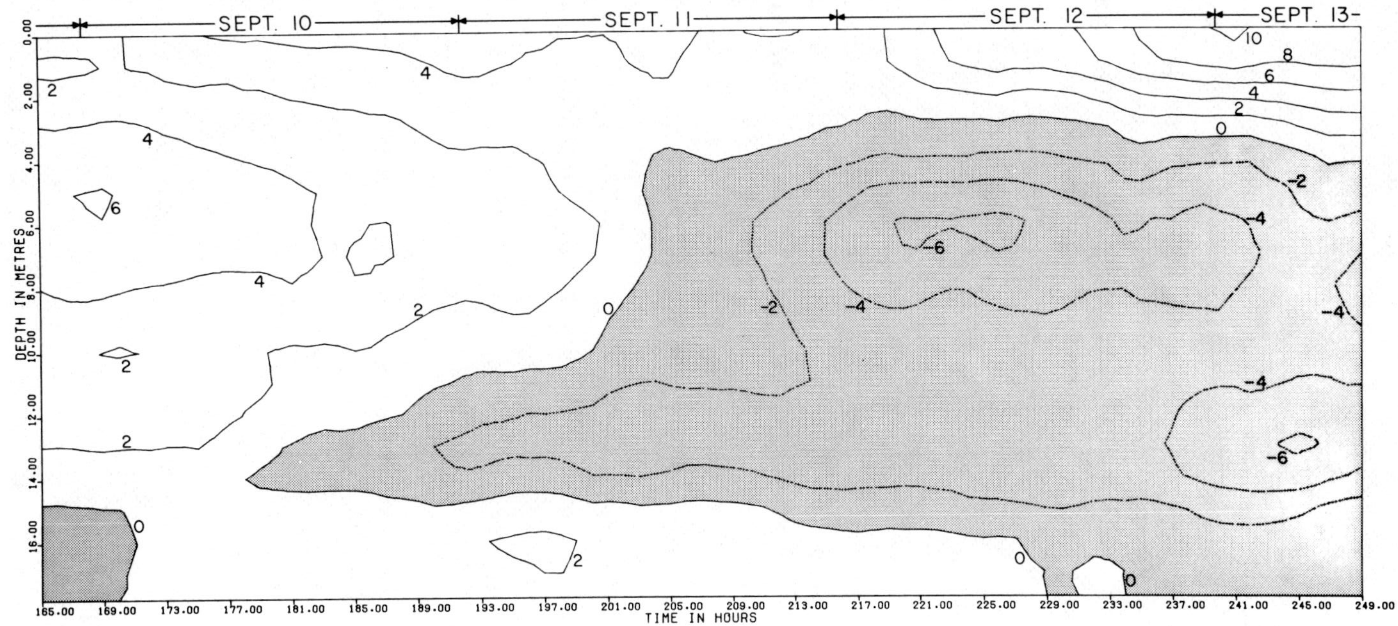

Fig. 8 Continuation of Fig. 7 during September 10-13, 1974.

hourly samples were taken across each transect using vertically profiling current meters and CTD sensors (Elliott and Hendrix, 1976). During the analysis of the data, however, it was discovered that the circulation was more complex than the steady two-layered flow investigated 25 years earlier in the James. Figs. 7-8 show the vertical structure of the low-pass currents which were observed at one of the transects. (The current data were first resolved into the longitudinal direction and then filtered with a 25 hour box-car filter.)

At the beginning of September 5 (hour 49) the net circulation was in the correct sense for a partially mixed estuary. There was a seaward flow of 6-8 $cm.s^{-1}$ near the surface, and a reverse landward flow of 16 $cm.s^{-1}$ near mid-depth. However, from hour 50 onward the bottom flow reversed its direction and started to flow seaward, reaching a maximum strength of 10-12 $cm.s^{-1}$ by hour 85. There appears to have been an upward phase propagation through the water column which resulted in the surface layers also reversing their direction of net flow. By hour 55 the surface water had started to flow landward and during hours 55-90 it had a negative velocity of 2-4 $cm.s^{-1}$. During most of the first measurement period the residual flow was in the reverse sense to that usually associated with a partially mixed estuary.

There was no sampling on September 7 and 8, coupled with the effects of the filter this caused a three day gap in the residual current records which continue in Fig. 8. By the end of September 9 (hour 168) there was a positive seaward flow throughout most of the water column. However, a three-layer flow then developed in which there was a landward flow near mid-depth. (This three-layered flow is in the reverse sense to the three-layered flow which can occur in tributary embayments, e.g. Baltimore Harbour, when a water of intermediate density flows seaward at mid-depth). The three-layered flow intensified and reached a peak early on September 12 when the landward flow at mid-depth had a magnitude of 6-8 $cm.s^{-1}$. Eventually, toward the end of the second period of observation, the flow reverted to the usual estuarine pattern. The time scale for the transient flow, from the initial reversal to the final recovery of the estuarine circulation, was of the order of 10 days. In general, the residual flow was highly coherent at the two transects and the profiling data agreed well with the results obtained from moored current meters. The intensive observations were deliberately made at locations which were near the long-term current meter mooring and the observations suggest a further mechanism which can cause current fluctuations in an estuary:

Prior to the start of the intensive measurements there had been a period of moderate down-stream winds in the Potomac (about 5 $m.s^{-1}$). These winds were also present during the first 24 hours of the measurements and local forcing is thought to have caused the strong estuarine flow observed at the beginning of September 5 (this was a Mode 1 response). Following this wind stress, however, the horizontal salinity gradients were observed to be in the reverse sense to those usually associated with an estuary, i.e. the salinity decreased towards the mouth. This suggests that the reverse circulation may have been driven by the perturbed density field. The preliminary results from a numerical model which can include the effects of a wind stress (Elliott, 1976) have suggested that the salinity reversal was caused by the combined effects of surface water being blown out of the estuary and by higher salinity water, which would be advected into the estuary by a compensating return flow, up-welling within the estuary. Following the relaxation of the wind, the internal density would tend to drive the net currents in the opposite sense to that usually associated with a partially-mixed estuary. Observational evidence that the salinity gradient in an estuary can be reversed following a period of down-stream winds has been reported for other estuaries by Elliott (1976) and Weisberg (1976). Therefore the intensive observations are considered to have been taken during a period when the estuary was responding to adjustments in the density field: this adjustment had a time scale of around 10 days.

DISCUSSION

The recent observational results, which have been summarized here, show that non-local forcing in the Chesapeake Bay due to the influence of the coastal ocean is an important mechanism which influences the distribution of non-tidal currents and elevations within the Chesapeake estuarine system. Previous work had concentrated on the analysis of the gravitationally driven internal circulation, based on data which had been collected for relatively short periods of time during calm weather. In contrast, the recent measurements show that the mean velocities associated with the meteorological forcing can be an order of magnitude larger than those associated with the gravitational circulation. Furthermore, the important time scales are now known to extend to at least 20 days which is significantly longer than was previously thought to be the case. The intensive observations and preliminary results from a numerical study also suggest that the internal adjustment of a perturbed density distribution, which has a time scale of the order of 10 days, may be an important mechanism influencing the estuarine circulation.

The following comments can therefore be made on the need for future studies:

1. Is the Chesapeake Bay system typical? We have presented results which we believe to be valid for the Chesapeake Bay but it needs to be resolved whether these mechanisms are also important in other types of estuaries.

2. Numerical models need to include Ekman effects at the open coastal boundary. Effort should also be directed towards modelling the fluctuations in the non-tidal flow; this requires models that are efficient enough to be run repeatedly for simulations of around 100 tidal cycles in duration without excessive computer costs. The models should also be able to include the interactions between branching tributaries. (Three-dimensional models are capable of satisfying the interaction requirement, but are unlikely to be economically feasible when run for long simulations.)

3. The recent observations have shown that field studies of limited durations (say 5-10 days or less) in the Chesapeake Bay are unlikely to lead to a true understanding of estuarine dynamics and that much longer records are required to resolve the forcing mechanisms. It will be necessary to collect long records in several types of estuaries before such questions will be fully answered.

ACKNOWLEDGMENTS

This study was made at the Chesapeake Bay Institute of The Johns Hopkins University and was supported by the National Science Foundation under grant OCE74 - 08463. We thank Dr D.W. Pritchard for the encouragement that he gave throughout the study.

REFERENCES

Elliott, A.J., 1976. A numerical model of the internal circulation in a branching tidal estuary. Chesapeake Bay Institute Special Rept. 54, Ref. 76-7, 85 pp.

Elliott, A.J., 1976. Response of the Patuxent estuary to a winter storm. Chesapeake Sci., 17:212-216.

Elliott, A.J., 1977. Observations of the meteorologically induced circulation in the Potomac estuary. Est. Coastal Mar. Sci., (in press).

Elliott, A.J. and Hendrix, T.E., 1976. Intensive observations of the circulation in the Potomac estuary. Chesapeake Bay Institute Spec. Rept. 55, Ref. 76-8, 35 pp.

Pritchard, D.W., 1954. A study of the salt balance in a coastal plain estuary. Jour. Marine Res., 13:133-144.

Pritchard, D.W., 1956. The dynamic structure of a coastal plain estuary. Jour. Marine Res., 15:33-42.

Wallace, J.M. and Dickinson, R.E., 1972. Empirical orthogonal representation of time series in the frequency domain. Part I: Theoretical considerations. J. App. Meteor., 11:887-892.

Wang, D-P and Elliott, A.J., 1977. Non-tidal variability in the Chesapeake Bay and Potomac River: evidence for non-local forcing. Submitted to J. Phys. Oceanogr.

Weisberg, R.H., 1976. The nontidal flow in the Providence River of Narragansett Bay: A stochastic approach to estuarine circulation. J. Phys. Oceanogr., 6:721-734.

LONG-PERIOD, ESTUARINE-SHELF EXCHANGES IN RESPONSE TO METEOROLOGICAL FORCING

NED P. SMITH
Johnson Science Laboratory, Harbor Branch Foundation
Fort Pierce, Florida 33450

ABSTRACT

Recording current meter data from two approximately one-month periods are used to investigate non-tidal exchanges between Corpus Christi Bay, Texas, and the northwestern Gulf of Mexico. The net transport in May and June, 1975, is an outflow which appears to be driven primarily by slowly falling coastal water levels. Rising coastal water levels in July, 1976, appear to be responsible for a net inflow into the bay. In each study, non-tidal current variations, occurring over time scales on the order of three to six days, are superimposed onto the very long period net transport. Related studies in the same area suggest that these estuarine-shelf exchanges are in response to meso-scale meteorological forcing. Dominant processes include the set-up and set-down of coastal water levels by cross-shelf windstress, a cross-shelf Ekman transport produced by longshore windstress, and an inverse barometer effect.

INTRODUCTION

The exchange of water between estuaries and the inner continental shelf occurs over a wide range of time scales and in response to a variety of astronomical, thermodynamical and meteorological forces. The relative importance of these forcing mechanisms varies greatly both in space and time. An investigation of estuarine-shelf exchanges should ideally extend over a time interval sufficiently long to include meteorological events lasting many days, and, in some areas, the fortnightly or monthly tidal periodicities associated with the principal tidal constituents cycling in and out of phase. Meteorological forcing of minor importance at times of spring or tropic tides may be relatively important under neap or equatorial tidal conditions.

The northwestern Gulf of Mexico is a region of characteristically low amplitude tides and strong winds. Zetler and Hansen (1970) have summarized available information on tidal harmonic constants for the Gulf of Mexico. For Port Aransas, Texas, in the northwestern corner (Site A in Figure 1), the tide is mixed but principally diurnal. The O_1 and K_1 diurnal constituents both have amplitudes on the order of 15 cm. The M_2 and S_2 partial tides are approximately 8 cm and 2 cm, respectively. Tidal ranges at times of tropic tides are generally less than 0.75 m.

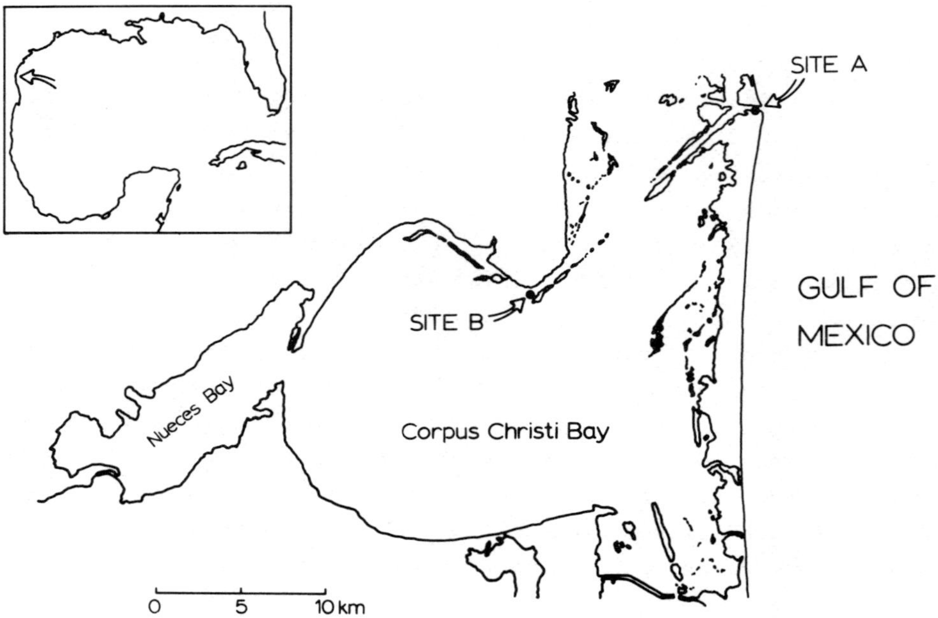

Fig. 1. Study Sites A and B at the coast at Port Aransas, Texas, and at the entrance to Corpus Christi Bay, respectively. Insert shows the study area in the northwestern Gulf of Mexico.

On the other hand, the National Weather Service lists the International Airport at Corpus Christi, Texas, as the third windiest in the country, with a multi-annual average wind speed of 23.5 km/hr. One might expect, therefore, that meteorological forcing is relatively important in exchanging water between the inner continental shelf and the intracoastal bays that lie just inside the northwestern rim of the Gulf of Mexico.

The purpose of this paper is to document long-period exchanges occurring specifically between Corpus Christi Bay, Texas, and the inner shelf of the northwestern Gulf of Mexico, using time series of direct current measurements; and to suggest some of the meteorological forcing mechanisms responsible for these exchanges.

THE OBSERVATIONS

Direct current measurements were obtained using a General Oceanics Model 2010 inclinometer recording current meter. Sampling was carried out just above the bottom in about 7.5 m of water at the edge of the ship channel connecting Corpus Christi Bay with the inner continental shelf of the northwestern Gulf of Mexico (Fig. 1).

The first study was over an approximately 30-day period from 13 May through 12 June, 1975, at Site A on the coast at Port Aransas, Texas. The second study covered approximately 31 days from 29 June to 3 August, 1976. Measurements were made at Site B, at the entrance to Corpus Christi Bay, approximately 19 km from the coast along the Corpus Christi Ship Channel. Currents were measured at half-hourly intervals in both studies.

The long-channel components of the recorded current velocity vectors were used in the analysis. Inclination angles were read to the nearest degree. Current speeds are non-linearly related to inclination angles, but within the speed range of 10 to 40 cm/sec the precision of the long-channel component current speeds is approximately ±1 cm/sec.

Water level measurements were obtained from a U.S. Army Corps of Engineers water level recorder located at Site A. Water level data were read relative to a datum of one foot (30.48 cm) below mean sea level, and have a precision of approximately ±3 mm.

RESULTS

Since estuarine-shelf exchanges are largely driven by variations in coastal water levels, it is appropriate to begin by computing an energy density spectrum to determine over what time scales coastal water level variations occur. Figure 2 shows the spectrum of hourly water levels recorded at Site A between 29 January, 1974, and 5 March, 1975. The energy density calculations utilize a fast-Fourier transform technique, after the linear trend has been removed from the data (Fee, 1969). The principal diurnal and semi-diurnal tidal constituents stand out as clearly isolated spectral peaks and are consistent with the amplitudes reported by Zetler and Hansen (1970). Also evident are the overtides and compound tides, reflecting shallow-water effects, at periodicities of approximately eight and six hours.

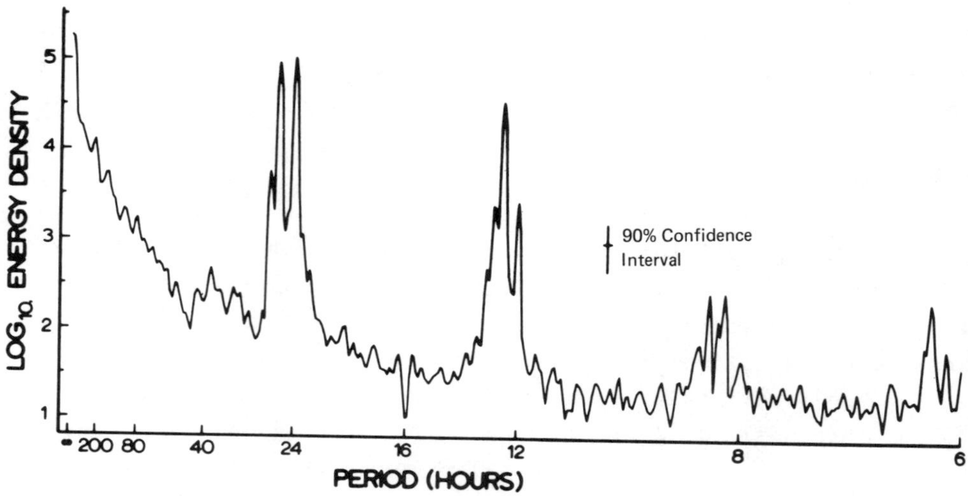

Fig. 2. Energy density spectrum of hourly water levels measured at Site A between 29 January, 1974, and 9 March, 1975. Vertical axis is in cm^2/c.p.h. Spectral resolution is 0.00042 c.p.h.

Of particular interest here, however, is the rise of computed energy density levels at periodicities longer than about two days. This may reflect many thermodynamic processes, and perhaps includes some of the long-period tidal constituents. However, it is felt that the increase in energy density levels in the long-period part of the spectrum, and thus long-period shelf-estuarine exchanges, are due primarily to meteorological forcing in various forms.

Two recent investigations have been carried out to monitor shelf-estuarine exchanges directly with recording current meter data obtained between Corpus Christi Bay and the inner continental shelf of the northwestern Gulf of Mexico. The data provide information both on the relative importance of tidal and non-tidal exchanges, and on the characteristics of non-tidal exchanges occurring over a wide range of time scales.

Figure 3 shows the half-hourly long-channel current components recorded from 13 May to 12 June, 1975, at Site A on the coast. Dominating the pattern is the approximately fortnightly cycle from tropic to equatorial and back to tropic tides, governed by the position of the moon in its orbit. Less apparent is the fact that the entire pattern is shifted into the ebb portion of the plot. This is brought out clearly, however, when the tidal

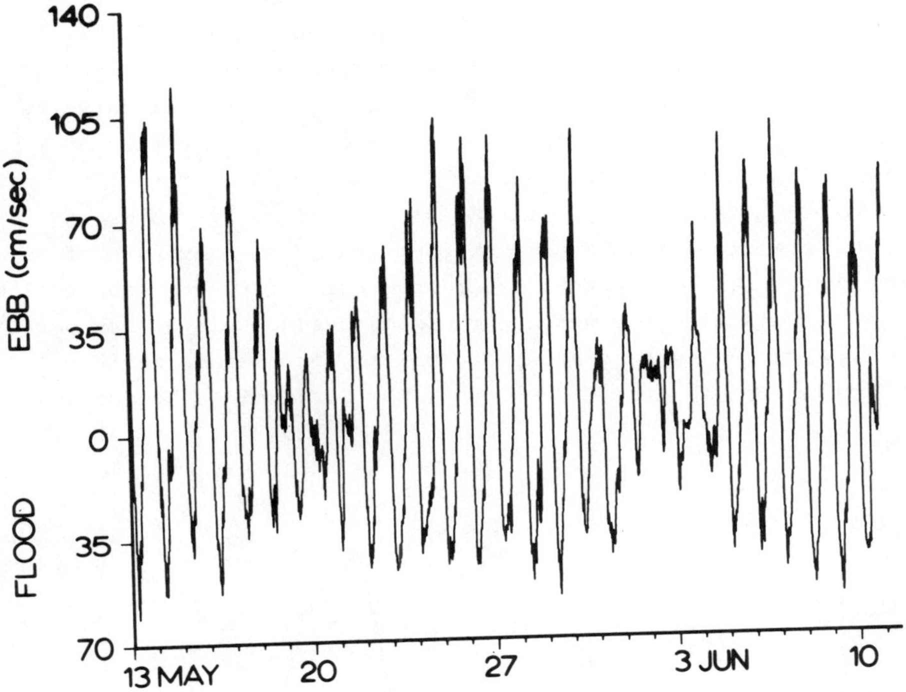

Fig. 3. Half-hourly, long-channel current components from Site A, in cm/sec, 13 May to 12 June, 1975.

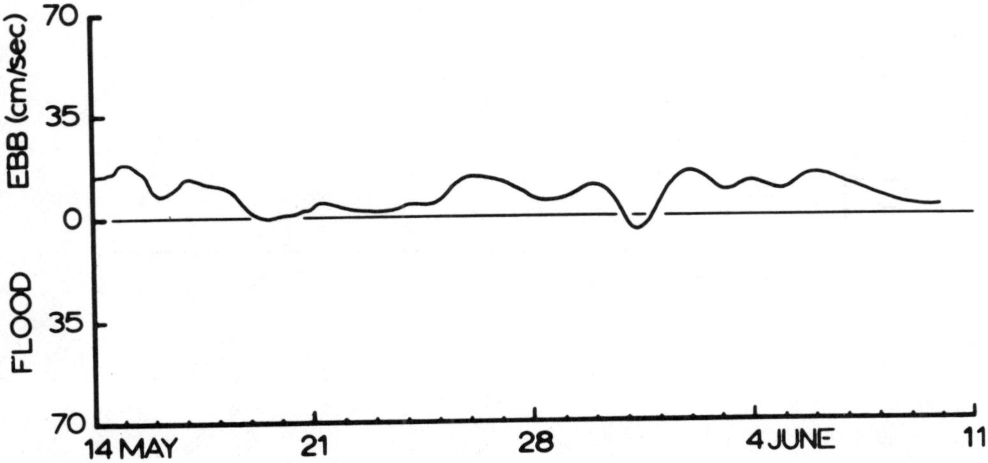

Fig. 4. Filtered long-channel current components from Site A, in cm/sec, 14 May to 11 June, 1975.

period components of the current record are removed by numerically filtering the raw data. Figure 4 shows the numerically filtered non-tidal long-channel components past Site A. Two points are particularly noteworthy in this figure. First, as noted above, non-tidal currents are nearly consistently in the ebb portion of the plot. Second, there are, superimposed onto the net ebb, quasi-periodic variations in the long-channel current components over time scales on the order of three to six days. Non-tidal currents are generally less than 20 cm/sec, and the average current past Site A over this time interval was an outflow of 8.4 cm/sec.

Although the estuarine-shelf exchanges occurring over intermediate time scales on the order of three to six days are of primary concern in this paper, a word regarding the very long period exchanges, explaining the net ebb recorded during this 30-day period, may be in order. Figure 5 shows the multi-year average monthly water levels computed for Galveston by the National Ocean Survey

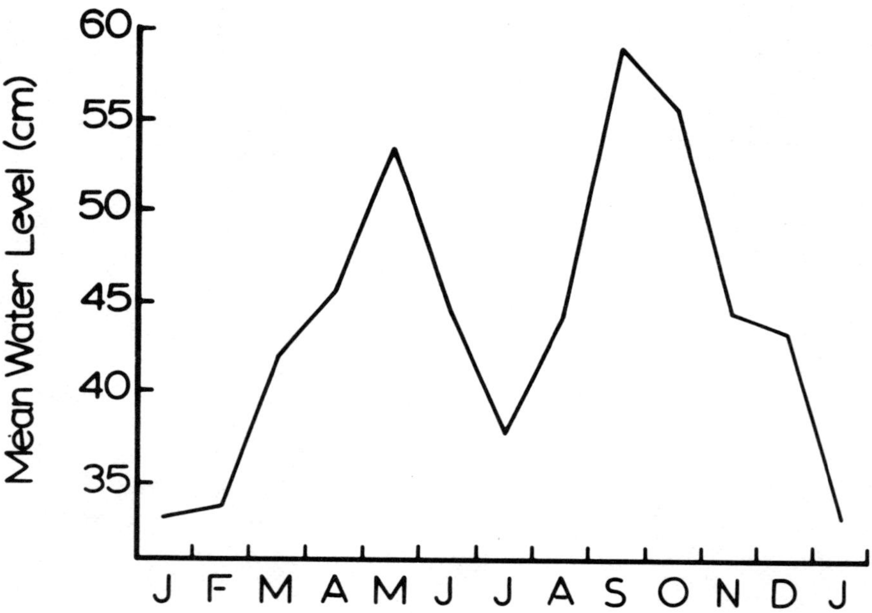

Fig. 5. Multi-year mean monthly water levels, relative to a datum one foot below mean sea level, for Galveston, Texas, 1958-1973.

between 1958 and 1973. The average annual cycle for the northwestern rim of the Gulf of Mexico is comprised of a semi-annual rise and fall, with high water occurring during the months of May and September, and low water during January and July. The causes of the semi-annual variations in coastal sea level have been suggested by Marmer (1954), Whitaker (1971) and Sturges and Blaha (1976), and appear to include both thermohaline and dynamic processes in the northwestern Gulf of Mexico. The spring run-off is primarily responsible for the relatively high coastal water levels in May. Seasonal warming and cooling produce a maximum expansion and contraction of the water column in September and January, respectively. Sturges and Blaha have suggested that the July minimum may be related to a temporary storage of water in a large anticyclonic gyre in the western Gulf maintained by the curl of the windstress.

Of particular importance here is the decrease in coastal water levels between the May high and the July low. The net ebb recorded during the 30-day study period at Site A reflects a slow draining of the intracoastal bays as water levels fall at the coast. There may be the added effect of the outflow of freshwater entering the coastal bays from rivers draining South Texas, but river inflow into coastal bays is small in this semi-arid region.

The relative importance of tidal and non-tidal exchanges past Site A and over this time interval can be investigated qualitatively by plotting the cumulative net displacements (Fig. 6). This is computed just like a progressive vector diagram, however since the calculations involve the scalar long-channel components

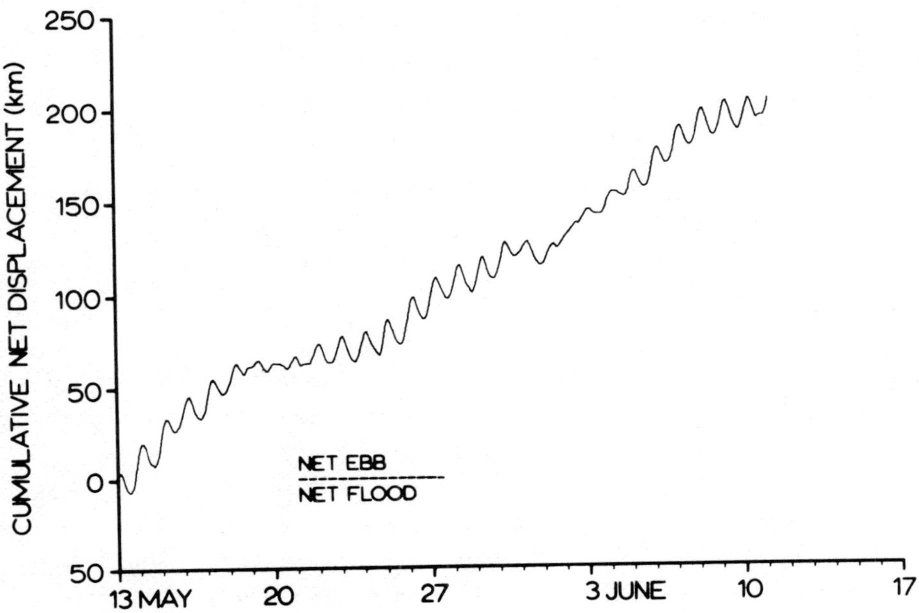

Fig. 6. Cumulative net displacement, in km, past Site A, 13 May to 12 June, 1975.

of the current, the cumulative displacements can be plotted against time, rather than in x-y coordinates. Figure 6 shows clearly the transition from tropic to equatorial tidal conditions, but it is apparent that the tidal exchanges constitute little more than small perturbations on the much more important long-period exchanges. Dominating the plot is the net outflow that extends through the entire study period.

The second study involved half-hourly direct current measurements at Site B (Fig. 1) between 29 June and 3 August, 1976. Figure 7 is a plot of the long-channel current components at the edge of Corpus Christi Bay, approximately 19 km from the coast

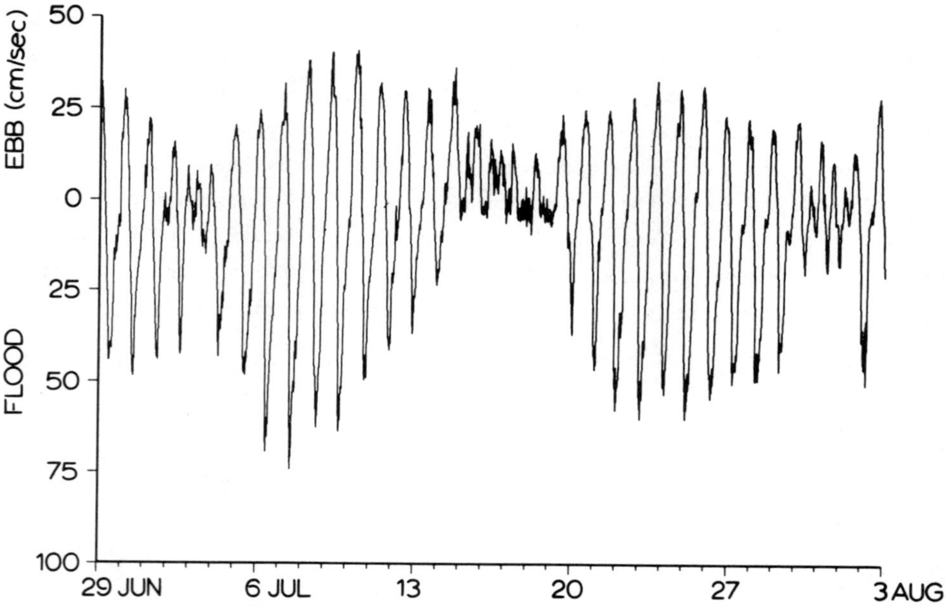

Fig. 7. Half-hourly, long-channel current components from Site B, in cm/sec, 29 June to 3 August, 1976.

along the ship channel. The same basic pattern appears, with a well-defined transition from tropic to equatorial tidal conditions. Current speed ranges are somewhat diminished, reflecting the frictional and constrictional effects of the shallow waters and channels.

The numerically filtered, non-tidal estuarine-shelf exchanges are shown in Figure 8. A markedly different pattern is apparent, however, contrasting with that recorded during May and June of the previous year. The approximately 31 days of data from the mid summer, 1976, study shows a net inflow past Site B. During the first

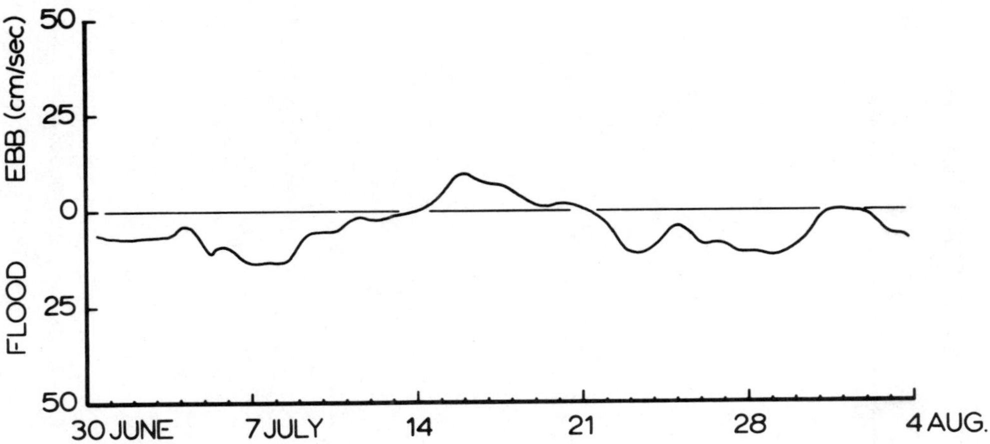

Fig. 8. Filtered long-channel current components from Site B, in cm/sec, 30 June to 3 August, 1976.

two weeks of the study there was a net flood into Corpus Christi Bay. This resumed for the final two weeks, following a period of about a week during which there was a net outflow. Intermediate time scale non-tidal variations are apparent superimposed onto the very long period exchanges. Non-tidal currents during this time interval varied between approximately ±10 cm/sec. The average flow was a flood of just under 5 cm/sec.

Figure 9 shows the cumulative net displacement past Site B during the 31-day study period. Again, the tidal motions appear as relatively small perturbations, with the most important exchanges occurring over much longer time scales. The dominant inflow during this time interval may be explained largely in terms of the very long period, semi-annual variations in coastal water levels. This study was conducted during the time of, and just following the July minimum, and as water levels began rising toward the October maximum. The net inflow thus reflects a slow flooding of the bays in response to rising coastal water levels. The brief period of net outflow may reflect the fact that the time period of the study fell so close to the time of lowest water that a quasi-steady rise had not yet begun.

DISCUSSION

In a recent paper, Weisberg (1976) discussed the need for sufficiently long current meter records, in view of non-tidal forcing occurring over time scales well in excess of the semi-diurnal

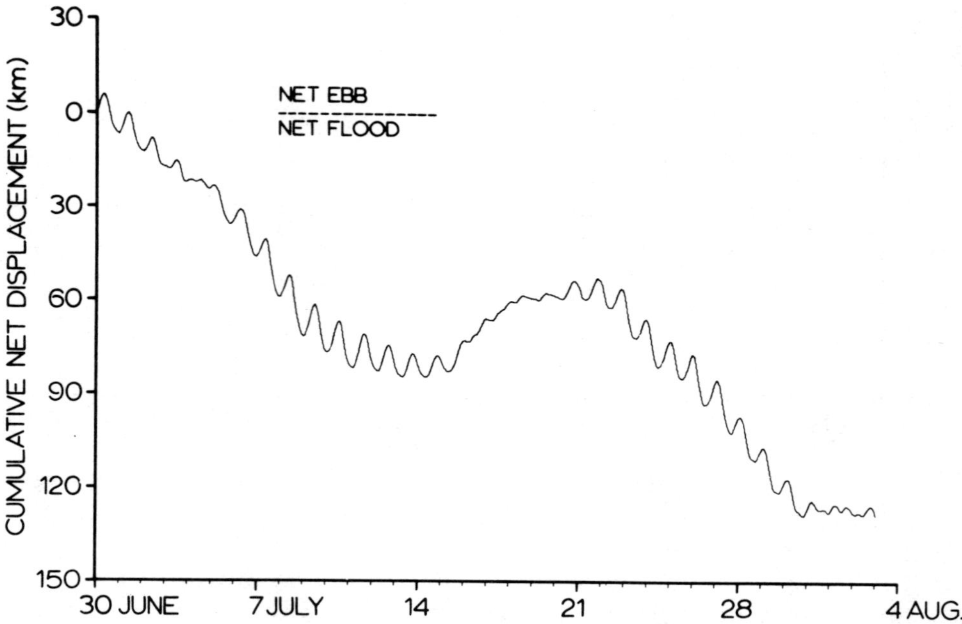

Fig 9. Cumulative net displacement, in km, past Site B, 30 June to 3 August, 1976.

and diurnal tidal periodicities. The results presented here support this suggestion, but indicate that in some areas at least seasonal variations in coastal water levels may result in correspondingly long period variations in the mean flow of water between estuaries and the inner continental shelf. In the northwestern Gulf of Mexico, with relatively small freshwater inflow into coastal bays from rivers draining South Texas, the semi-annual variations in coastal water levels appear to reverse the net estuarine-shelf transport over the same time scales. In other areas, the very long period variations may just alter the rate at which estuarine waters are exported onto the shelf.

Superimposed onto the semi-annual variations in estuarine-shelf exchanges, but occurring over time scales well in excess of tidal periodicities, are the quasi-periodic, meteorologically-forced exchanges, characteristically at time scales on the order of three to six days. Many other investigators have noted similar variations in current meter and water level records. Weisberg (1976) filtered out the tidal contribution to a 51-day current record from the Providence River, Rhode Island, and found quasi-periodic variations occurring over time scales on the order of four to seven days.

Beardsley, *et al.* (1977) have reported coastal water level variations and sub-surface pressure fluctuations over time scales on the order of several days. Groves (1957) documented non-tidal water level variations over intermediate time scales at many coastal and island stations in the Atlantic and Pacific, and discussed some of the most probable meteorological forcing mechanisms.

Two additional studies have recently been carried out in the northwestern Gulf of Mexico to investigate variations in coastal water levels and thus estuarine-shelf exchanges. In the first (Smith, 1977), regional pressure gradients were used to infer surface windstress over shelf waters. Statistically significant coherence-squared values were computed between variations in the volume of Corpus Christi Bay and variations in both the longshore and cross-shelf windstress components. Results indicated that the cross-shelf component of the windstress produces a set-up or set-down of coastal water levels over time scales on the order of two to four days, and thus forces a slow filling or draining of the bay. One may tentatively assign at least a part of the intermediate time scale variability noted in Figures 4 and 8 to the cross-shelf component of the surface windstress.

Over longer time scales, the volume of the bay is more coherent with variations in the longshore component of the windstress, suggesting coastal water levels rise and fall in response to a cross-shelf Ekman transport of shelf waters.

In a second study, just being completed, as yet unpublished data suggest that significant variations in coastal water levels may be forced by spatial variations in the surface pressure field. This inverse barometer effect seems to be particularly important over time scales on the order of two to six days, and water level variations estimated to be approximately ±5 cm are exceeded by only three astronomical tidal constituents in the northwestern Gulf of Mexico.

Analysis of non-tidal current or water level data and locally measured meteorological variables suggests that the estuarine-shelf exchanges occurring over time scales on the order of a few days do not occur as a local response to meteorological forcing. Coherence spectra (not shown) computed from the long-channel current components at Site B and both the longshore and cross-shelf windstress components computed from coastal wind data indicated statistically insignificant values over time scales associated with meteorological forcing. On the other hand, estuarine-shelf exchanges were found to be statistically significant when windstress values were computed from regional pressure gradients (Smith 1977). Similarly, the theoretical inverse barometer relationship of -1 cm/mb was very nearly matched when cross-Gulf atmospheric pressure differences were compared with cross-Gulf water level differences. Yet a comparison of local atmospheric pressure and water level variations measured at Port Aransas, Texas, resulted in a relationship of -0.82 cm/mb. This suggests that the estuarine-shelf exchanges observed at some point along a coast may be more a response to meso-scale meteorological forcing than a purely locally driven process.

CONCLUSIONS

One may conclude that where tidal processes are small, such as in the Gulf of Mexico, or in many estuarine areas sufficiently removed from the coast, meteorological forcing over time scales on the order of several days may play a significant role in estuarine-shelf exchanges. This is especially true in estuaries having little inflow of fresh water. Meteorological forcing may occur in several forms, with windstress and perhaps inverse barometric effects dominating. Studies repeated at various times of the year indicate that regions having substantial seasonal water level variations and small inflow of fresh water may undergo long-period reversals in the net transport between the estuary and the adjacent inner continental shelf lasting over periods of many weeks. The higher coherences between estuarine-shelf exchanges and regional meteorological forcing suggest that these exchanges do not occur at a response to purely local conditions.

ACKNOWLEDGMENTS

Mr. James C. Evans provided valuable help in the computer analysis of the current and water level data; Dr. J. S. Holland assisted in the installation and recovery of the recording current meters. Water level data used in the study were provided by Mr. D. T. Graham of the Army Corps of Engineers in Galveston, Texas.

Harbor Branch Foundation, Inc., Contribution Number 75.

REFERENCES

Beardsley, R., H. Mofjeld, M. Wimbush, C. Flagg and J. Vermersch, Jr. 1977. Ocean tides and weather-induced bottom pressure fluctuations in the Middle Atlantic Bight. Journ. of Geophysical Res. 82(21):3175-3182.

Fee, E. 1969. Digital computer programs for spectral analysis of time series. Univ. of Wisconsin, Milwaukee, Center for Great Lakes Research, Special Report No. 6, 17 pages.

Groves, G. 1957. Day to day variation of sea level. Meteorological Monographs 2(10):32-45.

Marmer, H. 1954. Tides and sea level in the Gulf of Mexico. In: Gulf of Mexico, its origin, waters and marine life. Fishery Bulletin, Fish and Wildl. Serv. U. S. 55(89):101-118.

Smith, N. 1977. Meteorological and tidal exchanges between Corpus Christi Bay, Texas, and the northwestern Gulf of Mexico. Estuarine and Coastal Marine Science 5(4):511-520.

Sturges, W. and J. Blaha. 1976. A western boundary current in the Gulf of Mexico. Science 192:367-369.

Weisberg, R. 1976. A note on estuarine mean flow estimation. Journ. of Marine Res. 34(3):387-394.

Whitaker, R. 1971. Seasonal variations of steric and recorded sea level of the Gulf of Mexico. Texas A & M University, Ref. 71-14T, 110 pages.

Zetler, B. and D. Hansen. 1970. Tides in the Gulf of Mexico--a review and proposed program. Bulletin of Marine Sci. 20(1): 57-69.

SURGE-TIDE INTERACTION IN THE SOUTHERN NORTH SEA

D. PRANDLE and J. WOLF
Institute of Oceanographic Sciences, Bidston Observatory, Merseyside, ENGLAND.

ABSTRACT

Observations of storm surges in the River Thames show that surge peaks tend to occur on the rising tide and seldom, if ever, occur on high tide. This tendency has been attributed to the interaction between tide and surge propagation as described by the non-linear terms in the associated hydrodynamic equations. A recent study by Prandle and Wolf (4) examined the mechanics of interaction within the River Thames and showed that an important component of it originates outside of the river; this component is investigated in the present paper.

A method of identifying interaction in the southern North Sea is developed involving the use of two hydrodynamic numerical models, one simulating tidal propagation and the other surge propagation. Operating these models concurrently, the coupling between tide and surge is introduced by perturbation terms which represent the influence in either model of sea levels and velocities computed by the other. This approach has been used to simulate the pattern of interaction which occurred during the disastrous storm surge of 30 January to 2 February 1953. It is shown that interaction in the southern North Sea results primarily from the quadratic friction term, developing significantly in the coastal region off Lowestoft as far south as the Thames estuary due to the high velocities associated with both tide and surge propagation in that area. Changes in the surface elevation of tide and surge due to the effects of interaction may develop rapidly in certain localised regions such as the Thames estuary. There may also be longer period changes of the order of the duration of the storm due to a systematic displacement of the M_2 tidal regime.

1. INTRODUCTION

The observed tendency of surge peaks to occur on the rising tide in the Thames has long intrigued researchers interested in tide and surge phenomena. It is also of major practical importance since the empirical formulae that have been derived to predict coastal flood levels arising from storm surges must take account of this tendency. Hence such questions arise as to whether it is possible for surge peaks to occur at high tide and, if so, under what conditions. Prandle and Wolf (4) have investigated the dynamics of tide-surge interaction

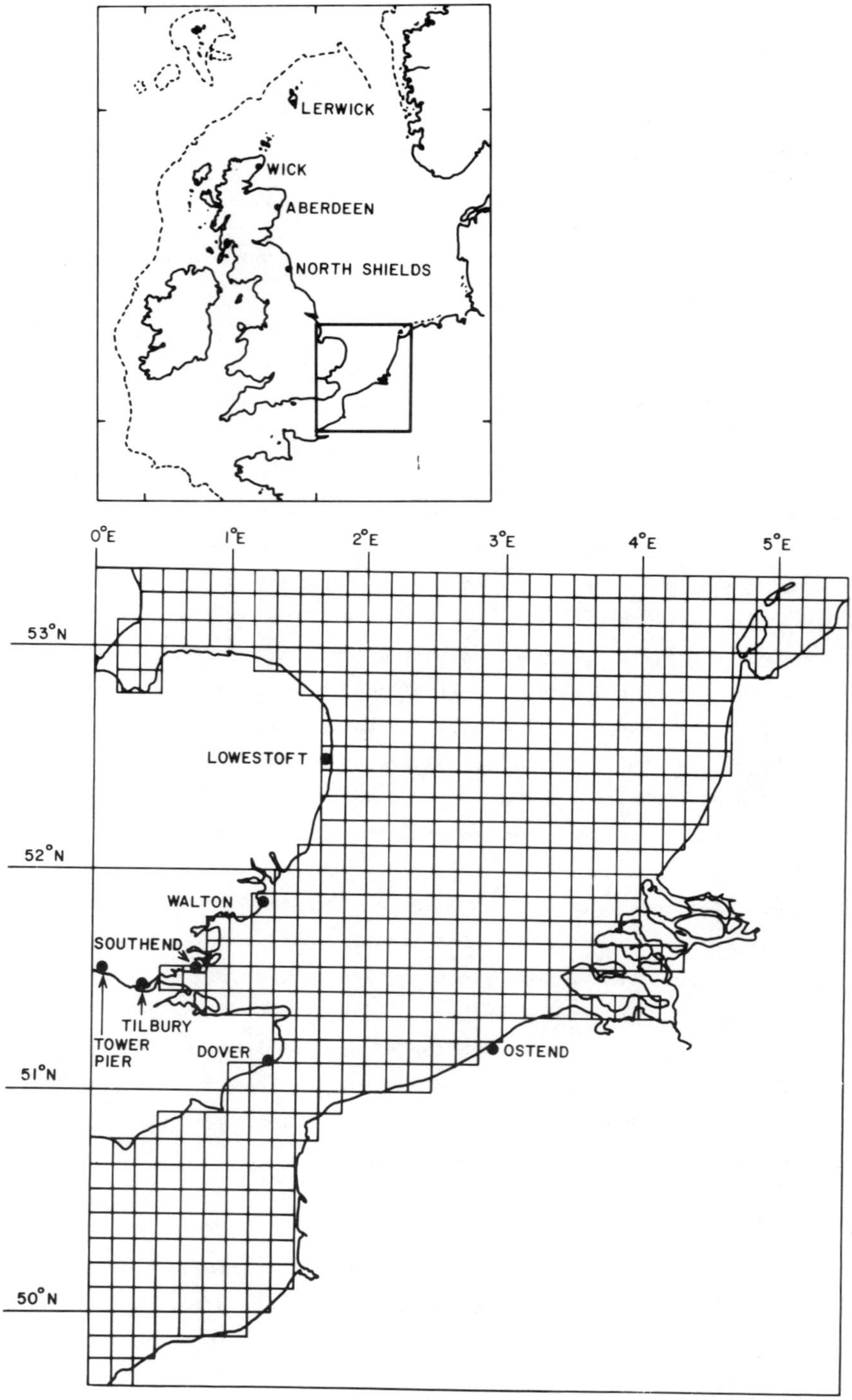

Fig.1 Schematic representation of the southern North Sea.

in the River Thames and have thereby attempted to provide some explanation for the observed distribution of surges. By analysing statistically the southward propagation of a number of discrete surge peaks along the east coast of Britain they showed that the peaks tend to occur on the rising tide in the Thames irrespective of the phase relationship between tide and surge in the northern North Sea. Using numerical models to simulate tide and surge propagation along the river from a seaward boundary at Walton the observed increase in interaction effects along the river was reproduced. However the predominance of surge peaks occurring on the rising tide was shown to be due, in part, to interaction effects which modify surge profiles prior to their arrival at the mouth of the Thames. In their statistical analysis of observed surges, Prandle and Wolf showed that this interaction, occurring outside of the river, developed significantly between Lowestoft and Walton; it is this feature of interaction that is of primary interest in the present paper.

The paper considers the propagation of tide and surge in the southern North Sea with the objective of examining the mechanics of their interrelationship in that region. The study is limited to the simulation of a particular surge event, namely the major surge of 31 January to 2 February 1977; Prandle (3) made a comprehensive study of this surge with the aid of a numerical model. The present approach employs two versions of this same model operated concurrently and referred to subsequently as parallel models. One model simulates tidal propagation and the other surge propagation. Non-linear interaction between tide and surge is introduced by perturbation terms which represent the influence of the surge on the tidal propagation and, likewise, the influence of the tide on the surge propagation. In this way, it is possible to account for the interaction between tide and surge while retaining the separate identities of the two phenomena. It is then possible, for instance, to determine the way in which the M_2 tidal amphidromic system in the southern North Sea changes during the course of the storm. It is also possible to variously include or omit certain of the perturbation terms so that the interaction due to a particular term can be explicitly identified.

The application of these parallel models has demonstrated that the major source of interaction arises from the quadratic friction term. In consequence, the areas where this interaction is most effective were identified by examining the spatial distribution of the velocity field for both tide and surge propagation.

Throughout this text the word interaction is used to denote both an actual physical process and its resultant effect on surface elevation, Z_I, where the latter is defined as the difference in surface elevation between the value computed from a simulation of tide plus surge combined, Z_C, and the sum of the separate components computed from simulations of tide alone, Z_T, and surge alone, Z_S; i.e. $Z_I = Z_C - Z_T - Z_S$

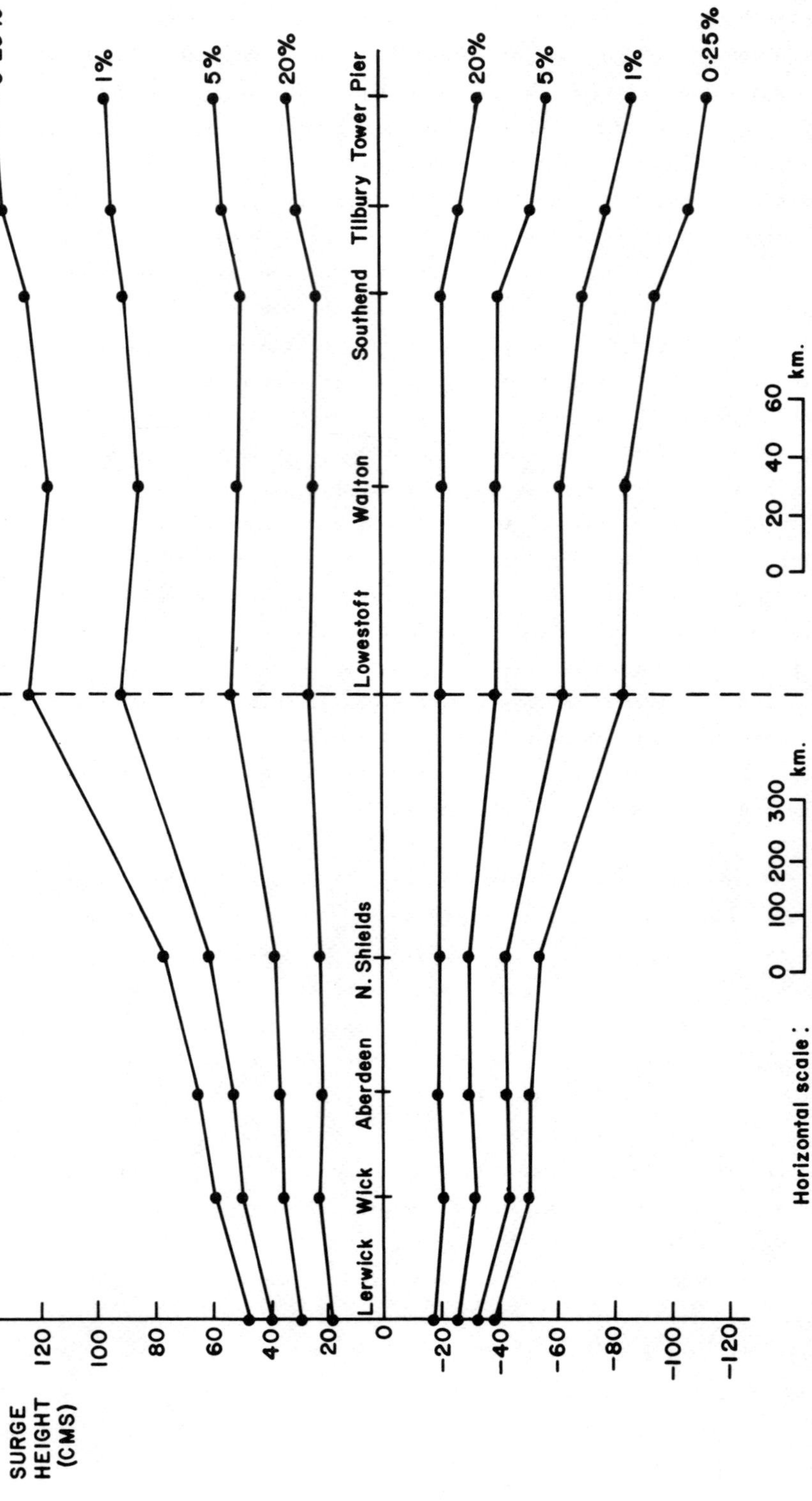

... location for R = 0.25, 1, 5 and 20%.

2. STATISTICS OF OBSERVED SURGES

An analysis was made of observed surges at the following ports: Lerwick, Wick, Aberdeen, North Shields, Lowestoft, Walton, Southend, Tilbury and Tower Pier. The locations of these ports are shown in figure 1 and extend from the northern North Sea southwards along the east coast of Britain and thence into the Thames as far as Tower Pier.

The observed data comprised 5 years of hourly recordings at each location for the years 1969 to 1973. At each port, hourly residual heights were calculated as follows:

$$R_t = O_t - P_t - M \qquad (1)$$

where R_t is the residual elevation, or surge (at time t);
O_t the recorded elevation;
P_t the predicted astronomical tide;
and M the annual mean of O_t, calculated for each year separately.

At each location, the surge data were analysed to compute probability distributions expressed in terms of percentage exceedances of a particular surge level. The percentage of surges exceeding a value Z_P was denoted by P_Z given by $P_Z = n/N \times 100$, where n is the number of surge values exceeding Z_P out of a total number of hourly surge values N (approximately 44,000 for the 5 year period). The analysis was performed for positive and negative surges separately. The results are shown in figure 2 in the form of the values of Z_P corresponding to P_Z = 0.25, 1, 5 and 20% respectively for the various ports along the path of propagation of the surge. The horizontal scale in figure 2 represents the distance between the ports. The values of Z_P for P_Z = 0.25% and P_Z = 1% are representative of peak surge levels at each location as might be observed during the propagation of a moderate to large storm surge. The variations in Z_P, for these two values of P_Z, show a steady and regular increase in amplitude between Lerwick and Lowestoft and thereafter remain reasonably constant between Lowestoft and Tower Pier.

The above analysis was then repeated but, rather than analyse the complete data set as a whole, the data was first separated into distinct subsets according to the timing of any particular observation relative to tidal high water. Each subset was then analysed separately as before. Figure 3 shows the mean surge level for each port at four tidal phases namely; (a) rising tide, $3\frac{1}{2}$ to $2\frac{1}{2}$ hours before high tide (HT); (b) high tide, HT $-\frac{1}{2}$h to HT $+\frac{1}{2}$h; (c) falling tide, HT $+2\frac{1}{2}$h to HT $+3\frac{1}{2}$h and (d) low tide, HT $-6\frac{1}{2}$h to HT $-5\frac{1}{2}$h. The divergence of the four curves is a measure of the degree of interaction at each location. The larger values indicated by the curves for surges on the rising tide clearly illustrate the increase in interaction as surges propagate southwards. The figure

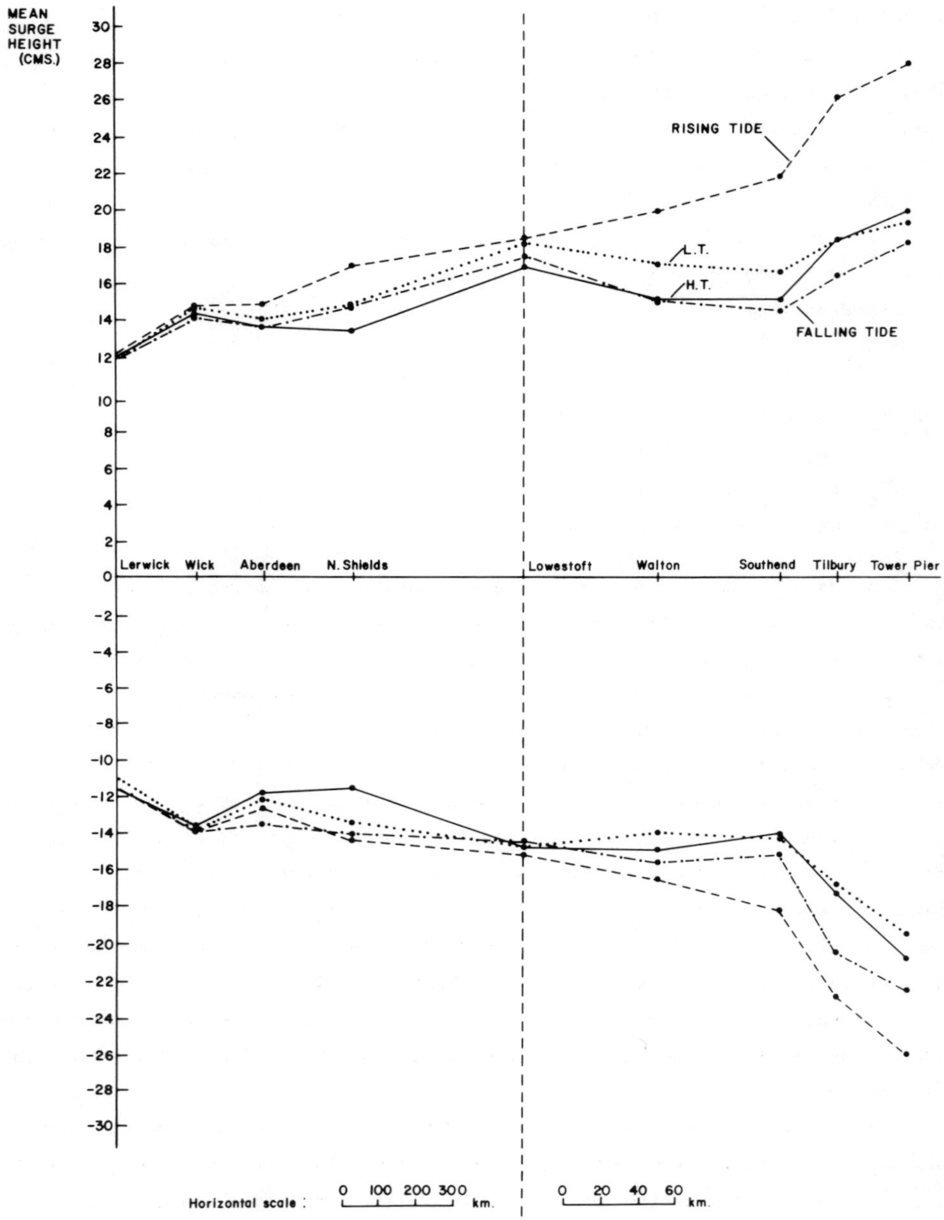

Fig.3 Recorded mean surge levels at four tidal phases.

indicates that interaction can be detected as far northwards as Wick and it then increases continuously as far as Tower Pier. An exception to this continuous increase is the small decrease between North Shields and Lowestoft; this may possibly be attributed to discontinuities in the tidal regimes between the central and southern North Sea and also the discontinuity in the coastline in the region of the Wash. However, the present feature of interest is the significant increase in interaction which occurs between Lowestoft and Walton.

The pronounced effect of interaction at ports south of Lowestoft is clearly illustrated in figure 4 which shows time-series of observed surge and predicted tide (the latter to one quarter of the vertical scale used for the surge) at Lowestoft and Southend during 1970. The time-series shown were considered to be representative of conditions at these two ports. The features of interest are: (a) at Lowestoft, there is an apparently random distribution of the timing of surge peaks relative to the phase of the tide, whereas (b) at Southend, the surge peaks almost always occur on the rising tide and never on high tide. Hence figure 4 emphasises the effects of interaction in this region as previously enunciated from the statistical analysis of recorded surges.

Prandle and Wolf [4] examined interaction in the Thames using the parallel model approach (§3) ; figure 5 shows some of the results from this study. Starting from a prescribed tide and surge input at the mouth of the model (Walton - Margate) the effects of interaction at Tower Pier are illustrated by the modification of the tide due to the influence of the surge and, likewise, the modification of the surge by the influence of the tide. The figure illustrates that the most significant effect of interaction is the reduction of the surge peaks through the influence of the tide, it was also shown that this reduction of the surge peaks was due to the quadratic friction term. An examination of figure 5 shows that the interactions $S'-S$ and $T'-T$, within the river, are not responsible for the peak of the net residual, $S'+T'-T$, occurring on the rising tide. This phenomenon must therefore be a function of the surge-tide phase relationship at the mouth. Since this phase relationship is evidently non-random it may then be concluded that the timing of surge peaks in the Thames is dependent on interaction occurring seawards of the line between Walton and Margate.

3. PARALLEL MODELS

The basis of the present modelling approach consists of a combined model of the southern North Sea and River Thames developed by Prandle [2] . This model comprises two parts dynamically interfaced, a one dimensional representation of the River Thames together with a two dimensional representation of that part of the North Sea south of latitude $53^{\circ}20'$ extending westwards into the English Channel as far as the Greenwich meridian. The model uses an

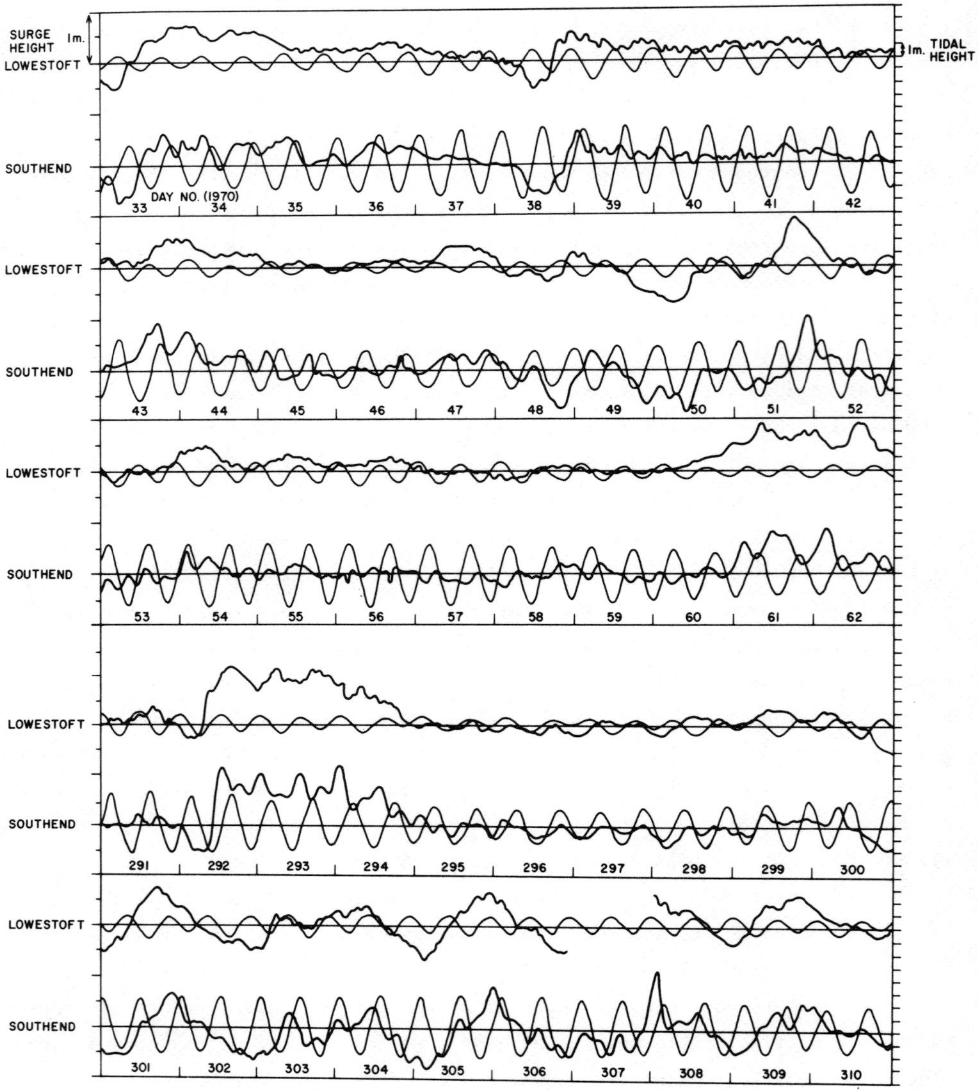

Fig.4 Time-series of predicted tide and recorded surge at Lowestoft and Southend during 1970.

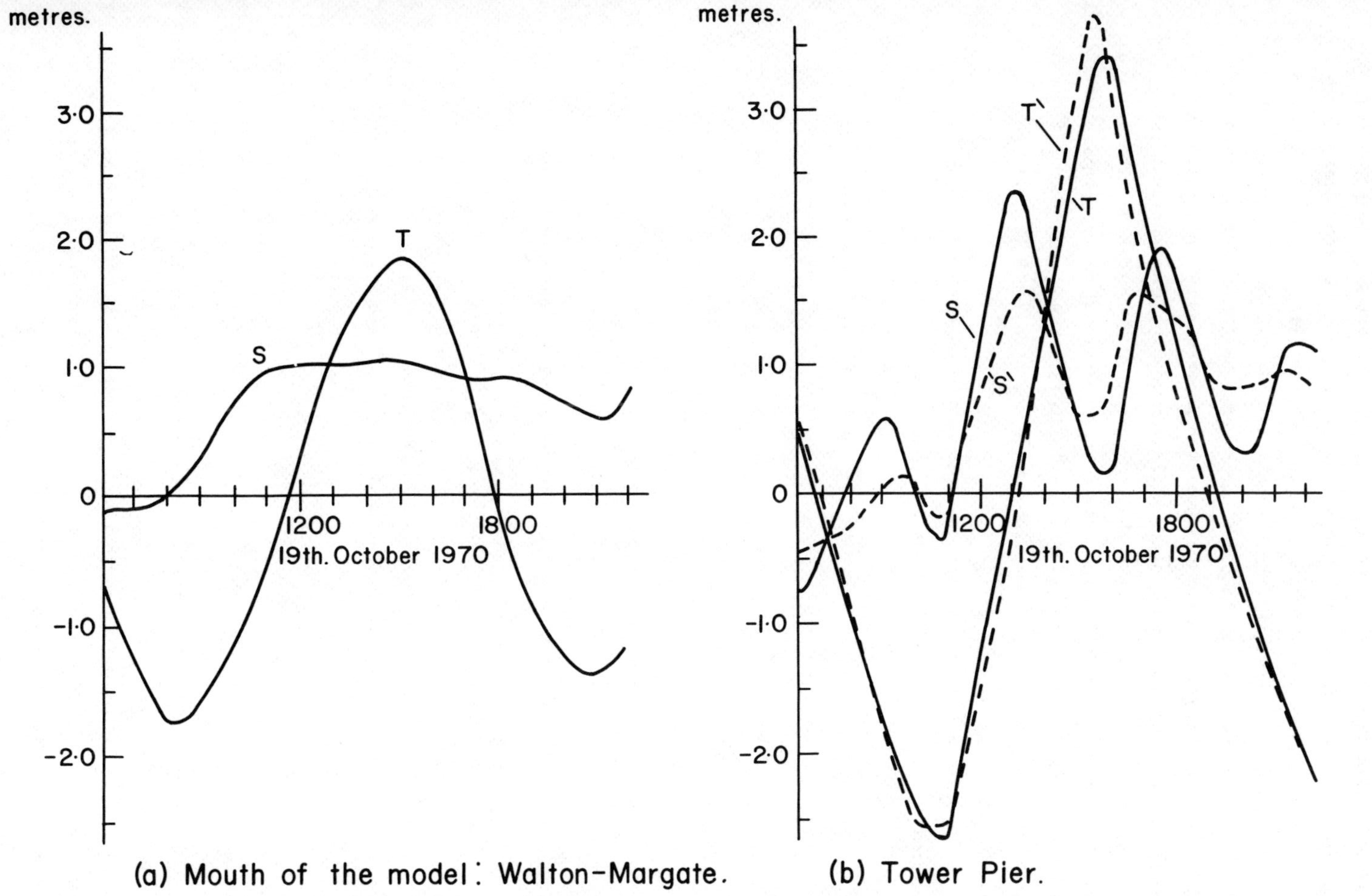

Fig.5 Propagation of tide and surge in the Thames computed by parallel models (Prandle and Wolf (4)); tide alone T, surge alone S, tide modified by interaction T', surge modified by interaction S'.

explicit finite difference scheme for solving, by means of a forward time-stepping procedure, the relevant equations of motion. The schematic representation of the southern North Sea is shown in figure 1.

Prandle (3) showed that this numerical model was able to accurately simulate the propagation of tide and surge throughout this region. The present objective is to gain an understanding of the properties of interaction by simulating tide and surge separately while introducing interaction between the two phenomena in the form of perturbations as already mentioned. The application of the parallel model approach to a one dimensional representation of the Thames has been described by Prandle and Wolf (4) , hence only the application to the two dimensional representation of the southern North Sea will be described here.

The relevant hydrodynamic equations may be expressed for space coordinates along lines of latitude and longitude :

$$\frac{\partial U}{\partial t} + \frac{\partial}{\partial x}\frac{U^2}{H} + \frac{\partial}{\partial y}\frac{UV}{H} + Hg\frac{\partial Z}{\partial x} + Ku(u^2+v^2)^{1/2} - \Omega V = 0, \qquad (2)$$

$$\frac{\partial V}{\partial t} + \frac{\partial}{\partial x}\frac{UV}{H} + \frac{\partial}{\partial y}\frac{V^2}{H} + Hg\frac{\partial Z}{\partial y} + Kv(u^2+v^2)^{1/2} + \Omega U = 0, \qquad (3)$$

$$\frac{\partial}{\partial t}Z + \frac{\partial}{\partial x}U + \frac{\partial}{\partial y}V = 0 , \qquad (4)$$

where x, y are orthogonal axes positive to the east and to the north,

t time,

g gravitational constant,

Z elevation of the water surface above a horizontal datum,

D depth of the bed below the same datum

u, v velocities (depth-averaged) in the x and y directions respectively,

K friction coefficient,

Ω Coriolis parameter,

and $H = D+Z$, $U = uH$, $V = vH$.

Numerical simulation of tide and surge propagation in the southern North Sea (Prandle (3)) has indicated that interaction is largely insensitive to the inclusion or omission of the convective terms (second and third terms in equations (2) and (3)) and hence, on this basis, it is justifiable, and also convenient, to omit these terms from further consideration.

The concept of the parallel model approach requires that, for all locations and at all times, the elevation and transports in the tidal model, Z_T , U_T and V_T respectively, together with the corresponding values Z_S , U_S and V_S in the surge model must satisfy the relationships :

$$Z_C = Z_T + Z_S \tag{5}$$

$$U_C = U_T + U_S \tag{6}$$

$$V_C = V_T + V_S \tag{7}$$

where the subscript C denotes values computed in a combined simulation of tide and surge. Inserting (5), (6) and (7) into equations (2), (3) and (4) produces the following equations for the combined propagation of tide and surge :

$$\frac{\partial}{\partial t}(U_T+U_S)+(D+Z_S+Z_T)\,g\frac{\partial}{\partial x}(Z_S+Z_T)+K(u_S+u_T)\{(u_S+u_T)^2+(v_S+v_T)^2\}^{1/2} - \Omega(V_T+V_S) = 0 , \tag{8}$$

$$\frac{\partial}{\partial t}(V_T+V_S)+(D+Z_S+Z_T)\,g\frac{\partial}{\partial y}(Z_S+Z_T)+K(v_S+v_T)\{(u_S+u_T)^2+(v_S+v_T)^2\}^{1/2} - \Omega(U_S+U_T) = 0 , \tag{9}$$

$$\frac{\partial}{\partial t}(Z_T+Z_S)+\frac{\partial}{\partial x}(U_T+U_S)+\frac{\partial}{\partial y}(V_T+V_S) = 0 \qquad 10)$$

In the friction term in (8) and (9) it is assumed that the relationships (6) and (7) for transports in the x and y directions respectively also apply for the associated depth-averaged velocities. It may be shown that this approximation is justifiable for small values of Z/D .

The open-sea boundary conditions employed in the simulation of the propagation of tide, surge and tide plus surge involve the specification of the sea surface elevation $Z_B(t)$ at every boundary grid square B .

Hence the boundary conditions to be satisfied in the parallel models are:

$$Z_{B,C} = Z_{B,T} + Z_{B,S} \tag{11}$$

The assumption made in the use of parallel models is that equations (8), (9) and (10) may be separated into two parts as follows :

for tide,

$$\frac{\partial}{\partial t} U_T + (D+Z_{\underline{S}}+Z_T)\,g\frac{\partial}{\partial x} Z_T + K\, u_T\{(u_{\underline{S}}+u_T)^2+(v_{\underline{S}}+v_T)^2\}^{1/2} - \Omega V_T = 0, \tag{12}$$

$$\frac{\partial}{\partial t} V_T + (D+Z_{\underline{S}}+Z_T)\,g\frac{\partial}{\partial y} Z_T + K\, v_T\{(u_{\underline{S}}+u_T)^2+(v_{\underline{S}}+v_T)^2\}^{1/2} + \Omega U_T = 0, \tag{13}$$

$$\frac{\partial}{\partial t} Z_T + \frac{\partial}{\partial x} U_T + \frac{\partial}{\partial y} V_T = 0 \quad , \tag{14}$$

with boundary conditions $Z_B = Z_{B,T}$,

and for surge,

$$\frac{\partial}{\partial t} U_S + (D+Z_S+\underline{Z_T}) g \frac{\partial}{\partial x} Z_S + K u_S \{(u_S+u_T)^2 + (v_S+\underline{v_T})^2\}^{1/2} - \Omega V_S = 0, \tag{15}$$

$$\frac{\partial}{\partial t} V_S + (D+Z_S+\underline{Z_T}) g \frac{\partial}{\partial y} Z_S + K v_S \{(u_S+\underline{u_T}) + (v_S+\underline{v_T})^2\}^{1/2} + \Omega U_S = 0, \tag{16}$$

$$\frac{\partial}{\partial t} Z_S + \frac{\partial}{\partial x} U_S + \frac{\partial}{\partial y} V_S = 0 \quad , \tag{17}$$

with boundary conditions $Z_B = Z_{B,S}$

In operating the tidal model with (12), (13) and (14), the surge parameters Z_S, U_S and V_S which appear in equations (12) and (13) are evaluated from the simultaneous operation of the surge model, while in operating the surge model with (15), (16) and (17) the tidal parameters Z_T, U_T and V_T which appear in (15) and (16) are obtained from the concurrently-running tidal model. Using this parallel model technique to simulate various surge events, it was shown that the results from the separate simulations of tide and surge could be combined to give values in close agreement with results obtained from the simulation of tide and surge combined, thus satisfying conditions (5), (6) and (7) as required. Hence, the additional terms underlined in equations (12), (13), (15) and (16) may be considered to represent the interaction between tide and surge. The magnitude of the shallow water interaction terms are proportional to a product of surge amplitude and tidal amplitude. However, the magnitude of the interaction associated with the frictional terms is a more complex function involving products of the surge amplitude and the tidal amplitude with the respective powers of these amplitudes varying according to both the instantaneous ratio of the tide and surge velocities and also their difference in direction.

The interaction terms in (12) and (13) involving the value of Z_S are subsequently referred to as the shallow water terms while the other interaction terms in (12) and (13) involving U_S and V_S are referred to as the quadratic friction terms; similar descriptions are used to refer to the corresponding terms in (15) and (16).

4. STORM SURGE OF 31 JANUARY - 2 FEBRUARY 1953

The loss of life and damage caused by the storm surge of 31 January - 2 February 1953 represents one of the worst natural disasters experienced in recent history along the shores of Britain and the North Sea coast of Continental Europe. The propagation of this surge in the southern North Sea was simulated using the parallel model technique outlined in § 3. Tide and surge elevations along the open-sea boundaries of the models were prescribed from the data used by Prandle (3) in an earlier comprehensive examination of this surge event. In addition, the effect of local wind forcing was incorporated by adding further terms to equations (15) and (16); the form of these wind forcing terms and the corresponding wind speed data were also extracted from the earlier study of Prandle.

One limiting factor present throughout the following discussion of the results of this simulation is the implicit assumption that there is no interaction between the tide and surge at the open-sea boundaries of the model. While the analysis of recorded surge data described in § 2 showed that some interaction does occur outside of the model region, it is considered that the essential features of the results obtained will remain valid.

Computer runs were carried out for

(1) model of tide alone (T),

(2) model of surge alone (S),

(3) model of tide with interaction from model (4), below, due to shallow water and quadratic friction (T'),

(4) model of surge with interaction from model (3), above, due to shallow water and quadratic friction (S'),

(5) model of tide plus surge combined (T+S).

The purpose of operating model (5) was to confirm that the values from models (3) and (4) satisfied the requirements of (5), (6) and (7) and hence the results from this model will not be discussed further.

(a) Interaction at Lowestoft and Southend

Elevations at Lowestoft and Southend computed from these four models are shown in figures 6 and 7 respectively, vertical lines drawn on these figures indicate the time of high tide at the particular location as computed by model (1). The modification of the surge through the influence of the tide is shown both by the divergence of the curves for S and S' and also by the separate curve for S'-S , similarly the modification of the tide through the influence of the surge is shown by the divergence of the curves for T and T' and by the curve of T'-T . The average magnitude of the interaction at Lowestoft is about half that at Southend, i.e. approximately in the ratio of the magnitude of the respective tidal ranges as suggested by the form of the interaction terms (§ 3). The time-series for S'-S at Lowestoft shows little correlation with the tidal phase whereas the corresponding time-series at

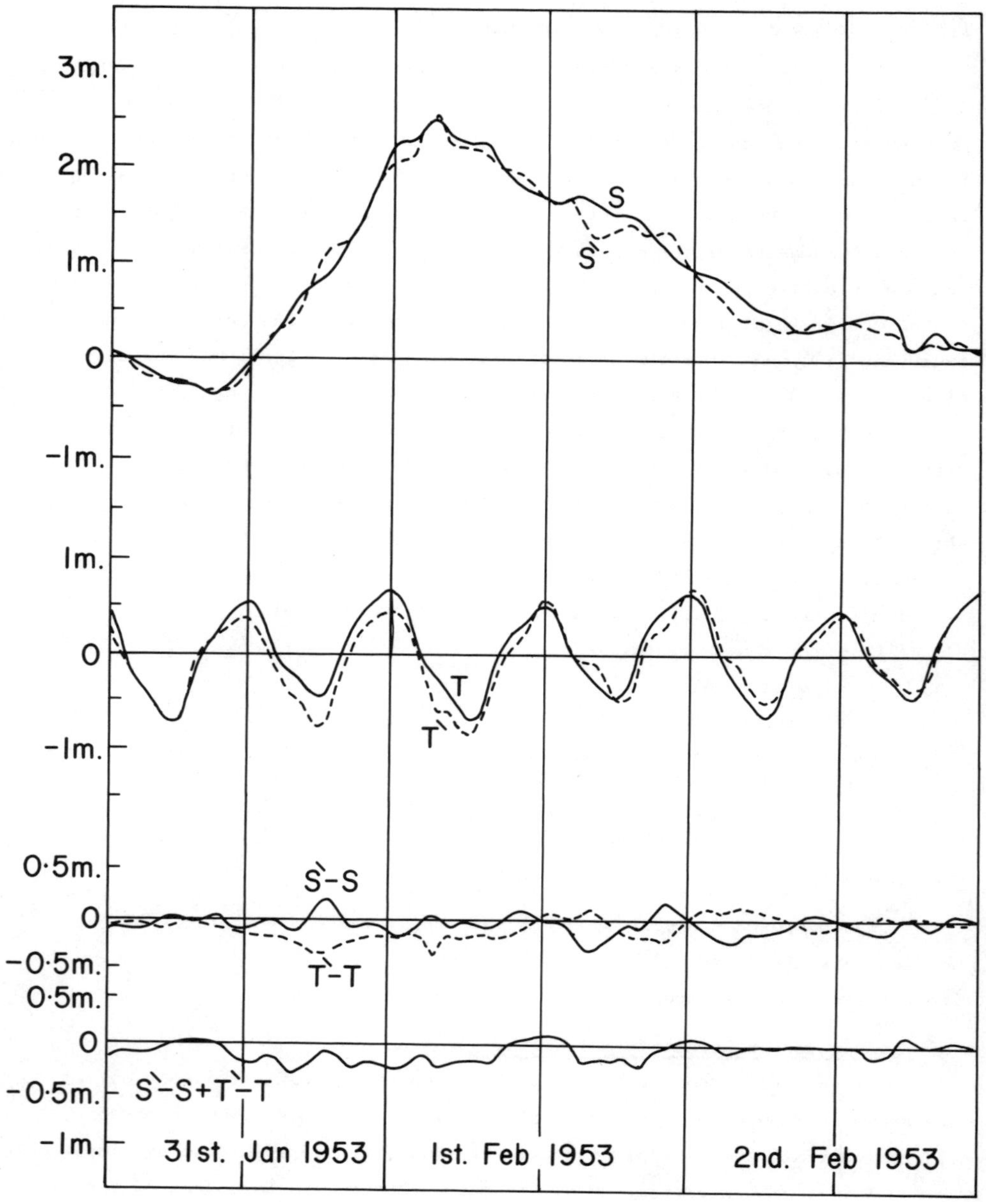

Fig.6 Computed values of tide, surge and interaction at Lowestoft.

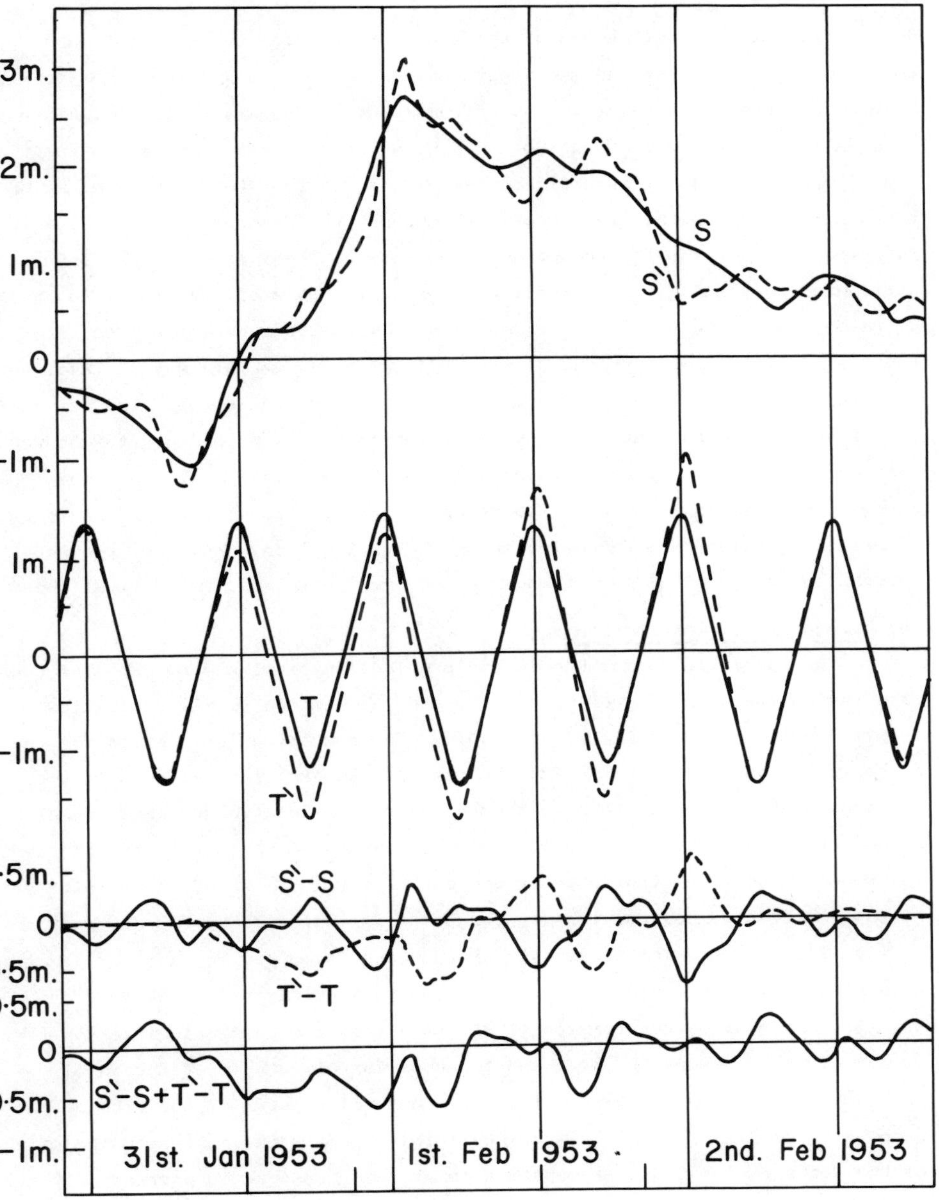

Fig.7 Computed values of tide, surge and interaction at Southend.

Southend is clearly influenced by the tidal stage. The curves for T' and T at both Lowestoft and Southend show that the influence of the surge tends to reduce mean water levels for about 24 hours from 12.00 31 January to 12.00 1 February. This is followed over the next 24 hours by an appreciable increase in the tidal range at Southend and a significant phase delay at Lowestoft; an examination of the modification of the tidal regime throughout the area of the model is described in § 4(e). The reduction in mean tidal level at both locations may be attributed to the negative surge levels preceding this effect, the long duration of these changes to the tidal regime accords with the suggestion made later (§ 4(e)) that a time constant of the order of a day or more may be involved in the response of the tidal regime to the influence of the surge. The curves for S'-S and T'-T show, at both Lowestoft and Southend, a tendency to counteract each other. The net interaction curve at Southend, S'-S+T'-T, shows some correlation with tidal phase, in particular it illustrates the tendency for interaction to increase surge levels on the rising tide. However, the net interaction curve at Lowestoft shows no obvious correlation with tidal phase.

(b) Components of interaction

The model runs described in the previous subsection were repeated but with the interaction between models (3) and (4) limited to (A) quadratic friction only and (B) shallow water only. The resulting interaction for the cases of (A) and (B) and also for the complete interaction examined in the previous subsection are shown in figure 8 for four locations. These results show that, at all four locations, the modification of the surge through the influence of the tide indicated by the curves for S'-S is of the same order as the modification of the tide by the surge as indicated by the curves for T'-T . The tendency for these two interactions to oppose one another is also evident again. The separate curves for interaction effects of quadratic friction and shallow water respectively show that the quadratic friction term is dominant throughout the area and accounts for almost all of the total interaction. The interaction effect due to shallow water of the tide on the propagation of the surge is almost negligible whereas the corresponding effect of the surge on the tidal propagation is of some consequence particularly at Southend. The time-series for this latter effect at Southend is highly correlated with the tidal phase and contributes towards the characteristic increase in surge levels on the rising tide. The lack of any obvious spatial coherence between the results at different locations is discussed further in the following subsection.

(c) Spatial distribution of interaction

The spatial distribution, at 06.00 1 February 1953, of the modified surge, S' , the modified tide, T' , tide plus surge combined, (T+S)(or (T') + (S'))

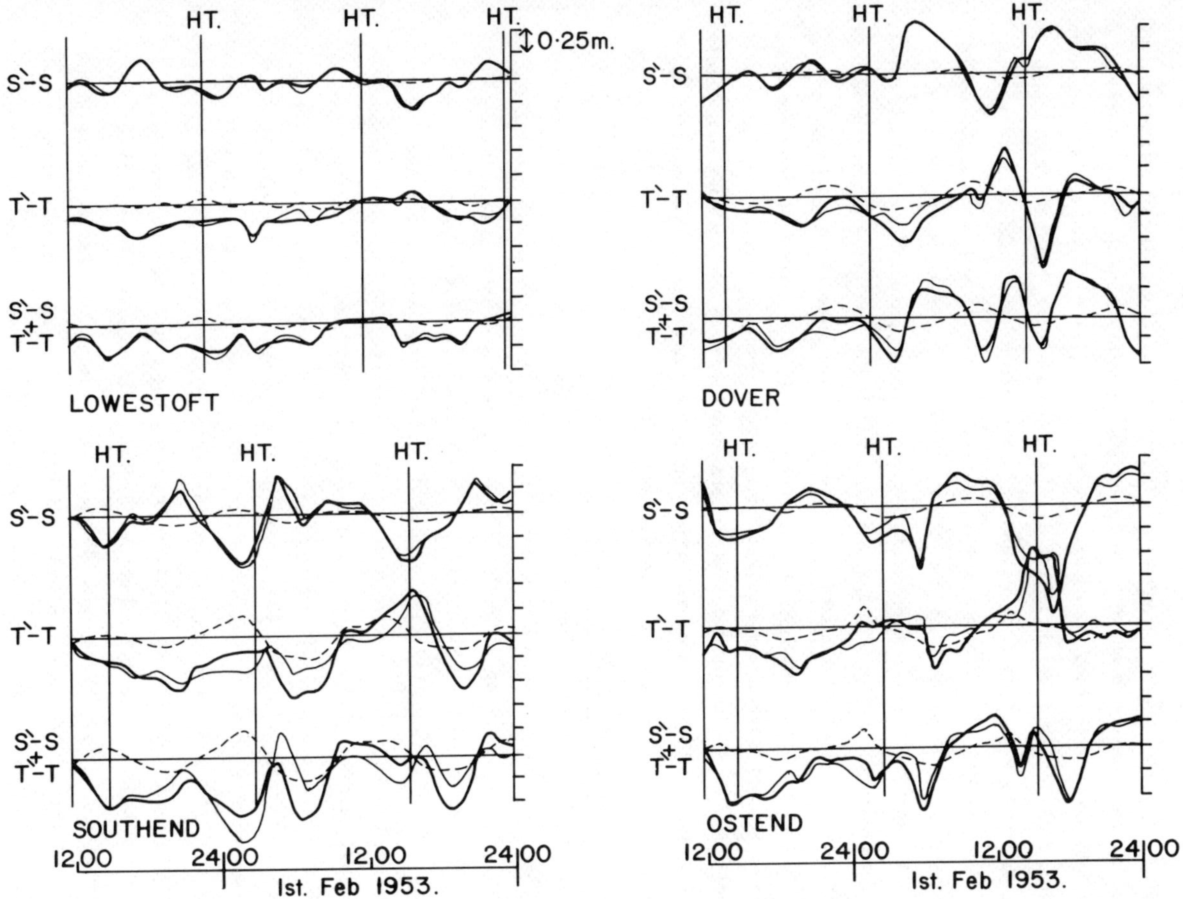

Fig.8 Components of interaction at Lowestoft, Southend, Dover and Ostend. ▬▬ complete interaction, —— quadratic friction only, --------- shallow water only.

together with all of the various components of interaction are shown in figures 9 and 10. The values shown for this specific time may be regarded as representative of values occurring during a large surge event. The values of interaction due to quadratic friction are shown to be everywhere in close agreement with the values for complete interaction. The values of interaction due to shallow water are shown to be much smaller in relation to the interaction due to quadratic friction. However, the shallow water interaction of the surge on tidal propagation is significant in the region of the Thames estuary and eastwards along a section from Ostend to the mouth of the Rhine. This region also coincides with the maximum total interaction; this may be attributed to the large magnitude of both tide and surge elevations in this region (§3).

The spatial distribution of the various components of interaction appear to be rather complex. An attempt was made to follow the development through time of these spatial distributions, however an examination of successive distributions of the type shown in figure 9 and 10 for values at hourly intervals did not reveal any clear trends in the changing distributions of interaction. It was concluded that interaction in this region does not develop in a slow and gradual fashion in either space or time but, on the contrary, develops rapidly and often in a localised sense.

(d) Spatial distribution of velocity

Since it has been shown that quadratic friction is the major cause of interaction in the southern North Sea, it follows that the areas where interaction develops most strongly will be those where both tide and surge velocities are largest. Figure 11 shows the spatial distribution of mean absolute velocities for (A) tide alone, T , (B) surge alone, S , and (C) tide plus surge, (T+S) , where these mean values were obtained by averaging values over a semi-diurnal period from 14.30 1 February to 03.00 2 February. The distribution of tidal velocities in this region is fairly well established and the results for figure 11(a) are in good agreement with the distribution of maximum values shown in the atlas published by Seehydrographischer Dienst, Rostock 1975 (5) . The distribution for surge alone shows high velocities along the east coast of Britain extending into the Dover Strait and English Channel. Similarly the distribution for tide plus surge confirms this concentration of high velocities. Hence, since almost all large surges which occur in the Thames estuary contain a significant component originating in the northern North Sea, this component will experience considerable interaction as it propagates in the coastal region off Lowestoft as far south as the Thames estuary.

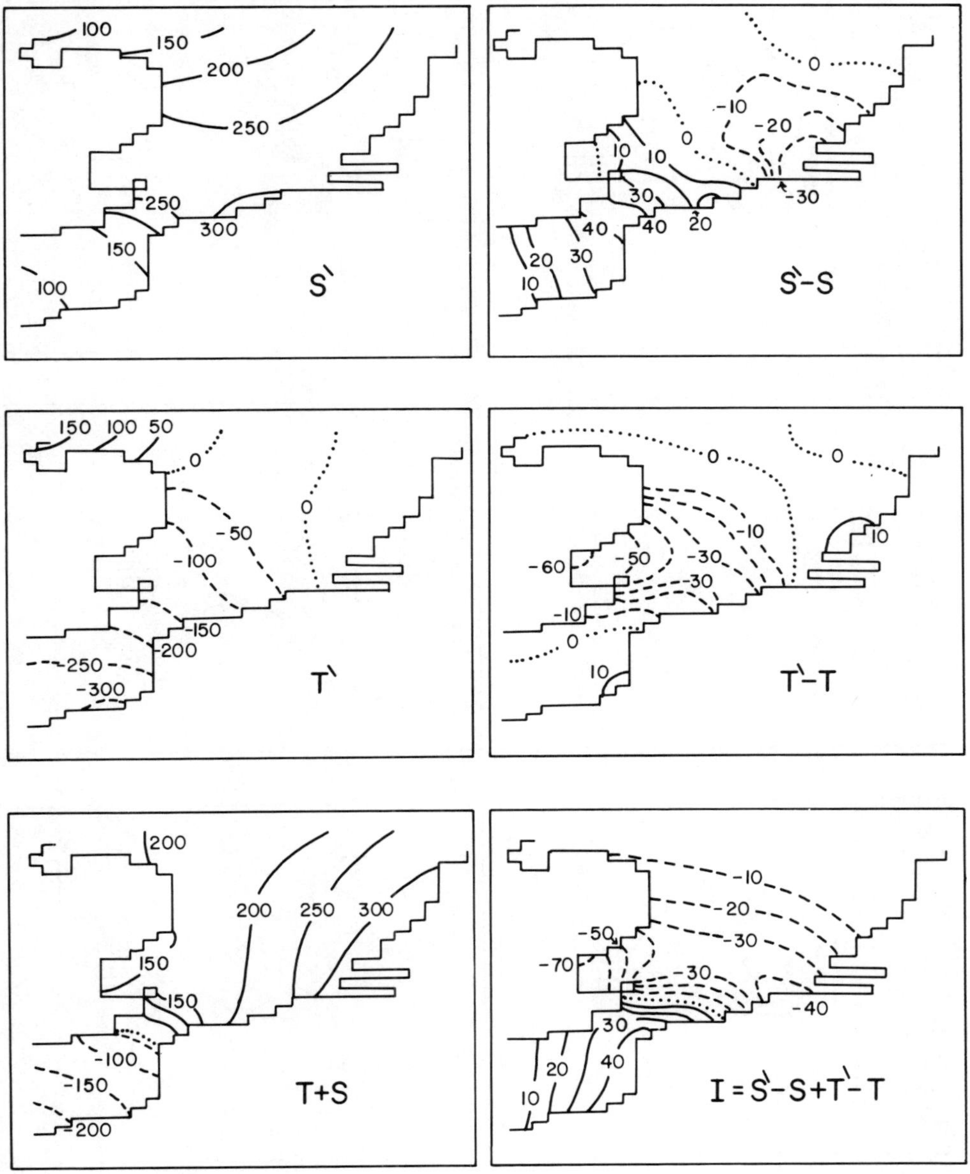

Fig.9 Computed tide, surge, tide plus surge and components of interaction; instantaneous values at 06.00 1 February 1953 in cm.

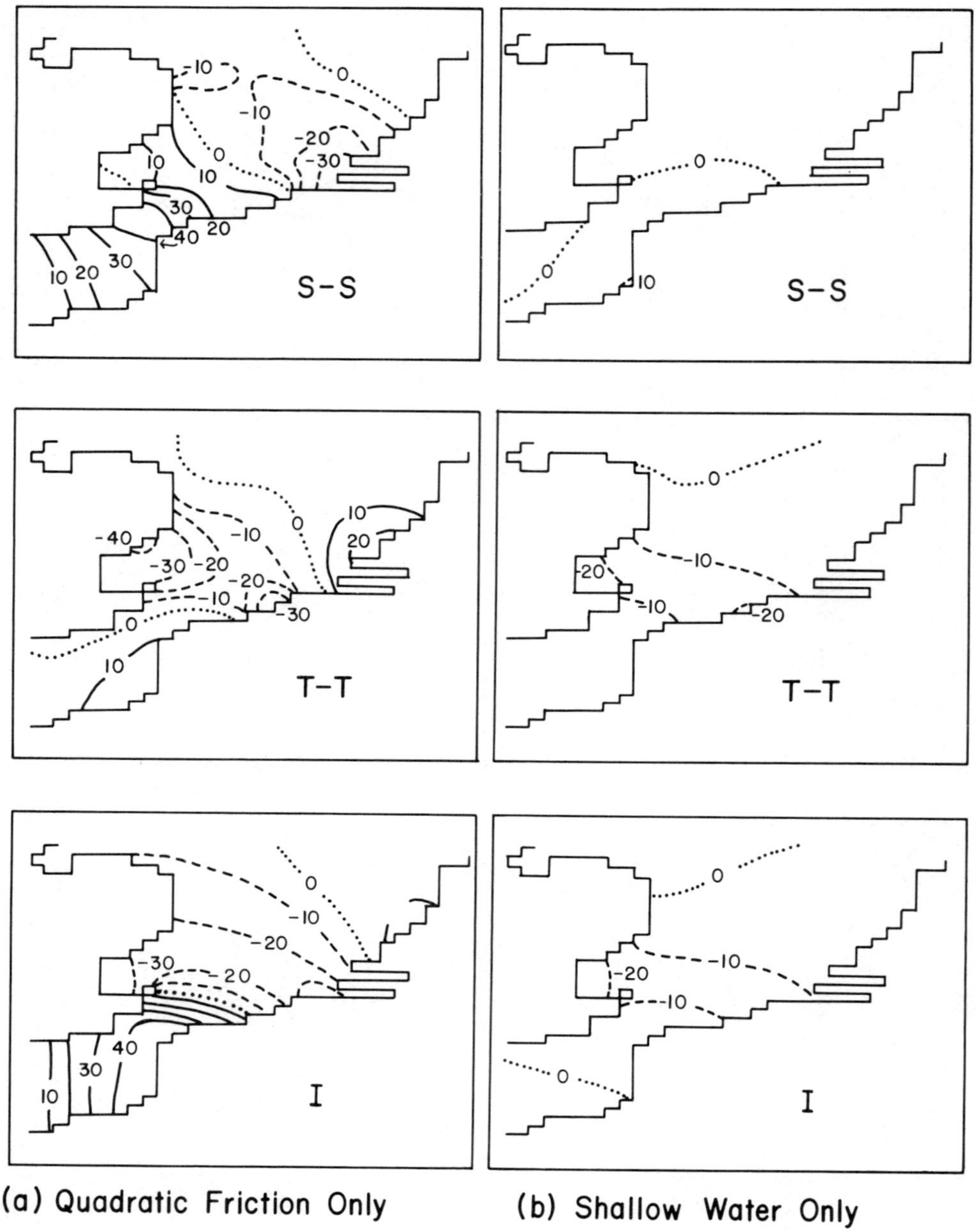

Fig.10 Computed components of interaction due to (a) quadratic friction only and (b) shallow water only; instantaneous values at 06.00 1 February 1953 in cm.

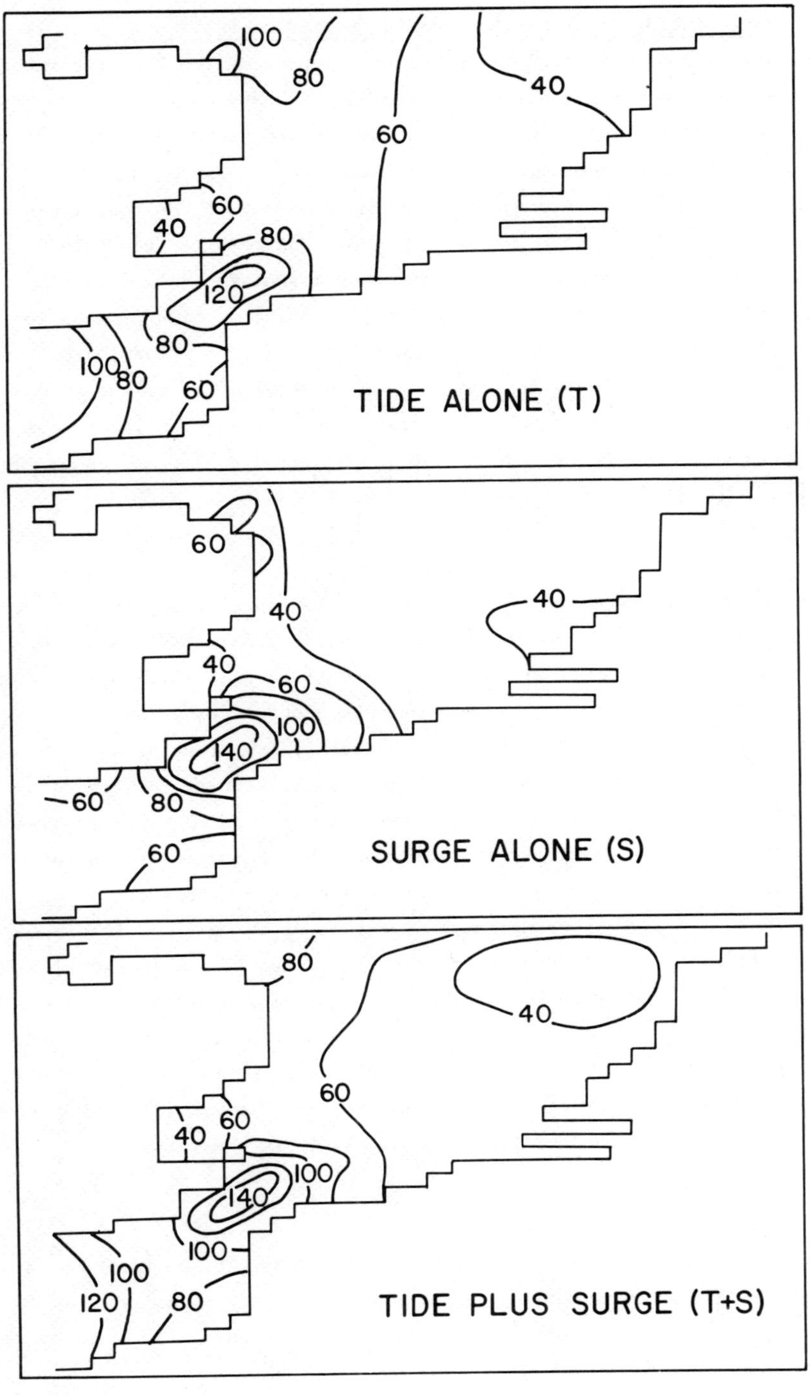

Fig.11 Distribution of mean velocities computed for tide alone, surge alone and tide plus surge; values shown are in cms^{-1} averaged over the period 14.30 1 February to 03.00 2 February 1953.

(e) Variation in the M_2 distribution over the period of the storm

The simulations described in §4(a) were repeated with the difference that the tidal distribution specified at the open boundaries in (A) model (1) of tide alone; (B) model (3) of tide with the influence of surge and (C) in model (5) of tide plus surge combined, was restricted to the single M_2 constituent. In the original case the tide was composed of the constituents for M_2, S_2, K_2, N_2, O_1, K_1 and M_4 (Prandle (2)). The results from this simulation of the '53 surge and the M_2 tide showed that the values for interaction of (A) tide on surge propagation, (B) surge on tide propagation and (C) the combination of (A) and (B) were, in all cases, almost identical to the values obtained for the simulation of the same surge with the more complete tidal regime. This indicates that the major surge tide interaction in this region is between the surge and the M_2 tide, confirming a similar result found by Banks (1) .

A particular advantage of restricting the simulation to surge plus the M_2 tide only is that it allows the distribution of the co-phase and co-range lines associated with the M_2 tide to be easily determined at any stage during the surge event. Figure 12 shows successive distributions of the M_2 tide, each determined from a Fourier analysis over one tidal cycle with the time of the mid-cycle as indicated, the interval between each distribution shown is equal to the period of M_2. In each case, the latest distribution is superimposed over the previous distribution in order to illustrate relative displacements between successive tidal distributions. The amphidromic system is shown to be initially displaced westwards and then to rotate in an anti-clockwise sense until it returns to the original distribution after approximately 3 days corresponding to the duration of the storm. This evidence of a relatively longer period displacement of the tidal regime is particularly interesting since it offers the possibility of including a systematic correction to the predicted tide in the course of a large surge event. However, the displacements shown by the present model are severely restricted by the assumption of fixed boundary conditions. An equivalent simulation using a model of the whole of the North Sea should prove extremely interesting.

5. CONCLUSIONS

1. An examination of storm surges recorded in the River Thames has shown that surge peaks tend to occur on the rising tide. This effect is attributed to interaction between tide and surge as described by the non-linear terms in the relevant hydrodynamic equations.

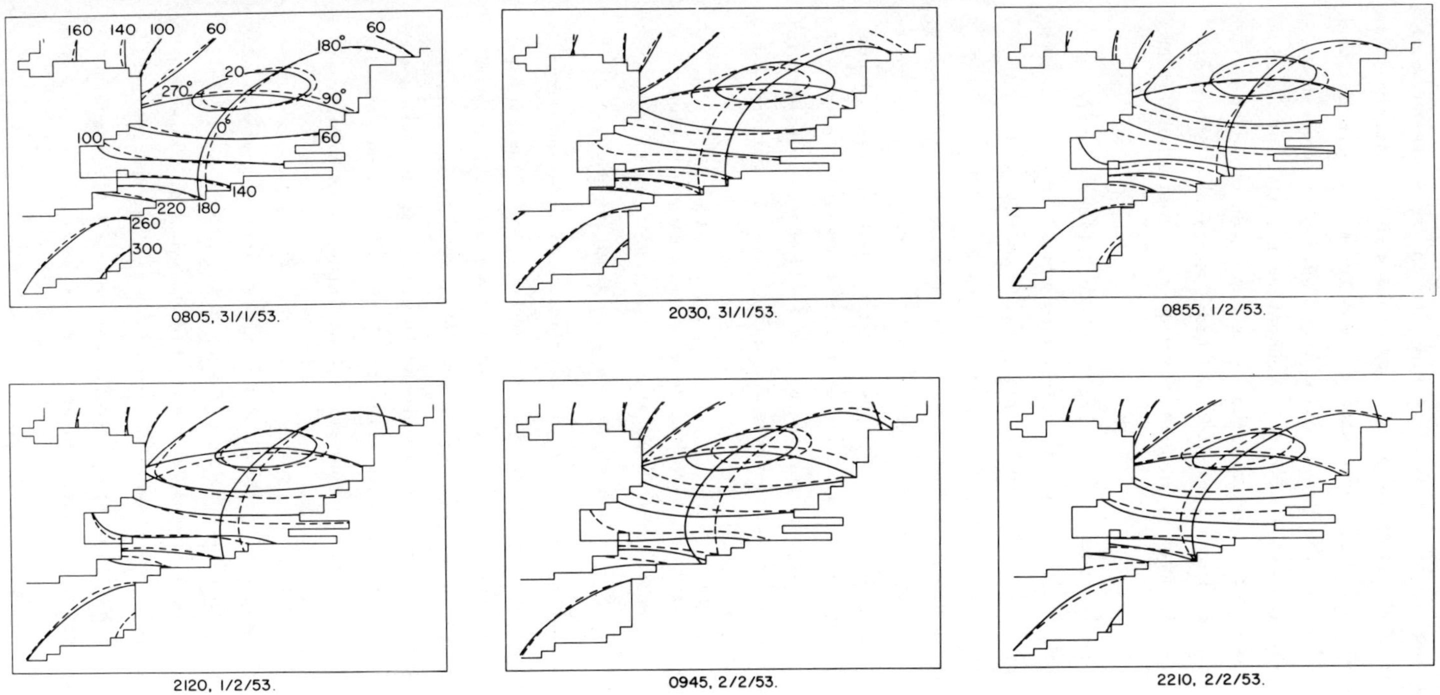

Fig.12 Variation in the M_2 tidal regime over the period of the '53 storm; continuous lines show the distribution at the times stated, dashed lines show distribution one (M_2) period earlier. Co-range lines show amplitude in cm.

2. A method of identifying the mechanics of interaction in the southern North Sea has been developed involving the use of two numerical models, one simulating tidal propagation and the other surge propagation. The two models are operated concurrently with cross linkage from perturbation terms which introduce the influence of the surge into the model of tidal propagation and the influence of the tide into the surge model. The magnitude of these interaction terms were shown to be a function of surge amplitude and tidal amplitude with the respective powers of these amplitudes, in part, dependent on instantaneous flow conditions.

3. This modelling approach was used to simulate the interaction occurring during the disastrous storm of 30 January to 2 February 1953. It was shown that interaction in the southern North Sea results primarily from the quadratic friction term and that the modification of the surge propagation by the tide is of a similar order of magnitude as the modification of the tidal propagation by the surge. The interaction from shallow water terms is generally restricted to the modification of tidal propagation by the surge and is only of significance in the Thames estuary and the region east of the estuary between Ostend and the mouth of the Rhine. However, the time-series for the shallow water interaction in the Thames shows that this term contributes to the increase in surge heights on the rising tide in the river.

4. An examination of the spatial and temporal developments of the various components of interaction suggests that changes in water level due to interaction can develop rapidly in time and may be localised in space.

5. A study of the spatial distributions of velocity for both tide and surge was made since the importance of the quadratic friction term suggests that interaction will develop most effectively in those regions where the velocities associated with both tide and surge propagation are greatest. These spatial distributions showed that the coastal region around Lowestoft as far south as the Thames estuary is an area of high velocities for both tide and surge and hence this accords with the important observation that interaction develops rapidly between Lowestoft and the Thames.

6. A simulation of the '53 surge with the M_2 tide only, showed that almost all of the surge-tide interaction may be accounted for by this constituent. This simulation also enabled the displacement of the M_2 tidal regime by the surge event to be traced. The displacement was found to consist of an orderly anti-clockwise rotation of the amphidromic system with an associated time-constant of about 3 days or, effectively, the total duration of the storm.

ACKNOWLEDGEMENTS

The work described in this paper was funded by a Consortium consisting of the Natural Environment Research Council, the Ministry of Agriculture Fisheries and Food, and the Departments of Energy, Environment, and Industry.

REFERENCES

1 J. E. Banks, Phil. Trans. R. Soc. Lond., A, 275 (1974) 567-609.
2 D. Prandle, Institute of Oceanographic Sciences, Bidston, Merseyside, England, 4 (1974).
3 D. Prandle, Proc. R. Soc. Lond., A, 344 (1975) 509-539.
4 D. Prandle and J. Wolf, "The Interaction of Surge and Tide in the North Sea and River Thames" (in press).
5 "Atlas der Gezeitenstrome fur die Nordsee, den Kanal und die Irische See" Dritte, verbesserte Auflage Seehydrographischer Dienst, der Deutschen Demokratischen Republik, Rostock 1975.

RESIDUAL PHENOMENA IN ESTUARIES, APPLICATION TO THE GIRONDE ESTUARY

R. BONNEFILLE
Electricité de France, Chatou (France)

ABSTRACT

Integration of equation of residual phenomena (velocity and salinity) in the case of an estuary with constant width and depth, shows the possibility to have some closed residual streamlines near the bottom.

In the general case, integration is more complicated, but the conclusion is the same.

Numerous data about the Gironde estuary are used to estimate the value of the three more important new coefficients introduced by the theory of residual phenomena : the longitudinal and vertical mixing coefficients of salinity and the vertical mixing coefficient of momentum.

BRIEF REVIEW OF RESIDUAL PHENOMENA

The theory of residual phenomena (Pritchard, 1956) is based on the division of physical events in two elements. The first one is independant of the time, there is the residual part ; it is the time-averaged value of the considered events during a given period, at the minimum a tide-period. The second element depends on the time ; it represents the fluctuation of the physical event, induced by the tide,relatively their mean values. For example let us considering the velocity vector $\vec{V}_i$ and the salinity s

$$\vec{V}_i = \vec{V}_o(x,y,z) + \vec{V}_m(x,y,z,t)$$

$$s = s_o(x,y,z) + s_m(x,y,z,t)$$

where $\vec{V}_o$ and s_o are the residual values on the period T_m

$$V_o = \frac{1}{T_m}\int_o^{T_m} \vec{V}dt \ , \ s_o = \frac{1}{T_m}\int_o^{T_m} sdt$$

and $\vec{V}_m$ and s_m are the fluctuante components, the time-averaged values of those are null.

If, we introduce these fonctions in the momentum and diffusion equations ; then, if this equations are time-averaged, as in the conventional theory of the turbulence, new terms are appearing ; they are introduced by the products of fluctuantes components $\vec{V}_m$ and s_m. This terms are modelised by using mixing coefficients. The horizontal components of these coefficients represent the dispersion by tide currents on areas the dimensions of whose are from 100 m to 10 km. Of course, the mixing coefficients are greather than the turbulent coefficients, about 100 times more, for example from 100 to 1000 m^2/s. Nevertheless the vertical mixing coefficients have about the same value of vertical coefficients of turbulence (1 to 10 cm^2/s).

The same method is used to define the width-averaged residual quantities from the time-averaged residual events ; for example $\vec{V}$ and $\bar{s}$ defined by the relationships.

$$\vec{V}_o = \vec{V}(x,z) + \vec{V}'(x,y,z)$$

$$s_o = \bar{s}(x,z) + s'(x,y,z)$$

with

$$\vec{V} = \frac{1}{b}\int_{-b/2}^{b/2} \vec{V}_o dy \ , \ \bar{s} = \frac{1}{b}\int_{-b/2}^{b/2} s_o dy$$

on which b(x,z) is the width of the estuary and Ox, Oy, Oz the longitudinal, lateral and vertical axes of coordinates. Their averaging operation over the width of the estuary introduces new terms due to the fluctuations of the velocity and salinity from a side to the other side of the estuary. These new terms are also modelised by new mixing coefficients (Ronday, 1975). But by the effects of the sides of the estuary the equation transformations are more complicated, and it is necessary to do some assumptions about the values of residual quantities on the sides. A preliminary study showed that it is usefull to assume that the residual velocity is null on the side :

$$V_o(x,\pm b/2,z) = 0$$

It is now possible to modelise the side effects as the mixing effects, and define the general mixing coefficients :

K_{dx}, K_{dz} for the momentum, from the momentum equation,

K_{sx}, K_{sz} for the salinity from the diffusion equation.

After elimination of negligeable terms, the equation are reduced at the following forms (Chatwin, 1976) :

- momentum equation :

$$g\frac{\partial\rho}{\partial x} + \frac{\partial}{\partial z}\left[\frac{\rho}{b}\frac{\partial}{\partial z}(K_{dz}\frac{\partial}{\partial z}bu)\right] = 0$$

- continuity equation :

$$\frac{\partial}{\partial x}(bu) + \frac{\partial}{\partial z}(bw) = 0$$

- diffusion equation :

$$u\frac{\partial\bar{s}}{\partial x} + w\frac{\partial\bar{s}}{\partial z} = \frac{\partial}{\partial x}(K_{sx}\frac{\partial\bar{s}}{\partial x}) + \frac{\partial}{\partial z}(K_{sz}\frac{\partial\bar{s}}{\partial z})$$

where

x is longitudinal seawards, z vertical upwards,

g is the gravity acceleration,

ρ is the specific mass of water,

u and w are the horizontal and vertical components of the residual velocity $\vec{V}$.

The continuity equation suggests to search a stream function Ψ_o, that is obtained easily if we do the following assumptions :

- the depth of the estuary is constant : h_o doesn't depend of x or z,
- the width b is constant in a cross-section : b(x,z) becomes b(x),
- the residual discharge

$$Q_R = \int_{-h_o}^{o} bu\ dz$$

is constant along the estuary,

- the mixing coefficient of momemtum K_{dz} is independant of z,
- the residual velocity is null on the bottom, as on sides,

$u(x,-h_o) = w(x,-h_o) = 0$

- the water specific mass ρ depends linearly on the residual salinity :

$\rho = \rho_o(1+a\overline{s})$

a is a constant, its value is approximatly 0,76 if the salinity is expressed in gr/liters,

- the residual salinity is assumed to be constant on the vertical ; we put

$\overline{s} = S_s(x)$

where S_s is the superficial width and time-averaged salinity ; this assumption consists on the first order solution of the diffusion equation.

Let we us define the non-dimensional stream-function $\Psi_o(x,Z)$ by :

$$u = -\frac{Q_R}{bh_o}\frac{\partial \Psi_o}{\partial Z}\ ,\quad w = \frac{1}{b}\frac{\partial}{\partial x}(Q_R\Psi_o)\ ,\quad Z = \frac{z}{h_o}$$

we obtain :

$$\Psi_o = \frac{1}{2A}(2Z-1)(Z+1)^2Z + \frac{1}{2}(Z^2-1)Z$$

with

$$A(x) = (24K_{dz}Q_R)/(agbh_o^4\frac{dS_s}{dx})$$

and for u and w

$$u(x,Z) = \frac{3}{2}u_m(1-Z^2) + u_s(1-Z-8Z^2)(1+Z)$$

$$w(x,Z) = \frac{h_o}{b}\frac{d}{dx}(bu_s)(2Z-1)(1+Z)^2Z$$

where there are introduced :

- the vertical-averaged horizontal residual velocity

$$u_m(x) = \frac{Q_R}{bh_o(x)}$$

- the horizontal residual velocity due to the salinity gradient

$$u_s(x) = \frac{agh_o^3}{48K_{dz}}\frac{dS_s}{dx}$$

This solution shows that the horizontal residual velocity u contains two proportional parts to u_m and u_s. The first one flows seawards ; which is more important near the mouth and near the end of the estuary in the zone where the variation of salinity is small. The second part is seaward near the free surface and upward near the bottom ; it is important on the central zone of the estuary near the inflexion point of the longitudinal distribution of salinity. This opposition between the two parts of the velocity u gives to the vertical distribution of the residual velocity its specifical aspect, particularly the change of direction near the bottom in the zone called the "neutral point" (Hansen and Rattray, 1965).

As Ψ_o is function of x by the parameter A, it is easy to trace the stream lines, defined for the constant values of Ψ_o, by the equation :

$$Z^4 + \frac{3+A}{2}Z^3 - \frac{1+3A}{2}Z^2 - A\Psi_o = 0$$

The figure 1 shows that the neutral point doesn't exist if $A > 1$ only ; that is happening on the mouth and the end of the estuary. On the contrary for $A < 1$ the direction of the residual velocity changes near the bottom. In this case we could have closed residual trajectories as it is shown in the figure 2, for which the stream lines are computed for a possible distribution of A(x) along the estuary.

Using this results it is easy to integrate the diffusion equation at the second order approximation ; it appears a small variation of the salinity along the verticale

$$s(x,Z) = S_s(x) - \frac{u_s ho}{20K_{sz}}\frac{dS_s}{dx}\left[(8Z^3+15Z^2-10)Z^2 - \frac{5u_m}{2u_s}(2-Z^2)\,Z^2\right]$$

EXTENSION TO THE ESTUARIES WITH VARIABLE DEPTH

Taking in account that the depth h is function of x and y, brings a set of complications in the equations. But it is possible to use the same method ; let we us define the functions $\eta(x,y)$, $\beta(x)$ and the mean depth $h_o(x)$, by the relationship :

$$\frac{1}{h(x,y)} = \frac{1}{h_o(x)}\left[1 + \eta(x,y)\right]$$

$$\beta(x) = \frac{1}{b}\int_{-b/2}^{b/2}\eta(x,y)\,dy$$

It is also possible to obtain results like with a constant depth ; if we introduce the function G(x) :

$$\frac{dG}{dx} = a\frac{dS_s}{dx} + \frac{\beta}{h_o}\frac{dh_o}{dx} - \frac{1}{1+\beta}\frac{d\beta}{dx}$$

which is used in the expression of u_s :

$$u_s(x) = \frac{gh_o^3}{48K_{dz}(1+\beta)^3}\frac{dG}{dx}$$

In the case of an estuary with a flat bottom, we have $\beta = 0$, and we find again the same expression for $u(x,Z)$ except that h_o depends on x. The expression of the vertical component w is more complicated by the fact of the slope of the bottom :

$$w(x,Z) = \left[(1+Z)\frac{dh_2}{dx} - Z\frac{dh_1}{dx}\right]u + \frac{Q_R}{2b}\frac{d}{dx}\left(\frac{1}{A}\right)(1-Z-8Z^2)(1+Z)$$

where $h_2(x)$ is the mean elevation of the free surface and $h_1(x)$ the elevation of the flat bottom, or

$$h_o = h_2 - h_1$$

APPLICATION TO THE GIRONDE ESTUARY

The numerous measurements of velocities and salinities made in Gironde (Bonnefille, 1971) (17 surveys from 1965 to 1975, with sometimes simultaneous explorations on 5 verticales) allowed us trying to confirm the theories of residual phenomena and to estimate the new mixing coefficients K_{dz}, K_{sx} and K_{sz}. The figures 3 and 4 show the mean characteristics of the estuary S_s, b and h_o. The figure 5 gives an example of comparisons of theoritical and measured distributions of residual velocity $u(Z)$ and salinity $\overline{s}(Z)$ for differents sections of the estuary. It has been also possible to compute for each set of measurements, respectively u_m, u_s, K_{dz}, K_{sz} and finally Q_R.

The analysis of data shows an interesting first result. The residual depth $h_o(x)$ and the horizontal gradient of salinity $\frac{dS_s}{dx}$ have about always the same value ; this fact increases the interest of residual phenomena, because the determination of $A(x)$ becomes very easy. The analysis of results carrier on the following conclusions.

Residual discharge Q_R (figure 6)

Theoretically, if β is null, Q_R is constant ; this assumption is not well confirmed by the results. Nevertheless a more interesting hypothese would be that the rapport Q_R/h is nearly non-dependant on x. In this case we have

$$u_m S_s = K_{sx}\frac{dS_x}{dx}$$

expression giving easily K_{sx}

Longitudinal mixing coefficient of salinity K_{sx} (figure 7)

This coefficient could be considered as constant

$$K_{sx} \sim 4000 \text{ m}^2/\text{s}$$

Vertical mixing coefficient of salinity K_{sz} (figure 8)

The distribution of Ksz along the estuary is not simple ; but this coefficient is not important because it appears only on the vertical distribution of salinity. That is a second order phenomena. We can admit

$K_{sz} \sim$ 3-6 cm^2/s for $h_o \sim$ 7 m

$K_{sz} \sim$ 17 cm^2/s for $h_o \sim$ 11 m

Vertical mixing coefficient of momentum K_{dz} (figure 9)

This coefficient increases with the size of the flow, and mainly with the size of the aera of the cross-section. K_{dz} is very important because it is a main parameter on A(x) which determines the general form of the residual circulation. The mean value of K_{dz} is

$K_{dz} \sim$ 1 to 20 cm^2/s

for the cross-section areas 30000 to 100000 m^2.

REFERENCES

Bonnefille, R., 1971. Remarques sur les écoulements moyens à l'aval de la Gironde. AIRH, Paris, 4 : 229-233.

Chatwin, D.C., 1976. Some Remarks on the Maintenance of the Salinity Distribution in Estuaries. Estuarine and Coastal Marine Science, 4 : 555-566.

Hansen, D.V. and Rattray, M., 1965. Gravitational circulation in straits and estuaries. Journal of Marine Research, 23 : 104-122.

Ronday, F.C., 1975. Etude de l'envasement et de la variation longitudinale du coefficient de dispersion dans les estuaires partiellement stratifiés. Annales des Travaux Publics de Belgique, 4 : 1-18.

Pritchard, D.W., 1956. The dynamic structure of a coastal plains estuary. Journal of Marine Research, 15(1) : 33-42.

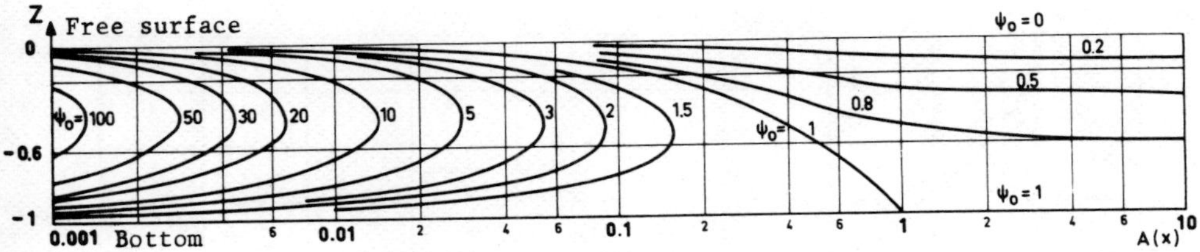

Fig.1 - Residual streamlines of a partially stratified estuary.

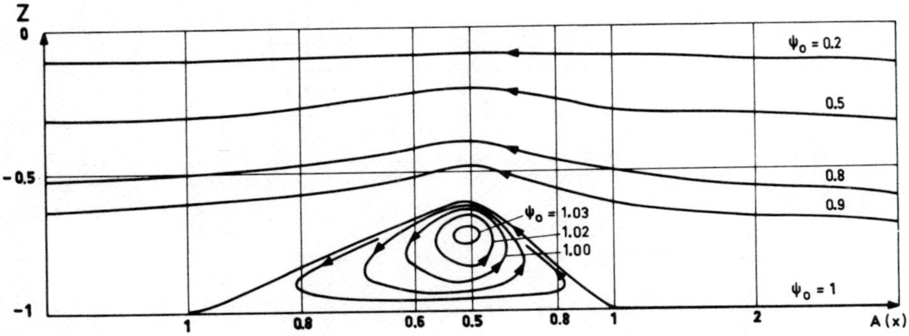

Fig.2 Example of residual streamlines in an estuary

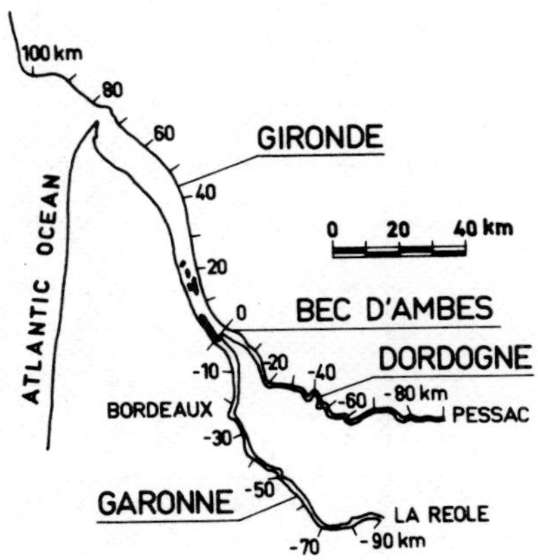

Fig.3 The Gironde Estuary

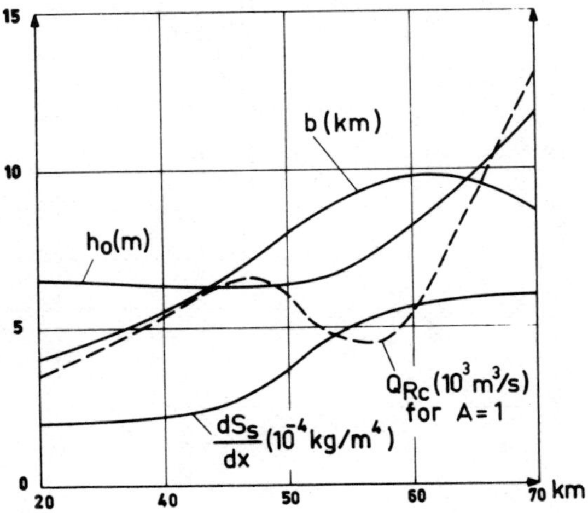

Fig.4 Values of h_o, b, $\frac{dS_s}{dx}$ along the Gironde Estuary

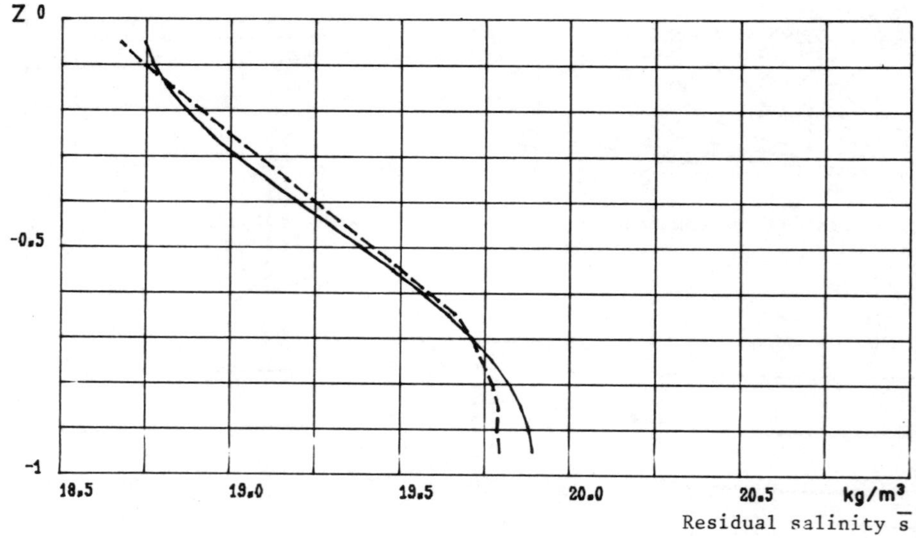

measured distribution

theoritical distribution

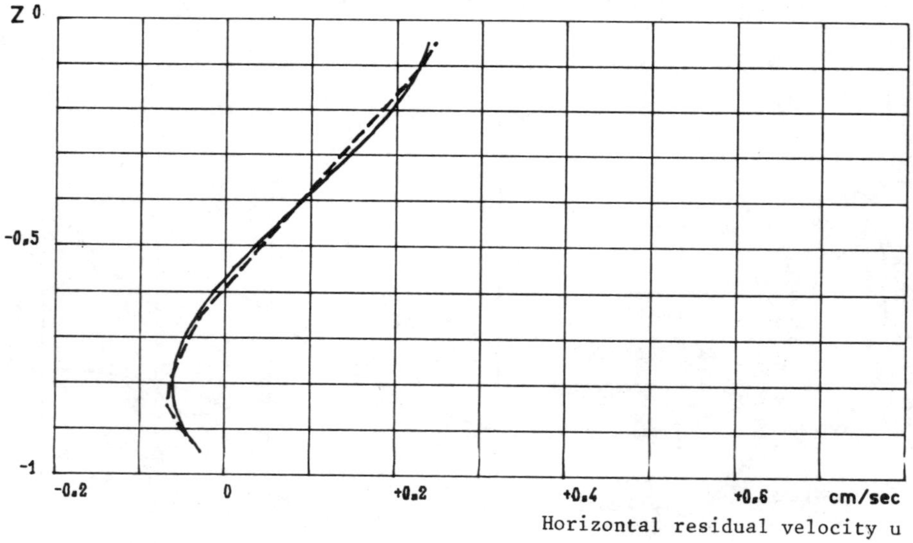

Fig.5 Example of measured and theoritical residual salinity and velocity at km 89 the 27 june 1968

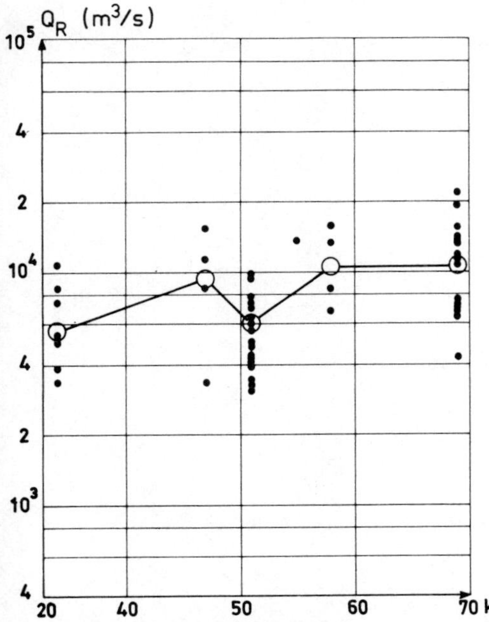

Fig.6 Residual discharge Q_r

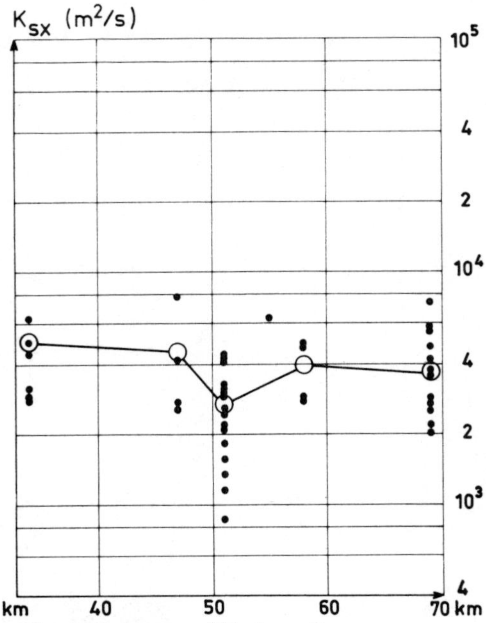

Fig.7 Mixing coefficient K_{sx}

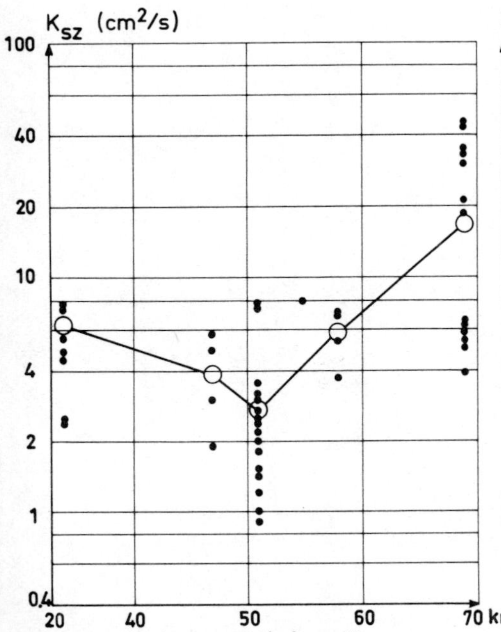

Fig.8 Mixing coefficient K_{sz}

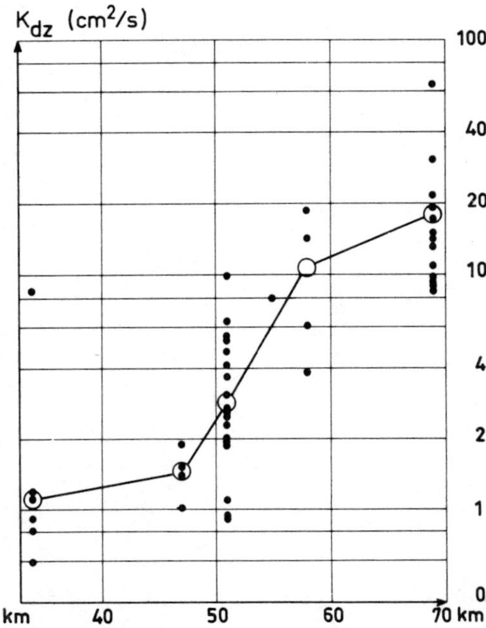

Fig.9 Mixing coefficient K_{dz}

ASYMMETRY AND ANOMALIES OF CIRCULATION AND VERTICAL MIXING IN THE BRANCHING OF A LAGOON-ESTUARY

Y. GALLARDO
O.R.S.T.O.M., Centre Océanologique de Bretagne, Brest (F.)

ABSTRACT

The lagoon Ebrié, in Ivory Coast, is formed of different bays and branches before communicating by an artificial canal with the Gulf of Guinea. The two principal lateral branches are quite different : the western channel forms a natural prolongation of the central channel while the eastern channel begins with a constriction. Observed at the eastern and western entrances of the lagoon, the circulation is statistically different, particularly in the upper layer and during ebb-tide. The eastern channel shows, at times, anomalies of residual velocity profiles which determine the relative asymmetry : there is often a seaward jet in the mid layer. The stronger residual anomalies are connected with sensible departures from the semi-diurnal period, involving the existence of beats between the tides and other subtidal frequencies. A decrease of Richardson number occurs during the anomalous profiles. The vertical mixing, its asymmetry and anomalies could be explained by a criterion for the maintenance of turbulence, depending on transient stages of river discharges and on the wind at the subtidal frequencies.

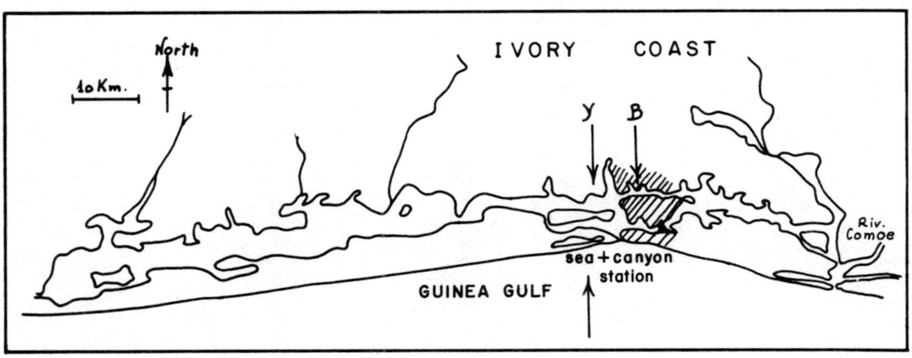

Fig. 1 - LAGOON EBRIE -Locations of temperatures - salinities - currents observations.

INTRODUCTION

The study of horizontal transport and of vertical mixing at the principal eastern and western gates of the lagoon Ebrié (fig. 1) was included in a larger and previously begun environmental program at the "Centre de Recherches Océanographiques d'Abidjan" in Ivory Coast.

Geophysical and physical descriptions (TASTET, 1974) of the whole lagoon have indicated its morphology, the monthly fresh water inflows, the tides and their currents. Because of the complicated topography with its sills, constrictions and bays and because of the variable fresh water inflow, the tides in the lagoon are much damped down and often out of phase with respect to the oceanic tides. TASTET (1974, p. 18) observes that these phase differences may reach respectively 40 minutes and 2.6 hours in the Central Channel, not far from the entrance of the eastern Channel. The seaward flow measured in the canal, during a tidal cycle of june 1966 was more than 400 $m^3 s^{-1}$, value which corresponds well to the mean fresh water inflow of June 1970, 1971, 1972 (TASTET, 1974, tab. 2, 3, 4).

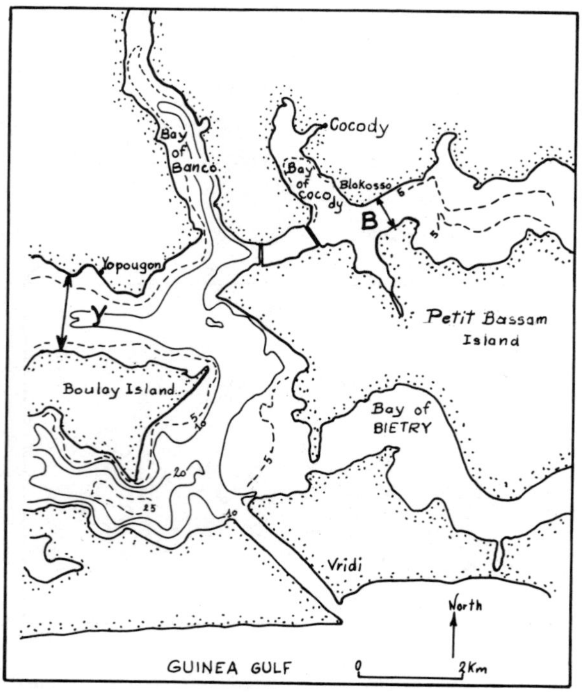

Fig. 2 - Geography and isobaths of the inner estuary (from J.P. TASTET, 1973)

From January to July 1976, current, salinity and temperature measurements were made, every month, at the locations B and Y of fig. 2 : until April, EKMAN current-meters, sampling five minutes, every two hours, were completed with a mooring of two AANDERAA currentmeters recording integrated velocity, direction, temperature and depth every ten minutes.

From August to December, the program was modified and concentrated in the eastern channel, in order to obtain some tangible results between the circulation, averaged over several tidal cycles, and the estimations of fresh-water inflow.

EAST-WEST ASYMMETRY

A glance at the fig. 2 indicates the complexity of the estuarine morphology ; however, the net flow coming from the canal must reach the area of branching without sensible loss because of the large and deep central channel. Since both branches represent approximately equivalent areas, and under the assumption that the water surface remains horizontal, it could be supposed that the velocities V_y and V_b would be inversely proportional to the surface of the vertical sections S_y and S_b ; that would lead to the following relationship

$$V_y = (S_b/S_y)\ V_b = 0.53\ V_b$$

V_y, S_y, V_b, S_b, are the tidal velocities and the vertical sections at the locations y and b.

The observed velocities, summed over ebb or flow periods, are distributed along a principal axis expressed by

$$\frac{y - \overline{y}}{s_y} = \frac{x - \overline{x}}{s_x}$$

where $\overline{x}$, $\overline{y}$ are the means, s_x^2 , s_y^2 the variances of the summed velocities x and y.

Fig. 3a, b, c, d, show our results at the depths 2 m and 4 m, for ebb (downstream velocities) and flow (upstream velocities) compared to the simple model described above ; the upper layer deviates eastwards during flowing tide (fig. 3 a) ; during ebb-tide there is an excess of velocities in the upper layer and a loss in the mid layer (fig. 3 b, d) of the western section ; the mid flow (fig. 3 c) is far from a linear partition and presents important fluctuactions in January, February and April.

When tidal amplitude is the highest (April) and when the fresh water inflow is maximum (June), the upper layer follows better the theoretical branching both during ebb and flow. When the tidal amplitudes are the lowest (March and May), the asymmetry seems to be the greatest. Nevertheless, the strong asymmetry of February has no obvious cause. On the contrary, in the mid layer, the asymmetry is great for the highest tides (April) ; however the asymmetries in January and February are obscure.

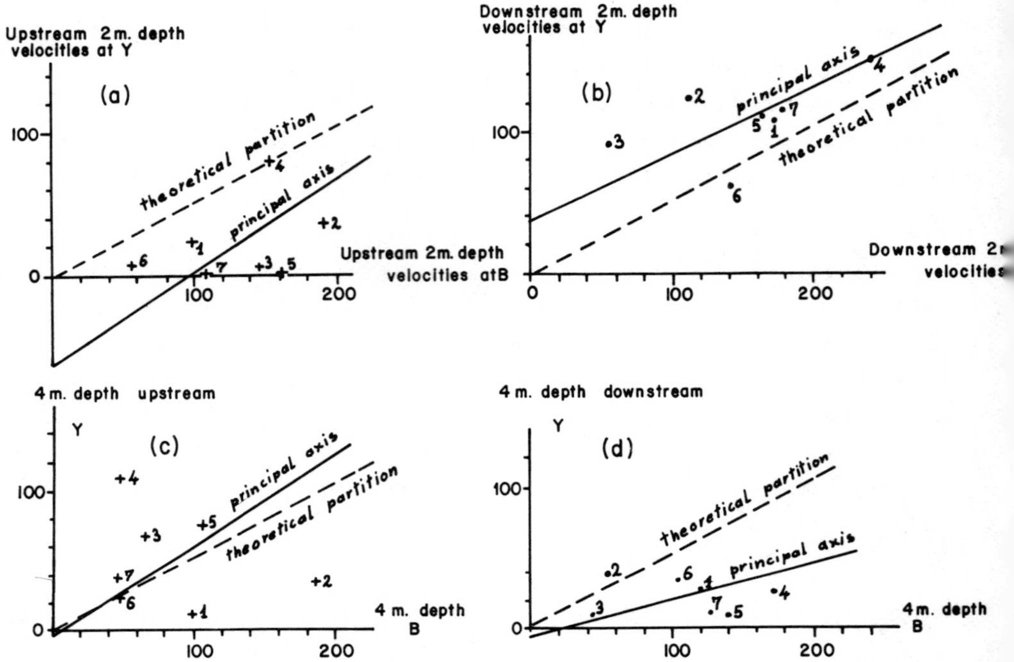

Fig. 3. Theoretical and observed partitions of the summed velocities of the channels B and Y.

Transverse motions, computed from AAnderaas during the period January-April, represent 20-25 % of the axial current in the West and only 10-15 % in the East. The effect of SW wind on the transverse motion is possible in the wide western estuary.

Richardson number too, computed between 2 m and 4 m depth at maximum ebb and flow are statistically different : greater in the eastern branch during ebb-tide but in the western branch during flow. On the whole, over a complete tidal cycle the Ri values are not significantly different in both branches.

The mean circulation observed during ebb and flow in each layer of B should correspond to a landward motion of the upper layer (about 5 cm s^{-1}) and to a seaward current in the mid layer (about 12 cm s^{-1}) : the spatial asymmetry existing from January to July, for variable tidal amplitudes, must be connected to a great anomaly of circulation. These results are confirmed and completed by the sequences of residual (i.e. averaged on a semi-diurnal tidal cycle) velocities and salinities in the eastern channel.

ANOMALIES OF CIRCULATION

Sequences of eight and six tidal cycles, respectively in August and October, were realized in the eastern channel. Tidal amplitudes were practically constant during each cycle. Constriction of the eastern channel greater tidal currents and the eastern origin of the fresh water inflow during the second part of the year, have fixed the choice of longer observations in this channel to the detriment of western channel.

Profiles of residual velocities (fig. 4,5) are variable and point out, at times, a persistent anomaly of the profile similar to those inferred above from the asymmetry : a maximum of seaward circulation at mid depth gives a jet profile.

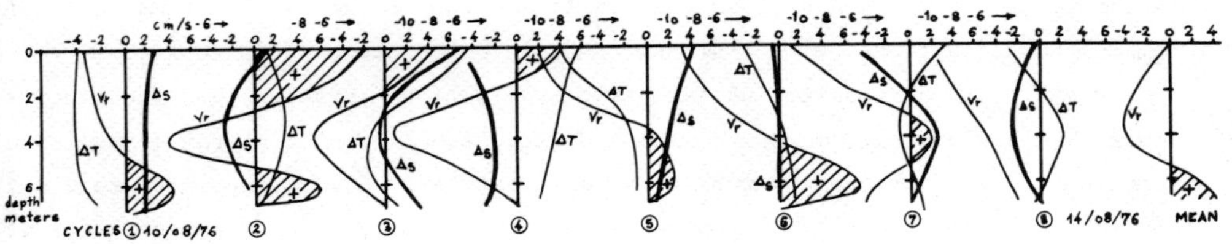

Fig. 4 - Sequence of 8 tidal cycles in August 1976 : residual velocities, deviations of temperatures and salinities from their mean profile. Numbers indicate the cycles.

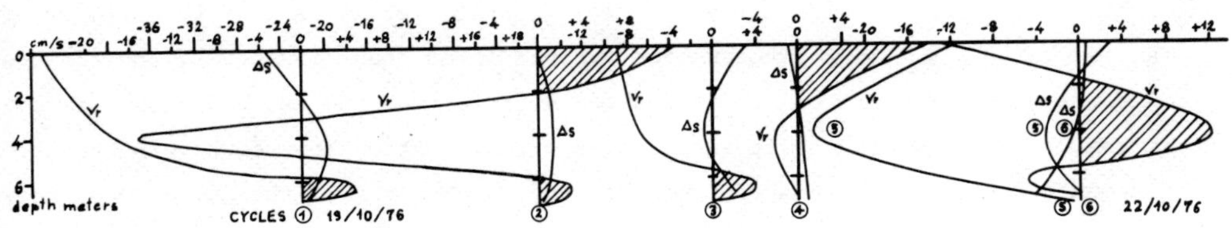

Fig. 5 - Sequence of 6 tidal cycles in October 1976.

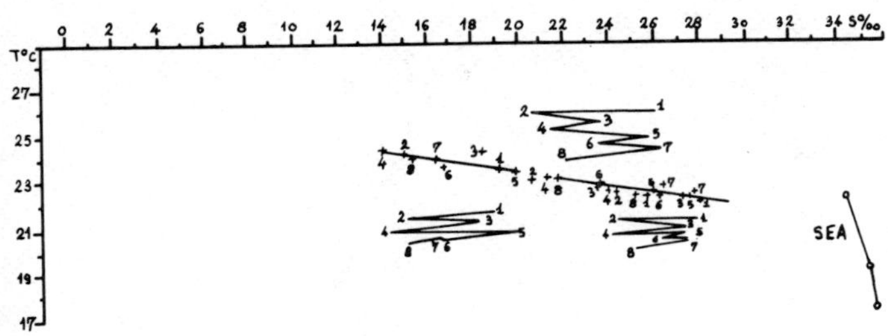

Fig. 6 - Diurnal oscillations of residual salinities during the sequence of August. Number indicate the cycles.

The comparison of fig. 4 with fig. 6, showing diurnal oscillations of residual salinities, indicates that strong changes of salinity occur with the appearance or disappearance of the jet profile, respectively between the cycles 1-2 and 4-5 ; moreover a landward jet, existing during the seventh cycle, disappears with a decreasing salinity.

The diurnal oscillation is not obvious on the horizontal circulation of fig. 4 ; however it becomes clear on the residual, relative, vertical motion of the maximum vertical salinity gradient. The diurnal oscillation of fig. 6 is still well marked in the maximum velocities during flow.

During October, we observe increasing periods between the appearances of the maximum flow velocities which coincide with the residual seaward anomalies. On the contrary, when the period decreases below that of the semi-diurnal tide, the residual anomaly is landwards in the mid layer. The frequency of maximum ebb velocit remains quasi-constant, with a slight tendency below the semi-diurnal.

In short, those results indicate that oscillations existing in a frequency range lower than the diurnal, may give rise to beats, from which arise the anomalies of circulation. In fact, the diurnal tidal oscillation in August, seems to reduce the lagging and, consequently, the anomalies with respect to the strong anomalies of October.

ANOMALIES OF VERTICAL MIXING

Is the gradient Richardson number a good indicator of vertical mixing or not ? generally the small tides (March and May) generate values of Ri frequently greater than 2, while high tides give numerous values lower than 2. However, the effects of vertical mixing for the same tides are different in the two branches : a glance at fig. 7 indicates the habitual stronger stratification in the western channel (segments Y are larger than segments B). But, on the whole, the differences between the eastern and western Ri values are not significant.

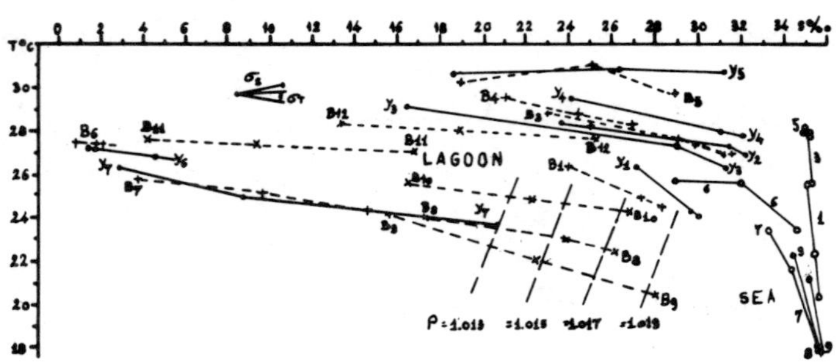

Fig. 7 - Temperature - Salinity - Depth diagrams (numbers indicate months)
+ and . = 2, 4, 6 m depth ∘ = 0, 10, 20 m sea depth
Lagoon observations averaged on a tidal cycle
σ_S and σ_T are the standard deviations on 8 successive tidal cycles.

We have pointed out a large time scale variability of velocity profiles in the eastern channel, with the appearance of jet profiles. We observe a decreasing tendency of the Ri values in presence of these anomalous profiles :

TABLE 1

Richardson's numbers at maximum velocities

		AUGUST							
tidal cycle n°		1	2	3	4	5	6	7	8
FLOOD	upper	0.5	1.3	5.4	7	556 →	2.5	2600 →	3.7
	lower	1.8	11	0.6	0.2	0.6	12	0.2	4.1
EBB	upper	59 →	5.3	10	1.8	0.6	58 →	0.7	4.8
	lower	0.5	0.5	0.9	0.0	11	1.3	∞	3.8
averaged Ri		15	4.5	4.2	2.2	142	18	-	4.1
		OCTOBER							
tidal cycle n°		1	2	3	4	5	6		
FLOOD	upper	49 →	0.5	123 →	5.3 →	0.8	22		
	lower	4.7	1.7	1.1	14	4.9	0.2		
EBB	upper	0.0	1.0	0.2	56 →	0.0	1.5		
	lower	0.4	0.0	0.4	7.0	6.4	1.9		
averaged Ri		13	0.8	31	21	3.0	6.4		

To the anomalies of cycles 2, 3, 4 in August and cycles 2, 4, 5, 6 in October often correspond averaged Ri significantly lower than in the other events ; more precisely, the process of destabilization between two consecutive cycles occurs in the upper layer, as indicated by the arrows on the table 1.

To our mind, the process of destabilization which appears at frequencies lower than the semi-diurnal tide, contributes moreover to the vertical mixing by changing the conditions of maintenance of turbulence. The ratio Az/Kz may define a critical value of the Richardson number (TAYLOR, 1931, PROUDMAN, p. 101, 1953) below which turbulent energy is supplied from the mean motion. Kz and Az, the coefficients of eddy-diffusion and of eddyviscosityare computed from the residual velocity and salinity fluctuations in August and October ; fluctuations of vertical velocities are estimated from the vertical oscillations of the salinity around its mean value at the depth 3 m. We obtain two different critical values, 2.1 in August and 6.5 in October ; that result suggests a variability of the turbulence which could explain some apparent anomalies of vertical mixing and the better mixing in the eastern estuary ; more precisely, the mean product $\overline{w's'}$ of vertical velocities and salinities fluctuations is the essential factor of the variability between August and October : when that product increases, the critical value decreases.

$\overline{w's'}$ increases tenfold in August with respect to October.
Lastly Kz is found larger in August than in October : their relative magnitude equal 8.

The ratio Az/Kz, estimated at a smaller time-scale, from the temperature and current AANDERAA measurements every ten minutes, gives the following results : during the lowest tidal amplitude (March) its values are 4.6 at location B and 3.0 at location Y ; during the highest tidal amplitude (April) its values are 0.10 and 0.11. For these very different tidal velocities (multiplicative factor of 3) the coefficient of eddy-diffusion Kz is in the range 0.01 - 1 $cm^2\ s^{-1}$ (low amplitude) and 5 - 50 $cm^2\ s^{-1}$ (high amplitude) : the highest value is found at Y for low amplitude, at B for high amplitude.

The principal theoretical and experimental results quoted by WELANDER (1968, p. 22-26) indicate that turbulence can be sustained when the flux RICHARDSON number, defined as R_f = (Kz/Az). Ri, lies generally below the mean value 0.3. From the local Ri observed during March, April, August and October it appears that R_f is about a few unities. Theoretically, and that is observed by comparison of March, August and October, the ratio Az/kz does not depend on the scales. The too large values of the observed R_f arises, to our mind, principally from the vertical scale of the local Ri observed : the vertical gradients should be estimated every 30 cm on the vertical, in order to obtain realistic values of Ri and hence of R_f.

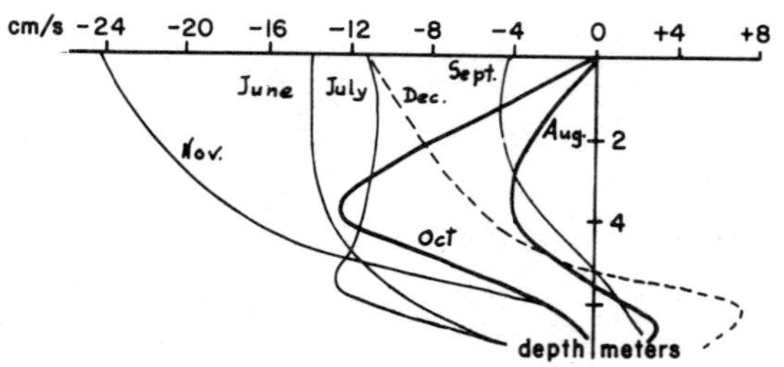

Fig. 8 - Means of the velocity profiles in the eastern channel.

DISCUSSION

The evolution of transports on fig. 8 is coherent with the habitual monthly fresh water inflows (TASTET, 1974) : the transport of about 600 $m^3\ s^{-1}$ in June corresponds well to the strong rainfalls in 1976. The anomalous profiles of August and October may represent transient stages of the river discharges for which the mean wind drift is opposite. The T-S diagrams of January (fig. 7) indicate more mixing than the consecutive months of the dry season : It is well known that atmospheric circulation is particular in January, with a seaward wind which gives rise to a coastal upwelling. COLIN (personal communication, 1977) shows a significa diurnal pike and an important variability around 4-6 days, for the annual wind spectrum in Abidjan. These scales correspond well to the changes of residual salinities and velocities observed in the eastern channel. WEISBERG (1976) demonstrates the effect of the wind variability on the estuarine circulation, and the necessity of measuring numerous tidal cycles, in order to obtain the "mean" circulation. Obviousness, the habitual S W wind has a very different effect on the residual circulation of the eastern and western channels : fig. 9 a indicates that anomalous profiles are often generated in the eastern channel, because the wind drives the circulation landward. On the contrary in the western channel, (fig. 9 b), the seaward circulation is favoured in the upper layer and, consequently, the typical estuarine circulation appears better.

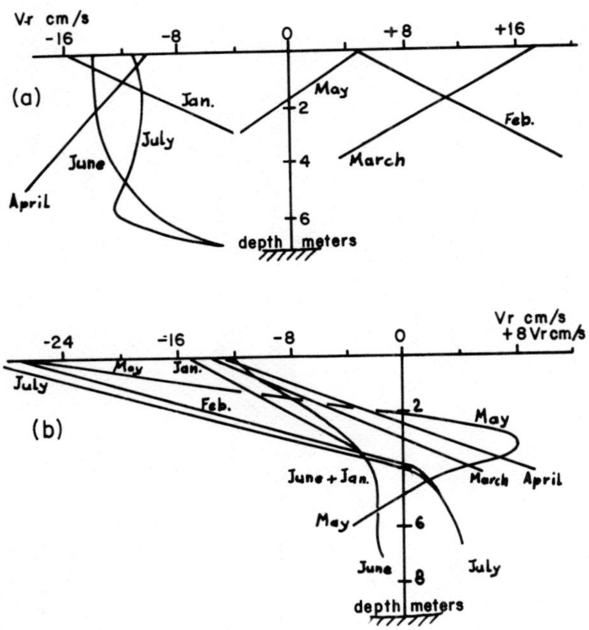

Fig. 9 - a - unsteadiness of the residual velocity profiles at station B.
b - steadiness of the residual velocity profiles at station Y.
(extrapolated from 1 or 2 meters depth to the surface).

CONCLUSION

We have observed a great variability of the residual circulation in a branching lagoon estuary. That variability may give rise to asymmetries between the eastern and western channels. The coefficient of eddy-diffusion Kz presents too a high range of variability which could be estimated from the dimensionless ratio Az/Kz. With respect to the general theoretical and experimental results which give a flux RICHARDSON number in the range 0.1-0.5, it appears that the gradient RICHARDSON numbers should be observed with a vertical distance of about 30 cm. The effect of the wind direction and velocity on the asymmetry is pointed out.

REFERENCES

PROUDMAN, J., 1953. Dynamical Oceanography. Methuen, London, 405 pp.
TASTET, J.P., 1974. L'environnement physique du système lagunaire Ebrié. Série doc. depart. sciences de la terre. Université d'Abidjan, 11 ; 28 pp, 58 fig., 4 cartes hors texte.
TAYLOR, G.I., 1931. Effect of variation in density on the stability of superposed streams of fluid, Proc. Royal Soc. (A), 132, London, 499 pp.
WEISBERG, R.H., 1976. The nontidal flow in the Providence River of Narragansett Bay : a stochastic approach to estuarine circulation. J. Phys. Oceanogr., 6, 721-734
WELANDER, P., 1968. Theoretical forms for the vertical exchange coefficients in a stratified fluid with application to lakes and seas. Acta R. Soc. Sci. Litt. Gothoburgensis Geophys., 1, 1-26.

DISPERSION BY TIDE-INDUCED RESIDUAL CURRENT VORTICES

J.T.F. ZIMMERMAN
Netherlands Institute for Sea Research, Texel, The Netherlands

ABSTRACT

It is shown that the nonlinear interactions of a tidal wave propagating over a complicated bottom topography produce a time-independent residual current velocity field which is a quasi-random function of space. The superposition of this Eulerian residual velocity field and the oscillatory tidal current gives rise to a dispersion process in the Lagrangian sense, the "diffusion coefficient" of which is related to the relevant parameters of both Eulerian components of the total velocity field. For diffusion time intervals small as compared with the Lagrangian velocity correlation time-scale, it appears that the gross transport of dissolved substances in a tidal channel should be described by the telegraph equation which takes into account the finiteness of the random Lagrangian particle velocities.

INTRODUCTION

Longitudinal dispersion (along the channel axis) in tidal inlets can be controlled by a manifold of physical processes (Fischer, 1976). If the effective longitudinal diffusion coefficient (K) is derived empirically, for instance from the salinity distribution, it is often difficult to say which of these processes is the dominant one. This impedes the expression of the diffusion coefficient in relevant physical parameters. However, for areas which are dominated by tidal currents, by first guess, Arons and Stommel (1951) expressed K in the amplitude of tidal velocity (U_1) and tidal displacement (l_1):

$$K = b_1 U_1 l_1 \tag{1}$$

It must be kept in mind, of course, that (1) arises primarily from dimensional reasoning and that, consequently, all of the relevant physics is buried in the factor b_1.

In the present paper we shall deal with a particular phenomenon which may be responsible for rather large diffusion coefficients (order $10^3 m^2 s^{-1}$) in tidal inlets, viz. the existence of a field of quasi-two dimensional (horizontal) irregularly distributed residual currents. These residual currents arise from nonlinear interactions of a tidal wave propagating over an area of complicated topography. They can be understood from the vorticity transfer between the tidal current and the mean flow (Sugimoto, 1975; Zimmerman, 1977), the irregular topography acting as a catalyst. Their pronounced existence in many areas, mainly in the form of vortices, has been demonstrated from field measurements (Sugimoto, 1975; Zimmerman, 1976) and by numerical (Tee, 1976) or hydraulic (Yanagi, 1976) modeling.

By their very existence residual eddies contribute to the longitudinal dispersion. Conceptually their dispersive action may be conceived in two different ways. First, the eddies produce an irregular distribution of residual current shear. It is well known that any shear, in collaboration with smaller scale (tide-induced) turbulence, enhances diffusion in the direction perpendicular to the shear direction. If regarded as such (Sugimoto, 1975) the longitudinal diffusion coefficient can be expressed in velocity (u_0) and length (l_0) scales connected with the residual eddy velocity field:

$$K = b_0 u_0 l_0 \tag{2}$$

Here, again, it is the factor of proportionality (b_0) that contains the relevant physics.

There is, however, another way in which a field of randomly distributed residual eddies produces dispersion on a large scale, even without the concurrent existence of smaller scale turbulence. This second effect arises from the fact that the Eulerian superposition of an oscillatory (tidal) motion on a field of residual currents, distributed randomly in space, can be transformed in a Lagrangian motion which is partly random in time. This effect is the subject of the present paper.

GENERATION OF RESIDUAL EDDIES

The large scale, quasi-two dimensional, flow produced by a tidal wave propagating through a non-rotating basin of uniform depth can be considered to be vorticity-free, the longitudinal velocity being at each moment only a function of the (x)-coordinate along the channel axis. Vorticity can, however, be produced by friction along the sides of the channel as well as by depth differences of the channel bottom. We shall concentrate here on the latter effect and outline the consequent production of residual vorticity, which is illustrated in Fig. 1. A tidal wave is running through a channel of depth H, producing an oscillatory current velocity U_1 along the channel axis. If now the bottom topography contains a depth perturbation h (x,y), the tidal velocity field will contain an oscillatory velocity perturbation $\{u_1(x,y),\ v_1(x,y)\}$ which, in general, will no longer be vorticity free. Let the associated vorticity be given by

$$\omega_1 = \frac{\partial v_1}{\partial x} - \frac{\partial u_1}{\partial y} \tag{3}$$

It may then be shown (Zimmerman, 1977) that, in linear approximation, in case of considerable bottom friction (small depth), ω_1 obeys the equation:

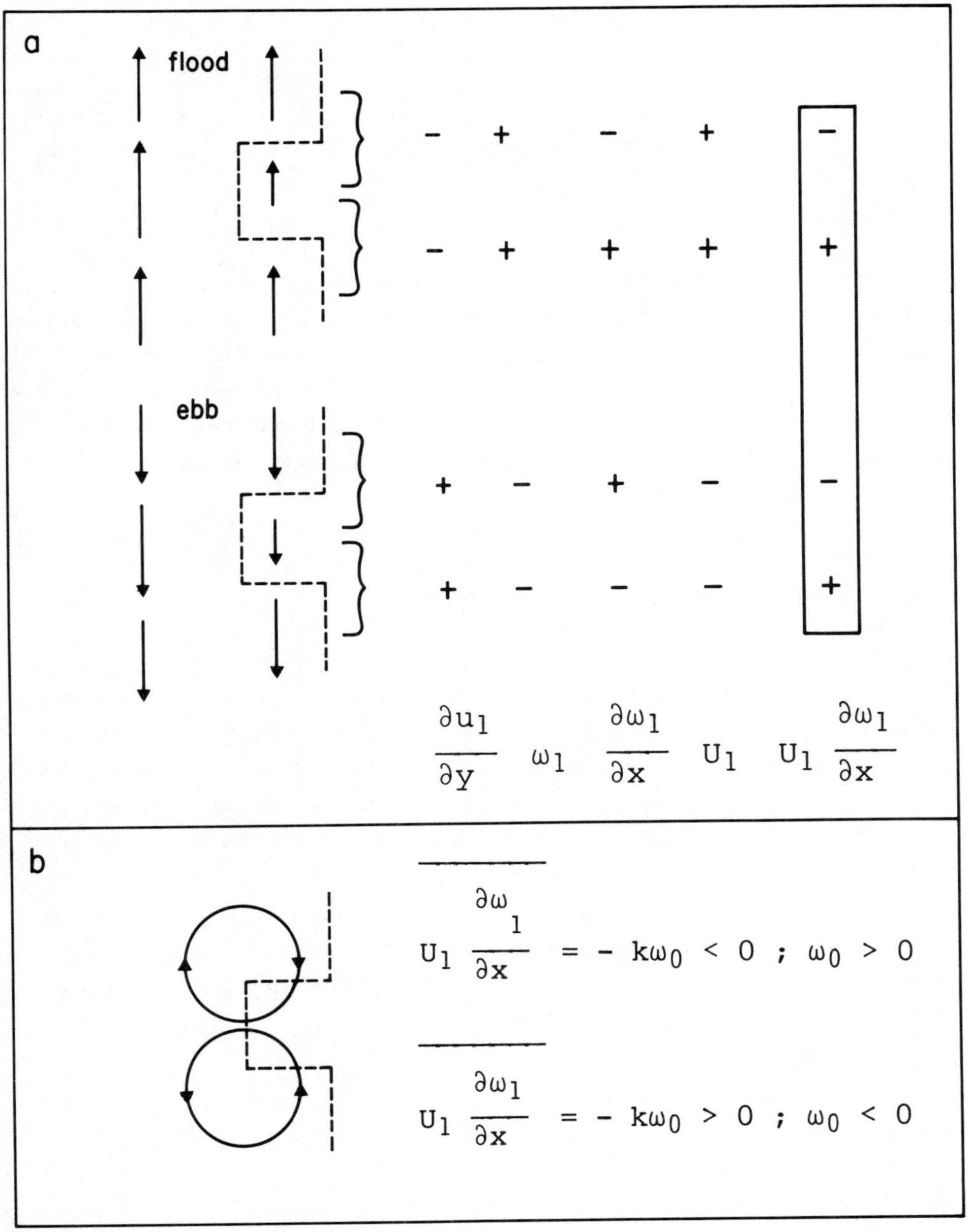

Fig. 1. a. Signs of various tidal parameters during the flood and ebb stages of the tide in a channel in which a shoal is present, shown by the dashed lines in left hand portion of the figure; the advective residual vorticity production term is outlined in the outermost column on the right. b. Residual circulation over the shoal due to the production of residual vorticity.

$$\frac{\partial\omega_1}{\partial t} = -\frac{k\omega_1}{H} - \frac{kU_1}{H^2}\frac{\partial h}{\partial y} \tag{4}$$

i.e. vorticity is produced by the bottom friction torque in the second term on the righthand side, k denoting the bottom friction coefficient, relating the bottom stress (τ_b) to the depth mean velocity by:

$$\tau_b = -k\rho U_1 \tag{5}$$

A localised depth perturbation, in Fig. 1 assumed to be a shoal, reduces the tidal velocity over the shoal by an increase in bottom stress, thereby creating vorticity. The sign of the vorticity as well as its longitudinal gradient reverses sign during the tidal cycle, as shown in Fig. 1. Hence, in linear approximation the tidal mean vorticity vanishes. However, residual vorticity can be produced if nonlinear advection terms are retained in the vorticity equation (Zimmerman, 1977). In that case, if $U_1 >> u_1$, we get:

$$\frac{\partial\omega_1}{\partial t} + U_1\frac{\partial\omega_1}{\partial x} = -\frac{k\omega_1}{H} + \frac{kU_1}{H^2}\frac{\partial h}{\partial y} \tag{6}$$

Averaging (6) over the tidal cycle gives:

$$\overline{U_1\frac{\partial\omega_1}{\partial x}} = -\frac{k}{H}\omega_0 \tag{7}$$

where ω_0 stands for the residual vorticity. As is shown in Fig. 1, the left hand side of (7) is of opposite sign on both sides of the shoal, but does not reverse sign during the tidal cycle. Hence, a cyclonic and an anticyclonic vortex are produced near the shoal (Fig. 1b). Now, in the real situation of many shallow tidal areas, the perturbations in water depth have a quasi random character, thereby randomizing the residual current velocity field. The latter, therefore, can be thought of as to exist of a random distribution of eddies of different strenght and size. Assuming now an ensemble of such random residual current fields, we may construct the spatial covariance $\langle u_0(x)u_0(x+r)\rangle$, from which a representative velocity scale $\langle u_0^2\rangle^{\frac{1}{2}}$ and length scale:

$$l_o = \int_0^\infty \frac{\langle u_0(x)u_0(x+r)\rangle}{\langle u_0^2\rangle}\,dr \tag{8}$$

are derived.

LONGITUDINAL DISPERSION

By definition the Eulerian residual current velocity field is a random function of space but time-independent. In a qualitative way it may now be shown that the Lagrangian residual velocity of a single waterparcel is a random function of time, if an oscillatory tidal current is added to the Eulerian residual current velocity field. The random structure in space of the residual current velocity field is shown in Fig. 2, together with the path of a particular waterparcel.

Starting at position O at the beginning of the flood tide, the parcel arrives at A' at slack water. If only the tide were acting it should arrive at A at that time. Hence the displacement A A' is due to the residual current velocity field. The same way of reasoning applies during its backward (ebb) motion from A' to B'. Here BB' is the displacement due to the residual current velocity field. Thus, during one tidal cycle the particle experiences a residual displacement OB'. The procedure may be repeated for all tidal cycles here after. Because of the random

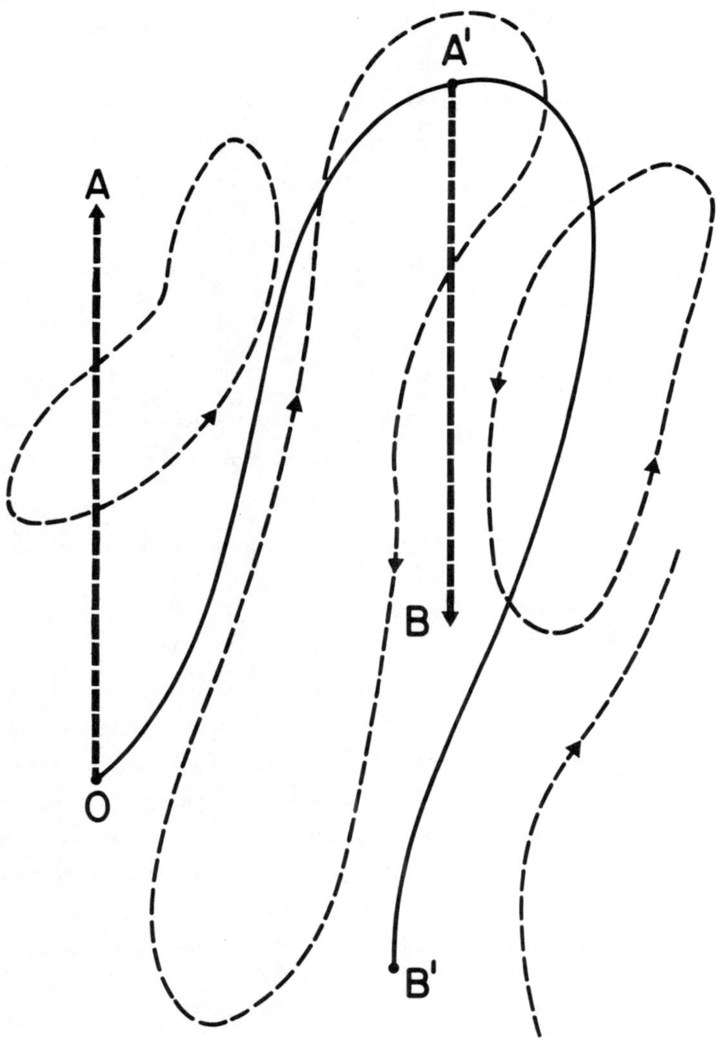

Fig. 2. Path of a particle during a tidal cycle shown by a solid line. The residual current velocity field is represented by dashed streamlines. The tidal motion proper, during the flood and ebb stages of the tide each, is shown by the thick straight dashed lines.

character of the residual current velocity it will be obvious that successive displacements AA', BB', ... form a random vector series, i.e. the particle experiences a "tidal random walk". Of course, neighbour vectors may be correlated, the magnitude of the correlation depending on the ratio of the r.m.s. step length of the tidal random walk and the integral length scale (l_0) of the residual current velocity field.

Considering now the problem of longitudinal particle dispersion as a random walk problem, we may use Taylor's (1921) classical theory of turbulent diffusion to express the effective longitudinal diffusion coefficient in the r.m.s. longitudinal step length, $\langle y_1^2 \rangle$, and the correlation coefficient of neighbour steps, c, provided that the successive displacements constitute a first-order Markov process. If now $y^{(n)}$ denotes the displacement due to the residual current velocity field during the n'th step and ε is the time-interval of each step, assumed to be half a tidal period here, then the mean-square Lagrangian residual velocity $\langle v^2 \rangle$ during each step is defined by

$$\langle v^2 \rangle = \frac{\langle y^2 \rangle}{\varepsilon^2} \tag{9}$$

whereas the correlation coefficient between the n'th and the m'th step is given by

$$\frac{\langle y^{(n)} y^{(m+m)} \rangle}{\langle y^2 \rangle} = c^m \tag{10}$$

Then, in the limit $\varepsilon \to 0$, assuming $\lim_{\varepsilon \to 0} \frac{\varepsilon}{1-c}$ to approach a finite value (τ), the mean-square particle displacement after time interval t reads:

$$\langle Y^2 \rangle = \langle v^2 \rangle \{2\tau t - 2\tau^2 (1-e^{-t/\tau})\} \tag{11}$$

from which an effective longitudinal diffusion coefficient can be defined by:

$$K = \tfrac{1}{2} \frac{d\langle Y^2 \rangle}{dt} = \langle v^2 \rangle \tau \{1-e^{-t/\tau}\} \tag{12}$$

For $t \gg \tau$, K attains the well-known form:

$$K = \langle v^2 \rangle \tau \tag{13}$$

Up to here K is expressed in Lagrangian quantities $\langle v^2 \rangle$ and τ. Since, in general, our information about the residual current velocity field is of Eulerian character we have to express the former quantities in the characteristics of the Eulerian velocity field. The latter is now supposed to consist of a homogeneous and normal random distribution in space of characteristic eddies represented by a stream function of Gaussian form (Zimmerman, 1976) upon which is superimposed an oscillatory tidal current. The Eulerian velocity field is then represented by the following set of parameters:

	tidal	residual	ratio tidal/res.
m.s. velocity	U_1^2	$\langle u_0^2 \rangle$	υ
length scale	l_1	l_0	λ

By using the assumption $U_1 >> \langle u_0^2 \rangle^{\frac{1}{2}}$ in a perturbation procedure for the Euler-Lagrange transformation (Zimmerman, 1976) it can be shown that the effective longitudinal diffusion coefficient may ultimately be expressed by:

$$K = b\ (\upsilon,\lambda)\ U_1 l_1 \tag{14}$$

where b (υ,λ) is a complicated function of the energy density spectrum of the residual current velocity field, weighted by functions which depend on υ and λ. Note that (15) resembles (1). However, here b (υ,λ) is not an empirical factor of proportionality but is theoretically related to the characteristics of the components of the Eulerian current velocity field. An equivalent expression, similar to (2), for K is:

$$K = c\ (\upsilon,\lambda)\ \langle u_0^2 \rangle^{\frac{1}{2}}\ l_0 \tag{15}$$

Here again c (υ,λ) is a complicated function of υ and λ. Formulas (15) and (16) show that K can neither be described by either only the tidal parameters $(U_1,\ l_1)$ or only the parameters (u_0, l_0) of the residual current velocity field. Although dimensionally the products of both sets of parameters produce a diffusion coefficient, it is shown by the dependence of the factors b and c on υ and λ, that it is the interaction of both field which does give rise to a dispersion process.

AN APPLICATION OF THE TELEGRAPH-EQUATION

The equation (11) for the second moment of the particle position corresponds to a transport equation of the form (Monin and Yaglom, 1971; Corrsin, 1974):

$$\frac{\partial^2 c}{\partial t^2} + \tau^{-1} \frac{\partial c}{\partial t} = \langle v^2 \rangle \frac{\partial^2 c}{\partial x^2} \tag{16}$$

where c is the cross-sectional mean concentration of the transported dissolvent This equation is the Telegraph equation which has a "wavelike" character for $t << \tau$, whereas it reduces to a "normal" diffusion equation for $t >> \tau$. The deviation from a "normal" diffusion equation for small t, due to the first term on the left hand side of (16), is a consequence of the finiteness of the particle velocity at all times. Since, in tidal areas, τ may be of the order of several tidal periods, the consequences of "diffusion with a finite velocity" in such areas should be briefly mentioned here.

In Fig. 3 we have shown the spread of a substance initially confined to a narrow band in the longitudinal direction, but homogeneously distributed over the channel cross-section. Assuming now that the residual current velocity field in the channel consists of a random distribution of residual eddies, the cross-sectional mean concentration of the substance develops according to (16). In Fig. 3 the cross-sectional mean concentration after one tidal period (T) is shown on the assumption that $T << \tau$. The "wavelike" diffusion over this time interval is demonstrated by the existence of two concentration maxima as compared to the distribution for "normal" diffusion if the finiteness of the particle

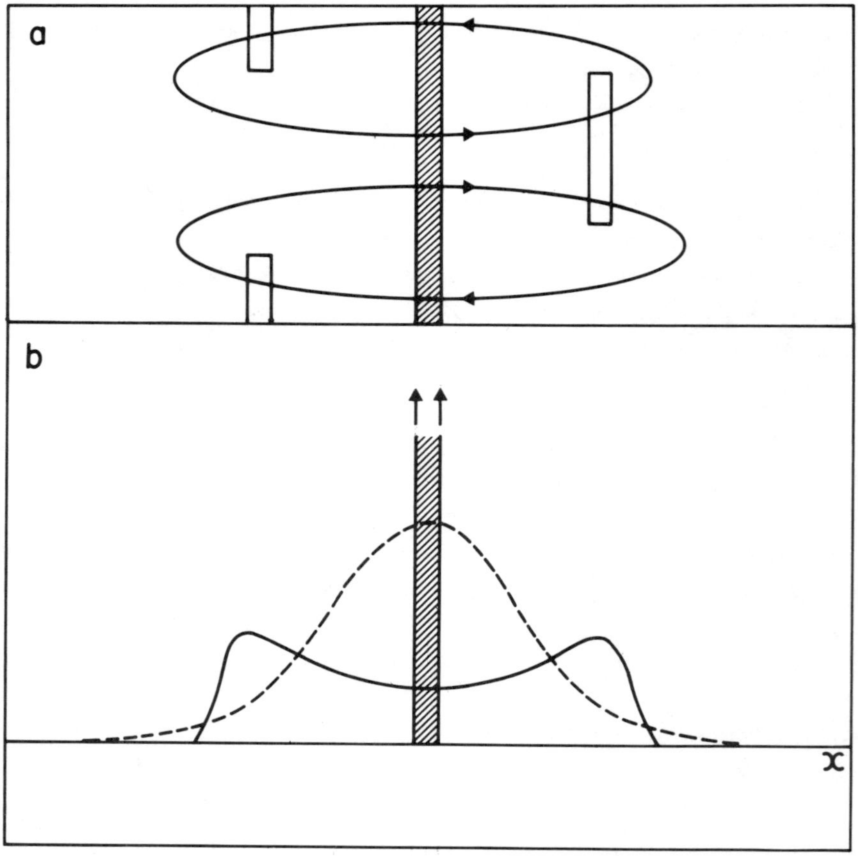

Fig. 3. a. The splitting of a band of dissolved substance in a tidal channel by the existence of a pair of large residual eddies. Both the initial (shaded) distribution and the distribution after one tidal cycle are shown. b. The distribution of cross-sectional mean concentration of the initial band (shaded) and after one tidal cycle (solid line) in the presence of residual eddies, the largest ones of which are shown in a. The dashed line shows the distribution of the cross-sectional mean concentration if, from the outset, the band-width of the substance cloud should have been large as compared with the scale of the residual eddies.

velocity is not taken into account. In this simplified example, the occurrence of two concentration maxima after a small time interval has a simple explanation. As the longitudinal dimension of the substance cloud is initially small as compared with the length-scale of the residual current vortices, the main effect of the large eddies is the splitting of the substance cloud in parts which are transported in opposite directions relative to their initial position (Fig. 3a). In the meantime, these parts are diffused to some degree by smaller eddies. For the cross-sectional mean concentration the result after one tidal period is a smoothed distribution with two maxima (Fig. 3b), the positions of which are located approximately at a distance of $\pm\langle v^2\rangle^{\frac{1}{2}}T$ from the initial position of the substance band. It might be that the complex patterns of tracer distribution in tidal areas with frequent occurrence of cloud splitting, as described for instance by Talbot and Talbot (1974), should be interpreted in this way.

A NUMERICAL EXAMPLE

To give an idea of the order of magnitude of parameters discussed before, we present here some values which are representative for the western Dutch Wadden Sea, a tidal area of about 130 km^2 in extent (for details see Zimmerman, 1976). Tidal current velocities in this area have a mean amplitude of 0.85 ms^{-1}, producing a mean amplitude of tidal displacement of 5.9 km. A representative scale velocity of residual eddies, which occur frequently in this area, is about 0.12 ms^{-1}. These eddies have an average length scale of 2.2 km. Hence the magnitude of the ratios υ and λ is, respectively, 2×10^{-2} and 0.4. This gives a correlation coefficient for neighbour steps (eq. 10, m = 1) of 0.85. Then the Lagrangian velocity correlation time scale (τ) amounts to 3.5 tidal periods. For the longitudinal diffusion coefficient for $t \gg \tau$, expressed by (14) we get a factor of proportionality, $b(\upsilon,\lambda)$, of 2.9. Hence, an average value of the longitudinal diffusion coefficient is about 800 m^2s^{-1}. Note, that the magnitude of the correlation time scale (τ) is such that for small time intervals the particular effects of "diffusion with a finite velocity" as described above should be taken into account in this area.

REFERENCES

Arons, A.B. and H. Stommel, 1951. A mixing length theory of tidal flushing. Trans. Am. geophys. Un., 32: 419-421.

Corrsin, S., 1974. Limitations of gradient transport models in random walks and in turbulence. Adv. Geophys.,18A: 25-60.

Fischer, H.B., 1976. Mixing and dispersion in estuaries. Ann. Rev. Fluid Dyn., 8: 107-133.

Monin, A.S. and A.M. Yaglom, 1971. Statistical fluid mechanics, vol. 1. MIT Press, Cambridge, Mass., 769 pp.

Sugimoto, T., 1975. Effect of boundary geometries on tidal currents and tidal mixing. J. Oceanogr. Soc. Japan, 31: 1-14.

Talbot, J.W. and G.A. Talbot, 1974. Diffusion in shallow seas and in English coastal and estuarine waters. Rapp. P.-v. Cons. perm. int. Explor. Mer, 167: 93-110.

Taylor, G.I., 1921. Diffusion by continuous movements. Proc. London math. Soc., (2) 20: 196-212.

Tee, K.T., 1976. Tide-induced residual current, a 2-D nonlinear numerical tidal model. J. Mar. Res., 34: 603-628.

Yanagi, T., 1976. Fundamental study on the tidal residual circulation I. J. Oceanogr. Soc. Japan, 32: 199-208.

Zimmerman, J.T.F., 1976. Mixing and flushing of tidal embayments in the western Dutch Wadden Sea-II: Analysis of mixing processes. Neth. J. Sea Res., 10: 397-439.

Zimmerman, J.T.F., 1977. Residual vortex formation by oscillatory (tidal) currents. Geophys. Astrophys. Fluid Dyn. (in press).

CORIOLIS, CURVATURE AND BUOYANCY EFFECTS UPON DISPERSION IN A NARROW CHANNEL

RONALD SMITH
Department of Applied Mathematics and Theoretical Physics, University of Cambridge

ABSTRACT

The work of Smith (1976) concerning buoyant contaminants in shallow channels is extended to include allowance for curvature of the channel and for the earth's rotation. For weak buoyancy the Coriolis and centrifugal effects lead to a transverse circulation and therefore reduce the longitudinal dispersion relative to that in a straight non-rotating channel. The response of the more dense fluid to these effects is inhibited by the proximity of the channel bed. Thus, it is the less dense fluid that tends to be moved to the outside curve of bends and to the right of the flow direction in the Northern Hemisphere. For stronger buoyancy the transverse density variations reduce or even reverse the circulation and can lead to an increase in the dispersion. Numerical results for the dispersion coefficient are presented for channels of parabolic cross-section.

INTRODUCTION

Much of engineering practice leans heavily upon the search for analogies to suggest empirical equations whose coefficients and range of applicability to the case in hand can be determined by laboratory experiments and field observations. For the dispersal of pollutants in a fluid environment the obvious analogy is with gaseous diffusion (Csanady 1973). Thus, if the dispersion is uni-directional and if the contaminant is conserved, then the appropriate model equation is

$$\partial_t(A\bar{c}) + \partial_x(A\overline{u}\bar{c}) - \partial_x(AD\partial_x\bar{c}) = 0 \tag{1}$$

Here A is the cross-sectional area, $\bar{c}$ the concentration, $\bar{u}$ the bulk velocity, and D is the effective diffusivity coefficient. There is a wealth of evidence to show the efficacy of the diffusion model. One noteworthy fact is that when there are currents D can be several orders of magnitude greater than the turbulent eddy-diffusivity (Fischer 1973). The explanation of this enhanced dispersion and an accurate theoretical prediction of the dispersion coefficient for neutrally-buoyant contaminants in laminar pipe flow were given by G.I. Taylor in 1953.

If equation (1) is compared with the cross-sectionally averaged diffusion equation, then D can be formally identified:

$$D = \bar{\kappa}_1 + \overline{u'c'}/\partial_x \bar{c} \tag{2}$$

Here κ_1 is the longitudinal laminar or turbulent diffusivity, $\overline{\quad}$ denotes cross-sectional averaging, and ' indicates departures from the average. Thus, to determine the shear-dispersion contribution to D it suffices to calculate the departures of the longitudinal velocity and the concentration from their cross-sectional mean values. Taylor's contribution was to recognise that after a sufficient length of time the steepening of transverse gradients due to the velocity shear would be approximately balanced by transverse diffusion (see Figure 1). In this asymptotic state it is comparatively easy to calculate c' and Taylor (1953) derived the result

$$\overline{u'c'}/\partial_x \bar{c} = a^2 \bar{u}^2 / 48\kappa_2 \, , \tag{3}$$

where a is the pipe radius and κ_2 the transverse diffusivity. It is the inverse dependence upon κ_2 which enables the shear-dispersion term to dominate the longitudinal diffusion in equation (2)

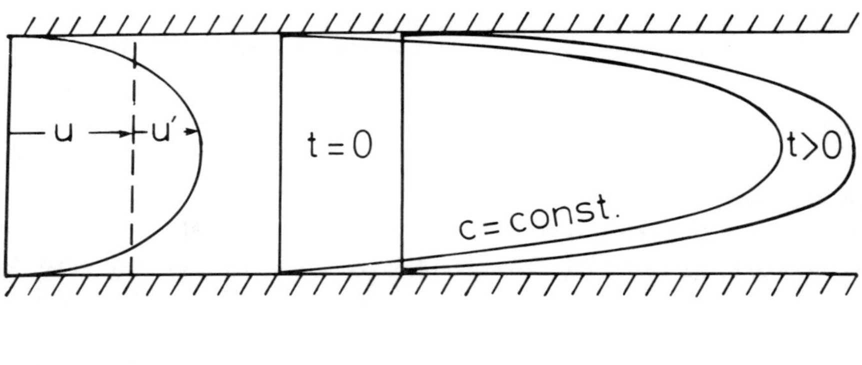

Fig. 1 Steepening of transverse concentration gradients by the velocity shear.

There are many ways in which Taylor's work has been extended. For open channel flow there is a preliminary asymptotic balance between vertical shear and vertical mixing, and Elder (1959) derived an equation of the same dimensional form as (3) but with the water depth h taking over the role of the pipe radius. At a later stage there is an asymptotic balance between lateral shear and lateral mixing, and the shear-dispersion is dramatically increased due to the channel breadth b replacing h (Fischer 1967). Holley, Harleman & Fischer (1970) demonstrated that for estuary flows the oscillatory nature of the flow interferes with the second asymptotic balance unless the time scale of lateral mixing is less than the period of oscillation.

Another class of extensions to Taylor's work is the inclusion of extra physical effects which modify one or both of u' and c'. Erdogan & Chatwin (1967) have shown that weak curvature and buoyancy effects can separately be accounted for by adding further terms to Taylor's formula (3) for shear-dispersion. Proceeding from the relationship (2) Fischer (1972) has shown that a temporal (i.e. mean and fluctuating) and a spatial (i.e. transverse and vertical) decomposition of u' and c' leads to a corresponding additive decomposition of the shear-dispersion. When there are several different physical effects modifying the flow, it would be extremely convenient if a phenomenological decomposition likewise led to an additive result for D. The present paper shows that this is not the case.

The chosen problem is an attempt to model the dispersion, on time scales of the order of one tidal period, of heat or salt in a narrow river or estuary. The physical effects accounted for are buoyancy, channel curvature, and the earth's rotation. Numerical results are presented to show the interactions between these effects. Studies which have considered one of the first two effects in the context of open channel flow include: Fischer 1969, Harleman & Thatcher 1974, Imberger 1976, Smith 1976. The inclusion of the Coriolis effect stems from a remark made in a review paper by Fischer (1976) noting the absence of any analyses of this potentially relevant effect. In the present paper confirmation is given of Fischer's speculation that, in the Northern Hemisphere, the earth's rotation tends to enhance the salinity to the left of the direction of flow, as observed by Pritchard (1952).

LIST OF HYPOTHESES

Fluid motion in rivers and estuaries is almost inevitably turbulent. As the underlying mathematical mode for the flow we represent the turbulent transports of mass and momentum in terms of eddy-diffusivity tensors with principal axes in the longitudinal, transverse and vertical direction. A more easily justified aspect of the mathematical modelling is the use of the Boussinesq approximation, in which we include the buoyancy effect due to density variations but neglect the corresponding inertia variations. Also, we shall neglect small-scale irregularities in the channel shape (Okubo 1973).

From this complicated starting point we seek those simplifications implied by the physical assumptions. These assumptions are:

i) Cross-sectional mixing takes place in much less than the tidal period:

ii) The dispersion time scale is comparable with the tidal period;

iii) Buoyancy, curvature and Coriolis effects all modify the dispersion at leading order;

iv) The channel is much shallower than it is wide.

The present work extends that of Smith (1976) in its more comprehensive nature of the third assumption. For ease of reference the author's earlier paper is henceforth designated as S .

NARROW CHANNEL APPROXIMATION

Assumptions (i) and (ii) together imply that the longitudinal length scale L of the pollutant distribution greatly exceeds a typical channel breadth scale B. Accordingly, the first step in S was to introduce the small parameter.

$\varepsilon = B/L$

and to specify the ε-ordering of the many terms in the equations of motion in relation to the basic dimensional quantitites B and U , where U is a typical longitudinal velocity. In the present paper the new assumption (iii) leads to a new scaling.

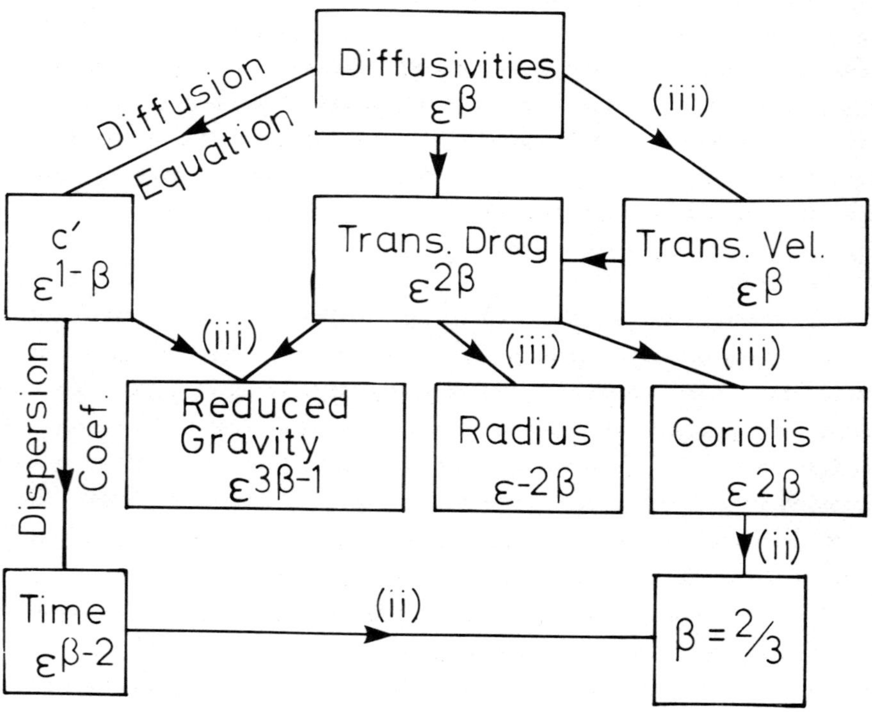

Fig. 2. Derivation of the scaling relationships from the physical assumptions.

To derive this scaling we follow through the mathematical consequences of the eddy diffusivities being of order ε^{β}, where β is to be determined. For brevity the full eqautions of motion are not presented, but use may be made of the simplified equations (4.5) given below. From the diffusion equation the assumed order of the diffusivities implies that c' is of order $\varepsilon^{-\beta}$ and that the dispersion time scale is of order $\varepsilon^{\beta-2}$. The transverse circulation can modify c' if the transverse velocity is of order ε^{β} (i.e. of the same order as the diffusivities). The associated transverse drag due to the turbulent stresses is of order $\varepsilon^{2\beta}$. By definition the longitudinal velocity is of order unity. Thus, Coriolis and centrifugal effects lead to transverse driving forces of the appropriate magnitude if the Coriolis parameter is of order $\varepsilon^{2\beta}$ and the radius of curvature is of order $\varepsilon^{-2\beta}$. Since the transverse density variations are only of order $\varepsilon^{1-\beta}$, the reduced gravity associated with density variations must be of order $\varepsilon^{3\beta-1}$, if the buoyancy forces as well are to contribute significantly to the transverse circulation. In geophysical applications the tidal frequency is of the same order as the Coriolis frequency. To satisfy assumption (ii) we need the dispersion time scale $\varepsilon^{\beta-2}$. This can be satisfied only if $\beta = 2/3$.

Because the contaminant is well-mixed across the channel it follows that the entire contaminant distribution tends to be carried along the channel with the flow at the bulk velocity $\bar{u}$. In order to focus attention upon less trivial aspects of the motion it is mathematically convenient to use axes moving at the bulk velocity. Proceeding as in S, a regular perturbation expansion

$$c = c^{(0)} = \varepsilon^{1/3}c^{(1)} + \varepsilon^{2/3}c^{(2)} + \ldots\ldots ,$$

leads to a preliminary simplification of the equations of motion. The main feature is that the transverse circulation can be described in terms of a stream function (see figure 3).

$$V^{(0)} = \partial_z\psi \quad , \quad W = -\partial_y\psi$$

Having derived simplified equations by the expedient of using moving axes, it is valid to transform the result back into the more familiar stationary axes:

$$\partial_z\psi\partial_y c^{(1)} - \partial_y\psi\partial_z c^{(1)} - \partial_y(\kappa_2\partial_y c^{(1)}) - \partial_z(\kappa_3\partial_z c^{(1)}) = (\bar{u} - u^{(0)})\partial_x\bar{c} \quad , \tag{4a}$$

$$\partial_z\psi\partial_y u^{(0)} - \partial_y\psi\partial_z u^{(0)} - \partial_y(\nu_{12}\partial_y u^{(0)}) - \partial_z(\nu_{13}\partial_z u^{(0)}) = -g\partial_x\zeta \tag{4b}$$

$$(\partial_y\psi\partial_z - \partial_z\psi\partial_y)\left[\partial_y^2\psi + \partial_z^2\psi\right] + (\partial_z^2 - \partial_y^2)\left[\nu_{23}(\partial_z^2 - \partial_y^2)\psi\right] + 2\partial_y\partial_z\Big[$$

$$(\nu_{22} + \nu_{33})\partial_y\partial_z\psi\Big] = \alpha g\partial_y c^{(1)} - 2R^{-1}u^{(0)}\,\partial_z u^{(0)} + f\partial_z u^{(0)} \quad , \tag{4c}$$

$$u^{(0)} = \psi = \partial_z\psi + \partial_y\psi\partial_y h = \partial_y c^{(1)} + \partial_y c^{(1)}\partial_y h = 0 \text{ on } z = -h, \tag{4d}$$

$$\partial_z u^{(0)} = \psi = \partial_z^2\psi = \partial_z c^{(1)} = 0 \text{ on } z = \zeta \tag{4e}$$

$$\partial_t(A\bar{c}) + \partial_x(\overline{A\bar{u}\bar{c}}) + \partial_x(\overline{A(u^{(0)} - \bar{u})c^{(1)}}) = 0 \ . \tag{5}$$

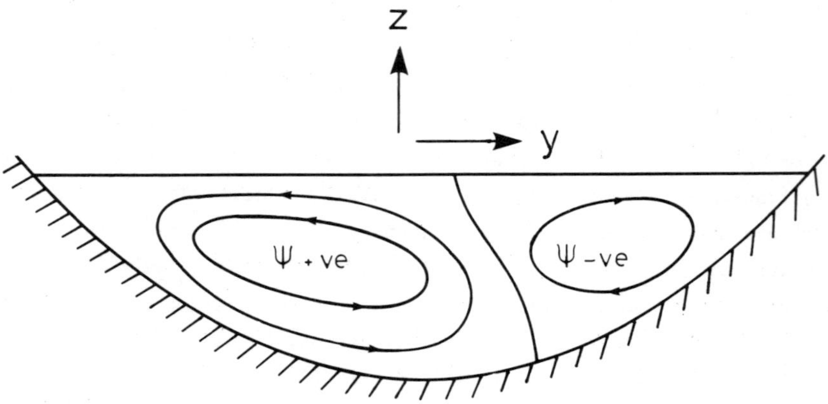

Fig. 3. Streamlines for the transverse circulation

Here κ_2 and κ_3 are the transverse and vertical eddy diffusivities for the contaminant, ν_{12} and ν_{13} are eddy viscosities for the longitudinal motion, ζ is the height of the free surface, g the gravitational acceleration, ν_{22} ν_{23} and ν_{33} are eddy viscosities for the transverse circulation, αg is the reduced gravity (positive for a buoyant contaminant), R is the radius of curvature of the channel, f the Coriolis parameter and h the water depth. Without loss of generality $c^{(0)}$ has been identified with $\bar{c}$, and $u^{(0)}$ is required to have cross-sectional average $\bar{u}$.

For any given channel it is not to be expected that the buoyancy, curvature and Coriolis effects should all significantly modify D . Thus, the specific ε-scalings derived and used above would not necessarily be appropriate. Fortunately, the final results (5.4) have a much wider range of validity than might be inferred from their derivation. The reason for this is that by retaining as many physical effects as possible the analysis has tended to retain negligibly small terms rather than to neglect potentially large terms. A careful reappraisal of the approximations enables us to show that if

$fB/U = O(\varepsilon^{f'})$, $R/B = O(\varepsilon^{-r'})$, $\alpha g/BU = O(\varepsilon^{\alpha'})$,

then it is necessary that at least one of the exponents should lie in the regions shown in figure 4, and the other exponents can lie to the right of the regions. The permitted range of the reduced gravity is more restrictive than that in S due to the present neglect of the longitudinal buoyancy-driven current.

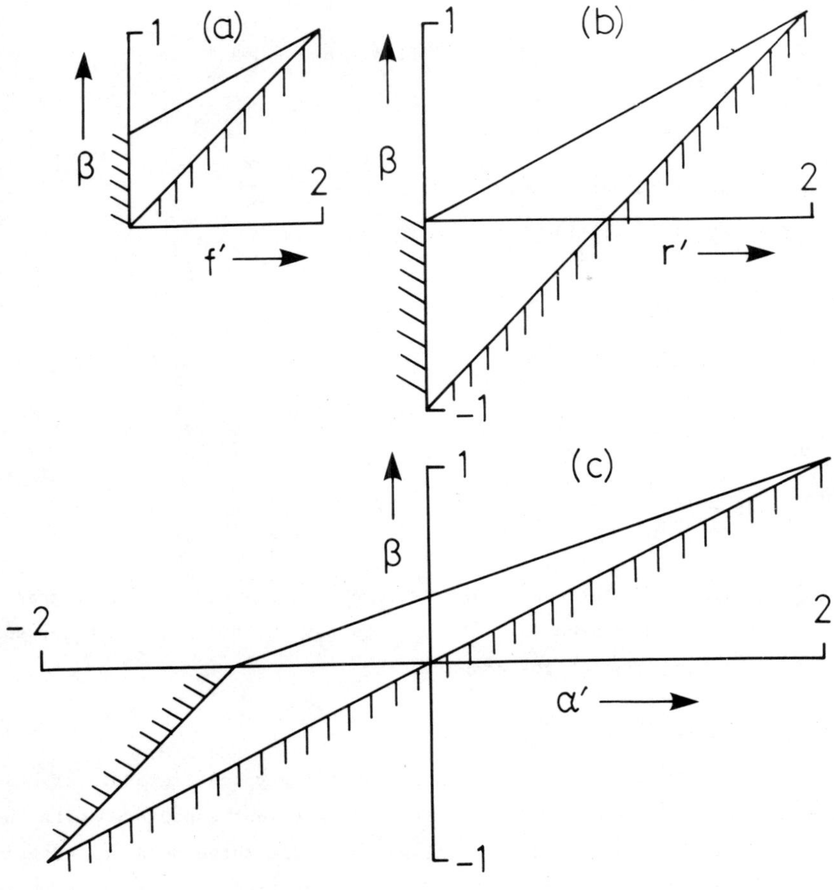

Fig. 4. Range of ε-exponents of (a) the Coriolis parameter, (b) the channel curvature, and (c) the reduced gravity, for the validity of equations (4,5).

SHALLOW WATER APPROXIMATION

Continuing as in S , we make a final simplification of the equations of motion by invoking assumption (iv). To do this we introduce a second small parameter

$$\delta = H/B$$

and we define a new vertical coordinate

$$z^* = \delta^{-1}z,$$

where H is a typical channel depth. The δ-scalings which are compatible with assumptions (ii) and (iii) are

$$\kappa_i = \delta\kappa_i^* \ , \quad \nu_{ij} = \delta\nu_{ij}^* \ , \quad c^{(1)} = \delta^{-1}c^{(1)*} \ , \quad \psi = \delta^{-1}\psi^* \ , \quad R = \delta R^* \ ,$$
$$f = \delta^{-1}f^* \ .$$

A regular perturbation expansion

$$c^{(1)} = c_o^{(1)} + \delta c_1^{(1)} + \delta^2 c_2^{(1)} + \ldots\ldots,$$

leads to the results

$$u_o^{(0)} = -g\partial_x\zeta\int_{-h}^{z}\frac{\zeta - z'}{\nu_{13}}\,dz' \ , \tag{6a}$$

$$\psi_o = \int_{-h}^{z}dz'\int_{-h}^{z'}\frac{dz}{\nu_{23}}\int_{z''}^{\zeta}dz'''\{\alpha g(\zeta - z)\partial_y c^{(1)} + \frac{u_o^{(o)2}}{R} - fu_o^{(o)} - \tilde{\psi}\} \tag{6b}$$

$$\partial_y c_o^{(1)}\{\int_{-h}^{\zeta}\kappa_2 dz + \int_{-h}^{\zeta}\frac{\psi_o^2}{\kappa_3}dz\} = \partial_x\bar{c}\int_{y-}^{y}dy'\int_{-h}^{\zeta}(u_o^{(o)} - \bar{u})\,dz. \tag{6c}$$

Here $u_o^{(0)}$ is required to have cross-sectional average $\bar{u}$, the integration constant $\tilde{\psi}(x,y,t)$ is chosen so that ψ_o is zero at the free surface, and $y_\pm$ denote the two sides of the channel.

QUALITATIVE ASPECTS OF THE FLOW

Equations (6a,b,c) admit of straightforward physical interpretations. The longitudinal current is driven by the pressure gradient associated with the slope of the free surface, and any influences of the three physical effects are relegated to higher order correction terms. In contrast, the expression (6b) for the transverse circulation is a linear sum of buoyancy curvature and Coriolis terms. Near the free surface the direction of these contributions to the circulation is respectively from less dense to more dense fluid, towards the outside curve of bends, and to the right of the flow direction. Near the channel bed there is a slower but thicker region of return flow. Finally, the variation in the concentration across the channel is determined by a balance between transverse diffusion as augmented by the transverse circulation and the steepening of the transverse gradients due to the velocity shear. As was noted by Fischer (1969), the augmented transverse diffusion agrees with Elder's (1959) results for shear dispersion in a transverse rather than a longitudinal context.

Our intention is to derive an equation involving only $\bar{c}$ and its partial derivatives. Thus, $\partial_y c_o^{(1)}$ is to be regarded as a derived quantity and, unless buoyancy is negligible, the linearity of equation (6b) for the transverse circulation becomes illusory. Indeed, if the expression for ψ_o is substituted into equation (6c) and the integrations are performed, then there results a cubic equation for $\partial_y c_o^{(1)}$. This means that $\partial_y c_o^{(1)}$ and hence ψ_o depend in a strongly nonlinear way upon f, R and $\partial_x \bar{c}$. In extreme circumstances the cubic can have multiple roots, but strictly this is beyond the range of the present analysis which implicitly assumes continuous dependence of all flow variables.

The stronger the transverse circulation the weaker is the transverse concentration gradient. This fact enables us to make several qualitative deductions concerning the transverse density variations. When buoyancy and rotation are the main effects then, in the Northern Hemisphere, concentration gradients are most pronounced when the contaminant is more dense to the left of the flow direction. This asymmetry tends to lead to the density being greater on the left bank. Similarly, when buoyancy and curvature are the main effects the contributions to ψ_o are in opposition to each other if the lighter fluid is towards the outer curve of the bend. Thus, the tendency is for the density to be greater on the inner curve. Dyer (1977) presents arguments to suggest that when there is significant vertical stratification the opposite occurs. The remaining combination, of rotation and curvature, has most pronounced concentration gradients for right hand bends. However, if the channel depth profile is symmetric then the density distribution is likewise symmetric.

DISPERSION COEFFICIENT

Substituting the result (6c) into equation (5), we find that $\bar{c}$ does indeed satisfy equation (1). The appropriate value of the dispersion coefficient is given by the formula

$$D = \frac{1}{A}\int_{y-}^{y+}\{\int_{y-}^{y} dy'\int_{-h}(u_o^{(o)} - \bar{u})dz\}^2\{\int_{-h}^{\zeta}\kappa_2 dz + \int_{-h}^{\zeta}\frac{\psi_o^2}{\kappa_3}dz\}^{-1} dy \qquad (7)$$

Two points are of immediate note. First, any transverse circulation reduces the dispersion relative to that of a neutrally-buoyant contaminant in a straight non-rotating channel (this explains the drop in D along an estuary where the salinity gradient is largest). Secondly, even if buoyancy is negligible the way that ψ_o occurs in equation (7) precludes D from being decomposed into separate contributions associated with curvature and rotation. If buoyancy is not negligible then the complexity is compounded by the nonlinear dependence of ψ_o upon the physical effec

Eddy diffusivities are at best an approximation to actual turbulent transports. Thus, there is necessarily some arbitrariness in the functional forms of the four eddy diffusivities κ_2, κ_3, ν_{13}, ν_{12}. Fortunately, D is given as an integral and is not sensitive to the details of the modelling of the turbulence. A particularly simple, and fairly realistic, model is given by the formulae

$$\kappa_i = \kappa_i' \, |\hat{u}_o^{(o)}| (h + \zeta) \quad , \qquad \nu_{ij} = \nu_{ij}' \, |\hat{u}_o^{(o)}| (h + \zeta) \; , \tag{8}$$

where $\hat{u}$ is the velocity at the free surface, and κ_i' ν_{ij}' are empirical constants (Talbot & Talbot 1974). Using equation (6a) we find that an equivalent model is

$$\kappa_i = \frac{3}{2}\kappa_i' \, \bar{u} \, (h + \zeta)^{3/2} \, A / \int_{y-}^{y+} (h + \zeta)^{3/2} \, dy \; .$$

A similar model in which the eddy diffusivities are proportional to the local depth was used by Fischer (1972).

For definiteness we restrict attention to channels of parabolic cross section:

$$\zeta = 0 \quad , \quad h = H\left[1 - (y \,/\, B)^2\right] \quad , \quad -B \leq y \leq B \; ,$$

and we specify the parameter values

$$\kappa_2' = .02 \qquad , \qquad \kappa_3' = \nu_{23}' = .005. \tag{9}$$

The appropriate non-dimensional measures for the longitudinal dispersion coefficient and for the curvature, buoyancy and Coriolis effects are

$$DH/B^2 |\bar{u}| \;\; , \;\; H/R \;\; , \;\; G = \alpha g H B \partial_x c / |\bar{u}| \bar{u} \;\; , \;\; fH/\bar{u}.$$

Figures (5a-e) show the dependence of the longitudinal dispersion coefficient upon both buoyancy and rotation for various values of the curvature. To obtain results for curvature of negative sign it suffices to reverse the sign of $fH/\bar{u}$. For non-symmetric depth profiles it would be necessary to reverse the sign of G .

It is noteworthy that while each effect separately reduces the longitudinal dispersion their interactions can increase the dispersion. For example, in figure (5a) with $fH/\bar{u} = .03$ the effect of increasing the buoyancy gradient is first to increase D and subsequently to reduce D . This, and several other qualitative features of figures (5a-e), can be explained in terms of the contributions of the three physical effects to ψ_o . From equation (6c) we see that the transverse density gradient reverses sign at the centre of the channel. Thus, buoyancy tends to lead to two circulation cells, as sketched in figure (2). This is unlike rotation and curvature which tend to lead to a single circulation cell. The effect of increasing buoyancy from zero is to increase ψ_0 at one side of the channel and reduce it at the other side. Whether this increases or decreases D depends upon whether the circulation or turbulence term dominates in the formula (7). Thus, in figure (5a) buoyancy decreases the longitudinal dispersion when the Coriolis effect is weak, but buoyancy increases the dispersion when the Coriolis effect is strong (e.g. with $fH/\bar{u} = .03$). In figures (5b–e) the centrifugal effect is progressively stronger and the region of weak circulation is moved upwards (i.e. in the upper half plane of figures (5b–e) curvature and rotation are in opposition to each other).

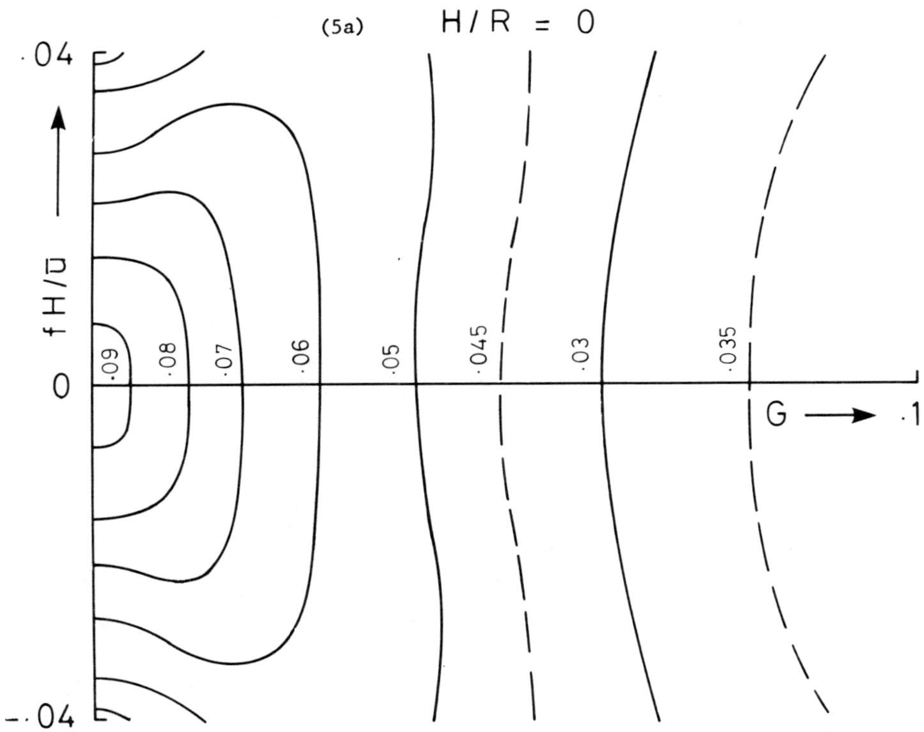

Fig. 5 Contours of the non-dimensional dispersion coefficient $DH/B^2|\bar{u}|$ for straight and curved channels of parabolic cross-section.

(5b) H/R = ·005

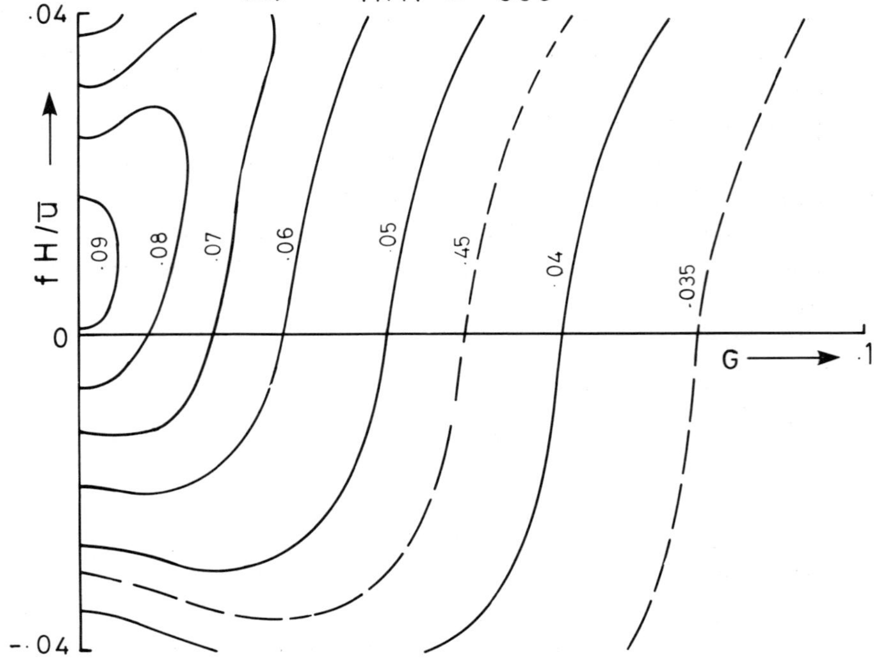

(5c) H/R = ·01

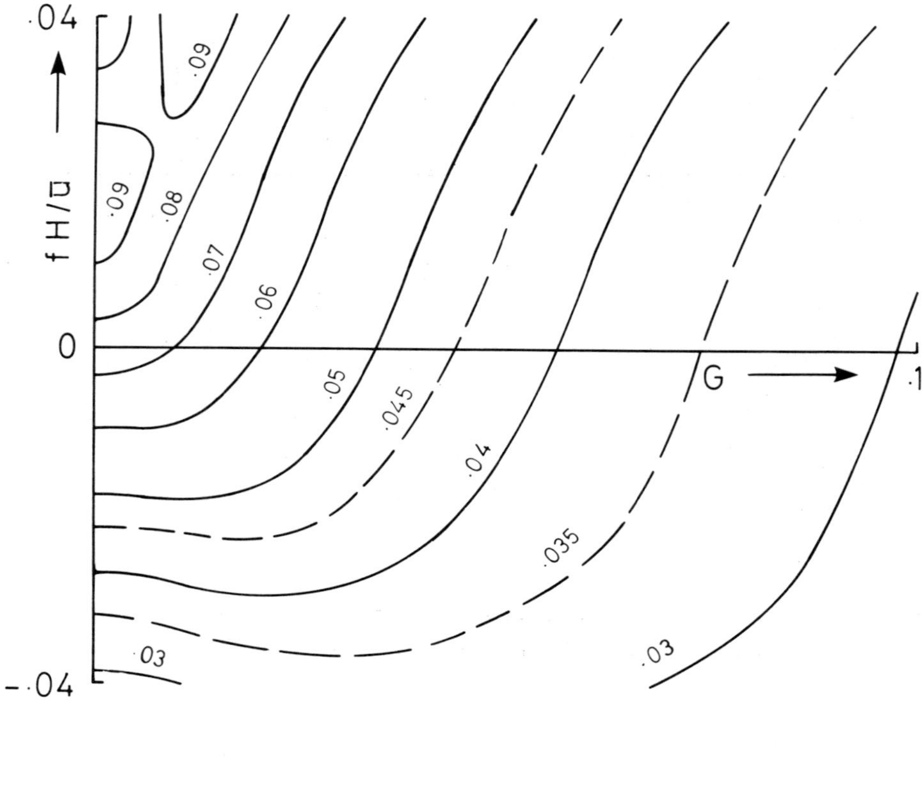

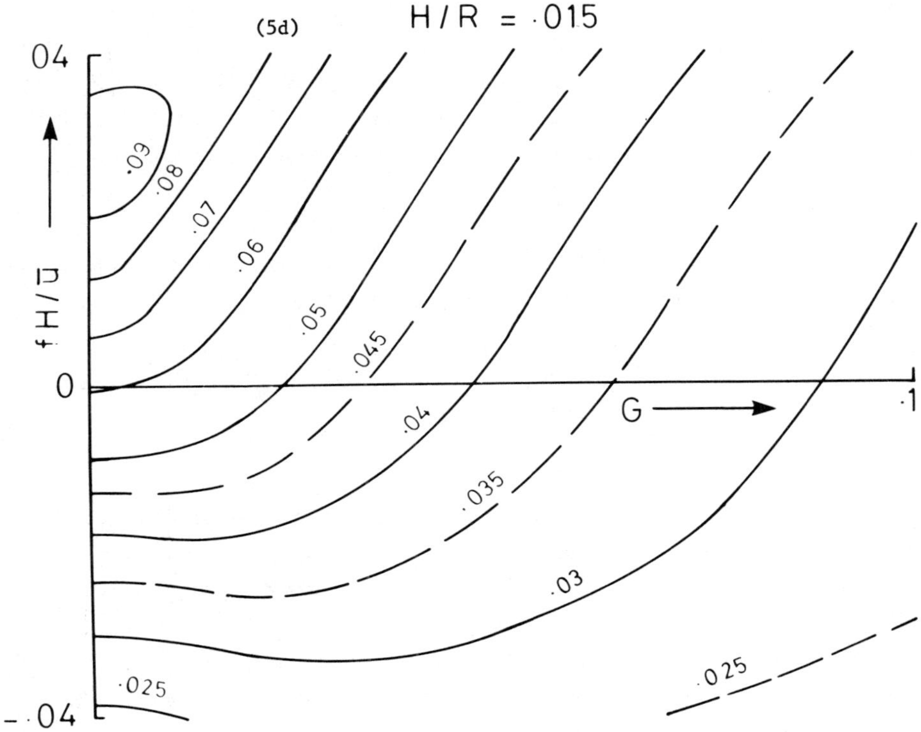
(5d)
H / R = ·015
·04
-·04
0
fH/ū
G
·1
·09
·08
·07
·06
·05
·045
·04
·035
·03
·025

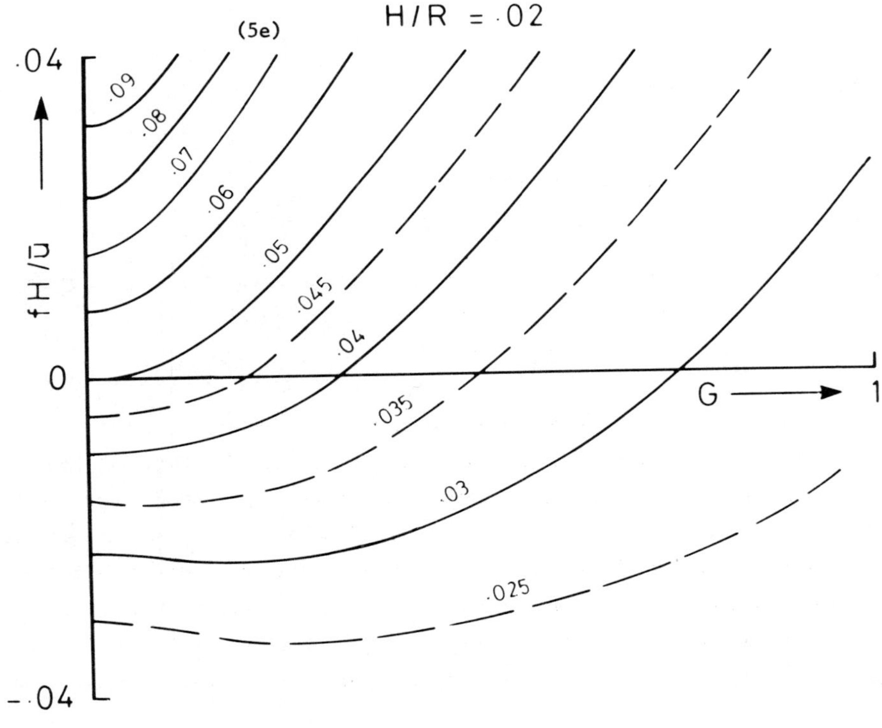
(5e)
H / R = ·02
·04
-·04
0
fH/ū
G
1
·09
·08
·07
·06
·05
·045
·04
·035
·03
·025

USE OF FIGURES 5a-e

If the above parameterisation (8.9) of the turbulence is accepted, then it is straightforward to use figures (5a-e) to get quantitative predictions of the dispersion coefficient. First, the depth-to-radius-of-curvature ratio H/R determines which of the figures is relevant. For example, with

$$H = 10 \text{ m} \quad , \quad R = 500 \text{ m}$$

it is figure (5e) that must be used. Secondly, the buoyancy gradient and the flow rate determine the appropriate position on the figure . For example, with

$$\partial_x \bar{c} = 10^{-4}\ {}^{o}\text{C m}^{-1} \ , \ B = 100\text{m}, \ \bar{u} = .1 \text{ m s}^{-1}, \ f = 10^{-4} \text{ s}^{-1},$$

the coordinate position in figure (5e) is

$$G = .01 \quad , \quad fH/\bar{u} = .01$$

Thus, for a right-handed bend the longitudinal dispersion coefficient is 5.8 $m^2 s^{-1}$. For a left-handed bend the sign of the Coriolis term has to be reversed and the dispersion coefficient is 3.9 $m^2 s^{-1}$. In a tidal flow the two different values of D would correspond to the two flow directions for the same bend.

If it is only the choice of the empirical constants κ_2', κ_3', ν_{23}' that is in question, then it is still possible to use figures (5a-e). The curvature parameterisation of the figure is with respect to

$$\left[2 \times 10^{4}\ \nu_{23}'(\kappa_2'\kappa_3')^{\frac{1}{2}}\right]^{-1} H/R$$

The contours are the values of

$$\left[50\kappa_2\right] DH/B^2|\bar{u}| \quad ,$$

and the coordinate axes are in terms of

$$\left[10^6\kappa_2'\nu_{23}'(\kappa_2'\kappa_3')^{\frac{1}{2}}\right]^{-2}\alpha g\partial_x\bar{c}BH/|\bar{u}|\bar{u} \quad , \left[2 \times 10^4\nu_{23}(\kappa_2'\kappa_3')^{\frac{1}{2}}\right]^{-1} fH/\bar{u} \ .$$

For more complicated models of the eddy diffusivities it would be necessary to calculate the dispersion coefficient directly from equations (6,7).

CONCLUDING REMARKS

A systematic derivation has been given of the flow and of contaminant dispersion in narrow channels. Allowance has been made for Coriolis, curvature and buoyancy effects. It has been found that there is a tendency for the fluid to be slightly more dense on the inner curve of bends and in the Northern Hemisphere to the left of the flow direction. Although the three physical effects contribute additively to the flow the same is not true of the dispersion coefficient. For a parabolic channel, with a simple model of the turbulent diffusivities, numerical results have been calculated. These results permit quantitative predictions of the dispersion coefficients and show that all three effects can significantly modify the dispersion.

ACKNOWLEDGEMENT

The author would like to thank C.E.G.B. for financial support.

REFERENCES

Csanady, G.T. 1973: Turbulent diffusion in the environment. D. Reidel, Dordrecht-Holland.

Dyer, K.R. 1977: Lateral circulation effects in estuaries. N.A.S. Symposium on Geophysics of Estuaries (In press).

Elder, J.W. 1959: The dispersion of marked fluid in turbulent shear flow. J. Fluid Mech. 5, 544-560.

Erdogan, M.E. & Chatwin, P.C. 1967: The effects of curvature and buoyancy upon the laminar dispersion of solute in a horizontal tube. J. Fluid Mech. 29, 464-484.

Fischer, H.B. 1967: The mechanics of dispersion in natural streams. J. Hydraul. Div. A.S.C.E. 93, 187-216.

Fischer, H.B. 1969: The effect of bends on dispersion in streams. Water Resour. Res. 5, 496-506.

Fischer, H.B. 1972: Mass transport mechanisms in partially stratified estuaries. J. Fluid Mech. 53, 671-687.

Fischer, H.B. 1973: Longitudinal dispersion and turbulent mixing in open channel flow. Ann. Rev. Fluid Mech. 5, 59-78.

Fischer, H.B. 1976: Mixing and dispersion in estuaries. Ann. Rev. Fluid Mech. 8, 107-133.

Harleman, D.R.F. & Thatcher, M.L. 1974: Longitudinal dispersion and unsteady salinity intrusion in estuaries. Houille Blanche 29, 25-33.

Holley, E.R., Harleman, D.R.F. & Fischer, H.B. 1970: Dispersion in homogeneous estuary flow. J. Hydraul. Div. A.S.C.E. 96, 1691-1709.

Imberger, J. 1976: Dynamics of a longitudinally stratified estuary. Proc. 15th Int. Conf. Coastal Eng. Hawaii (to appear).

Okubo, A. 1973: Effect of shoreline irregularities on streamwise dispersion in estuaries and other embayments. Neth. J. Sea Res. 6, 213-224.

Pritchard, D.W. 1952: Salinity distribution and circulation in the Chesapeake Bay estuarine system. J. Mar. Res. 11, 106-123.

Smith, R. 1976: Longitudinal dispersion of a buoyant contaminant in a shallow channel. J. Fluid Mech. 79, 677-688.

Talbot, J.W. & Talbot, G.A. 1974: Diffusion in shallow seas and in English coastal and estuarine waters. Rapp. P.-v. Reun. Cons. int. Explor. Mer. 167, 93-110.

Taylor, G.I. 1953: Dispersion of soluble matter in solvent flowing slowly through a tube. Proc. Roy. Soc. A 219, 186-203.

HOW SOME NEW FUNDAMENTAL RESULTS ON RELATIVE TURBULENT DIFFUSION CAN BE RELEVANT IN ESTUARIES AND OTHER NATURAL FLOWS

P.C. CHATWIN and PAUL J. SULLIVAN*
Department of Applied Mathematics and Theoretical Physics, University of Liverpool, Liverpool, England.

ABSTRACT

This paper deals with the relative turbulent diffusion of a cloud of passive contaminant, and in particular with the ensemble mean concentration $C(\underline{y},t)$ and the statistics of the fluctuation of concentration $c(\underline{y},t)$. Recent work by the authors (Chatwin and Sullivan, 1978a), using the fundamental equations (without eddy diffusivities), has led to results which are in many ways quite different from those presented e.g. in Csanady (1973). Among those results summarized here are new estimates of the magnitudes of C and $\overline{c^2}$ throughout the cloud, a proposal that there is a small core region surrounding the centre in which both C and $\overline{c^2}$ are much larger than elsewhere, a physical explanation of how the distributions of C, $\overline{c^2}$ and certain correlations and spectra can be self-similar outside the core in isotropic turbulence, and some speculations about the profound role of molecular diffusion. Some experimental confirmation of the physical picture leading to these results has also been obtained (Chatwin and Sullivan, 1978b). The main part of the present paper is an examination of how far these results can be used in the complicated velocity fields occurring in estuaries and other natural flows. In particular, certain time scales are obtained which measure departures of the velocity field in the neighbourhood of the cloud from isotropy, homogeneity and statistical stationarity. It is argued that for times since release less than the smallest of these, the effects of anisotropy, inhomogeneity and statistical unsteadiness of the velocity field on the dispersion of the cloud can be incorporated in the description summarized above. For the Tay estuary this minimum time is about 3 hours.

1. INTRODUCTION

Suppose that at a fixed time in a natural flow, specified for example in an estuary by the interval t_o since the last high tide, a quantity M of passive

*Permanent address: Department of Applied Mathematics, University of Western Ontario, London, Canada.

contaminant is released. Suppose further that the centre of mass of the cloud of contaminant is initially at $\underline{x}_0$ and that it is distributed in a specified way about $\underline{x}_0$. Let the distribution of concentration at any time $t>t_0$ be $\Gamma(\underline{x},t)$. Because the velocity field is turbulent the value of Γ is a random variable whose statistica properties depend both on $\underline{x}$ and t. Indeed these properties also depend explicitly on $\underline{x}_0$ and t_0, but, for brevity, the notation in this paper will not indicate such dependence.

The statistical properties of Γ considered in this paper will all be ensembl averages over the set of realizations of the dispersing cloud, in each one of which the release conditions, including the time and position of release, are identical. In each realization axes are taken whose origin coincides with the centre of mass of the cloud throughout the realization. When these axes are employed, the relevant flow dynamics is contained in the velocity $\underline{T}(\underline{y},t)$ of a point relative to that of the centre of mass, where $\underline{y}$ measures the position of the point relative to the centre of mass. The velocity field is determined by the momentum equations, which are independent of the contaminant when this is passive.*

The physical processes responsible for the dispersion of a cloud are summari in section 2. With the axes used, this description is in the framework of relative diffusion. Although some of the results apply when the less practically important framework of absolute diffusion is used, the latter is more difficult to discuss in detail because it is affected by the anisotropic large eddies of the turbulence whose structure is not governed by the inertial subrange. With relative diffusion however these large eddies are normally insignificant, at least initially. The relative diffusion of a cloud of contaminant is of basic physical interest and importance, not least because of the light it throws upon the structure of the velocity field. It is also of great practical significance. One example is the accidental spillage of a toxic chemical when the statistical properties of Γ, including its fluctuations, must be analyzed within the framework of relative diffusion (to avoid the artificial smearing which occurs with absolute diffusion). These properties determine the practical probabilities that are needed, for example the probability that, for a given period of time, the maximum concentration exceeds a given lethal value. The importance of the fluctuations in such a case is evident since their root mean square value is normally at least of the same order as the ensemble mean concentration and often much greater. A different practical problem is the way in which sound and light are absorbed and transmitted by the cloud, and this is naturally analyzed in terms of the spectral decomposition of Γ (Batchelor, 1959). Such factors may significantly affect the biological growth of a patch of plankton, for example. In view of the widespread occurrence of such

* Although some of the theoretical results apply also when it is not passive.

important practical problems, section 3 of this paper analyzes conditions under which the description given in section 2 can be used in the complicated flows existing in nature.

2. THE EVOLUTION OF A CLOUD IN INCOMPRESSIBLE FLOW

The equation governing Γ in each realization is

$$\frac{D\Gamma}{Dt} \equiv \frac{\partial \Gamma}{\partial t} + \underline{T}.\nabla\Gamma = \kappa\nabla^2\Gamma, \tag{1}$$

where $\underline{T} = \underline{T}(\underline{y},t)$ is the relative velocity field defined in section 1, and satisfies the incompressibility condition

$$\nabla.\underline{T} = 0. \tag{2}$$

In (1), κ is the molecular diffusivity. An immediate, and otherwise obvious, consequence of (1) and (2) is that the total amount of contaminant is constant, i.e.

$$\int\Gamma(\underline{y},t)\ d^3\underline{y} = M. \tag{3}$$

It is now useful to follow the normal practice and write both $\underline{T}$ and Γ as the sum of their ensemble means, $\underline{U} = \underline{U}(\underline{y},t)$ and $C = C(\underline{y},t)$ respectively, and fluctuations, $\underline{u} = \underline{u}(\underline{y},t)$ and $c = c(\underline{y},t)$ respectively. Thus, denoting an ensemble mean by an overbar,

$$\begin{aligned} &\underline{T} = \underline{U} + \underline{u}\ ;\ \overline{\underline{T}} = \underline{U}\ ;\ \overline{\underline{u}} = \underline{0}. \\ &\Gamma = C + c\ ;\ \overline{\Gamma} = C\ ;\ \overline{c} = 0. \end{aligned} \tag{4}$$

From (1) to (4) there follow in the usual way:

$$\int C\ d^3\underline{y} = M\ ;\ \int c\ d^3\underline{y} = 0\ ; \tag{5}$$

$$\frac{\partial C}{\partial t} + \underline{U}.\nabla C + \nabla.(\overline{\underline{u}c}) = \kappa\nabla^2 C\ ; \tag{6}$$

$$\frac{\partial \overline{c^2}}{\partial t} + \underline{U}.\nabla\overline{c^2} + \nabla.(\overline{\underline{u}c^2}) + 2\overline{\underline{u}c}.\nabla C = \kappa\nabla^2\overline{c^2} - 2\kappa\overline{(\nabla c)^2}. \tag{7}$$

Equations (6) and (7) are coupled through the terms in $\overline{\underline{u}c}$, in particular. In (7), the second and third terms on the left hand side are the effect of turbulent transfer while the fourth term represents production by feeding from the field of C. Of the two molecular terms on the right hand side, the first is ordinary diffusion but the second is everywhere negative and represents dissipation.

Results when $\kappa = 0$ - a cloud of marked fluid particles

Consider first a case when the effects of κ are negligible, which may be a good approximation in the early stages of the dispersion. With $\kappa = 0$, it follows from (1) and (2) that

$$2\Gamma\frac{D\Gamma}{Dt} = \frac{D\Gamma^2}{Dt} = 0 \text{ so that } \int \Gamma^2 d^3\underline{y} = \text{constant}. \tag{8}$$

Thus, using (4),

$$\int C^2\ d^3\underline{y} + \int \overline{c^2}\ d^3\underline{y} = \frac{M^2}{L_o{}^3}\ , \tag{9}$$

where L_o is a length determined entirely by the initial conditions. Now suppose that at time t the cloud is, on the average, spread over a volume of order $L^3(t)$, where

$$L^2(t) = M^{-1} \int |\underline{y}|^2 C \, d^3\underline{y}. \tag{10}$$

By mass conservation, as expressed by the first equation in (5), it follows that, at least in most of the cloud, C is of magnitude of order $M/L^3(t)$. Since it is obvious that $L(t) \to \infty$ as $t \to \infty$ it follows that $\int C^2 d^3\underline{y} \to 0$ as $t \to \infty$. Thus it follows from (9) that $\int \overline{c^2} \, d^3\underline{y} \to M^2/L_o^3$ as $t \to \infty$. Hence $\overline{c^2}$ cannot, in most of the cloud, have a magnitude of order M^2/L^6 as mistakenly assumed by Csanady (1973, p.236) and others. Rather (Chatwin and Sullivan, 1978a), the magnitude of $\overline{c^2}$ in most of the cloud is of order $M^2/L_o^3 L^3(t)$, consistent with (9). Granted these estimates of the magnitudes of C and $\overline{c^2}$, it follows that, _if_ their distributions are self-similar, they must be of the form (Chatwin and Sullivan, 1978a):

$$C = \frac{M}{L^3} f\left(\frac{|\underline{y}|}{L}\right) = \frac{M}{L^3} f(Y) \; ; \tag{11}$$

$$\overline{c^2} = \frac{M^2}{L_o^3 L^3} g\left(\frac{|\underline{y}|}{L}\right) = \frac{M^2}{L_o^3 L^3} g(Y). \tag{12}$$

The conditions under which self-similarity can be expected are discussed in Chatwin and Sullivan (1978a) and, from a different viewpoint, in section 3 of this paper. It is however important to emphasize that, even when there is no self-similarity, the _magnitudes_ of C and $\overline{c^2}$ are as in (11) and (12) over most of the cloud.

However these magnitudes cannot hold over the whole cloud. There is a core region of small volume surrounding the centre within which the magnitudes of C and $\overline{c^2}$ are at least one order greater than those in (11) and (12). Within this core the probability per unit volume of finding marked fluid is much greater than outside, and the production term in (7) is comparable with the other terms on the left hand side, in contrast with what happens outside the core when, according to (11) and (1 production of $\overline{c^2}$ is negligible compared with its transfer. The detailed structure of the core region depends on the small scale structure of the turbulence but in one case, examined in detail in Chatwin and Sullivan (1978a), the magnitudes of C and $\overline{c^2}$ within the core were of order M/L_o^3 and M^2/L_o^6 respectively, while its radius was of order L_o^3/L^2. It seems likely that the core radius will always be far too small to measure directly in experiments, whatever the small scale structure of the turbulence. However experiments on a plume emitted from a steady source (Chatwin and Sullivan, 1978b), while not being able to explore the core directly, showed that the dispersion was dominated by the central portion of the plume, consistent with the existence of a core, and that the dominant mechanism causing spreading was transfer from and to the central region by eddies of the velocity field with a length scale of order L (a wave number of order L^{-1}).

The effects of molecular diffusion

The dominant transfer mechanism by eddies with a length scale of order L is part of a cascade process in which interactions between eddies of the velocity field with components of the fluctuating concentration of comparable size produce smaller scale components and so on. Thus, as time evolves, more and more of the fluctuating concentration field is contained in components of small length scale (high wave number). One result of this process is the shrinking of the core radius with time. However this process also results in the increase of the surface area of the cloud and the increase of the gradients of Γ, and thus an increase in the effect of molecular diffusion with time. Eventually a balance is set up in which molecular diffusion effectively prevents the existence of components of the fluctuating concentration field with length scales smaller than the conduction cut-off length $(\kappa^2\nu/\varepsilon)^{\frac{1}{4}}$, where ν is the kinematic viscosity and ε is the rate of dissipation of kinetic energy per unit mass (Batchelor, 1952b). This balance occurs throughout the cloud in each realization, but, when ensemble means like C and $\overline{c^2}$ are considered, its effect is significantly greater in the core than elsewhere because of the far greater probability per unit volume of finding marked fluid there. Thus the primary effect of κ will be felt in the core, and through the dissipation term in (7), will be to reduce the magnitude of $\overline{c^2}$ there. The result of this effect is then transferred outwards to the region outside the core. In this region the direct effects of κ are negligible, and there is no reason why the shapes of the distributions of C and $\overline{c^2}$ there should be affected by it. Nor, by mass conservation, can the magnitude of C in most of the cloud be different from M/L^3. Thus, if (11) holds when the effects of κ are negligible, so does it also when they are not. However the magnitude of $\overline{c^2}$ can no longer be of order $M^2/L_o{}^3L^3$; the effect of κ transferred outwards from the core causes it to decrease faster than this. Thus, if (12) holds when the effects of κ are negligible, it must be replaced by

$$\overline{c^2} = \frac{M^2}{L_o{}^3L^3}g(Y) \times \eta(t,\kappa,\nu,\varepsilon), \tag{13}$$

when these effects are important. In (13), η tends to zero as $t \to \infty$ and depends strongly on κ. Evidence to reinforce this description is given in Chatwin and Sullivan (1978a, 1978b).

Useful correlations and their associated spectra

It is particularly valuable to use correlation functions and their spectra to describe the distribution of the fluctuating concentration field not least because of the practical advantage that experimental noise can be filtered out from the measured concentration records. Two natural correlations, both theoretically and practically, are $R(\underline{y},t)$ and $Q(\underline{y},t)$ defined by

$$R(\underline{y},t) = \overline{c(\underline{o},t)\ c(\underline{y},t)} \tag{14}$$

and

$$Q(\underline{y},t) = \int \overline{c(\underline{y}',t)\ c(\underline{y}+\underline{y}',t)}\ d^3\underline{y}'. \tag{15}$$

Thus R is a measure of the coherent structure of the fluctuating concentration field in relation to that at the centre while Q is a measure of coherent structure throughout the whole cloud. The correlations corresponding to R and Q in a plume were shown in Chatwin and Sullivan (1978b) to be self-similar, and of almost the same form, consistent with the existence of a core region. From the second of (5), it follows immediately that

$$\int R(\underline{y}, t)\ d^3\underline{y} = \int Q(\underline{y},t)\ d^3\underline{y} = 0, \tag{16}$$

and the corresponding results for a plume were verified experimentally in Chatwin and Sullivan (1978b).

It can be shown that both $\int R\ d^3\underline{y}$ and $\int Q\ d^3\underline{y}$ are finite so that R and Q have Fourier transforms, $\phi(\underline{k},t)$ and $\psi(\underline{k},t)$ respectively, defined by

$$\phi(\underline{k},t) = \int R(\underline{y},t)\ \exp(-i\underline{k}.\underline{y})\ d^3\underline{y} \tag{17}$$

and

$$\psi(\underline{k},t) = \int Q(\underline{y},t)\ \exp(-i\underline{k}.\underline{y})\ d^3\underline{y}. \tag{18}$$

From the inverse of (17) it follows that

$$\overline{c^2}(\underline{o},t) = (2\pi)^{-3} \int \phi(\underline{k},t)\ d^3\underline{k} = R(\underline{o},t), \tag{19}$$

so that $|\phi(\underline{k},t)|$ is the magnitude of the Fourier components representing $R(\underline{y},t)$. Similarly, from the inverse of (18),

$$\int \overline{c^2}(\underline{y},t)\ d^3\underline{y} = (2\pi)^{-3} \int \psi(\underline{k},t)\ d^3\underline{k} = Q(\underline{o},t). \tag{20}$$

Since, in each realization $\int c(\underline{y}',t)\ c(\underline{y}+\underline{y}', t)\ d^3\underline{y}'$ is a normal correlation function with a Fourier transform which provides an energy spectrum, it follows by linearity that $\psi(\underline{k},t)$ is the average energy spectrum. Both $\phi(\underline{k},t)$ and $\psi(\underline{k},t)$ contain far more detail than could reasonably be measured in an experiment, when one would probably attempt to find their values on the spheres $|\underline{k}|$ = constant. However the spectra corresponding to ϕ and ψ for a plume were measured in Chatwin and Sullivan (1978b) and found to be self-similar and nearly Gaussian.

An important application of Q and ψ is to the complete spectral decomposition of the cloud which is obtained by taking the Fourier transform of the relation (see (13) of Sullivan (1975)):

$$\int \overline{\Gamma(\underline{y}',t)\ \Gamma(\underline{y}+\underline{y}',t)}\ d^3\underline{y}' = \int C(\underline{y}',t)\ C(\underline{y}+\underline{y}',t)\ d^3\underline{y}' + Q(\underline{y},t).$$

This equation is extremely important for processes, like absorption and transmission of light and sound, that depend on the instantaneous concentrations, not only the fluctuations.

3. THE EFFECTS OF ANISOTROPY, INHOMOGENEITY AND UNSTEADINESS

Many features of the description just given seem certain to apply in any turbulent velocity field, in particular the estimates of the magnitudes of C and $\overline{c^2}$, the existence and importance of the core region, and the effects of κ. However the particular form of the velocity field determines the shapes of the distributions of C and $\overline{c^2}$, and how L(t) varies with time. For the special case of a statistically stationary Eulerian velocity field at high Reynolds number the inertial subrange may extend from metres to tens of kilometres (Richardson, 1926; Ozmidov, 1960). Provided L(t) lies within the inertial subrange and is significantly smaller than length scales typical of inhomogeneities in the velocity field (e.g. the depth and width of an estuary), the dispersion is governed by eddies (of length scales of order L(t)) whose statistical structure is isotropic and homogeneous, and determined entirely by ε. Under these circumstances the distributions of C and $\overline{c^2}$ outside the core must have the self-similar forms (11) and (13), and L(t) will be given approximately by (Batchelor, 1952a):

$$L(t) \simeq \alpha \sqrt{\varepsilon (t-t_0)^3} , \qquad (21)$$

where α is a universal dimensionless constant of order unity.

But in many natural flows these rather special conditions do not hold so that it is pertinent to enquire whether the results that they give can be extended to apply in such flows. The extensions considered here apply for times after release which are _not_ large compared with the time taken for mixing over the cross-section. Therefore they do _not_ lead to the description in terms of an effective longitudinal coefficient which was provided for absolute diffusion by Taylor (1954). In passing it is worth mentioning that in many natural flows it is doubtful in any case whether this description is ever properly justified (Chatwin, 1975, 1976; Dewey and Sullivan, 1977).

The essential point underlying the extensions to be discussed here is that from the point of view of an observer moving with the cloud, the correct viewpoint for relative diffusion, all the possible complications occurring in natural flows are observed as statistical unsteadiness in the relative velocity field $\underline{T}(\underline{y},t)$. There are four principal sources of such unsteadiness which are in general not independent. These are:

(i) unsteadiness in the mean Eulerian velocity field $\underline{V}(\underline{x},t)$ like that due to the tidal effect in estuaries - here $\underline{x}$ is measured relative to a fixed origin;

(ii) changes that arise from the cloud being convected into regions with different topography and density distributions;

(iii) the statistical sampling by the cloud of regions of different turbulence structure over the flow cross-section which is caused by convection by large eddies;

(iv) the convection of the cloud through regions of appreciable shear in $\underline{V}(\underline{x},t)$.

A time scale can be associated with each of (i) to (iv), and, provided each of these is large compared with the relative diffusion time scale T, given by

$$T = \left(\frac{1}{L^2}\frac{dL^2}{dt}\right)^{-1} , \qquad (22)$$

then it is reasonable to suppose that the effects of (i) to (iv) can be incorporated in the scales L(t) and $\varepsilon(t)$, but that otherwise the dispersion is as in isotropic turbulence provided the Reynolds number is sufficiently high for the inertial subrange to extend over length scales larger than L(t). This proposal seems particularly promising since observations of inertial subrange predictions like (21), reported by Richardson (1926), Ozmidov (1960) and others, have taken place in flows where to some degree all of effects (i) to (iv) are present. Under these circumstances the distributions of C and $\overline{c^2}$ will have the self-similar forms (11) and (13), where L(t) is determined (Batchelor 1952a) by

$$\frac{dL^2}{dt} \simeq 3\alpha^2\varepsilon(t)(t-t_o)^2 , \tag{23}$$

which is consistent with (21). When (23) holds it follows approximately from (22) that

$$T \simeq \frac{1}{3}(t-t_o). \tag{24}$$

Time scales associated with effects (i) to (iv)

The time scale T_1 associated with effect (i) can be defined by

$$T_1 = \left(\frac{1}{|\underline{V}|}\frac{\partial|\underline{V}|}{\partial t}\right)^{-1} , \tag{25}$$

and in many cases, including estuaries, it will be large with respect to the time scale T_2 associated with (ii). The simplest way of defining T_2 is in terms of a length scale $l(\underline{x},t)$ which measures the streamwise distance over which either the topography or the density distribution changes appreciably. In defining l, the value of $\underline{x}$ refers to the position of the centre of mass of the cloud and l may depend explicitly on t through effect (i). Having defined l, it is natural to define T_2 by

$$T_2 = \frac{l(\underline{x},t)}{V(\underline{x},t)} , \tag{26}$$

where V is the component of $\underline{V}$ in the streamwise direction.

A time scale associated with effect (iii), say T_3, is that taken for an element of contaminant to wander, by convection with large eddies, over that part of the cross-section containing appreciable changes in the turbulence structure. The shortest such time is relevant for the type of application being considered here and is obtained, in many cases including most estuaries, by determining the time taken for an element of contaminant to wander over the depth of the flow. If the local depth is $d(\underline{x},t)$ this gives (Sullivan, 1971b) an estimate of T_3 as

$$T_3 = \frac{10d(\underline{x},t)}{u_*} , \tag{27}$$

where u_* is the (local) value of the friction velocity which may also vary with $\underline{x}$ and t.

Although effect (iv), that of shear in the mean Eulerian velocity field, is clearly coupled with other effects, particularly (iii), it has been included separately because it provides a mechanism for the growth of L(t) that is rather different from that acting in isotropic homogeneous turbulence which was considered earlier, viz. the random encounters of the cloud with eddies whose length scales are of order L(t). For in the presence of mean shear there will be, on the average, a velocity difference $\Delta\underline{V}$ over the cloud giving a (vector) rate of extension of the cloud proportional to $\Delta\underline{V}$. Considering for simplicity the component ΔV of $\Delta\underline{V}$ in the streamwise direction gives $L/\Delta V$ as an estimate of T_4, the relevant time scale. In many cases the shear will be greatest in the vertical direction. Taking the z axis in the vertical allows ΔV to be estimated by $\Delta V \simeq L(\partial V/\partial z)$ provided second derivatives of V with respect to z are appropriately small, that is provided $|\partial^2 V/\partial z^2| L << |\partial V/\partial z|$. Using (21) this requires* $(t-t_o) << \{|\partial V/\partial z|/\alpha\varepsilon^{1/2}|\partial^2 V/\partial z^2|\}^{2/3}$. When this condition holds, $T_4 = L/\Delta V$ can be estimated by

$$T_4 \simeq |\partial V/\partial z|^{-1}. \tag{28}$$

Mean shear also affects the distribution of $\overline{c^2}$, for convection of the cloud by the largest eddies (especially the energy containing eddies) in regions of substantial mean shear produces fluctuations in concentration at a wave number of order $|\Delta V.T'|^{-1}$ where T' is a "turnover" time of these eddies. Of course, in the absence of mean shear, this effect is not present and the largest eddies affect the cloud only through the roles they play in effects (ii) and (iii).

The values of the time scales in some specific cases

Fischer (1972) quotes measurements by Bowden in the Mersey on the basis of which he considers that part of the estuary can be considered as a uniform channel 9 km long with a maximum depth of about 20 m and a root mean square tidal velocity of about 1 ms^{-1}. A study of the Tay estuary by Williams and West (1973) provides comparable data. The maximum depth was 25 m and the maximum tidal current velocity was about 2 ms^{-1}. Over a distance of about 10 km between Flisk and Newport the cross-sectional area of the Tay varies linearly between 5000 m^2 and 16000 m^2.

The value of T_1, defined in (25), is of course the same in both estuaries since it is due to the tide. Thus

$$T_1 \simeq 12 \text{ hrs} \simeq 4 \times 10^4 s. \tag{29}$$

For the Mersey the value of T_2, defined in (26), is infinite because there is no variation in the cross-section, while for the Tay, taking $l \simeq 10$ km and $V \simeq 1$ ms^{-1}, the estimate obtained is

$$T_2 \simeq 10^4 s. \tag{30}$$

To estimate T_3 in the absence of more detailed data from either estuary take $u_* \simeq 0.05V \simeq 0.05$ ms^{-1} and $d \simeq 20$ m. From (27) this gives

$$T_3 \simeq 4 \times 10^3 s, \tag{31}$$

for both estuaries. Unfortunately it is not possible to estimate T_4 for either estuary without detailed knowledge of the mean velocity profiles. But assuming $T_4 > T_3$, the smallest of the three times in (29) to (31), it follows from (24) that in the Tay the dispersion of the cloud should be essentially as in isotropic turbulence, extended in the way described above, provided $T << T_3$, i.e. provided $(t-t_o) << 12 \times 10^3 s \simeq 3\frac{1}{3}$ hrs.

*Note that ε is typically of order $10^{-7} m^2 s^{-3}$ (Sullivan, 1971a).

Finally it should be stressed that the estimates in (29) to (31) are very rough and based on inadequate data. In order to apply the ideas of this section in any particular case the time scales must be estimated taking into account both the details of the particular estuary and the history since release of the cloud.

REFERENCES

Batchelor, G.K., 1952a. Diffusion in a field of homogeneous turbulence. II. The relative motion of particles. Proc. Camb. Phil. Soc., 48: 345-562.

Batchelor, G.K., 1952b. The effect of homogeneous turbulence on material lines and surfaces. Proc. Roy. Soc. Lond., A213: 349-366.

Batchelor, G.K., 1959. Small-scale variation of convected quantities like temperatu in turbulent fluid. Part 1. General discussion and the case of small conductivity. J. Fluid Mech., 5: 113-133.

Chatwin, P.C., 1975. On the longitudinal dispersion of passive contaminant in oscillatory flows in tubes. J. Fluid Mech., 71: 513-527.

Chatwin, P.C., 1976. Some remarks on the maintenance of the salinity distribution in estuaries. Est. & Coastal Mar. Sci., 41: 555-566.

Chatwin, P.C. and Sullivan, P.J., 1978a. The relative dispersion of a cloud of passive contaminant in incompressible turbulent flow. Submitted to J. Fluid Mech.

Chatwin, P.C. and Sullivan, P.J., 1978b. Measurements of concentration fluctuations in relative turbulent diffusion. Submitted to J. Fluid Mech.

Csanady, G.T., 1973. Turbulent Diffusion in the Environment. D. Reidel Publishing Company, Dordrecht, 248 pp.

Dewey, R. and Sullivan, P.J., 1977. The asymptotic stage of longitudinal turbulent dispersion within a tube. J. Fluid Mech., 80: 293-303.

Fischer, H.B., 972. Mass transport mechanisms in partially stratified estuaries. J. Fluid Mech., 53: 671-687.

Ozmidov, R.V., 1960. The investigation of medium-scale horizontal turbulent exchang in the ocean using radiolocation observations of a floating buoy. Akad. Nauk SSSR Dokl. Earth Sci. 126: 536-538.

Richardson, L.F., 1926. Atmospheric diffusion shown on a distance-neighbour graph. Proc. Roy. Soc. Lond., A110: 709-737.

Sullivan, P.J., 1971a. Some data on the distance-neighbour function for relative diffusion. J. Fluid Mech., 47: 601-607.

Sullivan, P.J., 1971b. Longitudinal dispersion within a two-dimensional turbulent shear flow. J. Fluid Mech., 49: 551-576.

Sullivan, P.J., 1975. "The $^4/_3$rds law of relative diffusion". Mem. Soc. Roy. Sci. Liège, 6e série, 7: 253-260.

Taylor, G.I., 1954. The dispersion of matter in turbulent flow through a pipe. Pro Roy. Soc. Lond., A223: 446-468.

Williams, D.J.A., and West, J.R., 1973. A one-dimensional representation of mixing in the Tay estuary. In A.L.H. Gameson (Editor), Mathematical and hydraulic modelling of estuarine pollution. Water Poll. Res. Tech. Paper No. 13, Dept. Environ., London, pp. 117-125.

Professor Sullivan acknowledges the financial support of the National Research Council of Canada during the period when this work was done.

A ONE-DIMENSIONAL TIDAL MODEL FOR ESTUARINE NETWORKS

Langley R. Muir
Canada Centre for Inland Waters, Burlington, Ontario, Canada

INTRODUCTION

This paper is concerned with one-dimensional numerical modelling of unsteady flow in networks of canals, rivers, and estuaries. The model will allow the computation of water elevation and velocity in any network of open channels to which the following assumptions are applicable:

(1) flow is physically possible;
(2) flow is entirely subcritical (i.e. the Froude number is less than 1.0);
(3) flow is one-dimensional in space (i.e. laterally and vertically homogeneous);
(4) appropriate boundary conditions are available; and
(5) the section geometry of the channel is fixed (i.e. no deposition or scouring occurs).

Flow in open channels can be described by two equations, one expressing the conservation of mass (the continuity equation) and one expressing the conservation of momentum in the longitudinal direction (the momentum equation). In general terms, these equations form a set of non-linear partial differential equations. Depending upon the assumptions made, there are various methods available for the solution of these equations; but, since the development of the digital computer, numerical methods have generally been used.

This paper develops all of the theory necessary to construct a numerical model for simulating unsteady flow conditions in networks of open channels. The first section gives a formal method for describing the flow relationships in any network. The implicit finite difference method is then described and extended for use in networks. The equations of motion are given, with a description of a generalized equation solver, and the properties of the finite difference scheme are discussed. The final section describes one of many applications of the numerical model to show that the method does work.

THE NETWORK AS A GRAPH

The essential features of open channel flow in a network may be illustrated by considering an open channel network as if it were a mathematical entity known as a graph. A graph, in mathematical terms, may be defined as a connected set of lines on a plane surface. The points at which various lines meet or cross are known as nodes and, if direction may be determined, the graph is said to be directed. The relationships in the graph are purely topological in that distance relationships are not preserved.

The graph representation of a network of open channels consists of a number of lines called branches representing the elementary reaches of the open channel network and a certain number of nodes, each of which identifies the location at which two or more branches intersect. To make the graph more general and at the same time more applicable to river or estuarine systems, additional nodes are allowed on the boundaries of the graph or at arbitrary locations on the graph. For example, the river system shown in Fig. 1(a) may be schematized into the graph shown in Fig. 1(b). In the figure, the branches of the graph are shown as lines and the nodes are shown as dots.

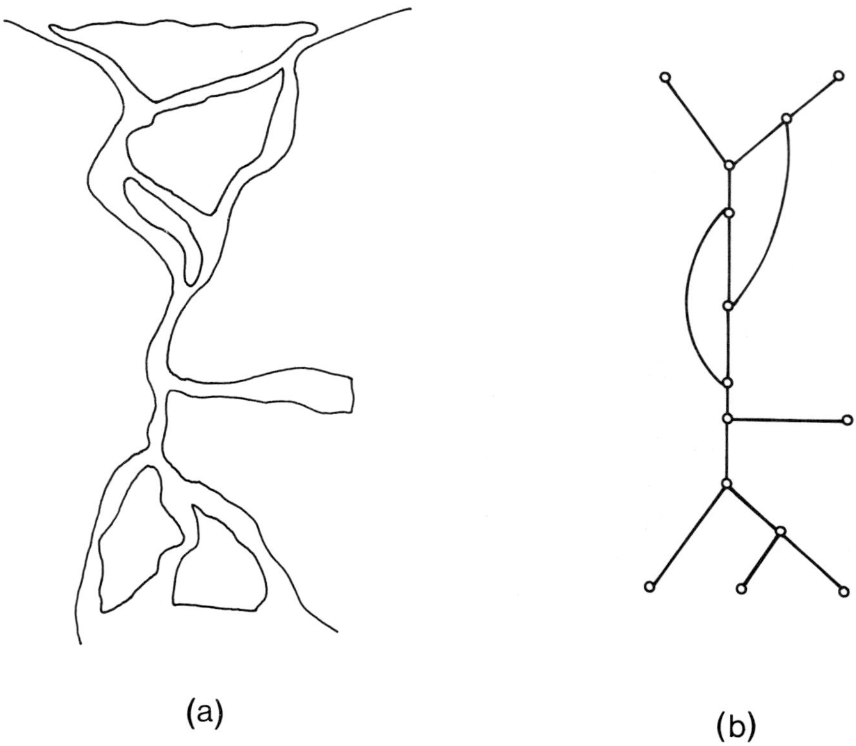

Fig. 1. A river network and its graph representation.

In modelling unsteady flow situations, each branch is considered equivalent to an elementary reach in the channel network. The nodes are placed at locations where flow properties are required or are known. At each node, it is necessary to know the cross-sectional geometry of the open channel and, during the solution of the equations of motion of the system, it is assumed that the system parameters will vary linearly between adjacent nodes. Therefore, in the schematization of an open channel network, it is necessary to take into account the physical parameters of a system, and these considerations are reflected in the additional nodes selected for inclusion on the graph.

It is possible to rearrange the graph so that there are, at most, three branches connected to each node. If more than three branches are connected to a particular node, as in Fig. 2(a), then the offending node may be split into two or more nodes, each of which has three branches, as in Fig. 2(b). The physical distance between the new nodes will be zero. The purpose of this schematization is to identify flow relationships and to simplify these relationships to an extent which allows simple bookkeeping for a computer program.

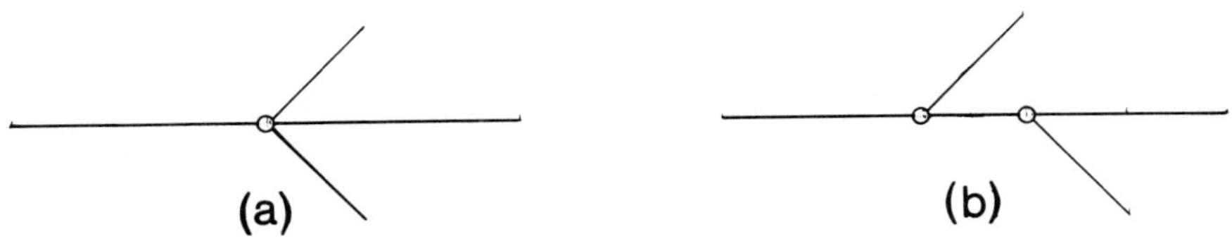

Fig. 2. Splitting of the junction node.

This splitting of nodes allows the identification of exactly four separate types of nodes. Each network can be schematized using only these four types of nodes. They are:

(1) a <u>bounding node</u> which is connected to only one branch;

(2) an <u>interior node</u> which is connected to exactly two branches;

(3) a <u>convergent node</u> which has two branches entering and one branch leaving; and

(4) a <u>divergent node</u> which has one branch entering and two branches leaving.

Types 3 and 4 are known collectively as junction nodes. In an unsteady flow problem, any given junction node could be either type 3 or type 4 depending upon the direction of flow. This situation may be automatically resolved by the computer program, if a node is classified according to its type when all flow is in the positive direction. Making this distinction between convergent and divergent nodes eases the computer programming difficulties without making the theoretical treatment any more difficult.

FINITE DIFFERENCE METHODS

These methods use finite difference approximations to the partial derivatives appearing in the equations of motion to transform the set of partial differential equations into a set of algebraic equations. It is the direct finite difference methods that hold the most promise for the solution of engineering problems when a digital computer may be used to solve large systems of algebraic equations. In this case, "direct" means the conversion of the partial differential equations to a finite difference formulation without using the characteristic equations. Two recent reviews of the various available methods are given in Price (1974) and Strelkoff (1970).

There are two finite difference methods, and the distinction between them lies in the method in which the finite differences are formulated and the resulting methods for the solution of the equations. The implicit method requires that all of the equations be solved simultaneously in order to advance the solution one time step. The explicit method proceeds down the open channel solving only one equation at one time. There are a large number of finite differencing schemes available for use with each method.

Because of the bookkeeping and equation solving requirements of the implicit scheme, the explicit method is much simpler to use. However, the explicit schemes are restricted in the size of the computational time step required to ensure a stable computational procedure. Numerical stability is achieved when small errors introduced in the computation diminish rather than increase in magnitude with succeeding computations. If too large a time step is used, the true solution to the equations may well be completely masked by the errors. The restriction is given by the well-known Courant condition, which is:

$$\Delta t < \frac{\Delta x}{u + \sqrt{gA/B}}$$

where:

B = width of water surface;

A = cross-sectional area;

g = acceleration due to gravity;

u = velocity; and

Δx = the distance interval used.

If friction is important, Garrison et al. (1969) have shown that the maximum Δt may be further limited by the following stability criterion:

$$\Delta t \leq \left[\frac{\Delta x}{u + \sqrt{gA/B}} \right] \left[1 - \frac{gn^2 |u| \Delta t}{2.2 (A/B)^{4/3}} \right]$$

where n is the Manning friction factor.

Examination of the stability criteria will show that, for typical river applications, time steps on the order of a few seconds may be required. For problems in large river systems which may involve tidal cycles or input hydrographs extending over several days, these small time steps cause the explicit method to be very wasteful of computer time.

The major restriction on the explicit method is the difficulty in handling flow in networks. As the computations proceed downstream and reach a junction, some arbitrary decision must be made about how much of the flow enters each branch. The profiles in each branch must then be computed separately and the results, at the end of the branches, compared. If these results are incompatible, then the computation must start again at the top of the branch and iterate in some fashion until the flows match properly. In the implicit method, this problem does not arise since all of the equations are solved simultaneously.

THE IMPLICIT METHOD

Of the two types of finite difference schemes available for the solution of the equations of motion, the implicit scheme has been chosen for reasons which have been discussed briefly and which will become clearer by the end of this section.

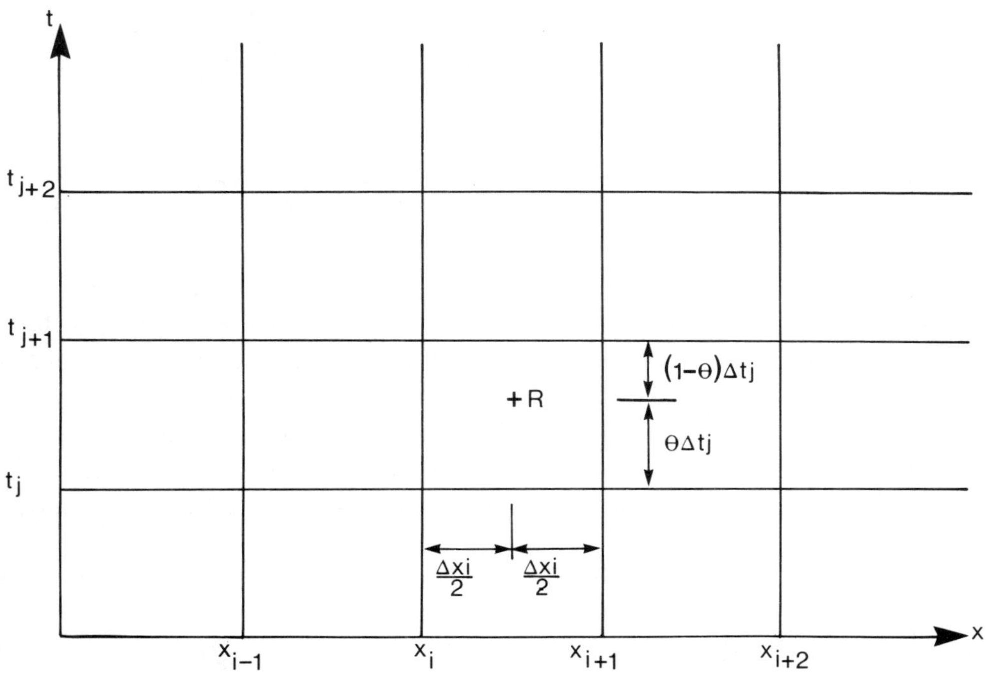

Fig. 3. The weighted four-point finite difference scheme.

Consider a rectangular grid, not necessarily uniform on the x-t plane as shown in Fig. 3. The function value and partial derivatives of a function at the point $x = x_i + \frac{\Delta x_i}{2}$ and $t = t_j + \theta\Delta t$ are given by:

$$K = \frac{\theta}{2}\,(K_i^{j+1} + K_{i+1}^{j+1}) + (\frac{1-\theta}{2})\,(K_i^j + K_{i+1}^j)$$

$$\frac{\partial K}{\partial t} = \frac{1}{2\Delta t}\,(K_i^{j+1} + K_{i+1}^{j+1} - K_i^j - K_{i+1}^j) \tag{1}$$

$$\frac{\partial K}{\partial x} = \frac{\theta}{\Delta x}\,(K_{i+1}^{j+1} - K_i^{j+1}) + \frac{(1-\theta)}{\Delta x}\,(K_{i+1}^j - K_i^j)$$

If the above are applied to a single-branched river or channel which contains only bounding or interior nodes, such as in Fig. 4, then at time = t_{j+1}, there are two unknowns for each node or grid point. A river system containing N branches, corresponding to N+1 nodes, has 2(N+1) unknowns at time t_{j+1}. For each pair of adjacent nodes at time t_{j+1}, a continuity equation and a momentum equation may be written in finite difference form relating the unknown values at time t_{j+1} to the known values at time t_j. Since there are two equations between each pair of adjacent nodes, there are a total of 2N equations in 2(N+1) unknowns. The addition of any two additional independent equations relating unknowns will produce a set of 2N+2 equations with 2N+2 unknowns which may be solved simultaneously to produce the solutions at time t_{j+1}.

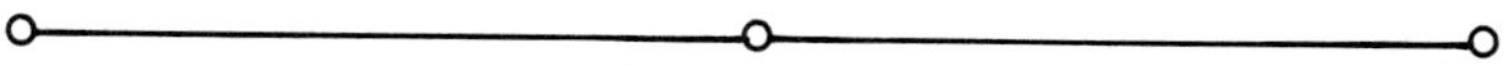

Fig. 4. An interior node.

The implicit method may easily be extended for use in a network of channels by considering the interior node shown in Fig. 4 and the junction node shown in Fig. 5(a). In Fig. 4, there are three nodes associated with the interior node, two bounding nodes and the interior node itself. The system is comprised of two branches for which four equations may be written. The two boundary conditions supply the other two equations necessary for the computation of the six unknown quantities.

Fig. 5. A junction node.

Consider the junction node shown in Fig. 5(a) and split this junction node into three nodes as shown in Fig. 5(b). There are now six nodes associated with the original junction node and hence twelve unknowns. The three branches provide six equations and the three boundary conditions provide an additional three equations, giving a total of nine equations for the twelve unknowns.

The three remaining equations necessary for a solution are provided fairly easily, since splitting the junction node into three parts has placed the actual junction itself between nodes where it may be treated as a simple input-output system and we may ignore the details of the physical mechanisms operating inside the junction. At any junction, a continuity equation will relate the amount of water entering the junction to the amount of water leaving the junction. This continuity equation provides one equation. Noting that the actual distance between the junction nodes is zero will provide two additional equations, since the water level at the three "junction nodes" must be equal. Hence, there are a total of twelve equations in twelve unknowns and the system of equations is solvable.

It may be shown, by simple mathematical induction, that the implicit method may be used in any arbitrary network of open channels that can be schematized as a plane graph. The major advantage of the implicit method is that, by solving all of the equations simultaneously, the flow in the various channels is computed automatically and no complicated iterative techniques are required to balance the various flows. The only disadvantage in the implicit method is the necessity of solving a large number of equations simultaneously.

EQUATIONS OF MOTION

Although both the bookkeeping system suggested by the graph method of describing the flow relationships and the equation solver may be programmed to be independent of the form of the equations used in the model, a choice of equations is necessary in order to proceed. The most generally useful set of equations to employ for describing flow in open channels is the non-linear St. Venant equations. If an open channel may be schematized as shown in Fig. 6, then the St. Venant Equations may be written as:

$$(b+b_S)\ \frac{\partial z}{\partial t} + \frac{\partial Q}{\partial x} - q = 0$$

$$\frac{\partial Q}{\partial t} + u\ \frac{\partial Q}{\partial x} + Q\ \frac{\partial u}{\partial x} + g\cdot A\ \frac{\partial z}{\partial x} + \frac{g\cdot K\cdot Q|Q|}{A\cdot h*} = 0$$

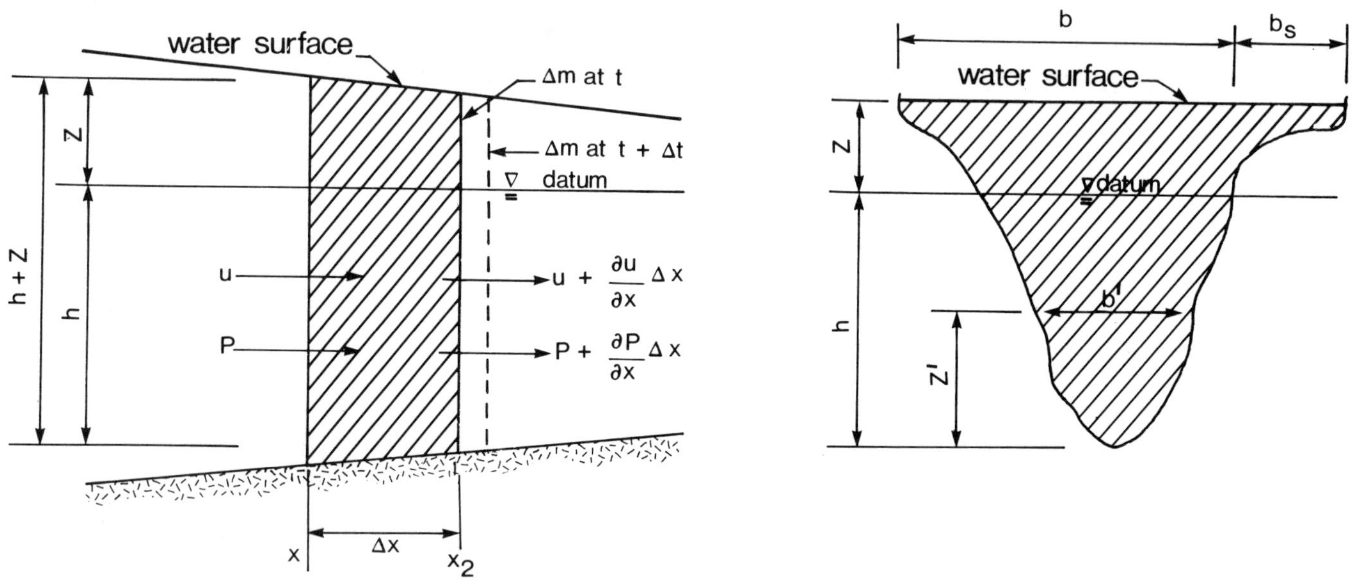

Fig. 6. Definition sketches for an irregular channel.

where:

Q = cross-sectional discharge;

u = cross-sectional mean velocity;

t = time;

x = distance along the channel;

g = acceleration due to gravity;

A = cross-sectional area;

z = water level above a horizontal datum;

K = friction coefficient, where

$K = 1/C^2$ (Chezy friction), or

$K = n^2/2.22(h^*)^{1/3}$ (Manning's friction);

h^* = total depth of water, or hydraulic radius;

b = width of conveyance section of channel;

b_s = width of storage section of channel; and

q = lateral inflow per unit length of channel.

In these equations, it is assumed that neither the water in the storage section of the channel nor the lateral inflow affect the momentum. A term involving the wind stress could easily be added to the momentum equation, and it will be seen that this would not affect the equation solver in any way.

The junction equations are fairly simple. Continuity requires that the inflow to a junction must equal the outflow and, since the length of the junction is effectively zero, no storage takes place in the junctions. Therefore, if nodes i, i+1, and k are the numbers associated with the junctions, the continuity equation is:

$$Q_i + Q_k = Q_{i+1}$$

The junction energy equations are simply that the water levels inside the junction are equal, or:

$$Z_i = Z_{i+1}$$

$$Z_i = Z_k$$

All of the various types of boundary conditions may be expressed in terms of the discharge and water level elevation at the node at which the boundary conditions are given.

By substituting the finite difference scheme into the equations of motion, we obtain a set of algebraic equations relating the properties of the system at time step j to the properties at time step j+1. In these equations, the only unknowns are the discharge and water level at the time j+1, and all of the equations may be written in the form:

$$f_1Q_i + f_2z_i + f_3Q_{i+1} + f_4z_{i+1} + f_5Q_k + f_6z_k = f_7$$

where the functions f_1 to f_7 are in terms of the section properties and the discharges and water levels at time j and at nodes i, i+1, and k.

Since the momentum equation is non-linear, some of these functions f_1 to f_7 are not constant but are functions of the unknown Q or z or both. For example, in the momentum equation, one of the non-linear terms is $u\partial Q/\partial x$. In the solution technique, this term is quasi-linearized to $U\partial Q/\partial x$ where U is the velocity computed from the last pass through the equation solver.

The system of equations is now in linear form and may be expressed as a matrix equation of the form:

$$[W] * [X] = [B]$$

where X is a solution vector. The matrix, W, is reduced in three steps:

(1) Zero diagonal entries in the matrix are filled. Any row, i, that has a zero diagonal entry has added to it a row, k, that has a non-zero entry in the ith column, where $i<k$.

(2) The matrix is upper-triangularized. Starting at the second row and working downwards, any non-zero entries to the left of the diagonal are eliminated through the use of the rows above.

(3) Once the matrix has been upper-triangularized, back substitution is used to fill the solution vector.

Each time the matrix, W, is reduced, the values for Q and z, extracted from the solution vector, are used to recompute the coefficients. The matrix consisting of the new coefficients must now be solved. This process is repeated until convergence to a predetermined number of significant digits is achieved in the Q and z values. Convergence to four significant digits usually takes less than 10 iterations.

Significant reductions in computing requirements can be made by avoiding storing or operating on the zero coefficients of the network matrix. This method, known as a sparse matrix technique, can easily be combined with Gaussian elimination to produce a reasonably efficient equation solver. The method used in this study is similar to that used by Chandrashekar et al. (1975, 1976) and is given in detail by Budgell (1977a). For a banded matrix, execution time increases linearly with the number of equations.

One of the most useful features of the equation solver, and of the bookkeeping scheme suggested by the graph method of describing the flow relationships, is that the numerical model may very easily be programmed to be completely independent of the exact form of the equations used. This can allow further saving in computer time if equations other than the full non-linear St. Venant equations may be used. Alternatively, the model may be easily expanded to take into account such complexities as more complicated junction equations, wind stress, or dispersion of a conservative tracer with minimum of programming changes.

PROPERTIES OF THE FINITE DIFFERENCE SCHEME

The set of algebraic finite difference equations is formally equivalent to the set of partial differential equations, describing the flow properties of the open channel network, only in the limit, as the time and distance steps approach zero. In any application of finite difference methods, it is necessary to examine the stability, computational error, and conservation properties of the finite difference scheme.

If the weighted, implicit four-point finite difference scheme given by (1) is applied to the linearized shallow water equations, an examination of the properties of the finite difference equation may be undertaken (Muir, 1975). The results are that:

(1) The growth factor for the propagation of error waves through time is given by:

$$|\lambda| = \frac{1 + (2\theta-2)^2 a + (\theta-1)b}{1 + 4\theta^2 a + \theta b}$$

where:

θ = weighting factor from the finite difference scheme;

$a = gh\ (\frac{\Delta t}{\Delta x})^2\ \tan^2\ (\frac{\sigma\Delta x}{2})$;

$b = K\Delta t$;

h = mean depth of water;

σ = frequency of wave;

$K = \frac{g \cdot u \cdot n^2}{2.22\ (h)^{4/3}}$;

u = mean flow velocity; and

n = Manning's roughness coefficient.

Hence, if $0.5 \leq \theta < 1.0$, the scheme is unconditionally stable and, if $0 \leq \theta < 0.5$, the scheme is conditionally stable depending on the relative magnitudes of Δx and Δt.

(2) Qualitatively, the error involved in the approximation of the partial differential equation to the finite difference equation is of the form:

$$E = (2\theta-1)0(\Delta t) + 0(\Delta t^2) + 0(\Delta x^2)$$

where:

0 = "the order of";

so when:

$\theta = 0.5$, the scheme has second-order accuracy, and the accuracy of the solution is independent of the relationship between Δx and Δt.

(3) The scheme conserves mass over one time step if $\theta = 0.5$. In general, over one time step:

$$\frac{\text{change in volume}}{\Delta t} = \theta(Q_1^{j+1} - Q_L^{j+1}) + (1-\theta)(Q_1^j - Q_L^j)$$

when the superscripts refer to time and the subscripts refer to the two ends of a branch. Hence, the departure from conservation is usually small, even if θ varies from 0.5.

All of the above results are obtained from an analysis of the linearized equation, but numerical experiments have shown that the results are also applicable to the full non-linear system of equations for a large number of practical examples.

APPLICATION

The theory presented in this paper has been programmed in FORTRAN IV, and the resulting computer program has been tested on a number of examples. Only one of these examples will be given here.

The St. Clair River joins Lake Huron and Lake St. Clair and is approximately 45 kilometres long, about 610 metres wide, and approximately 10 metres deep. The river has been extensively dredged and is a major navigational waterway. The lower end of the river, where it empties into Lake St. Clair, is a complicated delta region with a very large amount of swampy ground and very shallow channels. Historically, the flow has varied between 2830 and 6795 m^3/sec with a 50 per cent exceedence level of 5210 m^3/sec. The U.S. Army Corps of Engineers has maintained an extensive water level gauging network on the river for a number of years (P. Cox, personal communication, 1973).

A map of the river, giving the node numbering, is shown in Fig. 7. The geometrical data used in the model are given in Table 1. Examination of the hydrographic field sheets shows that the cross-sectional shape of the river may be approximated by a very wide rectangle and, as a result, the hydraulic radius (cross-sectional area/surface width) may be expressed as the mean depth. The river contains one major island, Stag Island, which is included in the model. The upper limit of the model is at Fort Gratiot and the lower limit is taken at Algonac, at the top of the delta region. No attempt was made to model the flow

TABLE 1.

Geometrical data for St. Clair River system. Datum is 173.74 M (IGLD)

Node Number	Topwidth (m)	Invert Level Below Datum (m)	Reach Length (m)	Manning's n	Water Level Gauge Name
1	304.8	8.59	2200	0.033	Fort Gratiot
2	396.2	8.63	1166	0.028	
3	874.8	4.97	1280	0.023	MBR
4	670.6	7.61	3871	0.023	
5	594.4	6.68	3734	0.023	Dry Dock
6	627.9	7.01	1189	0.020	Marysville
7	813.8	7.51	6187	0.020	
8	670.6	4.97	31	0.020	
9	670.6	6.19	5547	0.020	
10	493.8	7.43	1844	0.025	St. Clair
11	518.2	8.78	2743	0.025	
12	688.8	7.04	1890	0.025	
13	847.3	6.18	2896	0.021	
14	1005.8	5.08	3338	0.023	Marine City
15	868.7	6.48	2804	0.023	
16	688.8	8.03	2758	0.025	Roberts Landing
17	682.8	8.02	1859	0.022	
18	841.2	6.58	-	-	Algonac
19	457.2	3.09	3109	0.020	
20	466.3	3.13	1920	0.020	

in the delta region, since the section geometry of this region completely invalidates the hypothesis of one-dimensional flow and since there is not enough water level data to properly calibrate a model.

A major problem in modelling a natural river lies in the choice of a friction factor. In the case of the St. Clair River, a great deal of information is available which allows for a very fine calibration. A useful point in the calibration process is that a change in the friction factor in a given reach affects only the flow upstream of that reach.

The calibration period chosen for the St. Clair River was 19 June, 1973. On this date, discharge measurements were being taken in the river by the U.S. Army Corps of Engineers at St. Clair, Michigan. The mean flow on this day was 6987 m^3/sec. From this start, the calibration procedure was:

(1) use a discharge of 6087 m^3/sec and a known water level at Algonac;

(2) start with a Manning's friction factor of 0.020 which, from practical experience, is too low for this river;

(3) make a computer run and compare the calculated water levels with the observed water levels upstream;

(4) adjust the friction factor in the lowest sections of the model until the calculated water level at the second gauge upstream agrees with the observed

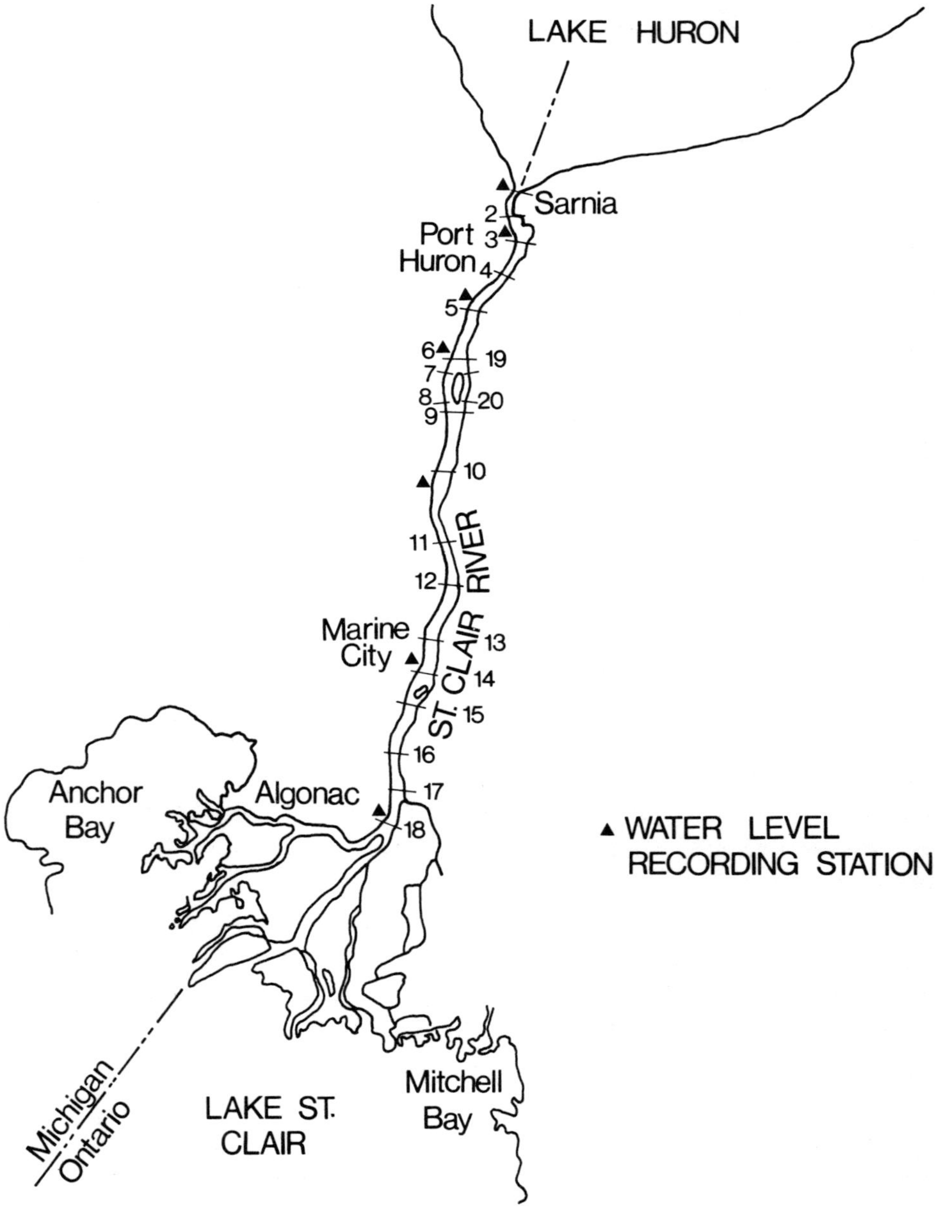

Fig. 7. The St. Clair River.

water levels;

(5) move upstream in this fashion, adjusting friction factors for one section at a time until all of the computed water levels agree with the observed water levels within a certain predetermined accuracy; and

(6) using two observed water levels and the determined friction factors, make another computer run to check that the discharge is correct. If it is not, adjust all friction factors up or down by the same relative amount until the computed and observed discharges match properly.

Using this procedure, the model was calibrated for 19 June, 1973, to within ±0.02 metres as the maximum error in water level. This accuracy is certainly within the expected limits of any one-dimensional model, assuming steady state flow conditions and considering the practical accuracy of discharge measurements. The calibration process used about 20 computer runs to match up the 9 water level gauges on the river. Although the process of calibration could easily be incorporated into the computer program, "intelligent guesses" allow a great saving in computer time.

After the model was calibrated for 19 June, 1973, two additional runs were made to show that the model would predict other flow conditions properly. The discharge for 19 June, 1973, was approximately 6087 m^3/sec, and this corresponds to an extremely high flow condition. The additional runs were: 18 July, 1968, which corresponds to a discharge of approximately 5181 m^3/sec, a "medium" flow condition; and 13 November, 1964, a "low" flow condition with a discharge of approximately 4162 m^3/sec. The observed water levels, computed water levels, and differences are shown in Table 2. Unfortunately, not all of the water level gauges were in operation all of the time. However, a good indication of the accuracy of the model is given by this table.

To show that the model properly predicts transients, it was run for a period during which a large transient passed through the St. Clair River. During 3-5 December, 1970, a storm was in progress, and this storm resulted in a one-metre storm surge on Lake Huron at Fort Gratiot. At that time, three gauges were in operation in the St. Clair River: one at Fort Gratiot, one at St. Clair, Michigan, and one at Algonac. Data are available from these gauges at hourly intervals.

A run was made using the observed water levels at Fort Gratiot and at Algonac as boundary conditions. The schematization was revised to use only 13 nodes, and all data from the steady state runs were retained. Fig. 8 shows the observed values compared with the calculated values for node 10 (St. Clair, Michigan). The agreement is quite close, considering that only hourly water levels are available. This time step is clearly too long to give adequate resolution in the phase of the transient.

Aside from the examination of transient flow in the St. Clair River, a very

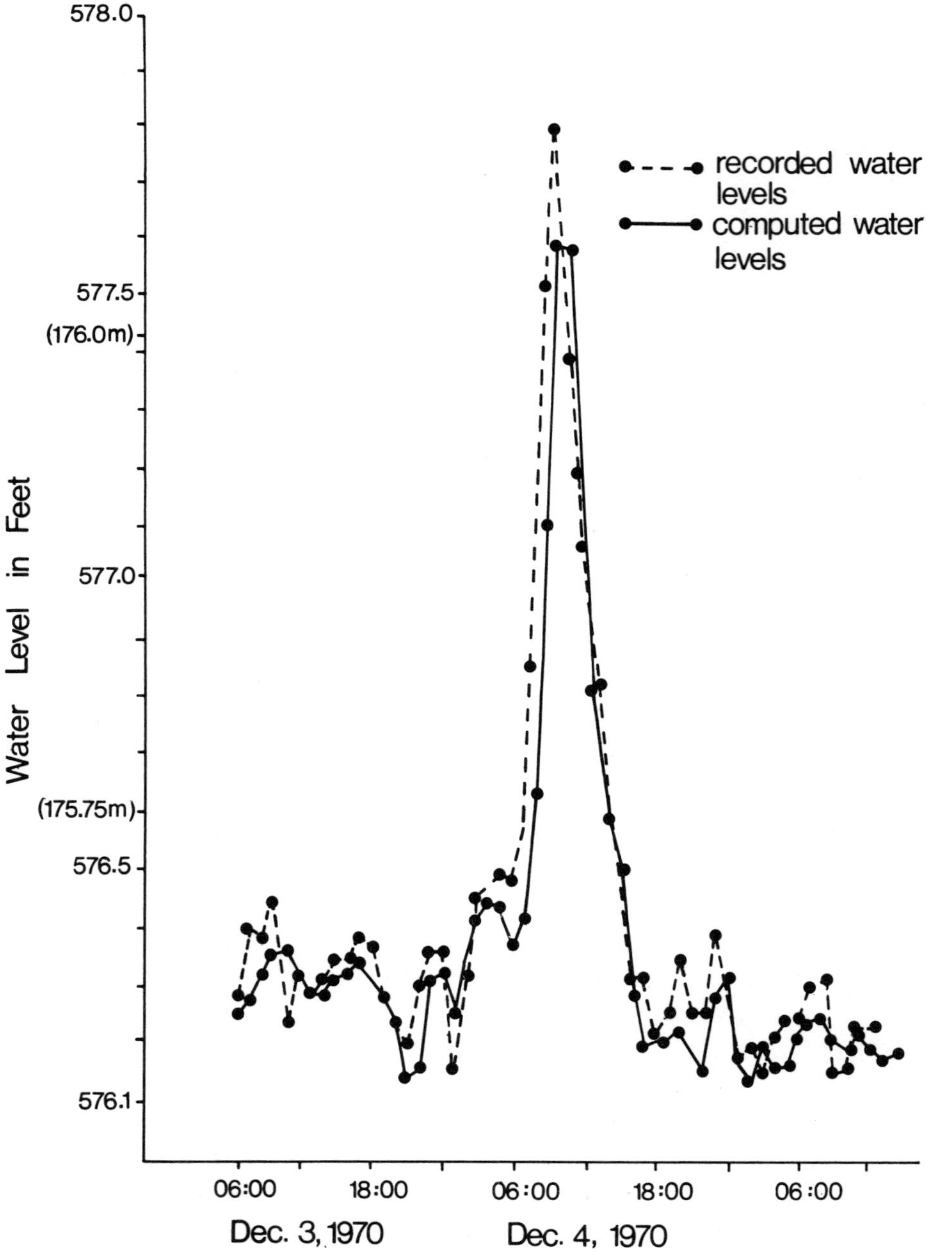

Fig. 8. Water levels at St. Clair, Michigan.

TABLE 2.

Observed and computed water levels for the St. Clair River under various discharges. Datum is 173.74 M (IGLD)

Node Number	19 June, 1973 Obs. (m)	Computed (m)	18 July, 1968 Obs. (m)	Computed (m)	13 November, 1964 Obs. (m)	Computed (m)
1	3.11	3.11	2.41	2.41	1.52	1.52
2		2.97		2.28		1.41
3	2.95	2.96	2.29	2.27	1.44	1.39
4		2.94		2.25		1.37
5	2.84	2.83	2.16	2.15		1.29
6	2.70	2.73	2.05	2.06	1.21	1.21
7		2.73		2.06		1.21
8		2.65		1.99		1.14
9		2.65		1.99		1.44
10	2.48	2.48	1.84	1.84	1.01	1.01
11		2.44		1.80		0.97
12		2.34		1.74		0.92
13		2.33		1.71		0.88
14	2.24	2.26		1.64		0.81
15		2.19		1.57		0.75
16	2.11	2.13	1.51	1.52		0.70
17		2.09		1.48		0.67
18	2.06	2.06	1.46	1.46	0.64	0.64
19	2.70	2.73	2.05	2.06	1.21	1.21
20		2.65		1.99		1.14
Computed Discharge		6087		5181		4162

extensive study of the tidal propagation in a large Arctic estuary has been carried out by Budgell (1977a). A part of the analysis consisted of an examination of the generation of shallow water tidal constituents using Power Spectral Analysis techniques. The results of this analysis were excellent and have been published by Budgell (1977b).

SUMMARY AND CONCLUSIONS

By describing a network of open channels in terms of graph theoretic concepts, the flow relationships in any network may be expressed in terms of only four types of nodes. This allows a computer program to be written which is totally independent of any particular network topology. By using the implicit method and a sparse matrix equation solver, the numerical model is independent of both the network topology and the particular form of field equations.

The model has been applied to a number of examples such as steady state river flow, transient river flow, and tidal propagation in large estuaries with excellent results. With minimum effort, the model may be rewritten to take into account the dispersion of a conservative tracer and, hence, could easily be used for water quantity studies as well as flow studies.

REFERENCES

Budgell, W.P. 1977a. Tidal Propagation in Chesterfield Inlet, N.W.T. Manu. Rep. Ser.: 3, Ocean and Aquatic Sciences, Canada Centre for Inland Waters, Burlington, Ontario.

Budgell, W.P. 1977b. Numerical Simulation of the Tides in Chesterfield Inlet. In: Proc. XVII Congress of the I.A.H.R., Baden-Baden, Germany.

Chandrashekar, M., L.R. Muir, and T.E. Unny. 1975. A Numerical Two-Dimensional Flow Model for River Systems. In: Proc. Modelling 75, Am. Soc. Civ. Eng. Specialty Conference, San Francisco, 407-425.

Chandrashekar, M., L.R. Muir, and T.E. Unny. 1976. A General Two-Dimensional River Simulator. In: Proc. Int. Symp. on Large Engineering Systems. The University of Manitoba. (in press).

Garrison, J.M., J.P. Granju, and J.T. Price. 1969. Unsteady Flow Simulation in Rivers and Reservoirs. J. Hydraul. Div. Am. Soc. Civ. Eng., 95:5:1559-1576.

Muir, L.R. 1975. Unsteady Flow in Networks of Open Channels. Manu. Rep. Ser.: 1, Ocean and Aquatic Sciences, Canada Centre for Inland Waters, Burlington, Ontario.

Price, R.K. 1974. Comparison of Four Numerical Methods for Flood Routing. J. Hydraul. Div. Am. Soc. Civ. Eng., 100:7:879-899.

Strelkoff, T. 1970. Numerical Solution of Saint-Venant Equations. J. Hydraul. Div. Am. Soc. Civ. Eng., 96:1:223-252.

A NEW APPROACH TO THE COMPUTATION OF TIDAL MOTIONS IN ESTUARIES

B.M. JAMART and D.F. WINTER
Department of Oceanography, University of Washington, Seattle, WA 98195

ABSTRACT

In this paper we present a summary of a new procedure for computing periodic tidal motion in estuaries with irregular boundaries and variable depth. The new method, described by Pearson and Winter (1977), appears to provide both computational speed and numerical accuracy in a wide variety of problems of practical interest. The starting point of the analysis is the standard set of vertically-integrated, time-dependent equations expressing conservation of horizontal momentum and mass in two dimensions. Since the motion in the estuary is assumed to be periodic, the dependent variables are Fourier decomposed. However, in order to avoid awkward Fourier decomposition of nonlinear terms and to decouple the different modes, advection and friction are evaluated by an iterative procedure. The time-dependent equations of motion are replaced by an equivalent set of modal equations, and it is shown that the boundary value problem consisting of the modal equations and appropriate boundary conditions can be rephrased in terms of a variational principle. The variational principle is used together with a finite element method to solve for the unknown flow variables. We describe here an application of the method to a segment of Hood Canal, Washington, where there is an interest in pollutant dispersal by tidal currents.

INTRODUCTION

Pearson and Winter (1977) recently described a new approach to the computation of periodic tidal motion in homogeneous estuaries and embayments with irregular shorelines and variable depth. In this paper we set forth the main features of the approach and describe some preliminary results of an application of the method to a segment of Hood Canal, a deep, weakly stratified inlet in Puget

Sound, Washington, U.S.A. The highlights of the method of Pearson and Winter can be summarized as follows: since shallow water theory is assumed to apply, the governing equations are the standard vertically-integrated, time-dependent equations expressing conservation of horizontal momentum and mass in two dimensions [see, for example, Hess and White (1974)]. The momentum equations include convective acceleration, Coriolis acceleration, and terms describing the effect of wind stress and bottom friction. Boundary conditions are specified along the shoreline boundaries, across the mouth of the estuary, and along other open boundaries. (The appropriateness of various boundary conditions is discussed by Pearson and Winter.) Since the motion in the estuary is assumed to be periodic, the dependent variables (water surface height and depth-averaged velocities) are Fourier decomposed. Consequently, the time-dependent equations of motion are replaced by an equivalent set of modal equations, with only x and y as independent variables. In order to circumvent the coupling between modes arising from the Fourier decomposition of the nonlinear terms, the latter are treated by an iterative procedure, similar to the one described below. Next, the boundary value problem consisting of the modal equations and appropriate boundary conditions is rephrased in terms of an equivalent variational principle. The variational principle is then used together with a finite element method to solve for the unknown variables, i.e., the coefficients of the Fourier series representations of the free surface elevation and of the depth-averaged velocities throughout the estuary. The actual space- and time-dependent currents and heights can then be reconstructed by Fourier synthesis.

The present discussion will be restricted to a special case of more general problem statements discussed by Pearson and Winter (1977 the reader is referred to that paper for further details.

In the following section, a somewhat simplified version of the governing equations is set forth and a review of the analytic procedure is presented. The next section is devoted to the numerical solution of the boundary value problem by a finite element technique. Preliminary results of tidal computations in Hood Canal, Washington, are presented in the last section.

FORMULATION OF THE PROBLEM

Denote by A the domain of interest in a homogeneous estuary bounded by an open section Γ_1, along which the periodic free surface height is prescribed, and a solid shoreline section Γ_2 along which the normal flow component vanishes. It is assumed that the effect of bottom friction on the horizontal momentum balance can be represented by a term proportional to $\underline{u}/D$, where $\underline{u}$ denotes the vertically averaged velocity and D(x,y) represents the time-mean depth. For convenience, in this analysis we neglect convective accelerations and wind stress. Since the estuary to which we intend to apply the analysis is deep and narrow, we neglect Coriolis acceleration and horizontal diffusion, and we linearize the equation for conservation of mass. With these simplifications, the equations of motion reduce to

$$\underline{u}_t + g\nabla h = -E\,\frac{\underline{u}}{D} \tag{1}$$

$$h_t + \nabla .\,(\underline{u}D) = 0 \ , \tag{2}$$

where h(x,y,t) is the elevation of the water above mean sea level, the subscripts denote derivatives, ∇ is the nabla operator, and E denotes a (constant) friction coefficient. The boundary conditions are

$$h = h^*(x,y,t) \quad \text{on} \quad \Gamma_1 \tag{3}$$

$$\underline{u}.\underline{n} = 0 \quad \text{on} \quad \Gamma_2 \tag{4}$$

where h* is a known function and $\underline{n}$ is the outward-directed unit normal vector. Let ω be the fundamental circular frequency of the periodic motion and let N be the number of modes required to represent the temporal variation of the dependent variables. In the absence of a "stationary" forcing function (as might be produced by persistent wind or river discharge), and since some of the nonlinear terms have been linearized and others neglected, there is no need to include a zeroth-order mode in the Fourier expansions of the dependent variables and forcing term. We can write, therefore,

$$\underline{u} = \text{Re} \sum_{n=1}^{N} \underline{U}_n\,(x,y)\, e^{-in\omega t} \tag{5}$$

$$h = \text{Re} \sum_{n=1}^{N} H_n\,(x,y)\, e^{-in\omega t} \tag{6}$$

$$h^* = \mathrm{Re} \sum_{n=1}^{N} H_n^* (x,y)\, e^{-in\omega t} \tag{7}$$

Substitution of (5) and (6) into (1) - (4) leads to coupled modal equations of the form

$$in\omega\, \underline{U}_n - g\nabla H_n = \frac{E}{D}\underline{U}_n \tag{8}$$

$$in\omega\, H_n - \nabla.\,(\underline{U}_n D) = 0 \tag{9}$$

with boundary conditions

$$H_n = H_n^* \quad \text{on} \quad \Gamma_1 \tag{10}$$

$$\underline{U}_n \cdot \underline{n} = 0 \quad \text{on} \quad \Gamma_2 \; . \tag{11}$$

Clearly, there is no coupling between modes in the linearized problem. Elimination of $\underline{U}_n$ between (8) and (9) yields an equation for H_n that can be written as

$$\nabla.(D\nabla H_n) + \frac{E^2 + in\omega ED}{E^2 + n^2\omega^2 D^2}\nabla D.\nabla H_n + \frac{n^2\omega^2 D + in\omega E}{gD} H_n = 0 \tag{12}$$

The boundary condition (11) becomes

$$\frac{\partial H_n}{\partial n} = 0 \tag{13}$$

Since we wish to compute H_n at a large number of points in the estuary, it is desirable for computational reasons (core memory limitation, essentially) to solve separately for the real and imaginary parts of H_n. However, it can be seen from Eq. (12) that the real and imaginary parts of H_n are coupled through two of the frictional terms (physically, this corresponds to a modification of the wave phase due to bottom drag). In order to affect the separation of the real and imaginary parts of H_n, we employed an iterative procedure in which the frictional terms responsible for the coupling are first placed on the right-hand side of the equations. Moreover, in order to guarantee the existence of a relatively simple equivalen variational principle, we also put uncoupled frictional terms on the right-hand side and treat them all in the global iterative scheme. Thus, we set $H_n = h_n + is_n$ in Eqs. (12) and (13), and assume that the dominant forcing excites h_n (i.e., $h_n^* \neq 0$ and $|s_n^*| \leq |h_n^*|$). Using superscripts to index the successive steps of the iteration, we write

$$\nabla.\left(D\nabla h_n^{2k+1}\right) + \frac{n^2\omega^2}{g} h_n^{2k+1} = f_1\left(h_n^{2k-1}, s_n^{2k}\right) \tag{14}$$

$$\nabla.\left(D\nabla s_n^{2k+2}\right) + \frac{n^2\omega^2}{g} s_n^{2k+2} = f_2\left(h_n^{2k+1}, s_n^{2k}\right) \tag{15}$$

where

$$f_1 = \frac{n\omega E}{gD} s_n + \frac{n\omega ED}{E^2+n^2\omega^2D^2} \nabla D.\nabla s_n - \frac{E^2}{E^2+n^2\omega^2D^2} \nabla D.\nabla h_n \tag{16}$$

$$f_2 = -\frac{n\omega E}{gD} h_n - \frac{n\omega ED}{E^2+n^2\omega^2D^2} \nabla D.\nabla h_n - \frac{E^2}{E^2+n^2\omega^2D^2} \nabla D.\nabla s_n \tag{17}$$

The boundary conditions for (14) and (15) are, respectively,

$$h_n = h_n^* \quad \text{on} \quad \Gamma_1 \qquad \frac{\partial h_n}{\partial n} = 0 \quad \text{on} \quad \Gamma_2 \tag{18}$$

$$s_n = s_n^* \quad \text{on} \quad \Gamma_1 \qquad \frac{\partial s_n}{\partial n} = 0 \quad \text{on} \quad \Gamma_2 \quad . \tag{19}$$

The iterative scheme proceeds as follows: as a first step (k = 0), compute $h_n^{(1)}$ by solving (14) subject to (18) with $f_1 = 0$; then substitute $h_n^{(1)}$ in (17), making $f_2 \not\equiv 0$, and thereby ensuring the existence of a nontrivial solution $s_n^{(2)}$ for the pair (15) - (19) even if $s_n^* = 0$. The next step (k = 1) is the computation of $h_n^{(3)}$ using $h_n^{(1)}$ and $s_n^{(2)}$ in (16), followed by a calculation of $s_n^{(4)}$ based on $h_n^{(3)}$ and $s_n^{(2)}$. The procedure is then repeated until sufficient convergence is achieved.

The problem formulation for h_n, i.e., solving (14) subject to (18) with known f_1, is equivalent to the variational condition $\delta J = 0$ for all δh_n vanishing on Γ_1, where

$$J = \int_A \left(\frac{D}{2} |\nabla h_n|^2 - \frac{n\omega^2}{2g} h_n^2 + f_1 h_n\right) dA \quad . \tag{20}$$

(The requirement that the normal derivative of h_n vanish on the variationally free boundary (Γ_2) is the natural boundary condition for this well-known variational principle.) A similar variational principle holds for s_n.

FINITE ELEMENT SOLUTION

Although the use of the finite element method (F.E.M.) in hydrodynamics is a fairly recent development, there are now available a number of texts and papers which set forth the theoretical basis of the method and examples of its use; presentations by Gallagher, Oden, Taylor, and Ziemkiewicz (1975), Connor and Wang (1973), Gray, Pinder and Brebbia (1977) are representative. We shall only briefly review the method as it pertains to the problem at hand, and refer the reader to the literature for more detailed accounts. The domain A is first divided into a number of finite elements (of triangular shape in this particular application) and we seek to compute the values of h_n and s_n at the element vertices. Let M denote the total number of vertices (or nodes), m_1 of which are located on the open boundary. The unknown variable (say, h_n) and parameters (e.g., D) are then approximated over each element by some interpolative representation in terms of their nodal values (h_{n_i}, D_i): we use the simplest possible basis (or shape) functions, i.e., linear interpolators and, as a consequence, the function J in Eq. (20), expressed as the sum of the integrals over each element, is a quadratic function of the h_{n_i} values. The variational condition requires that we set

$$\frac{\partial J}{\partial h_{n_i}} = 0$$

for all nodes except those belonging to Γ_1. This yields a set of ($M-m_1$) linear algebraic equations in M unknowns. This set, together with the m_1 equations expressing the boundary condition on Γ_1, completely determines the problem.

The numbering of the nodes can be done so as to minimize the bandwidth of the coefficient matrix, because the partial derivative of J with respect to a particular h_{n_i} involves only integrals over those elements having h_{n_i} as one of their nodes. The integration over each element is carried out either analytically (as for the contribution to the coefficient matrix of the first two terms of (20)) or numerically (as is necessary for the right-hand side vector, since f_1 is a nonlinear function of position). The L-U decomposition of the banded coefficient matrix was performed by an algorithm that utilizes row equilibration and partial pivoting. Since the matrix is the same for both variables h_n and s_n and does not change during the iteration process, its triangular decomposition need be done only once.

Finally, after the calculation of h_n and s_n has converged, the Fourier coefficients $\underline{U}_n$ of the velocities are computed according to Eq. (8). The differentiation of the H_n field, that is linearly interpolated over each triangle, gives constant element values that are taken as representative of the velocity coefficients at the centroid of each element.

APPLICATION TO HOOD CANAL

In an effort to ascertain the potential of the method for resolving efficiently the intricacies of natural tidal flows, we selected a segment of Hood Canal, Washington, as a test region. Hood Canal is the westernmost arm of Puget Sound (Fig. 1), which is a complex network of fjords and channels that communicates with the Pacific Ocean through the Strait of Juan de Fuca. The construction of a base for nuclear submarines at Bangor on the shore of Hood Canal makes the region of particular environmental interest and concern. Qualitative information concerning tidal currents in Hood Canal has been obtained from the Puget Sound hydraulic model and is available in various publications and technical reports from the University of Washington (e.g., Rattray and Lincoln, 1955; McGary and Lincoln, 1977). Although Hood Canal is not a truly homogeneous estuary, the intensity of freshwater runoff is usually low, so that the estuarine circulation mode does not modify substantially the barotropic tidal mode.

The test region has three open boundaries. The driving boundary condition (tidal height) is imposed along the northernmost boundary which is located off South Point on the seaward side of the sill. Rectangular basins of uniform depth whose volumes are roughly equivalent to those of Dabob Bay and the southern portion of Hood Canal are appended to the other two open boundaries, denoted Γ_3 and Γ_4, where a simple analytical solution is matched to the interior solution. In order to simulate the influence of Dabob Bay and southern Hood Canal on tidal currents in the Bangor region, we approximated water motion in the appendages by linear, frictionless long waves which are perfectly reflected at the closed ends of the

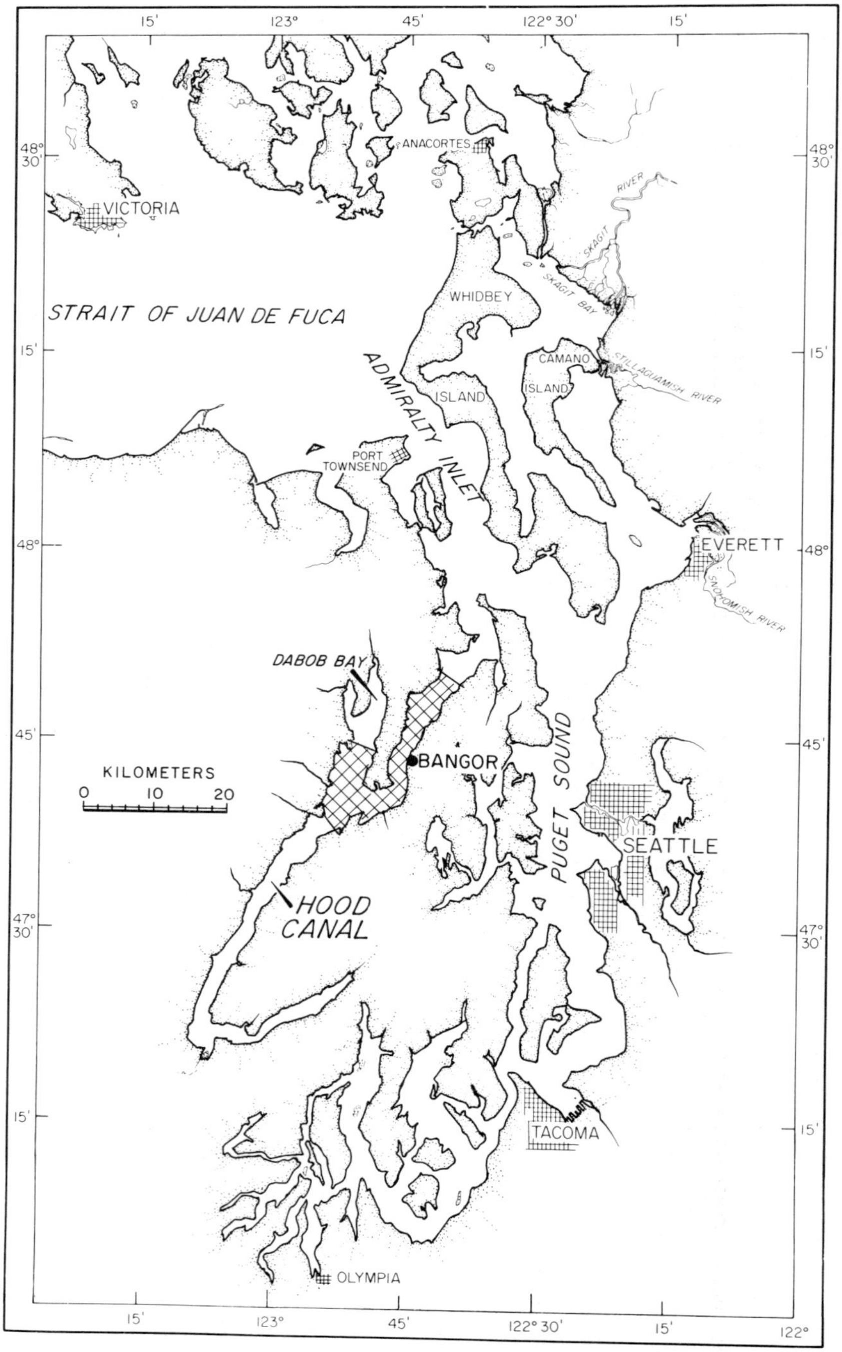

Figure 1. Map of Puget Sound. The hatched segment of Hood Canal is the test region.

rectangular basins. The matching condition relates the normal surface slope to the elevation on those boundaries. Such a condition can easily be incorporated in the variational principle. Alternatively, it can be implemented by direct manipulation of the algebraic system of equations in the following way: first, assume that the boundary elements along Γ_3 and Γ_4 are part of the domain in which the analytical solution is applicable; then replace the equations generated by the variations of the variable along such boundaries (e.g., $\partial J/\partial h_{n_k} = 0$, $h_{n_k} \in \Gamma_3$ or Γ_4) by expressions that relate the solution at each boundary node to that at a corresponding interior point in the direction of propagation of the one-dimensional wave. We performed numerical experimentation with both the methods for the case of a rectangular basin and each was found satisfactory. We used the latter (direct manipulation of the equations) in the Hood Canal production runs.

The finite element grid shown in Fig. 2 was prepared manually, and consists of 1495 triangles (853 nodes) of gradually varying area. The mesh resolution is finest in the region of principal interest between the seaward boundary and Hazel Point. Also, the resolution is finer along the sides of the channel as required for a fair approximation of the bottom topography (Fig. 3). The shallowest depth along the shoreline boundary is taken as 5 fathoms (about 9 m) so as to avoid unrealistically high friction coefficients and to make certain that linearization of the mass conservation equation remains a valid approximation. The bandwidth of the coefficient matrix is equal to twice the largest difference between node numbers of an element plus one, and in this case is 51: its minimization is hampered somewhat by the bifurcation of the channel at the southern end.

The next seven figures display selected results of a calculation corresponding to the hypothetical case of single mode forcing of period T_p = 12.5 hours, $h_1^* = 2$ m on Γ_1, and $E = 3.10^{-3}$ m sec^{-1}. Fig. 4 shows contours of the solution for $h_1(x,y)$, the cosine component of the elevation. Convergence within a tolerance of 0.5% for both h_1 and s_1 was achieved after the sixth iteration. The next figure shows the velocity field at the time of maximum flood, when the sole effect of friction is to slow the flow along the (shallow) shoreline. In this and subsequent figures, α denotes nondimensional time t/T_p. The velocity field is more varied during the transition from flood to ebb, i.e., around $\alpha = 0$. The cross-channel variations of phase due to "bottom" friction are such that,

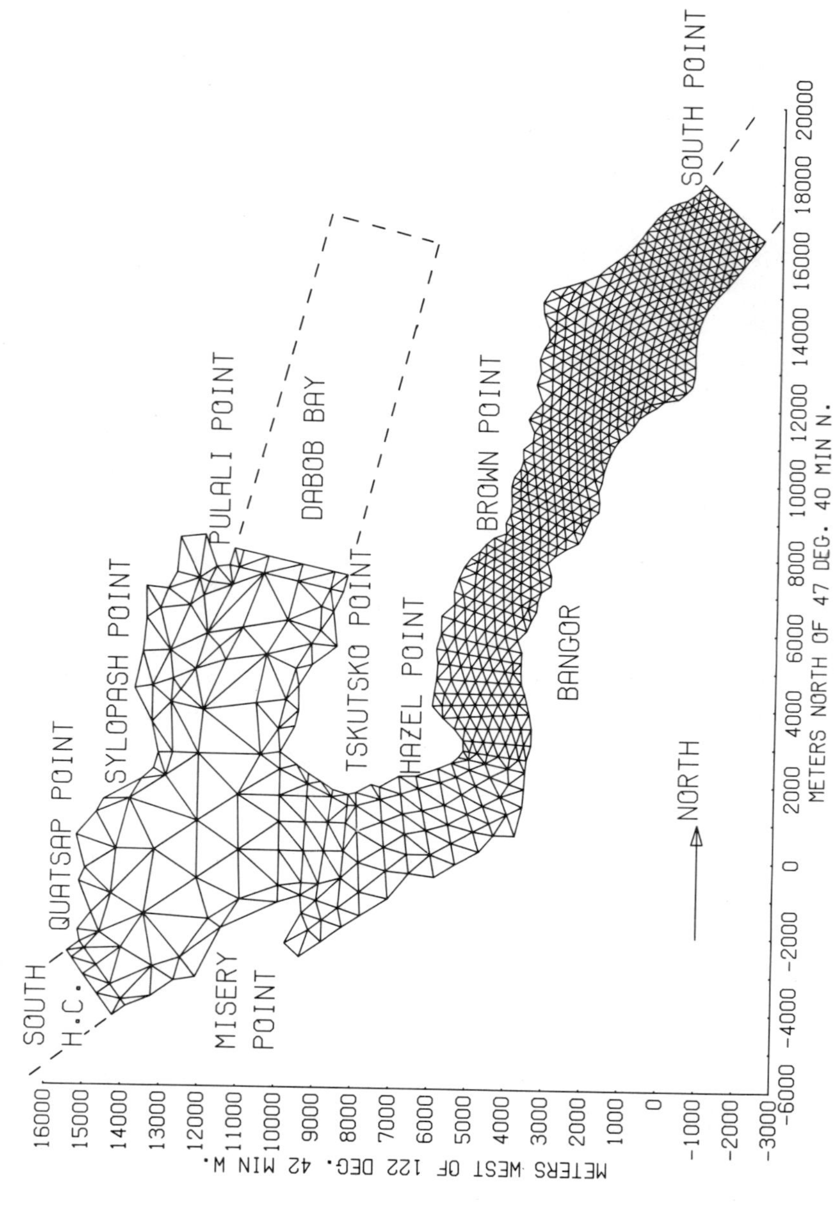

Figure 2. Finite element grid.

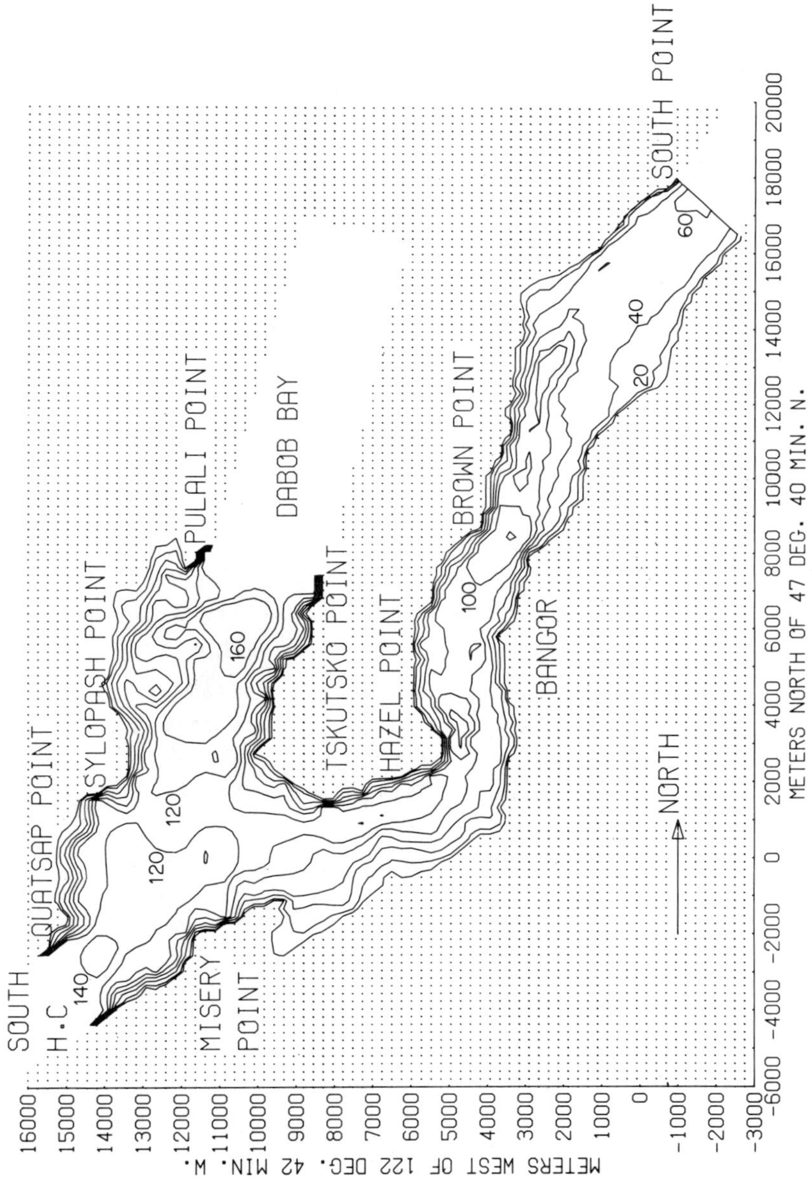

Figure 3. Topography of test region.

as observed in nature, the tide turns first along the shores. This topographical effect produces a shear that generates eddy-like structure during the transition period. Because of the several simplifications introduced in the equations and forcing function considered in this example, we have not made quantitative comparisons between our preliminary results and available hydraulic model data. However, it can be said that so far as their scale and position are concerned, the eddy patterns shown in these figures are similar to those observed in the Puget Sound hydraulic model.

For programming convenience, we used a CDC 7600 computer that has a large core memory. The program solving the problem described here utilizes a mass storage of about 125,000 words, is compiled in less than 3 CP seconds and executed in 6.5 CP seconds.

DISCUSSION

In this paper we have described preliminary steps and results of a study intended to demonstrate the feasibility of the method proposed by Pearson and Winter (1977) for the computation of tidal motion in homogeneous estuaries. The new method derives its efficiency, as compared with time-stepping procedures, from a combination of harmonic decomposition and an iterative scheme to treat nonlinearities. In solving the resulting elliptic problem, the method takes advantage of the superior flexibility introduced by the finite element approach. Lastly, the reformulation of the problem in terms of a variational principle, although not a necessary condition for the applicability of a finite element procedure, decreases the order of the highest derivatives to be approximated and simplifies the implementation of the boundary conditions.

Finite element methods are becoming increasingly popular among tidal modelers (e.g., Grotkop, 1973; Connor and Wang, 1973; Taylor and Davis, 1975; Brebbia and Partridge, 1976; Sündermann, 1977). Some workers, however, contend that they are economically unattractive, in comparison to finite difference methods, for practical calculations (e.g., Weare, 1976). Our experience with the present approach leads us to assert that the combination of harmonic decomposition and iteration is likely to provide computational efficiency in a wide variety of practical problems. We were pleased to learn at the Ninth Liege Colloquium that another group of researchers (Le Provost and Poncet, 1977) is currently working along similar lines.

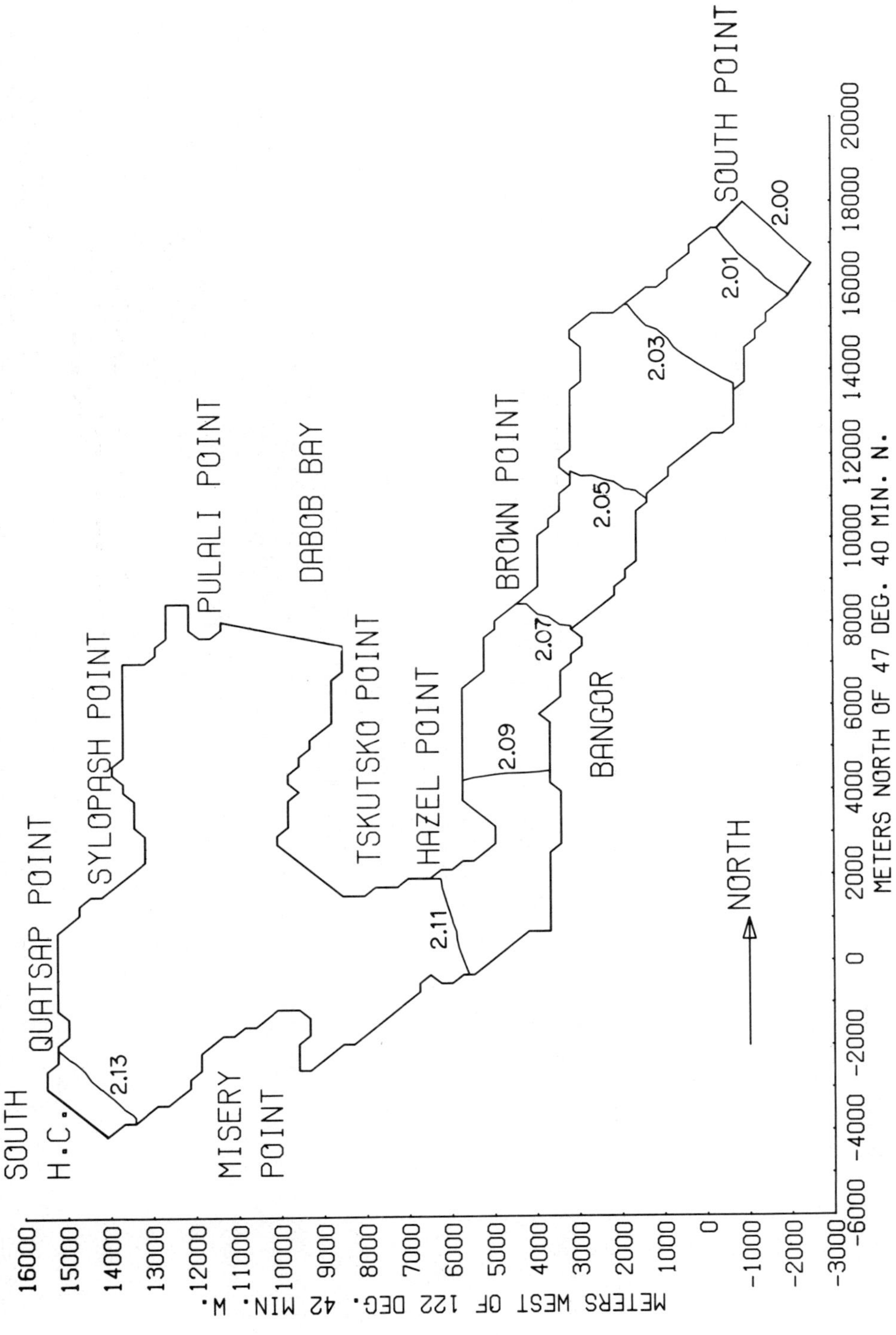

Figure 4. Isopleths of h_1, values in meters.

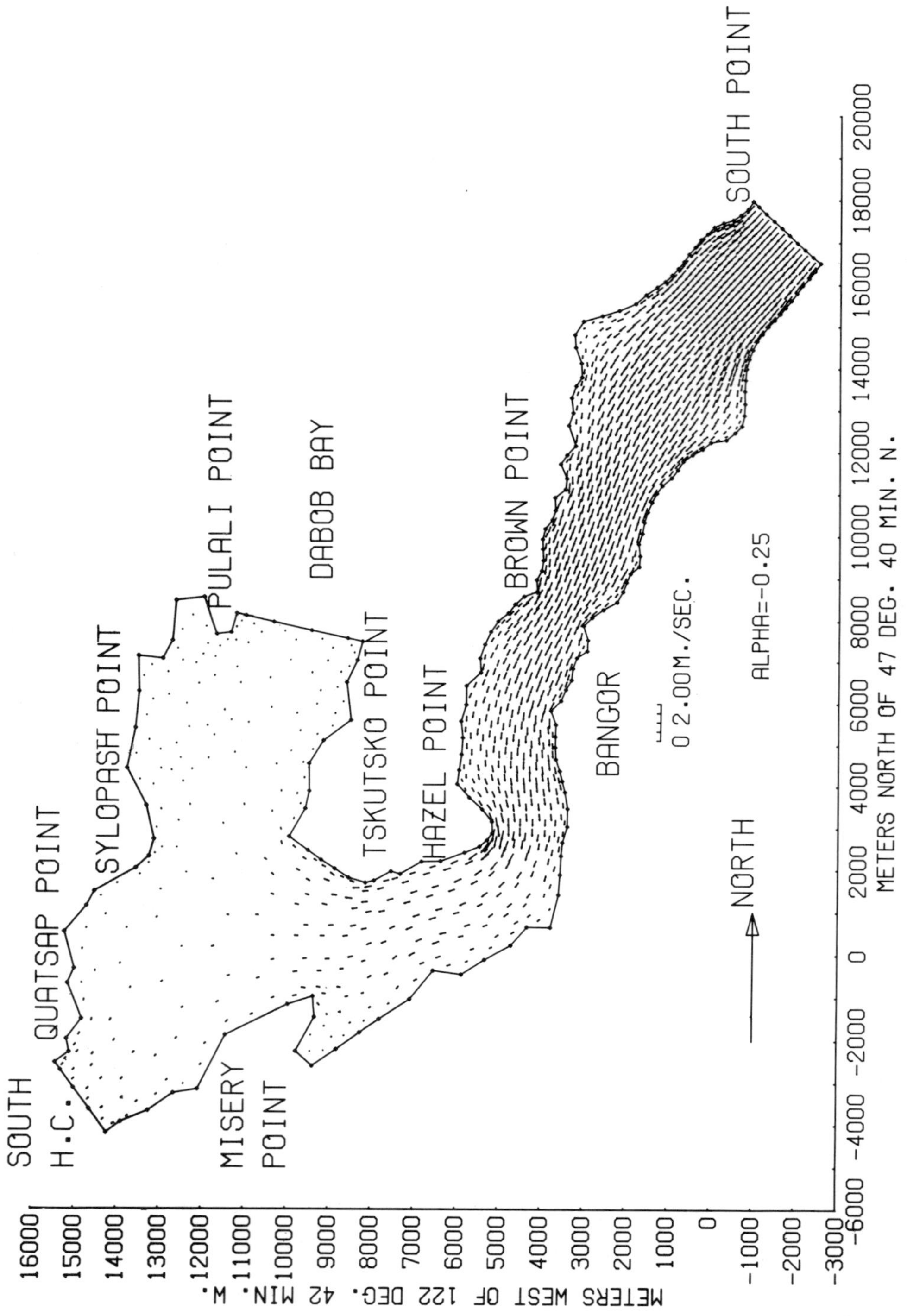

Figure 5. Velocity field at time $\alpha = -0.25$.

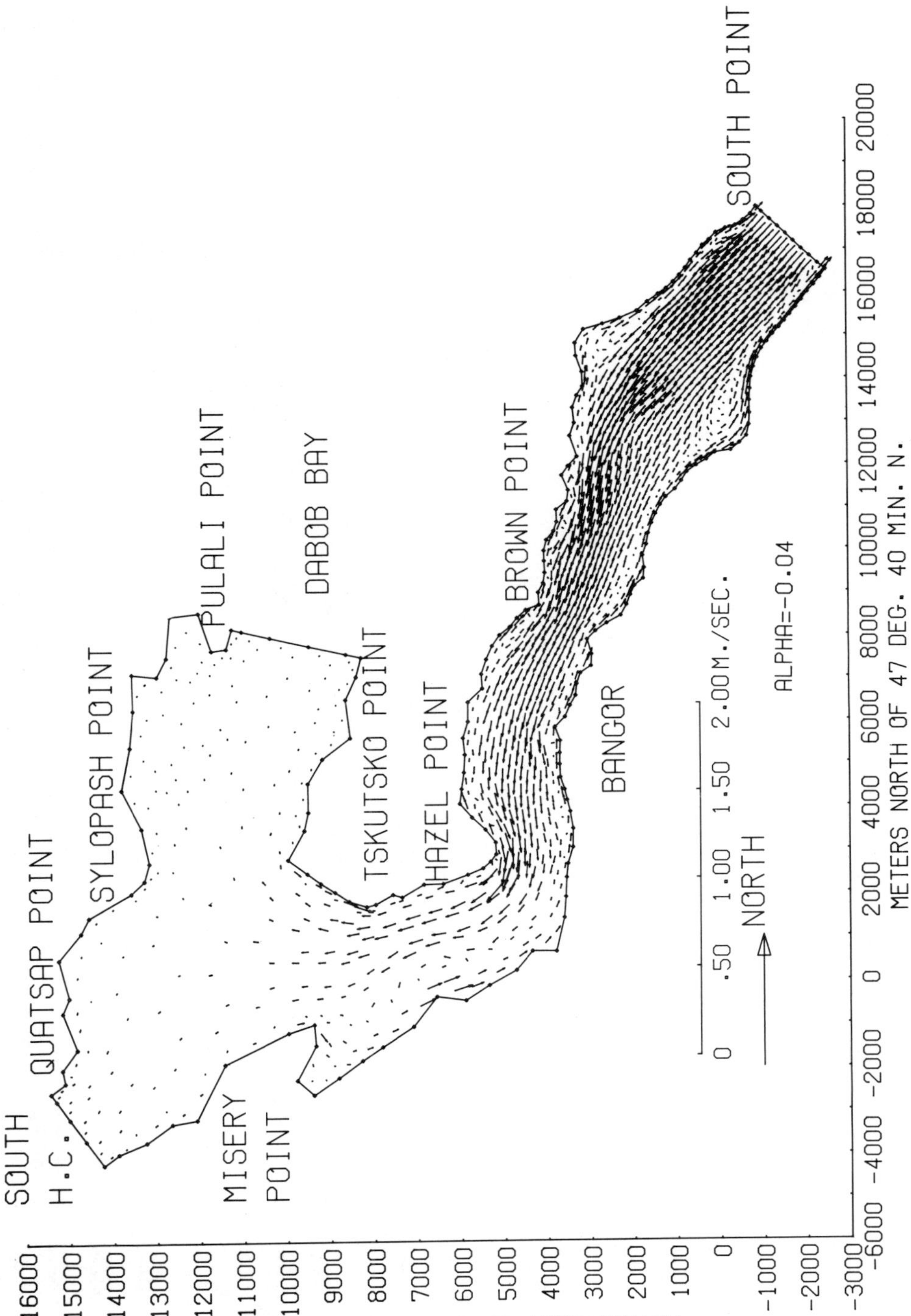

Figure 6. Velocity field at time $\alpha = -0.04$.

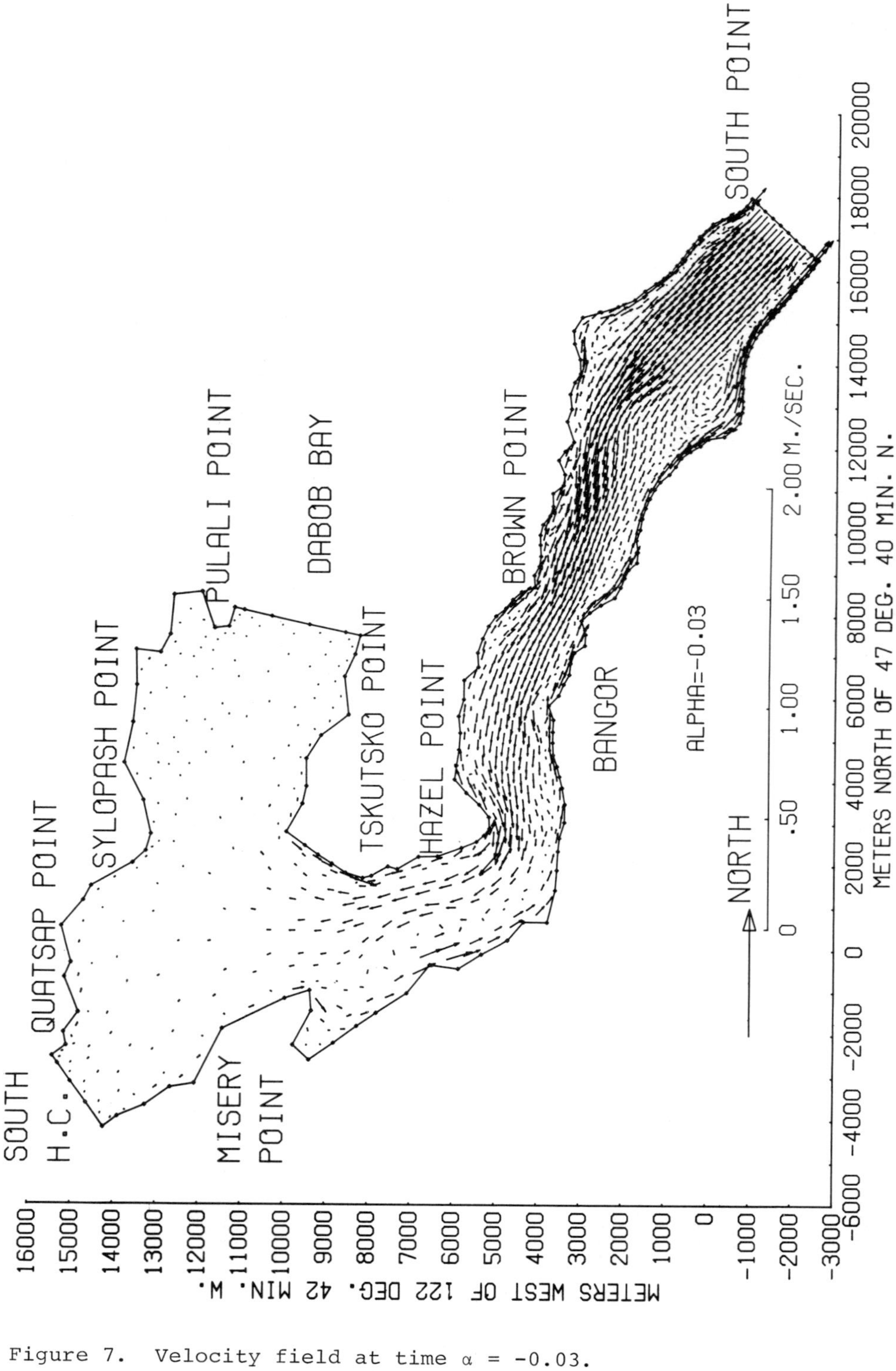

Figure 7. Velocity field at time $\alpha = -0.03$.

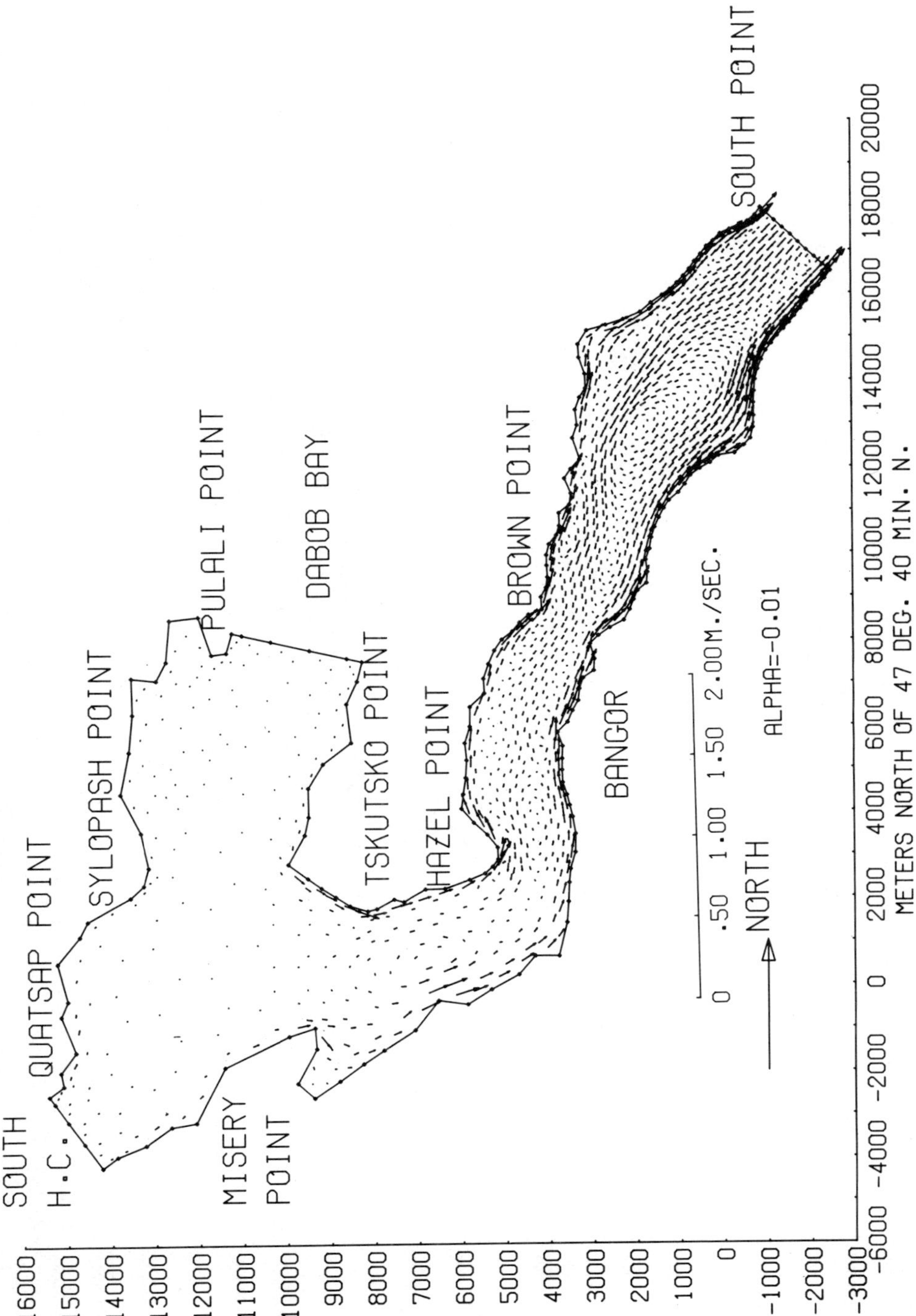

Figure 8. Velocity field at time $\alpha = -0.01$.

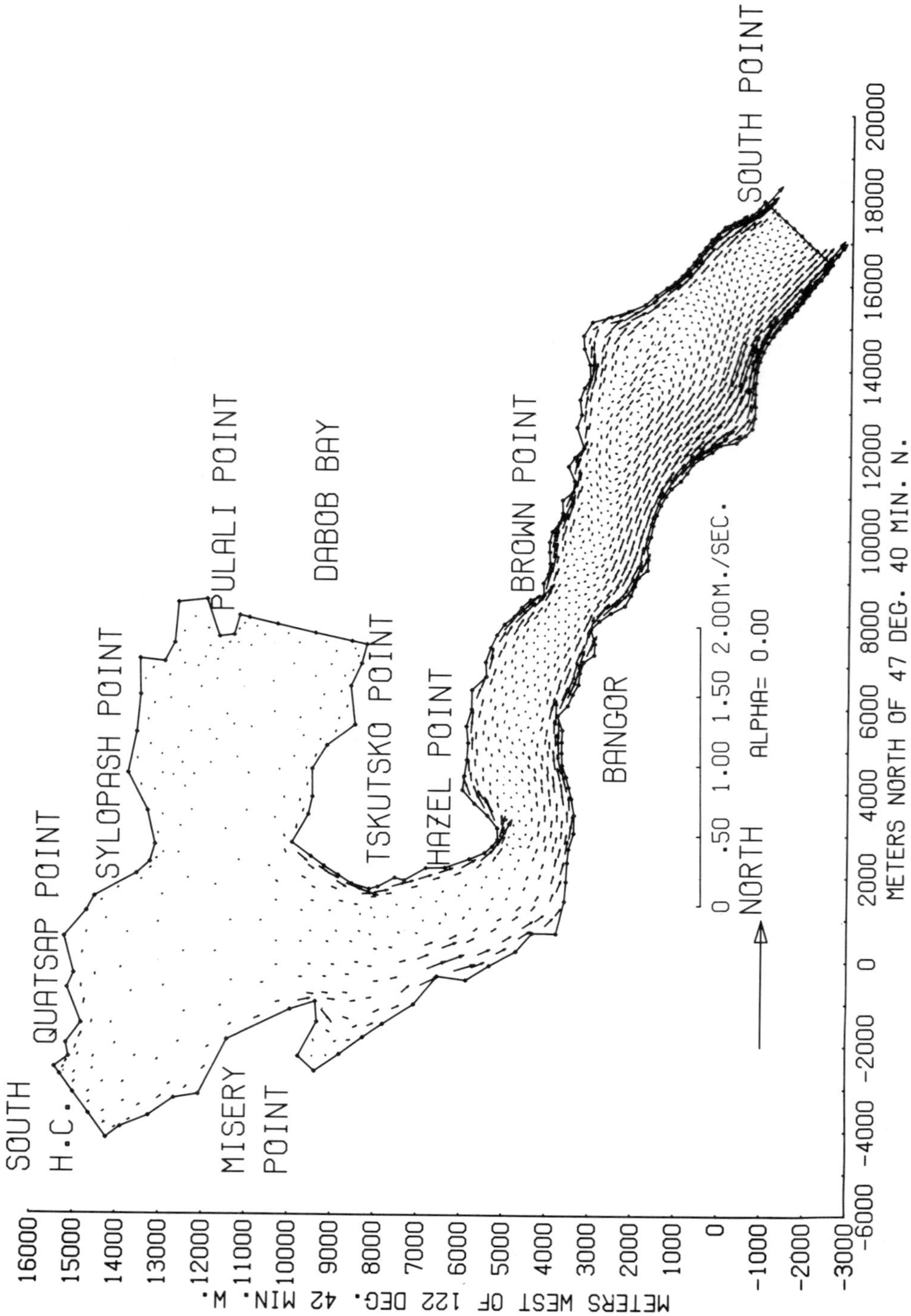

Figure 9. Velocity field at time α = 0.00.

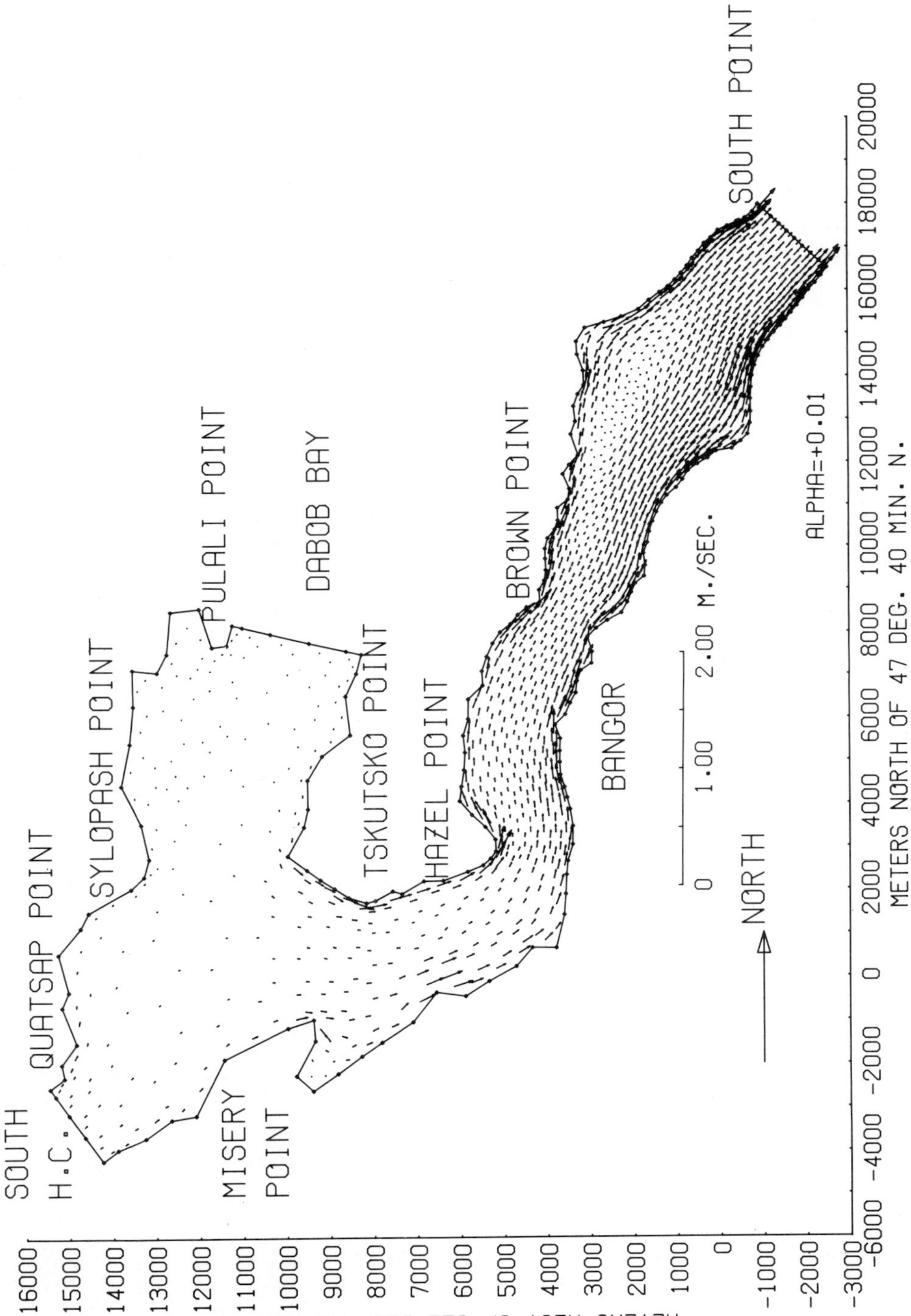

Figure 10. Velocity field at time $\alpha = +0.01$.

ACKNOWLEDGMENTS

The research reported here was supported in part by the Oceanography Section, National Science Foundation, under Grant OCE76-00406, and in part by the Washington Sea Grant Program, which is maintained by the National Oceanic and Atmospheric Administration, U.S. Department of Commerce. We are grateful to Ms. Jane M. Glass and Dr. Lawrence Lewin for their assistance in preparing the finite element grid of Hood Canal and the visual display of computer results.

REFERENCES

Brebbia, C.A., and Partridge, P.W., 1976. Finite element simulation of water circulation in the North Sea. Applied Mathematical Modelling, 1(2): 101-107.

Connor, J.J., and Wang, J.D., 1973. Mathematical models of the Massachusetts Bay. Part I: Finite element modeling of two-dimensional hydrodynamic circulation. Report No. MITSG 74-4, M.I.T., Cambridge, 57 pp.

Gallagher, R.H., Oden, J.T., Taylor, C., and Zienkiewicz, O.C., Eds., 1975. Finite Elements in Fluids, Vol. 1, Wiley - Interscience, London, 290 pp.

Gray, W.G., Pinder, G.F. and Brebbia, C.A., Eds, 1977. Finite Elements in Water Resources, Pentech Press, London, 1008 pp.

Grotkop, G., 1973. Finite element analysis of long-period water waves. Computer Methods in Applied Mechanics and Engineering, 2: 147-157.

Hess, K.W., and White, F.M., 1974. A numerical tidal model of Narragansett Bay. Marine Tech. Rept. No. 20, University of Rhode Island, Kingston, 141 pp.

Le Provost, C., and Poncet, A., 1977. Finite element method for spectral modelling of tides. In preparation. (*Le Provost's affiliation is with the Institut de Mécanique, Grenoble, France.*)

McGary, N., and Lincoln, J.H., 1977. Tide Prints: Surface Tidal Currents in Puget Sound. Washington Sea Grant Publication distributed by University of Washington Press, Seattle, Washington, 51 pp.

Pearson, C.E., and Winter, D.F., 1977. On the calculation of tidal currents in homogeneous estuaries. Journal of Physical Oceanography, 7(4): 520-531.

Rattray, M., Jr., and Lincoln, J.H., 1955. Operating characteristics of an oceanographic model of Puget Sound. Transactions, American Geophysical Union, 36: 251-261.

Sündermann, J., 1977. Computation of barotropic tides by the finite element method. In: Finite Elements in Water Resources, Pentech Press, London, pp. 4.51-4.67

Taylor, C., and Davis, J.M., 1975. Tidal propagation and dispersion in estuaries. In: Finite Elements in Fluids, Vol. 1, Wiley - Interscience, London, pp. 95-118.

Weare, T.J., 1976. Finite element or finite difference methods for the two-dimensional shallow water equations? Computer Methods in Applied Mechanics and Engineering, 7: 351-357.

Contribution No. 1002, Department of Oceanography, University of Washington, Seattle, WA.

A NUMERICAL MODEL OF BARATROPIC MIXED TIDES BETWEEN VANCOUVER ISLAND AND THE MAINLAND AND ITS RELATION TO STUDIES OF THE ESTUARINE CIRCULATION

P.B. CREAN
Institute of Ocean Sciences, Department of Fisheries and the Environment, Sidney,B.C.

INTRODUCTION

The waters between Vancouver Island and the mainland coasts of British Columbia and the State of Washington constitute a complex estuarine system in which the dominant source of fresh water is the Fraser River, Fig. 1. Before any realistic ecological models relating to the major fisheries of the region, including possible effects of domestic and commercial effluents or of oil spills, can be undertaken it is necessary to achieve a quantitative understanding of the physical processes that determine the essential character of the flow field and distributions of scalar properties. This involves an extended series of studies and it is the purpose of this paper to review briefly the present status of this work with emphasis on the roles played by the tides and tidal streams, then to describe, and present some results from, a numerical tidal model and lastly to indicate in general terms how these results are presently being used to introduce the effects of tides into an experimental "upper layer" model used to study the flow of fresh water from the Fraser River over the salt water in the Strait of Georgia.

The basic geographical features of the region, shown in Fig. 1, consist essentially of an outer strait (Juan de Fuca Strait) 140 km long and an inner strait (Strait of Georgia) of length 220 km. A complex system of islands (San Juan and Gulf Islands) and passages is located between the inner and outer straits. The inner strait is further connected to the open sea through a network of narrow channels, containing relatively shallow sills, which lead into Johnstone Strait, a single passage between Vancouver Island and the mainland.

Bottom topography in the region of the model is illustrated in Fig. 2 using depth contours in fathoms taken from existing hydrographic charts.

In the outer strait maximal depths decrease gradually from some 300 m near the entrance to about 100 m in the channels between the shoals in its inner end. In the region of the San Juan and Gulf Islands, two main conveying channels between the inner and outer straits may be distinguished, Haro and Rosario Straits. Of the two Haro Strait contains the deepest channel, characterized by depths of order 180 m - 250 m. Maximal depths along the Strait of Georgia are generally of order 150 m - 400 m. A topographical feature of particular interest in the present context is the large longitudinal change in cross-sectional area of the strait in the vicinity of the extensive shallow bank that delineates the seaward limits of the Fraser River delta.

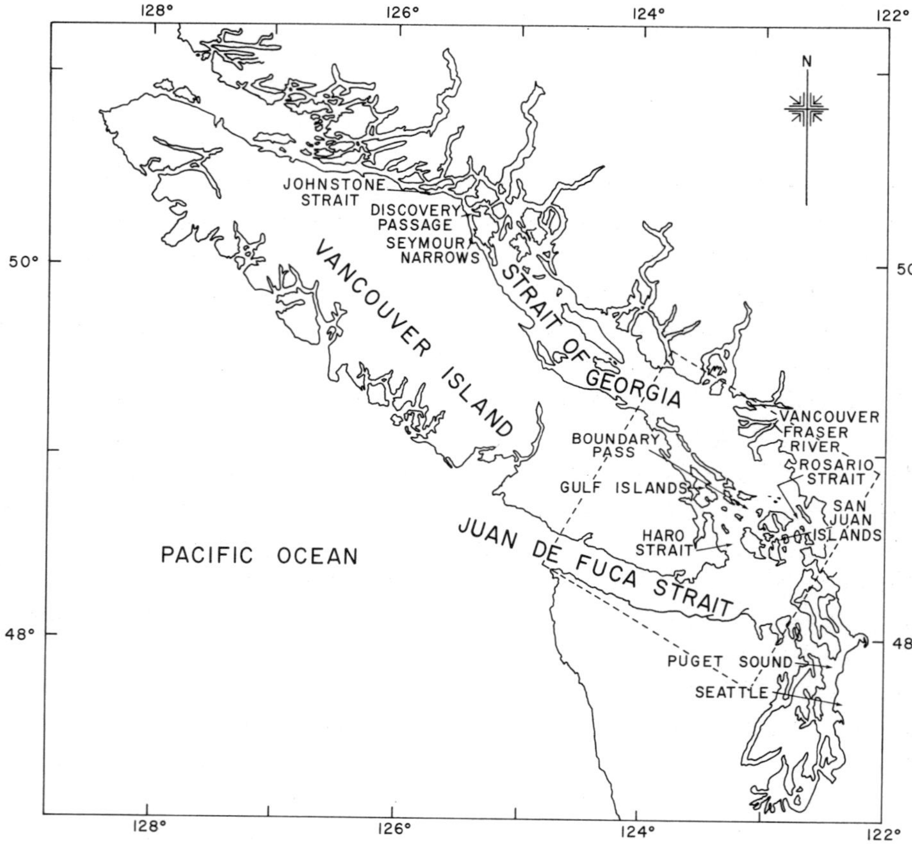

Fig. 1. The waters between Vancouver Island and the mainland coast. The dotted lines denote the boundary of the numerical model grid scheme.

A considerable number of studies pertinent to the physical oceanography of the system have been made over the past fifty years. Hutchinson and Lucas (1931) have described the water properties, nutrient concentrations and phytoplankton populations in the Strait of Georgia. The distribution and origins of the primary water masses within the Juan de Fuca Strait and the Strait of Georgia were described by Redfield (1950a). Waldichuk (1957) and also Tully and Dodimead (1957) have discussed the physical oceanography of the Strait of Georgia.

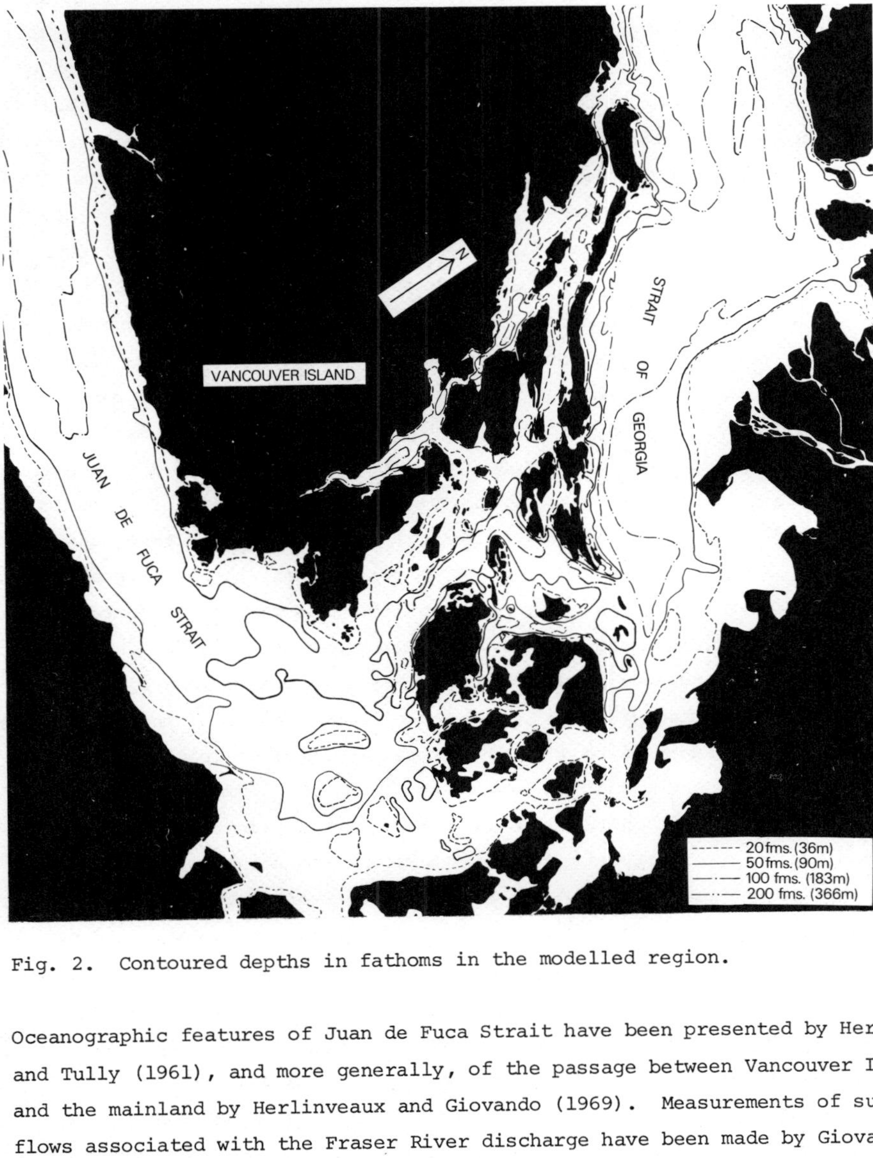

Fig. 2. Contoured depths in fathoms in the modelled region.

Oceanographic features of Juan de Fuca Strait have been presented by Herlinveaux and Tully (1961), and more generally, of the passage between Vancouver Island and the mainland by Herlinveaux and Giovando (1969). Measurements of surface flows associated with the Fraser River discharge have been made by Giovando and Tabata (1970), while a study of sub-surface current records, each of about one year duration, from a cross-sectional array of current meters located in the approximate vicinity of the Fraser delta in the Strait of Georgia has been reported by Chang, Pond, Tabata (1976). Good predictions of tidal elevations along the Fraser River under different discharge rates have been obtained using a one-dimensional numerical model (Ages and Woollard, 1976). Currents and scalar properties in Johnstone Strait, the single channel through which any exchanges of water between the northern part of the Strait of Georgia and the open sea must occur, have been discussed by Thomson (1976). Extensive data on water elevations and currents have been gathered by the Canadian Hydrographic Service. Data records from current meters moored over cross-sections in Juan de Fuca Strait, Strait of Georgia and Johnstone Strait are available, (Tabata et al. 1970,a,b,c, Tabata and Stickland 1972,a,b,c,d, Department of Environment 1972,a,b,c, 1973, Huggett et al. 1976a,b). Additional velocity measurements from vessels anchored in the general vicinity of the Fraser River plume have also been reported (Tabata et al. 1970). An extensive survey of tidal current measurements in Juan de Fuca and Rosario Straits has been presented by Parker (1977). Data on the distribution of scalar properties, has also been obtained, partly with a view towards the subsequent three-dimensional numerical simulation of processes dominating the estuarine circulation. Thus, in the course of twelve oceanographic cruises over a year, the basic seasonal changes in water properties throughout the Strait of Georgia, the vicinity of the San Juan Islands and Juan de Fuca Strait were determined (Crean and Ages, 1971). Further scalar data in Juan de Fuca Strait, where the seaward boundary of a three-dimensional model might plausibly be located, and over a period when an extensive array of moored current meters were in position over a cross-section of the Strait, have also been obtained (Crean and Miyake, 1976, Crean and Lewis, 1976). The first of these data sets includes simultaneous time-series observations by two vessels in collaboration with the Department of Oceanography of the University of Washington. In addition to the above, numerous local oceanographic studies have been made of the various mainland inlets.

The dominant estuarine features of the system may be illustrated by winter and summer salinity distributions over vertical sections taken through the median lines along the Strait of Georgia, Haro and Juan de Fuca Straits, Fig. 3 (Crean and Ages, 1971). The major source of salt water is located at depth at the entrance of Juan de Fuca Strait. Between winter and summer there occurs a marked increase in the salinity of the deep water in Juan de Fuca Strait as shown by the increased shaded area, representing water having a salinity in excess of 32.5‰, which is thought to be associated with upwelling processes on the continental shelf. The major single source of fresh water entering the system is the Fraser River which is estimated to account for some 70-75% of the total fresh water input to the inshore passage between Vancouver Island and the mainland (Herlinveaux and Tully, 1961). There exists a strong seasonal variation, determined by the thaw in the mountainous hinterland, in the surface dilution in the general vicinity of the river mouth. Of particular interest is the region of the "sills" between the inner and outer straits where the horizontal salinity gradients are indicative of strong mixing processes. The intermediate salinity water which results can either move seaward at the surface in Juan de Fuca Strait or sink to form deep water in the Strait of Georgia. As will be shown below, the tidal streams in this region are strong, and clearly exert a major influence on the fundamental estuarine circulation.

Annual variations in the water salinities at depth 100 m are evident throughout these coastal waters (Pickard, 1975). Following the summer salinity maximum in Juan de Fuca Strait there is evidence of a fall maximum in Boundary Passage followed by a winter maximum in the Strait of Georgia. Long-term variations of annual mean values are small. Reference below to a three-dimensional numerical simulation of the main features of the estuarine circulation within the system may be taken to imply the seasonal variations about the normative features of the winter and summer salinity distributions as sectionally shown in Fig. 3.

It is of interest to note that in recent years, the possibility of controlling the Fraser River discharge has been discussed. In such a case, the major intrusion of salt water present at depth in Juan de Fuca Strait during the summer would not be accompanied by maximal dilution of the surface waters in the Strait of Georgia. Any significant associated change in the estuarine circulation might effect primary food production critical to the major salmon fishery associated with the Fraser River. Thus one of the objects of these studies concerns the prediction of any such change likely to result from the imposition of such a control.

Both tides and winds exert a major influence on the motions of fresh water entering the Strait of Georgia from the Fraser River. Satellite pictures such as that shown in Fig. 4, can sometimes graphically illustrate the general extent of the shallow river plume that forms in the southern part of the Strait of Georgia when sharp contrasts occur between the silty water in the plume and sea water in the Strait. Reference to Figs. 1 and 2 permit ready recognition of such features as the Strait of Georgia, the Gulf and San Juan Islands, and the Fraser River Delta, just to the north of which lies the greater district of Vancouver. The Main Arm of the Fraser River, which carries about 80% of the total discharge, is clearly distinguished and the southward veering flow is partly brought about by an angled training wall, not evident in the picture, which extends out to the edge of the deep water.

Unfortunately such pictures do not afford a ready means of following the changing geographical extent of the plume, since clear skies and such clarity of contrast are generally associated with north-westerly winds. The present picture was taken during a period of 10-13 m/sec winds from the northwest and coincided with a lower-low water in the Strait.

Based partly on observations, which have been taken recently in connection with an "upper layer" model of the plume described briefly below, a provisional general description of the formation of the plume under moderate river-flow conditions is suggested as follows. As the water level falls at the river mouth and the tidal streams are ebbing in the Strait the salt wedge in the river

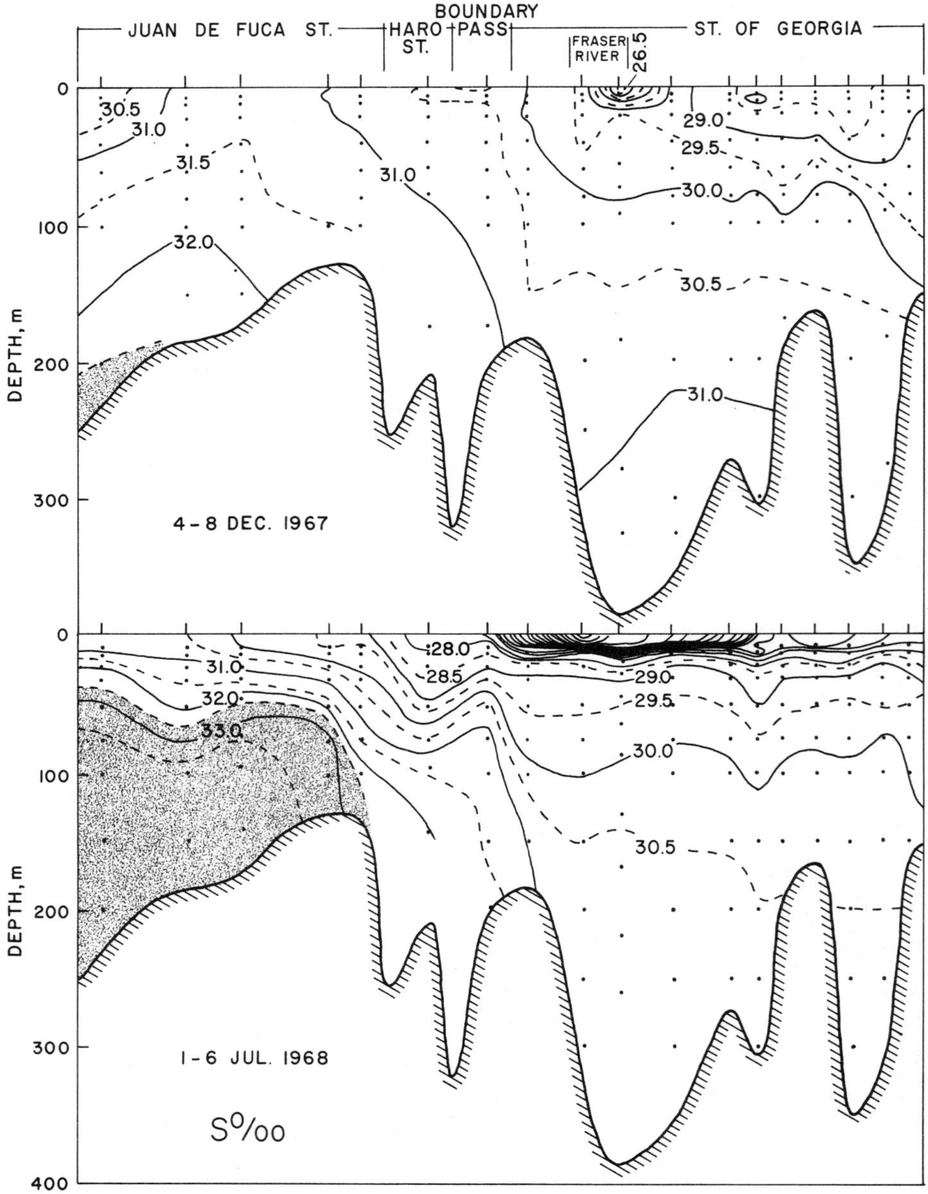

Fig. 3. Winter and summer contoured vertical sections of salinities ‰ through the median lines of Juan de Fuca Strait, Haro Strait, Boundary Passage and the Strait of Georgia (Crean and Ages, 1971).

retreats and the rate of discharge from the river increases. The silty water spreads out over the salt water to form a southward veering plume. Roughly an hour before lower-low water the discharge rate attains its maximum. As the streams turn to flood and levels rise in the Strait the discharge rate decreases, the salt wedge intrudes into the river and the shallow plume just formed, moves northward and outward away from the river mouth. The process is then repeated with new plume forming over the remnants of its predecessors except in such a case as shown in Fig. 4, where winds have swept such remnants southward towards the strong tidal-mixing region of the San Juan Islands.

Fig. 4. Satellite picture (LANDSAT-1 imagery from 1100 km) of the southern Strait of Georgia and the region of the San Juan and Gulf Islands showing the Fraser River plume. The land mass in the lower lefthand corner is Vancouver Island.

Under freshet conditions, it appears that the river flow holds its entrant direction in a strong narrow stream for a considerable distance across the Strait and then turns northward. This stream can maintain its identity over a distance of some 90 km from the river mouth, eventually following the line of the mainland coast and then dispersing (Giovando and Tabata, 1970). Clearly there exists much diversity in the possible motions of the 'plume' deriving from variations in the rate of discharge from the river and the mixed nature of the tides. Over and above these considerations there remain the effects of wind stress acting over the surface of the 'plume'.

It is evident that the tides and tidal streams play an important role in determining the estuarine circulation within the system, not only with respect to mixing processes, but also with regard to the relative proportions of the Fraser River discharge finding seaward egress respectively through Juan de Fuca or Johnstone Straits. It is also possible that there may exist significant residual flows induced by the barotropic tide.

Basic features of the main semi-diurnal (M_2) and diurnal (K_1) tides are illustrated by cophase and corange diagrams obtained from an earlier model described briefly below, Fig. 5. Of particular interest is the semi-diurnal degenerate amphidromic system in the inner part of Juan de Fuca Strait and rapid changes in amplitude and phase through the region of the San Juan Islands. Tides are of the mixed type and it will be noted that both constituents are of similar amplitude in the northern part of the Strait of Georgia.

The earliest theoretical work on tides in the system was undertaken by Redfield (1950b). Assuming a linear superposition of damped incoming and reflected waves he concluded that the frictional dissipation was considerably greater than that found in applying the same technique to Long Island Sound and the Bay of Fundy. Studies using an overall one-dimensional model of all the major conveying channels in the system, operated by prescribing elevations at the entrance to Juan de Fuca Strait and near the northern end of Vancouver Island, satisfactorily reproduced the main features of the M_2 and K_1 constituents, but required an unusually large frictional dissipation to optimise the agreement between the calculated and observed tidal constants around the interior of the modelled region. It was concluded that the inclusion of the Coriolis effect was essential before any further progress could be made (Crean, 1969). A two-dimensional, vertically-integrated numerical tidal model with multiple lateral boundary openings simulating connections with the ocean and the entrances to the various mainland inlets was developed. Excellent agreement was obtained between the observed and computed M_2 and K_1 tidal harmonic constants using an overall coefficient of friction of .0025 as commonly applied in work on coastal seas. Flows at the boundary openings in the mainland coast of the Strait of Georgia proved, however, to be quite unrealistic. The addition of further one-dimensional models to the existing two-dimensional scheme reduced the number of external openings to two, each simulating connections with the open ocean (Crean 1976). Numerical experiments with this model permitted a proper

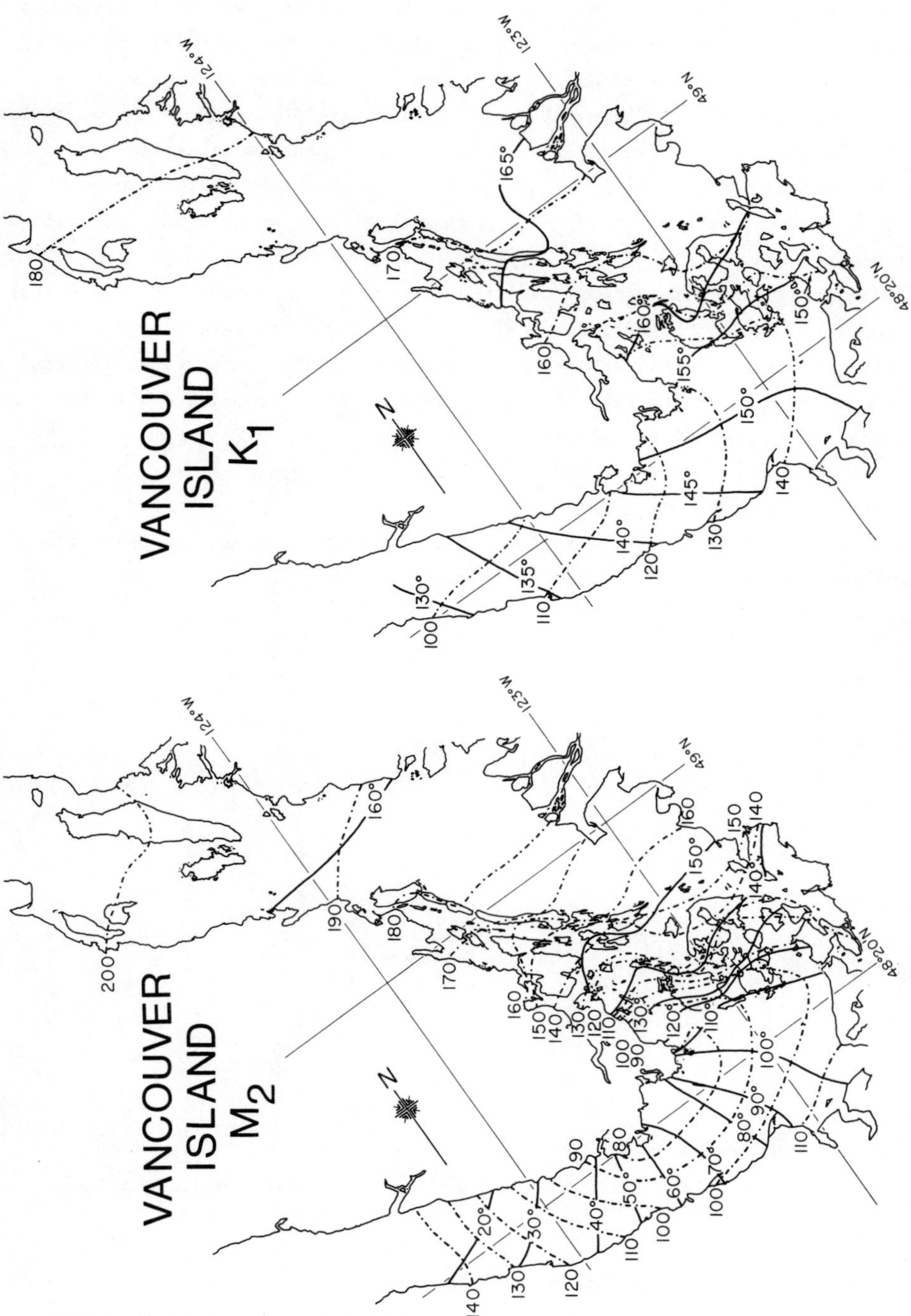

Fig. 5. Corange (cms) and cophase diagrams of the main semi-diurnal (M_2) and diurnal (K_1) tidal constituents derived from earlier studies employing an overall numerical model of the system.

distribution of friction to be determined which in turn revealed significant non-linear interactions between the major tidal constituents in the region of the San Juan Islands. Further, it was shown that the 4-km mesh size of the two-dimensional part of the model could not adequately simulate the propagation of long waves through Haro Strait unless a line of contiguous meshes having depths corresponding to maximum channel depth, rather than to the averaged depth over each mesh, was included. A summary of this work is being submitted for publication.

At an early stage of the work described above, a configuration of the coarse (4 km) grid model with boundary openings conforming to those employed in the present fine grid (2 km) mesh model was attempted. This attempt was eventually abandoned because of the highly unrealistic flow field associated with the open boundary in the Strait of Georgia.

FUNDAMENTAL HYDRODYNAMICAL EQUATIONS

The vertically integrated equations of continuity and momentum applicable to tidal motions in a flat rotating sea are:

$$\frac{\partial \zeta}{\partial t} + \frac{\partial U}{\partial x} + \frac{\partial V}{\partial y} = 0 \qquad (1)$$

$$\frac{\partial U}{\partial t} + \frac{\partial}{\partial x}\left[\frac{U^2}{\zeta+h}\right] + \frac{\partial}{\partial y}\left[\frac{UV}{\zeta+h}\right] - fV + \varepsilon\left[\frac{\partial^2 U}{\partial x^2} + \frac{\partial^2 U}{\partial y^2}\right]$$

$$+ g(\zeta+h)\frac{\partial}{\partial x}(\zeta) + \frac{kU\sqrt{U^2+V^2}}{(\zeta+h)^2} = 0 \qquad (2)$$

$$\frac{\partial V}{\partial t} + \frac{\partial}{\partial x}\left[\frac{UV}{\zeta+h}\right] + \frac{\partial}{\partial y}\left[\frac{V^2}{\zeta+h}\right] + fU + \varepsilon\left[\frac{\partial^2 V}{\partial x^2} + \frac{\partial^2 V}{\partial y^2}\right]$$

$$+ g(\zeta+h)\frac{\partial}{\partial y}(\zeta) + \frac{kV\sqrt{U^2+V^2}}{(\zeta+h)^2} = 0 \qquad (3)$$

where the notation is as follows:

x, y	Cartesian coordinates in the plane of the undisturbed sea surface
t	Time
$\zeta(x,y,t)$	Elevation of the sea surface
$U(x,y,t)$ $V(x,y,t)$	Components of vertically integrated velocity in directions x and y
$h(x,y)$	Depth of water below the undisturbed surface
f	Coriolis parameter, assumed constant over the region

k(x,y)	Coefficient of friction
g	Acceleration due to gravity
ε	Horizontal eddy viscosity

In the derivation of these equations it is assumed that the velocities are essentially independent of depth.

Equations (1), (2) and (3) are solved numerically for the sea area. The finite difference scheme is spatially similar to that used by Hansen (1961) but a modified numerical procedure in time is employed (Flather and Heaps, 1974).

The area is represented on a system of orthogonal grid lines. The grid notation is shown in Figure 6. Each mesh is of side 2 km and is characterized by a central elevation point while the normal components of vertically integrated velocity are evaluated at the mid-points of the sides.

To calculate the elevation $\zeta_i(t+\Delta t)$ at the centre of a mesh equation (1) is replaced by:

$$\left[\zeta_i(t+\Delta t) - \zeta_i(t)\right]\frac{1}{\Delta t} = -\left[U_i(t) - U_{i-1}(t) + V_{i-n}(t) - V_i(t)\right]\frac{1}{\Delta \ell} \tag{4}$$

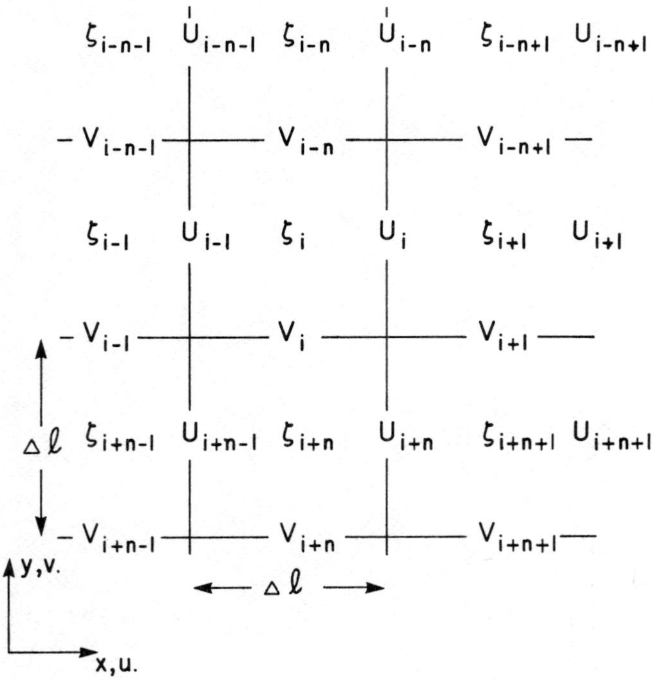

Fig. 6. Notation used in the computational grid.

The momentum equations (2) and (3) are replaced by:

$$\left[U_i(t+\Delta t)-U_i(t)\right]\frac{1}{\Delta t} = f\bar{\bar{V}}_i(t)$$

$$-\,g\frac{\bar{H}_i^x}{\Delta \ell}\left[\zeta_{i+1}(t+\Delta t)-\zeta_i(t+\Delta t)\right]-kU_i(t)\frac{\left[U_i^2(t)+\bar{\bar{V}}_i^2(t)\right]^{\frac{1}{2}}}{(\bar{H}_i^x)^2}$$

$$-\left[\frac{\left[\bar{U}_i^x(t)\right]^2}{\frac{1}{2}\left[\bar{H}_i^x+\bar{H}_{i+1}^x\right]}-\frac{\left[\bar{U}_{i-1}^x(t)\right]^2}{\frac{1}{2}\left[\bar{H}_{i-1}^x+\bar{H}_i^x\right]}\right]\frac{1}{\Delta \ell}$$

$$-\left[\frac{\bar{U}_i^y(t)\bar{V}_{i-n}^x(t)}{\bar{\bar{H}}_i}-\frac{\bar{U}_{i+n}^y(t)\bar{V}_i^x(t)}{\bar{\bar{H}}_{i-n}}\right]\frac{1}{\Delta \ell}+\frac{\varepsilon_i}{\Delta \ell^2}\left[U_{i-n}(t)+U_{i+n}(t)+U_{i-1}(t)+U_{i+1}(t)-4U_i(t)\right] \quad (5)$$

$$\left[V_i(t+\Delta t)-V_i(t)\right]\frac{1}{\Delta t} = -\,f\bar{\bar{U}}_i\,(t+\Delta t)-g\frac{\bar{H}_{i-n}^y}{\Delta \ell}\left[\zeta_i(t+\Delta t)-\zeta_{i-n}(t+\Delta t)\right]$$

$$-\,kV_i(t)\frac{\left[\bar{\bar{U}}_i^2(t)+V_i^2(t)\right]^{\frac{1}{2}}}{\left[\bar{H}_{i+n}^y\right]^2}$$

$$-\left[\frac{\bar{U}_{i+n}^y(t)\bar{V}_i^x(t)}{\bar{\bar{H}}_{i+n}}-\frac{\bar{U}_{i+n-1}^y(t)\bar{V}_{i-1}^x(t)}{\bar{\bar{H}}_{i+n-1}}\right]\frac{1}{\Delta \ell}$$

$$-\left[\frac{\left[\bar{V}_i^y(t)\right]^2}{\frac{1}{2}\left[\bar{H}_i^y+\bar{H}_{i+n}^y\right]}\;\frac{\left[\bar{V}_{i+n}^y(t)\right]^2}{\frac{1}{2}\left[\bar{H}_{i+n}^y+\bar{H}_{i+2n}^y\right]}\right]\frac{1}{\Delta \ell}+\frac{\varepsilon_i}{\Delta \ell^2}\left[V_{i-n}(t)+V_{i+n}(t)+V_{i-1}(t)+V_{i+1}(t)-4V_i(t)\right] \quad (6)$$

where the averaging of a grid variable with the neighbouring variable of that type in the direction x or y increasing is denoted respectively by expressions such as:

$$\bar{U}_i^x = \tfrac{1}{2}\ (U_i + U_{i+1})\ , \qquad \bar{U}_i^y = \tfrac{1}{2}\ (U_i + U_{i-n})$$

To denote the value at a point of a variable obtained by averaging the four values of that variable surrounding the point, the notation is:

$$\bar{\bar{H}}_i = \tfrac{1}{4}\ (H_i + H_{i-n} + H_{i-n+1} + H_{i+1}) \qquad \text{where } H_i = \zeta_i + h_i$$

$$\bar{\bar{U}}_i = \tfrac{1}{4}\ (U_i + U_{i+n} + U_{i+n-1} + U_{i-1}) \qquad \bar{\bar{V}}_i = \tfrac{1}{4}\ (V_i + V_{i-n} + V_{i-n+1} + V_{i+1})$$

DESCRIPTION OF THE MODEL

The grid scheme used in the model is shown in Fig. 7 where each square corresponds to the four sides of a mesh as illustrated in Fig. 6. Coastlines are simulated by mesh sides where the normal component of velocity is set equal to zero. Similarly, narrow islands or land barriers are depicted by closed mesh sides within the interior of the sea. In some instances ambiguities exist as to which side of a mesh best represents a part of the coastline. Such ambiguities are resolved by ensuring that regional surface areas in the model accord with those obtained from hydrographic charts.

The average depth over a mesh is specified at each elevation point within the interior of the sea. In regions where banks may be uncovered at low water, notably near the mouth of the Fraser River, an arbitrary depth of 5 m has been assigned. This is considered adequate at this stage of development of the model. Any realistic numerical simulation of flows over the shallow banks near the Fraser River mouth must also take account of the river discharge and its variations, and also the motion of the salt wedge.

A number of passes are included within the modelled region such as, for example, those leading from the region of the Gulf Islands into the Strait of Georgia. In some cases these have been simulated by one-dimensional channels inserted into the two-dimensional scheme. Earlier work with the overall tidal model showed that the comparatively small flows through these passes are not critical to the proper simulation of the tidal streams in the main conveying channels of the system, though they do possess considerable local importance. These features of the model will not be discussed here. Detailed adjustments of the flows through these passes will be undertaken when pertinent field observations, presently planned, are completed.

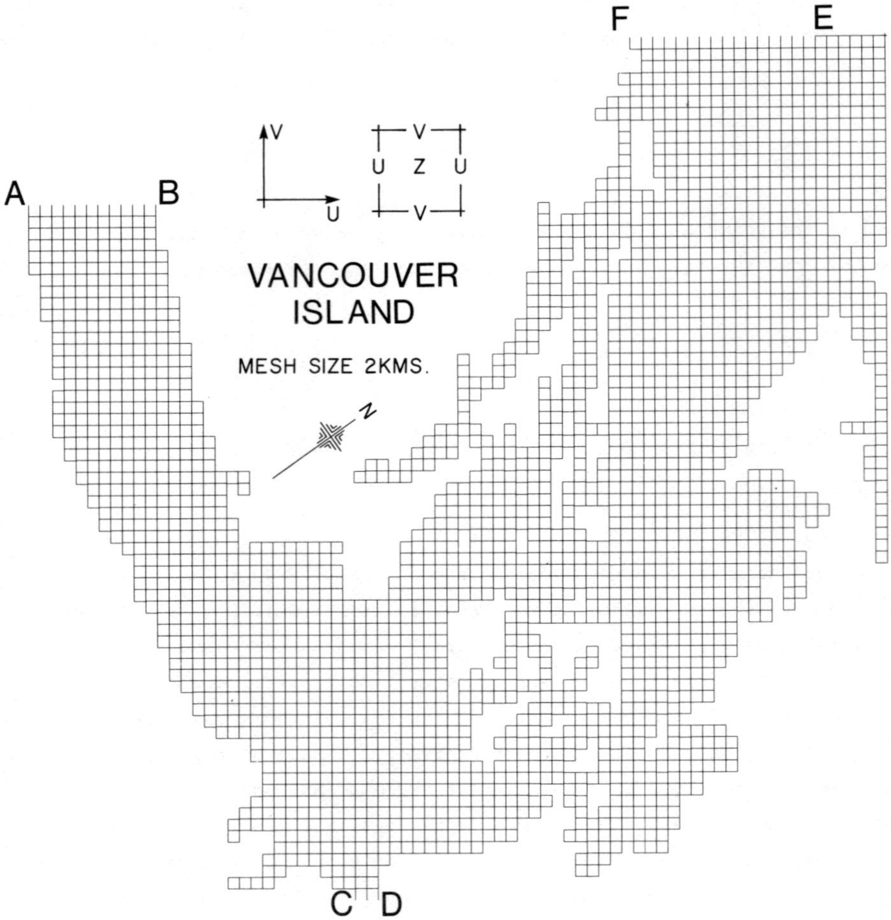

Fig. 7. The numerical model grid scheme. Rows of meshes where elevations are prescribed on open boundaries are denoted by AB, CD and EF.

The value of the friction coefficient used in the Strait of Georgia and in Juan de Fuca Strait is 0.003. In the region of the channels between the San Juan and Gulf Islands this value was increased to 0.03. The latter value appears unusually high when compared with those normally employed in work on coastal seas, for example Flather and Heaps (1974). The basic argument for the use of such high a value in the island region derives from earlier work with the overall model of the waters between Vancouver Island and the mainland. It was found that satisfactory agreement between the observed and computed tidal elevations around the interior of the modelled sea could only be achieved by the use of such high frictional dissipation, in particular with respect to the proper proportioning of the tides between the Strait of Georgia and the Puget Sound system. A corroborative feature of the argument concerns the existence of a significant non-linear interaction between the primary diurnal (K_1) and semi-diurnal (M_2) constituents which was shown to be due to the high frictional dissipation in the topographically-complex channels of the island region.

When the model is operated in the absence of lateral stresses small grid-scale fluctuations in the directions of velocity vectors ("noodling") occur in regions where the advective accelerations are important. The value of the horizontal eddy coefficient used to eliminate this feature of the model was 10^6 cm^2/sec, which, in view of the mesh size and timestep, implies a relatively small degree of lateral averaging (Kuipers and Vreugdenhil, 1973). The problem of numerically evaluating second derivatives near coastal boundaries was avoided by setting the coefficient equal to zero in such circumstances. The implied assumption is that generally the bottom stress acting on the water in a relatively shallow mesh contiguous to the coast is much greater than the lateral stress.

The maximum permissible time step yielding stable solutions is obtained from the Courant-Friedricks-Lewy criterion,

$$\Delta t \leq \frac{\Delta \ell}{\sqrt{2gh_{max}}} .$$

The time step that results for the present scheme is 23 secs.

Elevations were prescribed along the three open boundaries of the model shown by the lines, AB, CD, EF in Fig. 7, by interpolation of values from the overall model of the system which used a 4 km mesh. The boundary conditions in the case of the latter were predicted tides based on 61 harmonic constituents. As noted above, earlier attempts to operate part of the 4 km mesh model by prescribing elevations at these locations failed, firstly, because of the insufficient sensitivity of the model to permit proper adjustment of the frictional dissipation by optimising computed and observed elevations around the interior of the modelled sea and secondly because of the inability of tide gauges at either end of a long open boundary to adequately resolve the slopes of the water surface required to balance the geotrophic acceleration of the tidal streams moving normal to the open boundary. To overcome these difficulties in the present case, the distribution of frictional dissipation and the elevations of the water surface presented at open boundaries, were taken from the original overall model of the system. Subsequent comparisons of the velocity vector fields obtained at comparable instants from the two models showed excellent agreement except where modified by the changed representation of the coastal boundaries.

RESULTS

Illustrative of the performance of the model, some comparisons of computed elevations with those predicted on the basis of 61 harmonic tidal constituents over a seven-day period are now presented for representative locations around the interior of the modelled sea, Fig. 8.

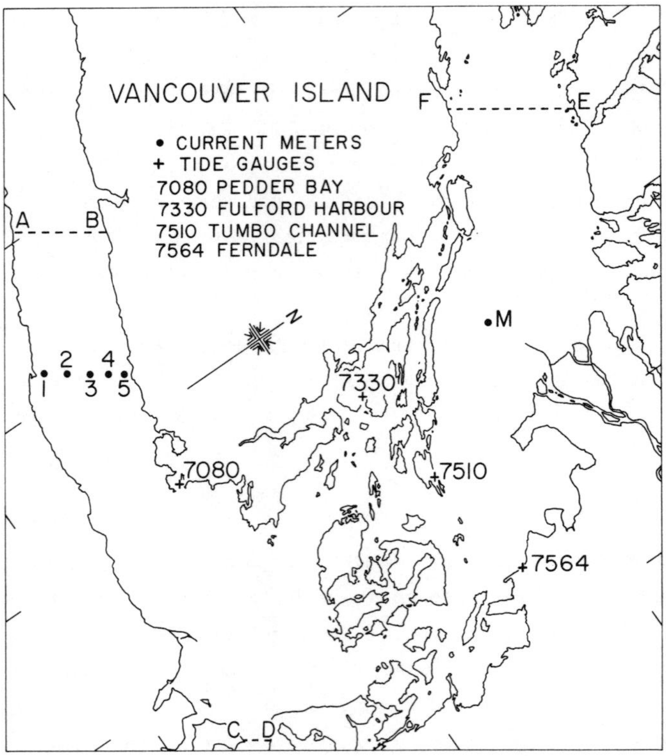

Fig. 8. Locations of current meters and tide gauges referred to in Figs. 9 and 10.

At Tumbo Channel (7510) and Ferndale (7564) the excellent agreement obtained is typical of the results obtained throughout that part of the Strait of Georgia included within the model, Fig. 9.

The most difficult area of the model in which to obtain good agreement is in the region of Haro Strait and in the degenerate semi-diurnal amphidromic region in the inner part of Juan de Fuca Strait, regions characterized by rapid changes in the spatial distribution of the harmonic constants Fig. 5.

A comparison of the predicted elevations with those computed by the model for Fulford Harbour (7330) shows good agreement though the curves from the model

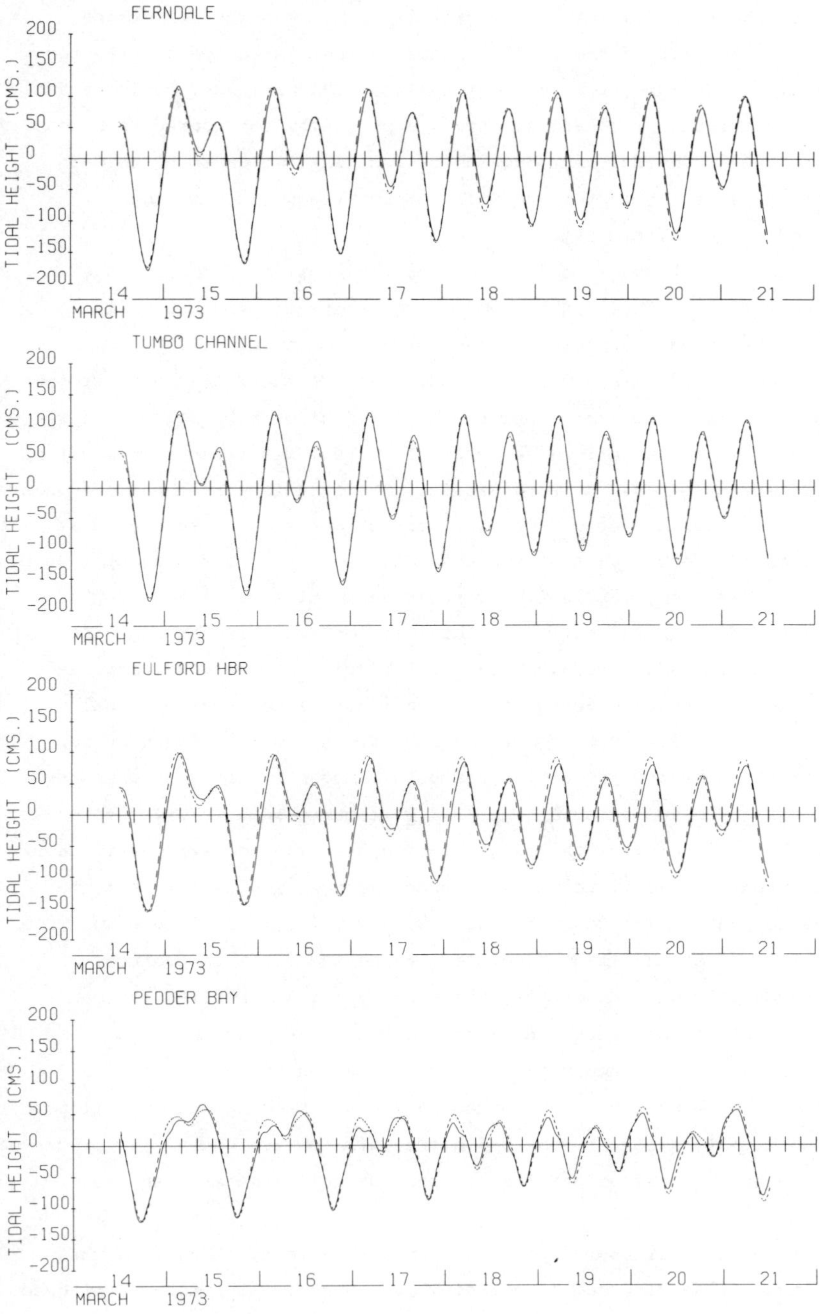

Fig. 9. Comparisons of the elevations computed by the model (continuous line) for some typical locations with those predicted (dashed line) on the basis of tidal harmonic constants.

show a small phase lead throughout the record and their exists a modest discrepancy in the tidal ranges. More detailed adjustments of frictional dissipation in this part of the model, including that in the passes leading into the Strait of Georgia from the region of the Gulf Islands, could probably improve the comparison. Such adjustments are presently considered of secondary importance to the successful reproduction of the tidal streams in the main conveying channels of the system and will not be attempted until the acquisition of further field data.

The Pedder Bay (7080) gauge is located near the semi-diurnal amplitude minimum. The agreement obtained is good though there is evidence of more accentuated higher tidal harmonics in the values from the model.

The primary object of this numerical study concerns the reliable prediction of the barotropic tidal streams. During the interval of time selected for this trial of the model an extensive array of current meters was in position over a cross-section of Juan de Fuca Strait. The locations of the moorings are shown in Fig. 8. Four Aanderaa recording current meters were suspended at each mooring in the deeper part of the channel. Comparisons representative of the agreement between the vertically averaged velocities predicted by the model and measured tidal velocities in this section are shown in Fig. 10. The measured velocities are resolved in the direction of the model axes and at this location may be considered as the longitudinal and cross-channel components of flow. (For purposes of the comparison the low frequency non-tidal part of the measured velocities has been removed by first suppressing the tidal signal by means of an $A^2_{24}.A_{25}/24.^2_{25}$ filter (Godin 1972), and then subtracting the residual from the original record.) The comparisons of the observed and predicted longitudinal V components show excellent agreement with respect to phase. The range of the observed velocities at 15 m and 50 m is appreciably greater than that predicted by the model. This feature, however, accords well with expectation based on observed profiles in open channel flow. Unfortunately current meter malfunctions precluded the presentation of a further comparison closer to the sea bottom. Such comparisons in the base of the more extended run of the original overall coarse grid model show that indeed at some 5-6 metres off the sea bottom the predicted, vertically-averaged, tidal velocities are considerably greater than those observed.

The smaller measured cross-channel components of velocity show much more variability than those directed longitudinally. Occasional intervals of good agreement between the values predicted by the model and those observed are thought to conform essentially to barotropic tidal oscillations, while the more irregular motions are associated with the intermittent excitation of internal modes. It will be recalled, Fig. 4, that the region where these

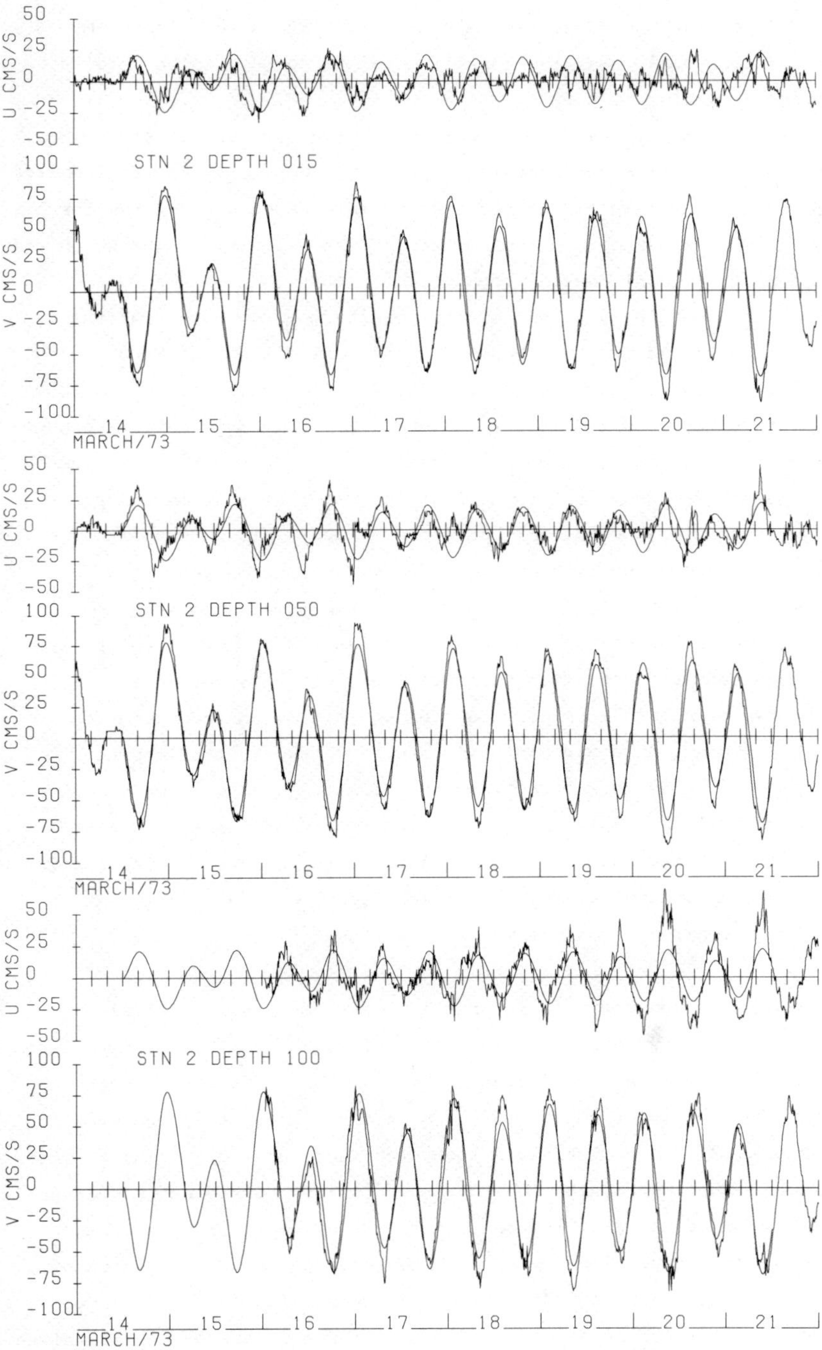

Fig. 10. Comparisons of the computed cross-channel (U) and longitudinal (V) tidal velocity components (smooth curves) with those observed (irregular curves) at three different depths (15, 50 and 100 m) at Station 2 in Juan de Fuca Strait (Fig. 8).

observations were made is stratified. During the period of the measurements, time series STD observations (Crean and Miyake, 1976) indicated displacements in the depths of isopycnals near the coastlines of up to 50 m which were approximately of semi-diurnal period.

In addition to the above, comparisons were also made between the elevations and velocities obtained from the overall coarse grid and limited-area fine grid models respectively. No significant differences were noted between the results obtained from each model except in those parts of the velocity fields which are strongly influenced by local coastal boundaries.

Fig. 11 shows the tidal elevations and the longitudinal component of the tidal streams, as obtained from the model for the period 1200, 15 March - 1600, 16 March 1973, at a central location in the Strait of Georgia off the mouth of the Fraser River, denoted by the letter M in Fig. 8. The range between higher-high water and lower-low water conforms approximately to the range contained between mean higher-high water and mean lower-low water at this location. Corresponding vector plots of the ebb and flood velocity fields when the streams are maximal are shown in Figs. 12 and 13. Small crosses denote the centre of a mesh where the velocity components, respectively averaged over those obtained from opposite mesh sides, are plotted. When the line denoting the velocity magnitude and direction exceeds 30 cms/sec, the vector is represented by two parallel lines of equal length, and so on for succeeding multiples of that speed, to avoid confusion due to excessive overlapping. Throughout most of Juan de Fuca Strait and the Strait of Georgia the tidal streams are essentially rectilinear. In the inner part of Juan de Fuca Strait there occurs some rotatory character to the streams though proper analyses in terms of the tidal constituents awaits a trial of the model of sufficient duration to permit separation of the major constituent tides by harmonic analysis.

The velocity vectors suggest some interesting aspects of the tidal flows near coastal features off the southernmost part of Vancouver Island where large speeds and marked changes in flow direction occur. The formation of back eddies changing in location from ebb to flood is clearly evident. Examination of a time sequence of such vector plots indicates that the eddy begins to form when the stream at that particular location has attained sufficient magnitude; the eddy then increases in spatial extent until the stream turns. A further point of interest concerns the differences between strength and direction of the ebb and flood streams in Haro Strait, again emphasizing the importance of local topography. It is of interest to note that these flow features lie along what is probably the primary route followed by water from the Fraser River moving seaward through Juan de Fuca Strait. Unfortunately, the 2 km grid scale used in this model is too coarse to permit other than qualitative indications of these dynamical features, which are pertinent to local navigation and which are held by some informants familiar with these waters to be real. Field observations are planned to test their authenticity.

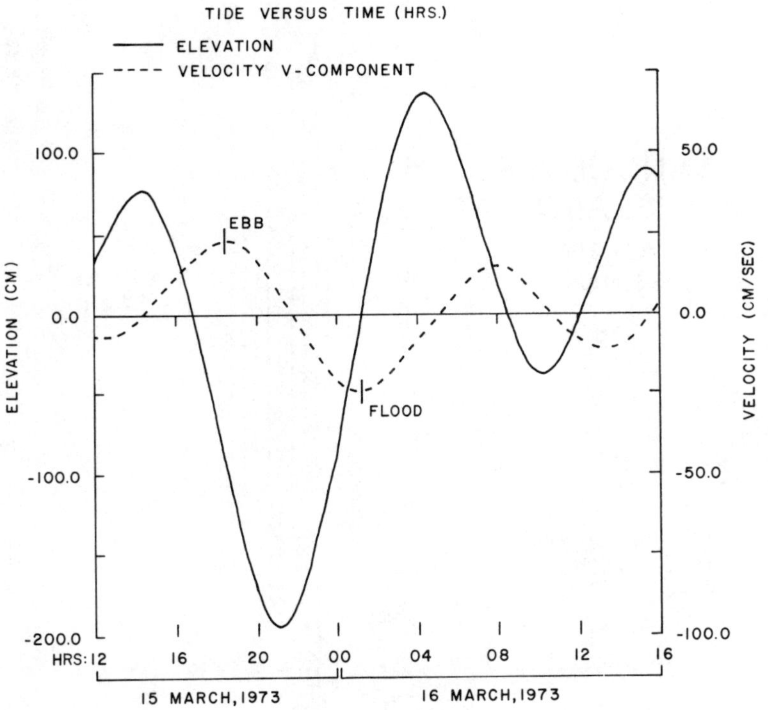

Fig. 11. Tidal elevations and the longitudinal component of velocity computed by the model for a location in mid-strait off the mouth of the Fraser River (M in Fig. 8). The words EBB and FLOOD denote the instants at which the velocity vector diagrams in Figs. 12 and 13 obtain.

Of particular relevance to the movement of the Fraser River discharge is the character of the tides and streams in the southern part of the Strait of Georgia. Considering the flow field in the general vicinity of the irregular line of closed mesh sides denoting the Fraser River training wall which extends over the shallow banks to the edge of the deep water in the strait, it is evident that streams decrease markedly in magnitude with northerly progression along the longitudinal axis of the strait, and also undergo a significant change in direction. More locally the velocity vectors show the streams moving off or onto the shallow banks. (No attempt has been made as yet to include the Fraser River discharge in this model since this is small compared to the tidal fluxes occurring in the Strait itself and cannot be adequately simulated in a barotropic model.)

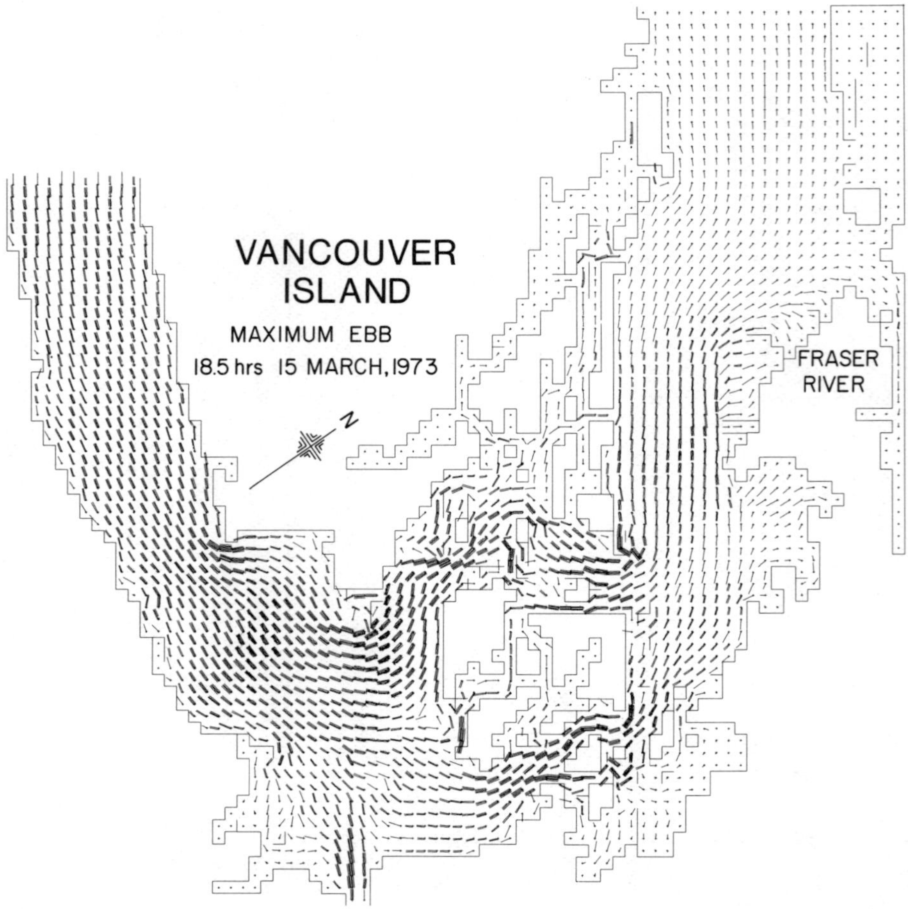

Fig. 12. Ebb velocity vectors obtained from the model.

A number of interesting dynamical inferences may be made from slopes of the water surface as illustrated by three-dimensional plots such as those shown in Fig. 14. In crude terms, all land has been removed above a base plane corresponding to the lowest elevation of the water surface computed in the modelled sea at the particular instant in question. Thus, islands are represented as 'holes' in the sea surface. The intersection of two lines in the sea area corresponds to the centre of a mesh in the model where elevations are calculated while a single line indicates the elevations along a channel represented in the model by an isolated sequence of contiguous meshes. The fields of elevation coincide in time with the ebb and flood velocity fields shown in Figs. 12 and 13 respectively.

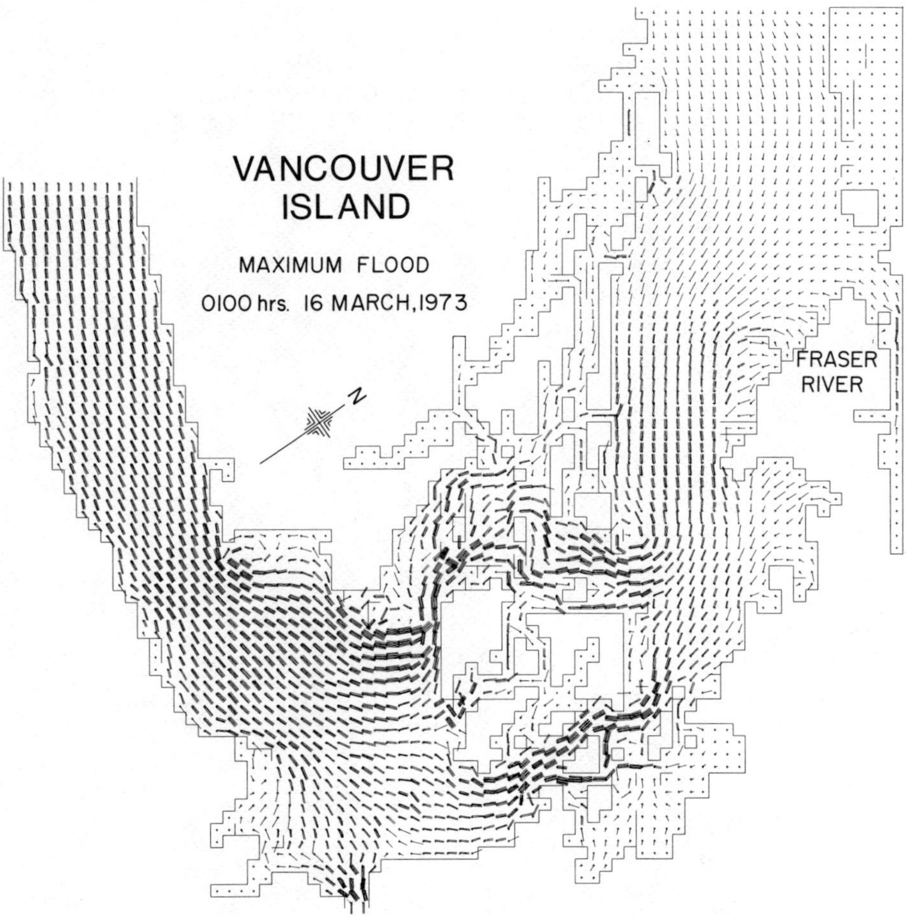

Fig. 13. Flood velocity vectors obtained from the model.

As noted above, the proper proportioning of tidal elevations and times of high water between the Strait of Georgia and the Puget Sound system required the introduction of high frictional dissipation in the region of the San Juan Islands. The marked slopes of the water surface through this region, which are more dramatically instanced in the case of the flood tide figure, are clearly evident. Elevations in the area contained between Vancouver Island and the Gulf Islands are similar to those in the northern part of Haro Strait and can differ markedly from those in the Strait of Georgia. The resulting strong tidal flows through the passes leading into the Strait of Georgia are well known to local navigators.

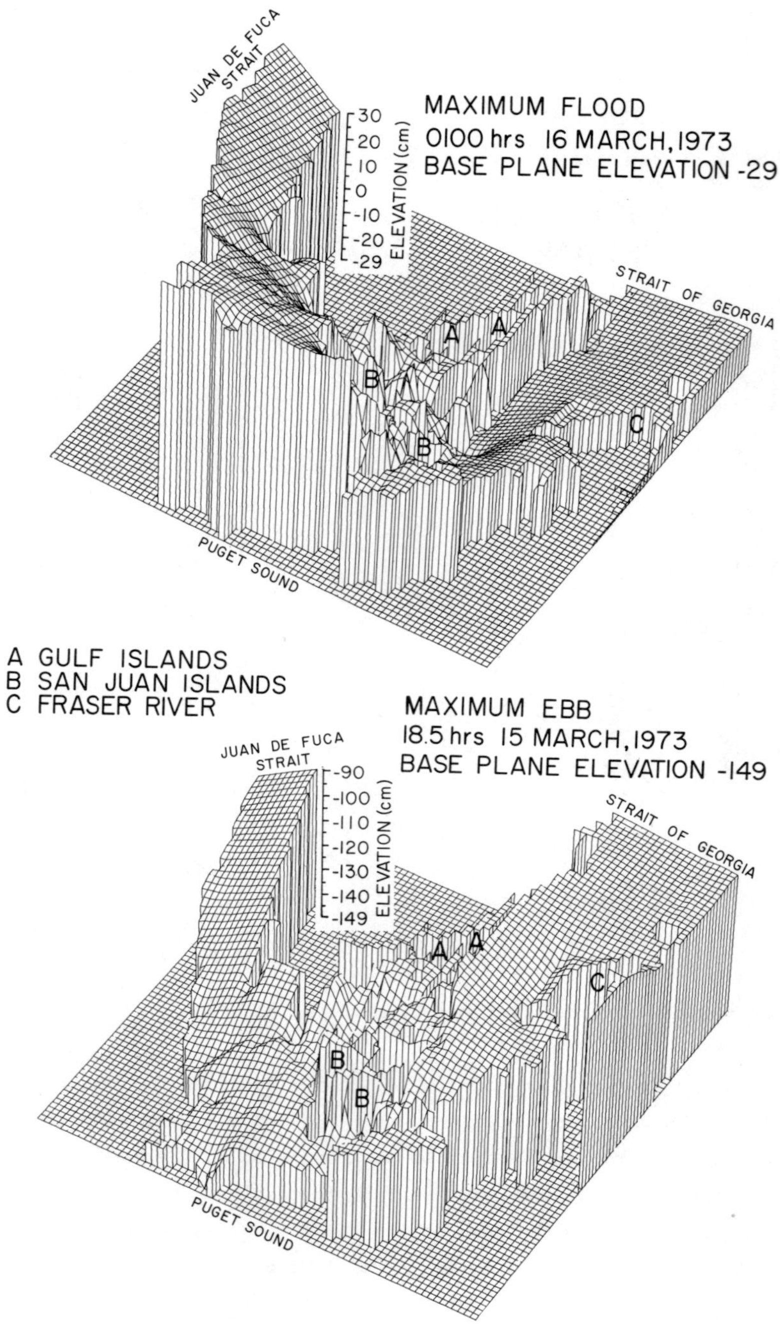

Fig. 14. Three-dimensional diagrams showing the shape of the water surface at the same instants respectively as the preceding diagrams of ebb (Fig. 12) and flood (Fig. 13) velocity vectors.

In the Strait of Georgia, there are cross-channel slopes of the water surface which balance the geostrophic accelerations due to the tidal streams moving parallel to the major axis of the strait, while in the vicinity of the river mouth local slopes are associated with the motion of water respectively off, or onto, the shallow banks.

A further feature of these diagrams concerns secondary undulations of the water surface in the inner part of Juan de Fuca Strait. These are more strongly in evidence when the streams are running full than at slack water and will be investigated further. At present it appears that they may be associated with depressions in the water surface of some 10-20 cm near the centre of the large eddies near coastal features which have been referred to above.

PRESENT AND FUTURE WORK

In collaboration with Mr. S. Huggett of the Canadian Hydrographic Service, velocity fields derived from the model are being used as the basis for an atlas of tidal currents in Juan de Fuca Strait and in the region of the San Juan Islands. Specific details of the tidal streams associated with coastal features, which are poorly resolved by the 2 km mesh size of the model, are being individually checked by tracking surface floats from an aircraft equipped with a special sighting device, inertial navigation system, radar altimeter, and magnetic tape data recorder (Grasty et al. 1977).

An important aspect of this work concerns residual currents induced by the barotropic tide and their possible significance to the estuarine circulation. It is evident from the computed velocity fields that relatively large net flows occur in regions characterised by strong tidal streams and complex topography. Using the vorticity equation derived from the depth mean flow momentum and continuity equations, the terms generating such secondary flows have been discussed by Kuipers and Vreugdenhil (1973). A short review of the problem of residual flows has been presented by Nihoul (1975). In the context of the present work there exists the possibility of examining residual flows obtained from models of 2 km and 4 km mesh scales of the same complex geographical region. Analyses of these results and comparisons with observed residual flows obtained from moored current meters are not as yet completed.

As described above, the eventual object of this work concerns a resolution of the main dynamical features of the estuarine circulation. The question arises as to how best the results obtained from the barotropic model presented above may be employed to further this end. Clearly a major concern is the motion of fresh water in the southern part of the Strait of Georgia, which may move southward into the strong tidal mixing area of the San Juan Islands or northward into the Strait of Georgia where the tidal flows are relatively weak. In the latter case, it may eventually move seaward through Johnstone Strait or the accumulated discharge over a lengthy period may be returned southward by the action of winds. The problem is immensely complex. The thickness of the plume is commensurate with the tidal range at the river mouth. There exist extensive areas of drying banks. The intrusion of the salt wedge into the Fraser River is related to the state of the tide and to variations in the strength of the river discharge. There is also the problem of simulating propagating shallow fronts between river water and sea water, or the remains of preceding "plumes", in the Strait. Any attempt to include all these features in a single three-dimensional model appears likely at present merely to proliferate dynamical confusion at heroic expense. The simplest approach holding promise of simulating the dominant aspects of the shallow upper layer motions in the Strait of Georgia, and which utilizes data from the above baratropic tidal model, will be summarized briefly. This is a collaborative project with co-workers at the Institute of Oceanography of The University of British Columbia, (Stronach, 1978).

Consider initially a system in which a lighter layer of fluid is superimposed upon a denser lower layer and in which the motions within each layer may be described by the vertically integrated equations of motion and continuity as applied to flat rotating sea (Proudman 1952). In the present case, the top layer is confined to the upper few metres of the water column which is generally some 150 - 300 m deep. Consistent with observations from moored current meters over a cross-section of the strait near the river mouth, the presence of this upper layer does not significantly affect the motion in the lower part of the water column which to a first approximation is dominated by the barotropic tidal streams. If now that part of the pressure gradient in the lower layer which derives from the presence of the upper layer, is set equal to zero, equations of motion and continuity applying to a single upper layer may be obtained in which the thickness of the layer now replaces the elevation of the water surface as one of the dependent variables, and in which additional terms in the momentum equations effectively induce "buoyant spreading". Empirical terms are also included to approximate entrainment of water from the lower layer into the upper. A particle of water in this upper layer will experience a pressure gradient associated with those slopes of the water surface giving rise to the barotropic tidal motions in the much thicker lower layer. At any point there will also exist a stress associated with the difference between the upper layer velocity at that point and the barotropic tidal velocity underneath. Both the components of the free surface barotropic tidal slope and underlying velocities are known throughout the strait from the original tidal model. The addition of a salt conservation equation, with an empirical expression to include entrainment, permits computation of the spatial distributions of upper layer salinity in the southern strait. These salinities may then be used to approximate the field of density. There is now at hand a simple and relatively inexpensive system which permits numerical experiments in connection with the changing surface slopes of the barotropic tide, varying river discharge, entrainment, mixing, vertical and lateral stresses, wind stresses, field accelerations and the formulation of adequate boundary conditions. A major intent of this study is the establishment of significant quantitative relations between the relative volume transports of water leaving the Strait of Georgia, respectively through the northern and southern openings and the river discharge rates, winds and tides.

An important feature of this approach is the relaxation of the stringent limitation on the time step imposed by the propagation speed of free surface waves in a region where a small mesh size is required and the water is relatively deep. This work, together with an appropriate program of field measurements, which includes STD casts, velocity profiling, drogue tracking, and also the recording of surface velocities at the river mouth, is presently in progress. The results obtained so far are most encouraging.

Though, to date, the model assumes dependence of the velocity components and salinity in the upper layer, it is proposed to take account of the vertical gradients, which are in fact an important feature of the plume, at least near the river mouth, by using the similarity that exists in the depth-profiles at increasing distances from the river mouth.

With regard to future work, two studies are planned. The first of these is an observational program to determine the character of deep salt water intrusions into the southern Strait of Georgia and possible relations to external forcing. The second involves the inception of three-dimensional numerical model studies with particular emphasis on maximal utilization of data deriving from the two-dimensional barotropic tidal and upper layer buoyant spreading models referred to above.

It would presently appear that these studies will culminate in a three-dimensional numerical simulation of the major estuarine motions and which might include such local processes on the continental shelf as could be shown to significantly affect the inflow of salt water. This model could, in principle, be coupled to an atmospheric model simulating the meso-scale orographic modifications of the geostrophic wind field by the mountains on the mainland and Vancouver Island. The realization of such an overall model, however, will require an extended program of field observations and numerical experiments to sufficiently resolve the essential dynamical features of the system.

ACKNOWLEDGEMENTS

The computer programming was carried out by Mr. P.J. Richards. I thank Dr. S. Pond and Dr. R. Thomson for critically reviewing the paper.

REFERENCES

Ages, A. and Woollard, A., 1976. The tides in the Fraser River. Pacific Marine Science Report 76-5: 100 pp. Institute of Ocean Sciences, Patricia Bay, Victoria, B.C.

Chang, P., Pond, S., and Tabata, S., 1976. Subsurface currents in the Strait of Georgia, west of Sturgeon Bank. J. Fish. Res. Board Can., 33: 2218-2241.

Crean, P., 1969. A one-dimensional hydrodynamical numerical tidal model of the Georgia-Juan de Fuca Strait system. Fish. Res. Board Can. Tech. Rep. No. 156: 32 pp.

Crean, P., 1976. Numerical model studies of the tides between Vancouver Island and the mainland coast. J. Fish. Res. Board Can., 33: 2340-2344.

Crean, P. and Ages, A., 1971. Oceanographic records from twelve cruises in the Strait of Georgia and Juan de Fuca Strait, 1968. Dep. Energy, Mines and Resources. Mar. Sci. Branch. 1-5: 389 pp.

Crean, P.B. and Lewis, A.G., 1976. Monthly sections - Strait of Juan de Fuca, June-Dec. 1973. Data Report 39 (Provisional) Institute of Oceanography, University of British Columbia.

Crean, P.B. and Miyake, M., 1976. STD sections Strait of Juan de Fuca, March-April 1973. Data Report 38 (Provisional) Institute of Oceanography, University of British Columbia.

Department of the Environment, 1972a. Data record of current observations, Strait of Georgia, section 2, Cape Lazo to Grief Point, 1970. Water Manage. Serv., Mar. Sci. Dir. Pac. Reg., Manuscr. Rep. Ser. 9: 88 p.

Department of the Environment, 1972b. Data record of current observations, Strait of Georgia, section 4, Gabriola Island to Gower Point, 1969-1972. Water Manage. Serv., Mar. Sci. Dir. Pac. Reg., Manuscr. Rep. Ser. 10: 153 p.

Department of the Environment, 1972c. Data record of current observations, Strait of Georgia, section 5, Porlier Pass to Sand Heads, 1969-1972. Water Manage. Serv., Mar. Sci. Dir. Pac. Reg., Manuscr. Rep. Ser. 11: 124 p.

Department of the Environment, 1973. Data record of current observations, Strait of Georgia, section 6, Samuel Island to Point Roberts, 1969-1970. Water Manage. Serv., Mar. Sci. Dir. Pac. Reg., Manuscr. Rep. Ser. 12: 96 p.

Flather, R.A. and Heaps, N.S., 1975. Tidal computations for Morecambe Bay. Geophys. J. R. Astr. Soc. 42: 489-517.

Giovando, L. and Tabata, S., 1970. Measurements of surface flow in the Strait of Georgia by means of free-floating current followers. Fish. Res. Board Can. Tech. Rep. No. 163: 69 pp.

Godin, G., 1972. The analysis of tides. University of Toronto Press, Toronto, 264 pp.

Grasty, R.L. and Gower, J.F.R., 1977. Inertial navigation for flight path recovery. Geol. Survey Can. 76-30, 12 pp.

Hansen, W., 1962. Hydrodynamical methods applied to oceanographic problems. Proc. Symp. Math. Hydrodynamical Methods of Phys. Oc. 1961. Institut für Meereskunde der Universität Hamburg. 1962: 25-34.

Herlinveaux, R.H. and Tully, J.P., 1961. Some oceanographic features of Juan de Fuca Strait. J. Fish. Res. Board Can. 18: 1027-1071.

Herlinveaux, R.H. and Giovando, L.F., 1969. Some oceanographic features of the inside passage between Vancouver Island and the mainland of British Columbia. Fish. Res. Board Can. Tech. Rep. No. 142: 48 pp.

Huggett, W.S., Bath, J.F. and Douglas, A., 1976. Data record of current observations, Juan de Fuca Strait, 1973. Institute of Ocean Sciences, Patricia Bay, Victoria, B.C.

Huggett, W.S., Bath, J.F. and Douglas, A., 1976. Data record of current observations, Johnstone Strait, 1973. Institute of Ocean Sciences, Patricia Bay, Victoria, B.C.

Hutchinson, A.H., and Lucas, C.C., 1931. The epithalassa of the Strait of Georgia: salinity, temperature, pH, and phytoplankton. Can. J. Res., 5: 231-284.

Kuipers, J., and Vreugdenhil, C.B., 1973. Calculations of two-dimensional horizontal flow. Report on basic research S 163, Part 1, Delft Hydraulics Laboratory.

Nihoul, J.C.J., 1975. Modelling of marine systems. Elsevier, Amsterdam, 272 pp.

Parker, Bruce B., 1977. Tidal hydrodynamics in the Strait of Juan de Fuca - Strait of Georgia. U.S. Dept. of Commerce, NOAA Technical Report NOS 69.

Pickard, G.L., 1975. Annual and longer term variations of deepwater properties in the coastal waters of southern British Columbia. J. Fish. Res. Board Can. 32: 1561-1587.

Proudman, J., 1953. Dynamical Oceanography. Methuen, London. 409 pp.

Redfield, A.C., 1950a. Proceedings of the colloquium on the flushing of estuaries. Woods Hole Oceanographic Institution Ref. No. 50-37: 175-177.

Redfield, A.C., 1950b. The analysis of tidal phenomena in narrow embayments. Papers in physical oceanography and meteorology, Vol. XI, No.4, Massachussetts Institute of Technology and Woods Hole Oceanographic Institution.

Stronach, J., 1978. Observational and modelling studies of the Fraser River plume. Ph.D. Thesis, University of British Columbia.

Tabata, S., and Stickland, J.A., 1972a. Summary of oceanographic records obtained from moored instruments in the Strait of Georgia -- 1969-1970: current velocity and seawater temperature from station H-06. Dep. Environ. Water Manage. Serv., Mar. Sci. Branch, Pac. Reg. Pac. Mar. Sci. Rep. 72-7: 132 p.

Tabata, S., and Stickland, J.A., 1972b. Summary of oceanographic records obtained from moored instruments in the Strait of Georgia -- 1969-1970: current velocity and seawater temperature from station H-16. Dep. Environ. Water Manage. Serv. Mar. Sci. Branch, Pac. Reg. Pac. Mar. Sci. Rep. 72-8: 144 p.

Tabata, S., and Stickland, J.A., 1972c. Summary of oceanographic records obtained from moored instruments in the Strait of Georgia -- 1969-1970: current velocity and seawater temperature from station H-26. Dep. Environ. Water Manage. Serv., Mar. Sci. Branch, Pac. Reg. Pac. Mar. Sci. Rep. 72-9: 141 p.

Tabata, S., and Stickland, J.A., 1972d. Summary of oceanographic records obtained from moored instruments in the Strait of Georgia -- 1968-1970. Current velocity from stations F-11, M-10 and I-31. Dep. Environ. Water Manage. Serv., Mar. Sci. Branch, Pac. Reg. Pac. Mar. Sci. Rep. 72-10: 22 p.

Tabata, S., Giovando, L.F., Stickland, J.A., and Wong, J., 1970a. Current velocity measurements in the Strait of Georgia -- 1967. Fish. Res. Board Can. Tech. Rep. 169: 245 p.

Tabata, S., Giovando, L.F., Stickland, J.A., and Wong, J., 1970b. Current velocity measurements in the Strait of Georgia -- 1968. Fish. Res. Board Can. Tech. Rep. 178: 112 p.

Tabata, S., Giovando, L.F., Stickland, J.A., and Wong, J., 1970c. Current velocity measurements in the Strait of Georgia -- 1969. Fish. Res. Board Can. Tech. Rep. 191: 72 p.

Tabata, S., Stickland, J.A., and De Lange Boom, B.R., 1971. The program of current velocity and water temperature observations from moored instruments in the Strait of Georgia -- 1968-1970 and examples of records obtained. Fish. Res. Board Can. Tech. Rep. 253: 222 p.

Thomson, R.E., 1976. Tidal currents and estuarine circulation in Johnstone Strait, British Columbia. J. Fish. Res. Board Can. 33: 2242-2264.

Tully, J.P. and Dodimead, A.J., 1957. Properties of the water in the Strait of Georgia, British Columbia, and influencing factors. J. Fish. Res. Board Can. 14: 241-319.

Waldichuk, M., 1957. Physical oceanography of the Strait of Georgia, British Columbia. J. Fish. Res. Board Can. 15: 1065-1102.

NUMERICAL INVESTIGATIONS OF THE INFLUENCE OF COASTAL STRUCTURES UPON THE DYNAMIC OFF-SHORE PROCESS BY APPLICATION OF A NESTED TIDAL MODEL

H.-G. RAMMING
Institut für Meereskunde der Universität Hamburg

SUMMARY

By application of a TIDAL NORTH SEA MODEL with high resolution in costal areas, several numerical investigations were carried out dealing with predictions on dynamical processes in off-shore shallow-water regions as subsequence of structural measures. One of these investigation series and their results will be presented and discussed here in the following.

The subject deals with the effect of the structural change of an existing breakwater and about the effect of a wash-up of a sand deposit in the shoal area upon the water levels, velocities, tidal residual currents and the mean transports of one tidal cycle in the off-shore range.
The size of the investigation area is determined by the approximate effect in the remote region. Here it must be ensured that the boundary conditions of the selected part area obtained from the NESTED NORTH SEA MODEL (RAMMING, 1975) will not be influenced by the structural changes.

The smallest mesh size of the applied model is 51 m. The areas of a differing mesh size are in an interaction to each other.

In the course of this article certain technical, practical details of the numerical treatment of the dynamic processes in shallow water with a very complex morphology will be stated so as used by the author. Numerical models are only <u>one</u> possibility for prognosticating alteration tendencies after structural measures in the coastal area. Numerical models have certainly great advantages in comparison to other investigation methods, but they also have their limits, however which should not be misunderstood. The obtained results must be particularly critical evaluated in respect of their physical importance.

The numerical investigations exclude a simultaneous assertion about a possible transport of solids since it was started from an unchangeable morphology apart from structural measures given into the model. Therefore, only forcasting for short periods of time will be possible. Long-time approximations with the aid of numerically determined residual currents and their alterations are only conditionally permissible.

INTRODUCTION

In coastal research work as well as in coastal engineering, numerical models for reproduction and estimation of dynamic processes are used additionally in an increasing number for some years already apart from the hydraulic models.

In both model types and investigation methods, the preparation of a similarity to nature will be a necessary condition for further investigations. This conformity between the process of nature - mostly determinded by measurements - and the obtained model results is limited in the case of a hydraulic model - a.o. by the technique - as well as in the case of the numeric model - a.o. by the approximation of the topography and by the given solution methods. The recognition and knowledge of these borders should be included in the considerations when evaluating and judging the results. The application of numerical methods for the reproduction of the movement processes in various depths of the sea by the use of at first course-mesh screen grids has led to remarkable results, and has given considerable impulses to hydrographical work. Therefore it was quite obvious to develop a model for the coastal range by a screen-grid solution under observation of the interaction between the areas of different mesh widths which could also be useful for coastal engineering. The prediction of an lateration of the dynamical processes as subsequence of building-constructional measures within the coastal range (breakwaters, wash-ups, deepenings of waterways, etc.) is a task for which a numerical model with a high screen-grid solution, when expertly used, can be at least indicate possible tendencies. The depth distribution will go into the numerical model. The topography as well as the coast line will be all the better approximated if the screen-grid distance is less. The applied method permits such a screen-grid solution and, thus, represents once more its flexibility for investigations in the coastal range.

For the numerical investigations, the method developed by HANSEN (HANSEN 1956, 1961) was used and as a base model, the already classically North Sea model was applied.

Here it may be permitted to abstain from a detailed description of the method, the discretisation, etc. in this connection and to refer to the appropriate literature (BRETTSCHNEIDER 1967; HANSEN 1956, 1961; RAMMING 1966, 1970, 1972, 1973; SÜNDERMANN 1966).

RAMMING (1975) gives further methodical hints towards the peculiarity of the method in regard to the screen-grid dissolution in the coastal range.

The following vertically integrated differential equations were used:

$$\frac{\partial u}{\partial t} + R^{(x)} - fv + u\frac{\partial u}{\partial x} + v\frac{\partial u}{\partial y} + g\frac{\partial \zeta}{\partial x} + A_h \Delta u = 0 \qquad (1)$$

$$\frac{\partial v}{\partial t} + R^{(y)} + fu + v\frac{\partial v}{\partial y} + u\frac{\partial v}{\partial x} + g\frac{\partial \zeta}{\partial y} + A_h \Delta v = 0 \qquad (2)$$

$$\frac{\partial \zeta}{\partial t} + \frac{\partial}{\partial x}(Hu) + \frac{\partial}{\partial y}(Hv) = 0 \qquad (3)$$

SOME REMARKS ON NUMERICAL HANDLING OF DYNAMICAL PROCESSES IN SHALLOW WATER AREAS

In previous models with course resolution and traditional quadratic bottom friction the simulation of physical processes producing water levels and velocities in extremely shallow water near the coast was not sufficiently accurate enough for various practical purposes.

It is well known that well established in the meanwhile vertically integrated equations of motion are not valid for describing the dynamical processes in shallow water areas.

The refined grid near the coast and in the estuaries permits a better approximation of the coast, the islands and the morphology in general. It has to be considered that as a consequence the agreement between computed and observed water levels and velocities at selected points has improved considerably. Moreover one has to consider the fact that sands and flats fall dry for some hours during a tidal cycle. It must also be considered that the main transport of water masses takes place in narrow channels on the flats. The refined model considers these facts explicitly but still the accuracy seems not to be good enough to encourage predictions, e.g. on effects of off-shore constructions. In particular the results show unsatisfactory phase lags at grid points located in areas with a very small water depth. Since the work of TAYLOR (1920) and DEFANT (1929) it is known that in numerical models the phase can be influenced by variations of the bottom friction. The results have been applied for the present investigations.

The application of the usual friction term is sufficient for water depths of more than 10 m. But in extremely shallow water areas with actual water depths below 3 or 4 m there may be some problems with numerical stability as a consequence of the numerical handling of the difference equation of motion.

Let us consider the usual vertically integrated hydrodynamical equations in a simple one-dimensional case:

$$(4) \quad \frac{\partial u}{\partial t} + \frac{r\, u|u|}{H} + \frac{\partial \zeta}{\partial x} = 0$$

$$(5) \quad \frac{\partial \zeta}{\partial t} + \frac{\partial}{\partial x}(Hu) = 0$$

From equation (4) follows, partially written in difference from

$$(6) \quad u' = u\left(1 - \frac{2\,\Delta t\, r\, |u|}{H}\right) - 2\,\Delta t\, g\, \frac{\partial \zeta}{\partial x}$$

Let us only consider the first part of the equation (6) which describes the energy dissipation due to bottom friction. This term with paranthesis is responsible for numerical stability, because it exists the following inequality

$$(7) \quad 1 - \frac{2\,\Delta t\, r\, |u|}{H} > 0 \qquad \text{or} \qquad (8) \quad H > 2\,\Delta t\, r\, |u|$$

otherwise one wins energy in a numerical way. It is obvious that this condition is very important for numerical experiments in shallow water and one has to find a solution to modify the difference equation in such a way that a numerical instability will not occur and for all that a good agreement between observations and numerical results can be expected. In other words: the difference equations must be applicable for shallow water without violation of physical laws.

In past experiments the equation (6) had been reduced to

$$(9) \quad u' = 2\ \Delta t\ g \frac{\partial \zeta}{\partial x}$$

whenever the condition (7) had been violated.

By means of the following arguments one can show that this neglect leads to a non-acceptable result:

a) the neglect of the whole term is not correct, because in that case the velocity is only determinated by the gradient term $g \frac{\partial \zeta}{\partial x}$ without consideration of the velocity in the time step before.

b) if we consider the equation of continuity (5) and substitute u by $g \frac{\partial \zeta}{\partial x}$ then it follows

$$(10) \quad \frac{\partial \zeta}{\partial t} + \frac{\partial}{\partial x} (g\ H \frac{\partial \zeta}{\partial x}) = 0$$

After some simple transformations and neglects of terms which have a second deviation it comes out the equation

$$(11) \quad \frac{\partial \zeta}{\partial t} + g \frac{\partial h}{\partial x} \frac{\partial \zeta}{\partial x} \sim 0$$

The variation of depth at the flats and sands is very small that is equivalent to $h_x \sim 0$ and one can say as a first approximation

$$(12) \quad \frac{\partial \zeta}{\partial t} \sim 0$$

This means a stop of the water motion at those areas or in those discrete points of the model where the bottom friction term had been neglected because of the small actual water depth. The sands and flats do not fall dry because the given condition is not correct. The results which are coming out are not satisfactory. In nature these water masses flow with relative high velocity into the narrow channels - so called "Priele". In the numerical model these water masses are missing, they are not present at other discrete points which now have a phase displacement (see fig. 5).

It is possible to solve this problem by means of varying the time step because the condition (7) depends also on t, but this method seems to be very complicated. One can avoid these complications if one varies the quotient $\frac{1}{H}$ of the friction term in such a form that numerical stability will not be voilated. The following treatment has been selected:

$$(13) \quad R = \frac{H + H_o e^{-pH}}{(H + H_1)^2} \quad \text{with } H_o = H_1 = 1 \quad \text{and} \quad p = 0.5;\ 1;\ 2;\ 3.....10$$

The investigations with the aim to prove this treatment have been made in a simple tidal model for an open channel with the following geometry:

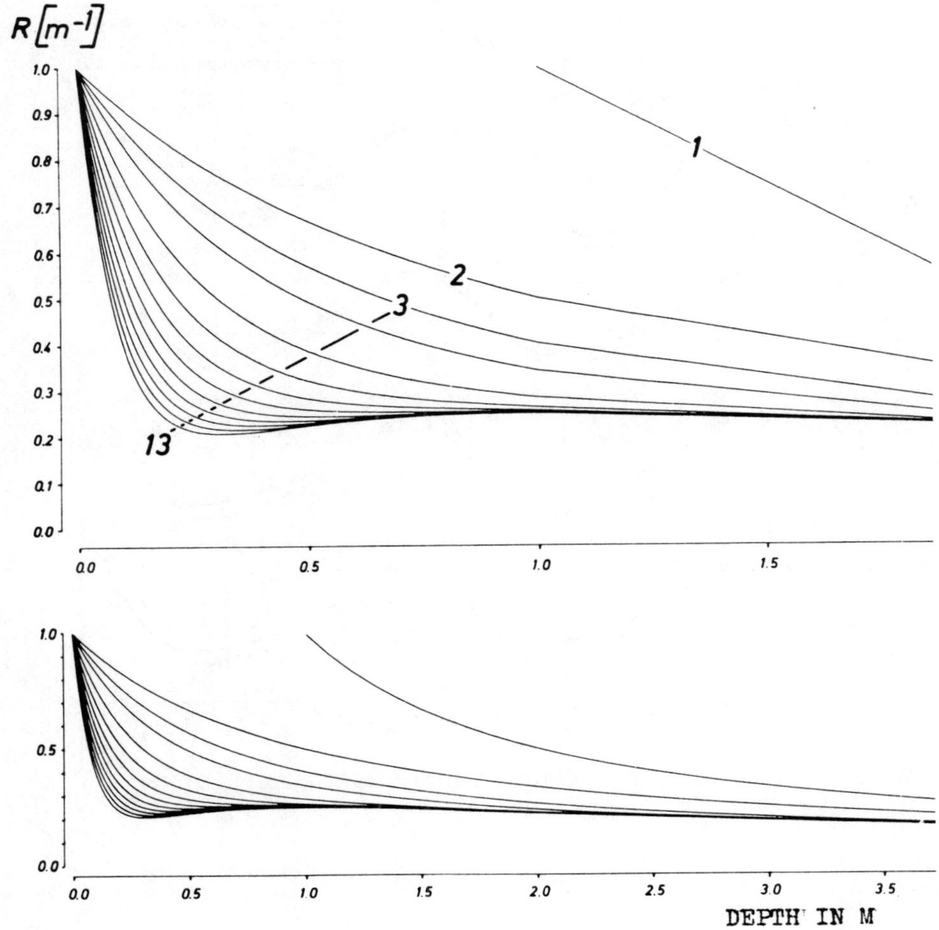

1) $\frac{1}{H}$

2) $\frac{1}{H+H_1}$ with $H_1=1$, $H=h+\zeta$

3) – 13) $\frac{1}{(H+H_1)^2} \times (H+H_0 e^{-pH})$ with $H_1 = H_0 = 1$
$p = 0.5;\ 1;\ 2;\ \cdots\cdots 10$

FIG. 1 COMPARISON OF DIFFERENT BOTTOM FRICTION TERMS

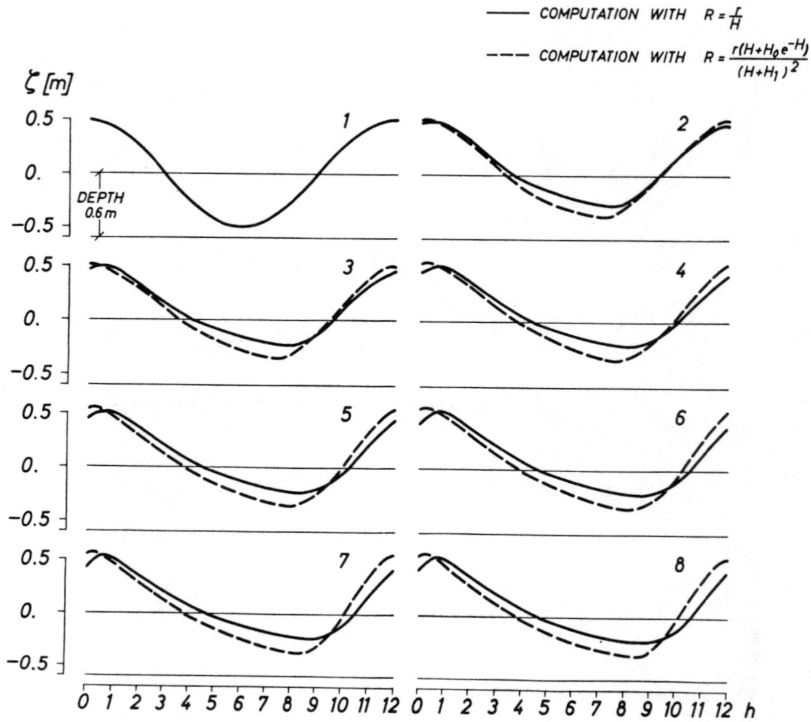

FIG. 2 COMPUTED WATER LEVEL VERSUS TIME AT DISCRETE POINTS 1 - 8

a) constant depth $h = 0.60$ m

b) constant width

c) total length of the channel: 4 000 m

d) 8 inner points

and the boundary conditions:

a) at the open boundary $\zeta = 0.5 \cos(2\pi/T)\, t$ with $T = 43\ 200$ s

b) at the closed boundary: $u \equiv 0$

The convective term $u \frac{\partial u}{\partial x}$ has been considered.

At fig. 1 the curves versus h with different factor p are presented. It is easy to see that the function 1/H is not suitable for shallow water. The best results have been obtained with p = 1. The advantage of this treatment is the convergence to the traditional bottom friction term with increasing water depth. Fig. 2 and fig. 3 show the results of the investigations with different friction terms.

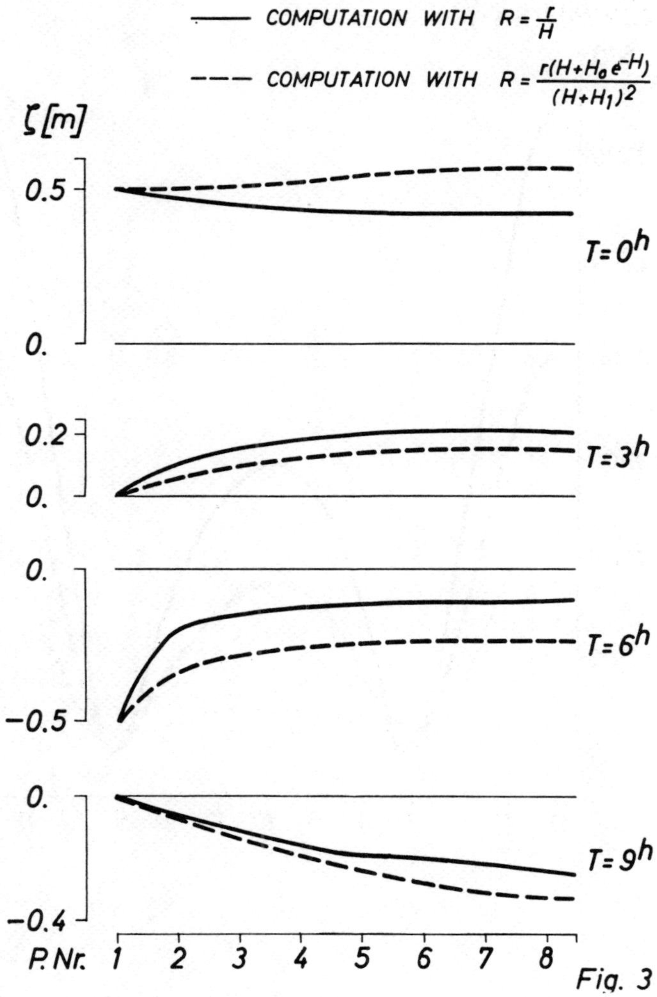

FIG. 3 COMPUTED SURFACE SLOPE VERSUS X

It can be definitely stated:

a) a phase lag from 30' at point 2 up to 50' at the points 7 and 8,

b) the water level at low water time is lower (about 14 cm) by means of the special treatment against the usual,

c) the water level at high water time decreases in the same case about 4 cm,

d) the smaller friction term allows an increasing velocity in shallow water.

Fig. 4 shows the computed bottom friction energy versus time of this discussed one-dimensional tidal channel. Three computations with different bottom friction treatments have been made.

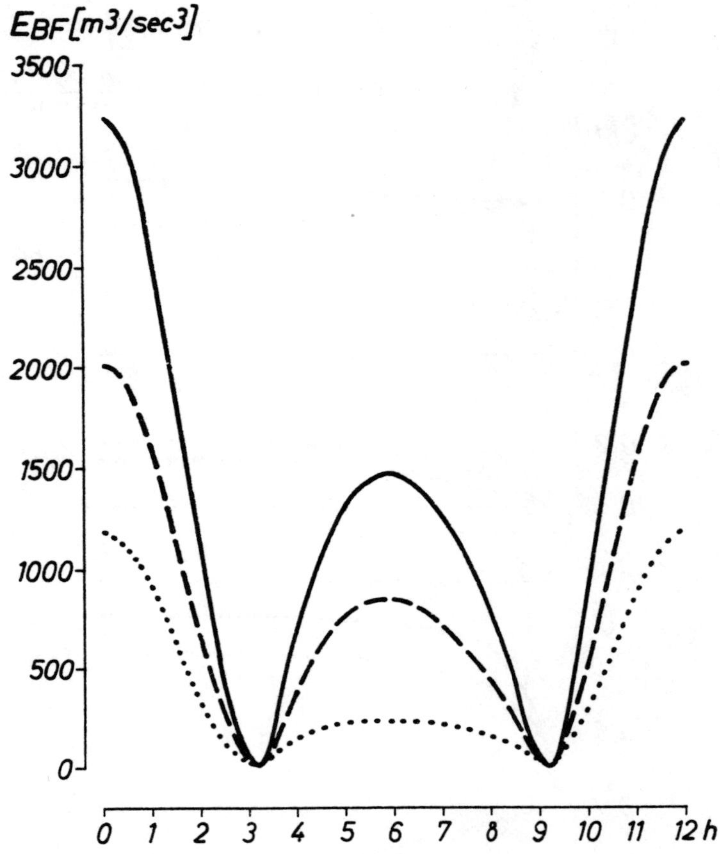

Computation with

.... $R = \frac{1}{H} r\ |u|\ u$

\- - - - - $R = \frac{1}{H + H_1} r\ |u|\ u$

———— $R = \frac{H + H_o e^{-H}}{(H + H_1)^2} r\ |u|\ u$

FIG. 4 ENERGY OF BOTTOM FRICTION VERSUS TIME

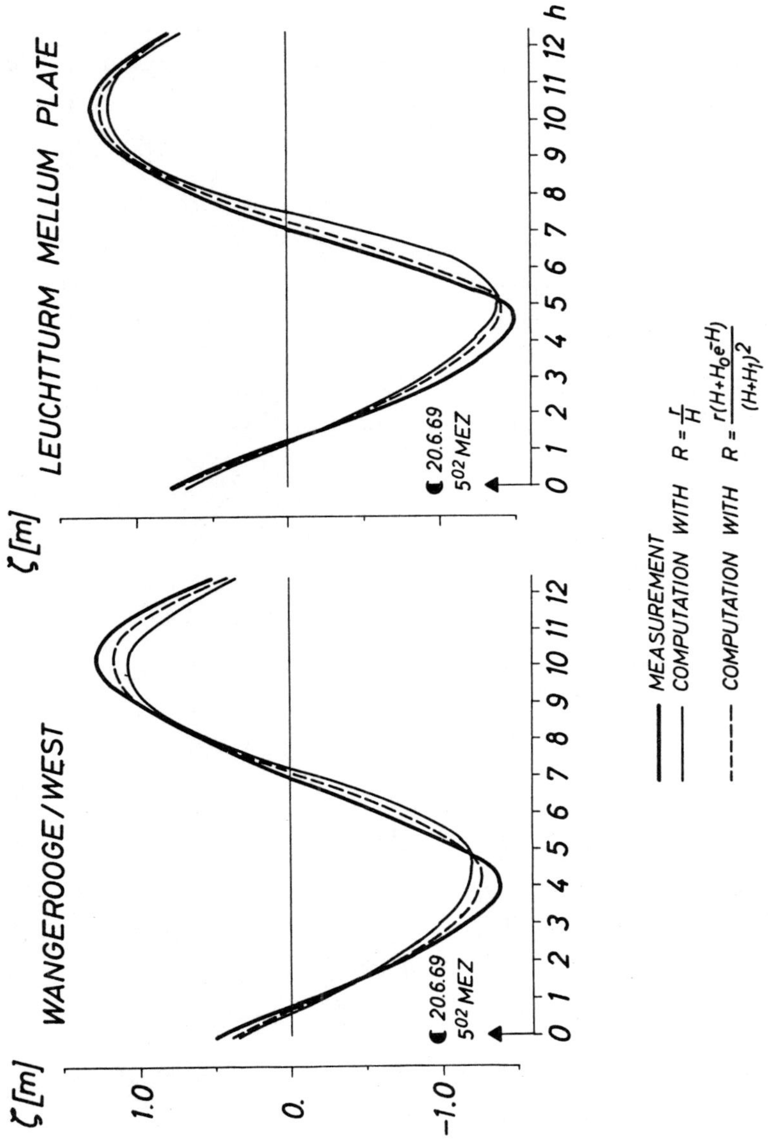

FIG. 5 COMPARISON BETWEEN MEASURED AND COMPUTED WATER LEVEL AT TWO SELECTED POINTS IN THE GERMAN BIGHT FOR TWO DIFFERENT TREATMENTS OF BOTTOM FRICTION

Fig. 5 shows a comparison between observed and differently computed water levels. These two examples are selected out of 80 points. This reproduction is not the best one but it is to comprehend a statistically average.

In conclusion it can be stated that by means of the special bottom friction treatment the vertically integrated hydrodynamic equations in difference form can be modified to be applicable for shallow water without violation of physical laws and one can achieve good results in a tidal model for an open channel and in a two-dimensional model with shallow waters areas.

Naturally, this equation is an empiric one which containes a numerical part as well as a physical part.

The advective terms as an important part of the description of the dynamical processes in shallow water in the numerical model can not be neglected. Unfortunately, they cause disturbances of a numerical type in the model which can be stabilised by using a time-step and screen-grid dependent coefficient A_h, which describes the horizontally turbulent eddy viscosity.

Shoal areas are often passed through by narrow channels with depths up to 5 m and more. On the grid points on the edges of these channels as well as on points where the shallow water area borders on navigable channels or waterways respectively, occur instabilities on account of the large $\frac{\partial h}{\partial x}$. Here also, the above mentioned coefficient will have a stabilizing effect.

If one considers the following equation

$$(14) \quad A_h = \frac{1-\alpha}{4} \cdot \frac{(2\,\Delta l)^2}{2\,\Delta t} \qquad \text{(BRETTSCHNEIDER, 1967)}$$

so it can be generalized by a dependency to the bottom slope to

$$(15) \quad A_h = \frac{1-\alpha e^{-\frac{\Delta h}{\Delta l}}}{4} \cdot \frac{(2\,\Delta l)^2}{2\,\Delta t}$$

and simplifying to

$$(16) \quad A_h = \frac{1-\alpha\,(1-\frac{\Delta h}{\Delta l})}{4} \cdot \frac{(2\,\Delta l)^2}{2\,\Delta t}$$

($2\,\Delta l$ = repective grid distance in x- or y-direction)

In the submitted investigations, the smoothing factor $\alpha = 0.98$ was set, A_h takes up values between 2.5 to 170 m^2/s according to grid distance. It is easily recognisable that by medium formation at a strong bottom slope a stronger dampening will occur.

The advective terms $u\,\frac{\partial u}{\partial x}$, $v\,\frac{\partial u}{\partial y}$, $v\,\frac{\partial v}{\partial y}$ and $u\,\frac{\partial v}{\partial x}$ will be treated on the boundaries of the numerical models as follows:
on the open boundary applies $\frac{\partial u}{\partial x}$ or $\frac{\partial v}{\partial y} = 0$;
on the closed boundaries the terms $u\,\frac{\partial u}{\partial x}$ and $v\,\frac{\partial v}{\partial y}$ cause no difficulties.
$v\,\frac{\partial u}{\partial y}$ and $u\,\frac{\partial v}{\partial x}$ are treated in the same way as in the center of the area since the components of velocity will disappear at the respective boundaries in accordance to the presuppositions, therefore, they will also be outside of the area equal to 0 (see fig. 6).

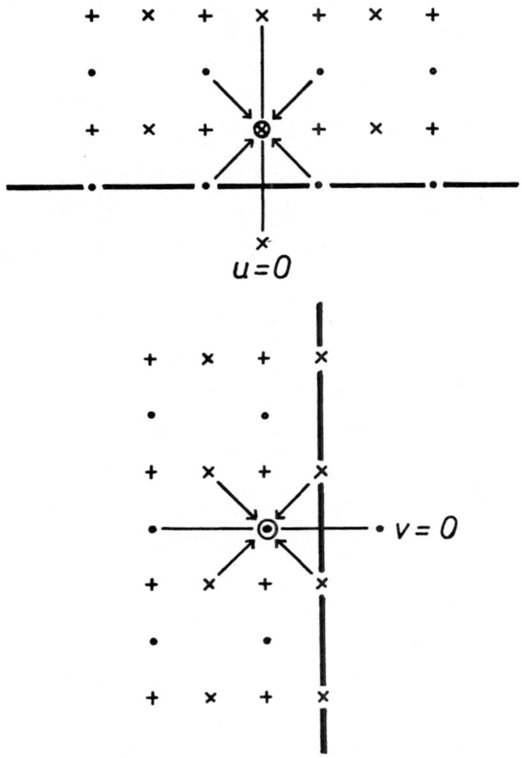

FIG. 6 THE GRID

The mean velocities of a tidal cycle will be determined as follows:

$\bar{u} = \frac{1}{T}\int_0^T u\,dt$; $\bar{v} = \frac{1}{T}\int_0^T v\,dt$; $\bar{u}$ and $\bar{v}$ are middled on the central ζ-point and form the components of the mean velocity $u\,(M)_m = \sqrt{\bar{u}^2 + \bar{v}^2}$ in the point M with the direction $\varphi = \text{arc tan}\,\frac{\bar{v}}{\bar{u}}$.

The components of the residual currents of a tidal cycle and its direction are defined as

$$u_r = \frac{\int_0^T Hu\,dt}{\int_0^T H\,dt} \quad \text{and} \quad v_r = \frac{\int_0^T Hv\,dt}{\int_0^T H\,dt} .$$

These components are also middled on the central ζ-point. The residual current velocity in point M is the $u\,(M)_R = \sqrt{u_r^2 + v_r^2}$ and the direction $\varphi = \text{arc tan}\,\frac{v_r}{u_r}$.

The residual current circulation is superimposed to the harmonical movement and enables a detailed insight into the long-time movement process which can be very informative for the judgement of the effectiveness of a structural measure.

Since overflowings of sandbanks as well as dry-falling sands occur in the shallow-water areas, these movement processes must also be reproduced by numerical methods which are physically interpretationable.

It is necessary to distinguish all thinkable cases, because it is possible that from time to time in some points there will be no water. One has to check the neighbourhood of each gridpoint from time step to time step with regard to the actual water depth, the depth distribution and the physical possibility of transports and directions of transports. It is also necessary to pay attention to the velocity of the overflow and the draining processes of extreme shallow water in order to make sure that the process in nature is in a good agreement with the process in the hydrodynamical-numerical model. The covered distance of water is computed and it will be checked if the next grid point has been reached. With this simple method one can find out the motion of the water line.

During the treatment of the continuity equation it should be checked whether $h + \zeta$ is ≤ 0. It will be sensible to select 0.02 m as a coefficient data. If $h + \zeta$ is ≤ 0.02, then the area represented by this point shall be considered as dry and $\zeta = -h$ is applied. The values $h + \zeta \leq 0.02$ will be stored by points and are then again transported to the point if $h + \zeta$ is > 0.02, i.e. if the actual water depth exceeds this value. Thus it will be possible that mass deviations may occur in time steps, but over a tidal cycle, the masses will be maintained.

THE EFFECT OF THE STRUCTURAL CHANGE OF A BREAKWATER ON MOVEMENT PROCESSES

The breakwater has a length of approx. 1 400 m and a present height of NN - 1.40 m.

The following three cases were investigated:

Model	I	breakwater, actual state
Model	II	breakwater, height NN + 1.50 m
Model	III	breakwater, actual height, half length

These marks are also valid for all figures within this paper.

The model

Before the beginning of the investigations it had to be checked at first:

a) how large should the model area be selected?

b) Where is a further screen-grid dissolution to be effected in accordance to tasks to be handled?

c) How far can or should a reduction of the screen-grid distance be carried out so that the peculiarities of the morphology and the breakwater can be well approximated?

d) How great will be the time step depending on the screen distance and the depth? Will the, thus, caused computation expenditure be supportable?

e) What tide should be selected for the investigations and what boundary values will be available?

to a)

The size of the model area to be selected will depend at first on the numerical investigations to be carried out and upon given questions. In the present case, the influence of a building-construction upon the dynamical processes must be investigated. It was also to be checked, which alterations of these vertical and horizontal movement processes will have to be expected in case of a possible constructional interference. Therefore, the size of the investigation area should be so selected that the open boundary of the numerical model can not be influenced by the changes of processes of movement to be expected. Based on the available floating measures and the morphology North of Wangerooge and Spiekeroog, the model area was selected as shown in figures 7 and 8.

From the later obtained results it can be seen that the open borders are sufficiently far away form the range in influence.

to b)

The selection of the screen grid and the degree of solution depends on the morphology and to the same extent upon the necessary accuracy of reproduction of the changes of movement to be expected.

Available were a part model with a mesh size of 457 m. The area North of Wangerooge offers itself for a further screen-grid solution. On the one side there are float measurements available and on the other side, changes in the processes of movements can be expected for certain in the socalled remote range in relation to the breakwater because of building-constructional measures which had to be investigated.

A screen distance of 153 m was selected in this part area. In this manner, the topography could be reproduced in a fineness which conforms to the available maps. This part area is illustrated in the figures 7 and 8. A further solution - the one-third-refinement method - was naturally required in the close range of the breakwater. A screen distance of 51 m between points of equal importance (ζ/ζ, u/u, v/v) has for the approximation of the morphology in the numerical model the subsequence that here a depth is indicated every 25.5 m since the ζ- as well as the u- and v-points are available. Unfortunately the breakwater does not run parallel to the screen.

The necessary deviations in the reproduction of the location in the numerical model should be considered as unimportant.

to c)

From the explanations under b) it can be concluded that the solution is an optimum one. The results which will be later explained add to this a multiple number of hints. In this connection it should also be mentioned, that to my knowledge, for investigations of this principal importance for the first time such extreme screen solutions have been applied for the first time and, apart from this, an interaction between areas of different grid distance were ensured.

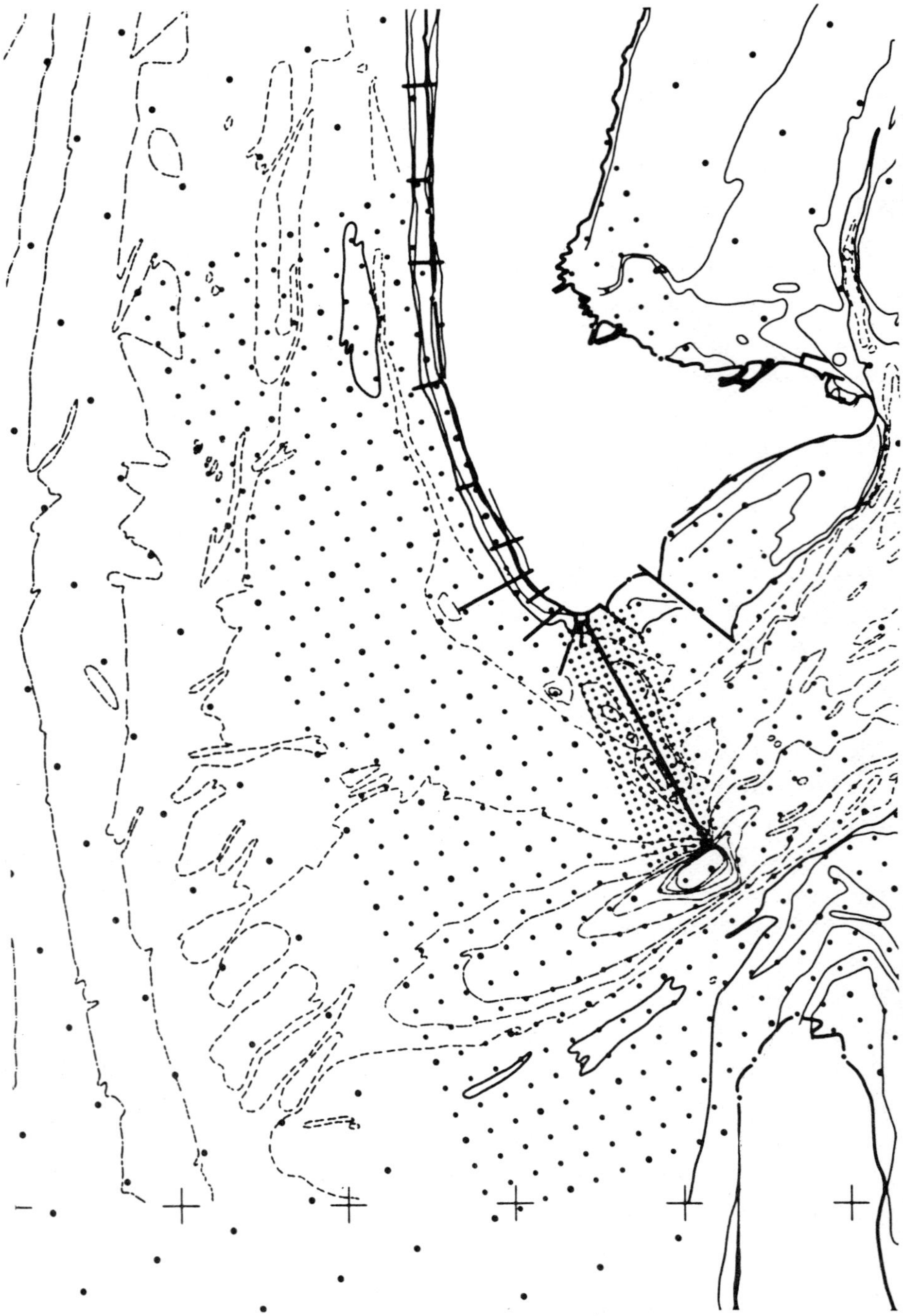

FIG. 7 MAP AND THE NESTED MODEL IN THE AREA OF THE BREAKWATER

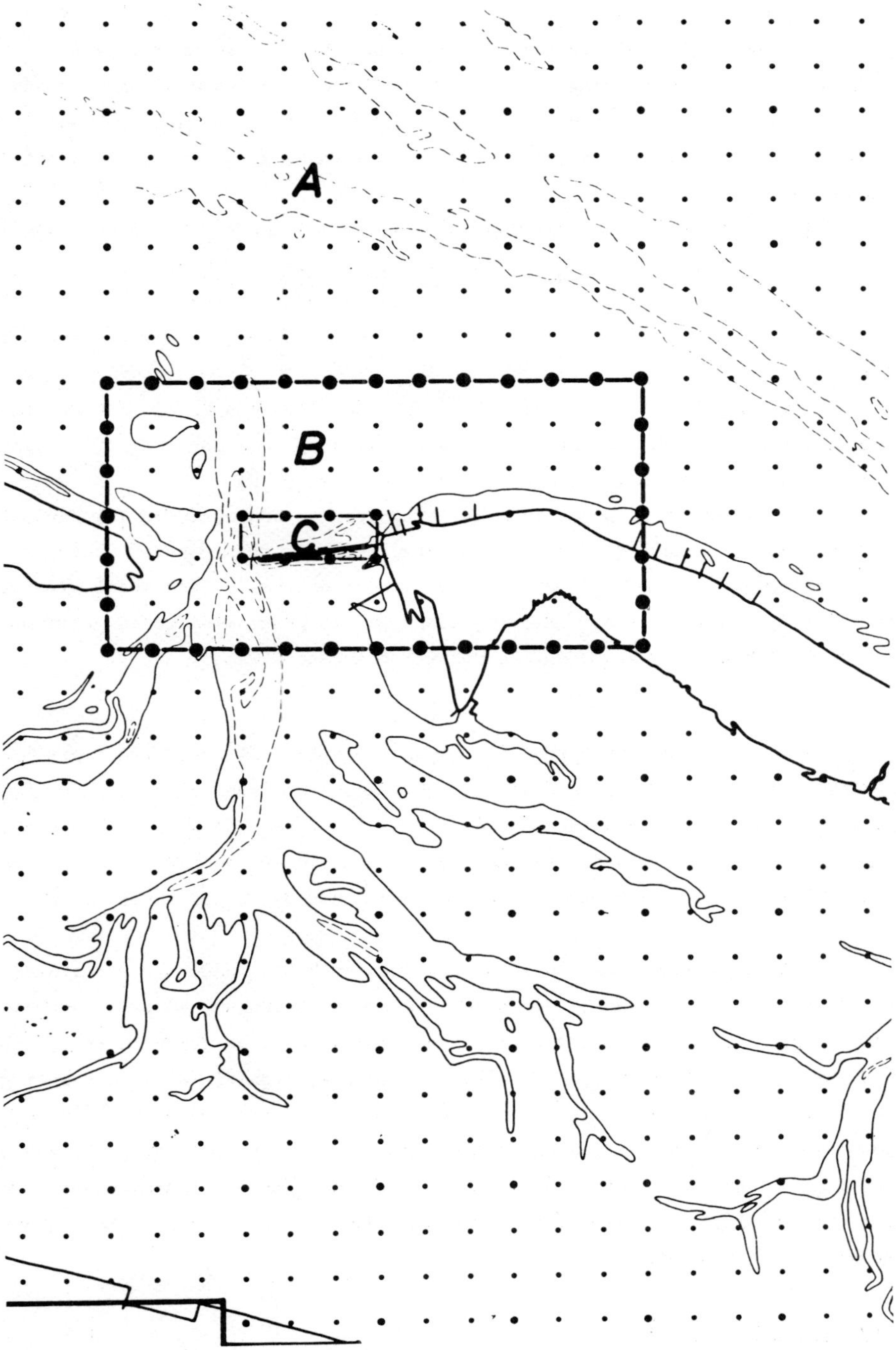

FIG. 8 AREAS A, B AND C WITH DIFFERENT GRID DISTANCES

to d)

The area with the lowest screen distance is bordered in the West by the breakwater's head. The further in the West situated deep channel has already a grid distance of 153 m. A further screen solution also in this range would lastly bring no other results and would led to a not anymore supportable computation expenditure.

The condtion for the numerical stability to Courant-Friedrichs-Lewy (1928)

$$2\ \Delta t\ \text{(time step)} < \frac{2\ \Delta l\ \text{(smallest grid distance)}}{\sqrt{2\ g\ H_{max}}\ \text{(greatest depth)}}$$

leads through the great depths in the above mentioned case to a very short time step.

With the given screen refinements and the given morphology follows the necessity for the numerical model to use a time step of 3 seconds. This means for a tidal period of 44700 s = 12^h 25 min = 14900 time steps.

This computation expenditure is considered as supportable. The periodical stationarity was obtained after two computed periods, this however, only because an approximated initial distribution were at our disposal. This model will be available for other possible investigations and is programmed to an optimum. The obtained results justify this computation expenditure.

to e)

An extensively undisturbed i.e. a not by wind influenced tidal period was selected for the investigations. The author had water-level observations of various levels out of an investigation series from the June 20, 1969 in hand. Naturally, any other tide can be used, but it should as free as possible from wind influences and the measuring period should not be in the range of the nipp- or spring time.

Since this model area is a part of the North sea model, by the aid of which the movement processes in the Weser-Jade-Regime were determined to a tast-orientate screen solution, the necessary water level values could be taken in the points at the open model boundary as f(t) from the available model. After completing the timely, linear interpolation and the subsequent 5-time overlapping middle structure for smoothening the curvature (this is absolutely permissible because of the very short time step) the water level values for the intermediate points were also obtained by linear interpolation - this is also permissible because of the short screen distance.

Natural similarity

The conformity of the determined velocity- and water level values to measurements and ovservations is an unalterable condition for prognostical investigations. This applies to the hydraulical model as well as to the numerical model. Measurements are mostly only available in a few discrete points. If the numerical model renders the timely course of the horizontal and vertical movements in these points or if the average deviations remain within the scope of the numerically possible accuracy, then it can be supposed that the reproduction of the processes of movement has succeeded within the scope of permissible error limits. This will not exclude, however, the critical areas will always exist which require a more close investigation and examination. The figures 10, 11, 12 show the agreement between observations and computations of water levels in the basic model.

Within the scope of the investigations which are the subject of this report, the numerically determined results were compared with the water level values on the tide gauge in Wangerooge West (June 20, 1969) and, thereby, it was determinded:

1. High water (model)	+ 122 cm
High water (nature)	+ 126 cm
Difference model/nature	- 4 cm
2. Low water (model)	- 130 cm
Low water (nature)	- 138 cm
Difference model/nature	+ 8 cm
3. Setting-in time high water (model)	
- Setting-in time high water (nature)	+ 20 min

Since the tide gauge Wangerroge West is extremely unfavourable situated, the deviations must be considered as negligible and absolutely permissible. It can be supposed that the screen solution with all its consequences is the cause of the result from the reproduction of these movements. Unfortunately the author had not other observations from this period for the model range in hand for comparing the natural similarity further.

For the reproduction of the movements in extremely shallow water with periodically dry-falling sands and shoal areas as well as water bearing channels, the treatment for bottom friction was well proved during the previous investigations as used again.

THE RESULTS

Velocities

Model I

Here the remark should be permitted that the solution of the grid within this area - as far as illustrated in the area A - is insufficient. With a point distance of 457 m, the locally extremely complicated current movements can not be reproduced satisvactorily. The illustrations of these processes in the areas B (153 m) and C (51 m) will provide a much better information over the in- and overflowing process.

1. WITTDÜN	20.6.69	5. SCHILLING	16.9.65
2. CUXHAVEN	20.6.69	6. VOSLAPP	16.9.65
3. WANGEROOGE/WEST	20.6.69	7. VAREL	16.9.65
4. MELLUM PLATE	20.6.69	8. DANGASTER SIEL	16.9.65

FIG. 9 MAP OF SPECIAL POINTS IN THE NORTH SEA MODEL IN WHICH THE COMPUTED AND OBSERVED TIDAL CURVES HAD BEEN COMPARED

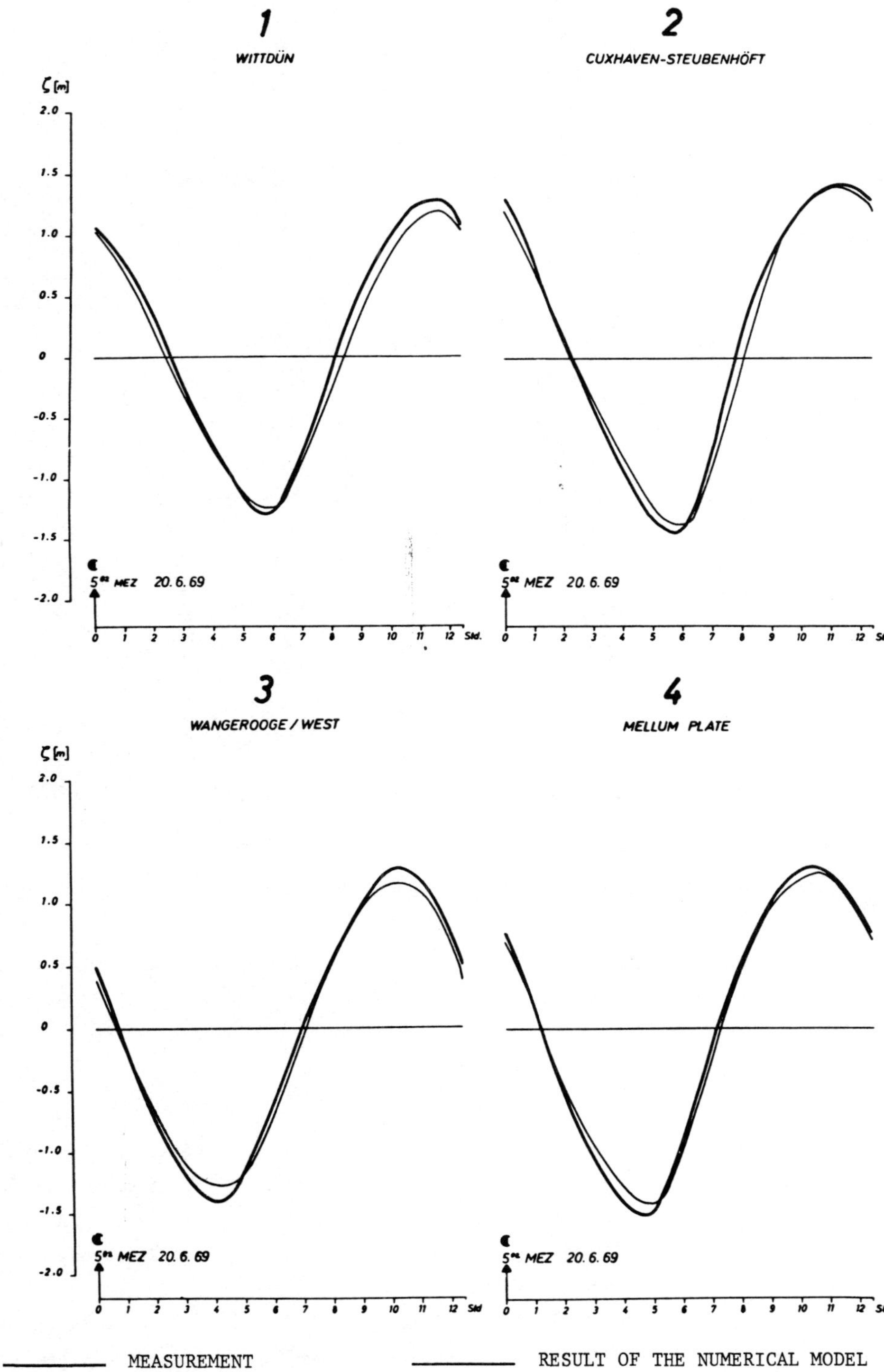

FIG. 10 COMPARISON BETWEEN OBSERVED AND COMPUTED TIDAL CURVES

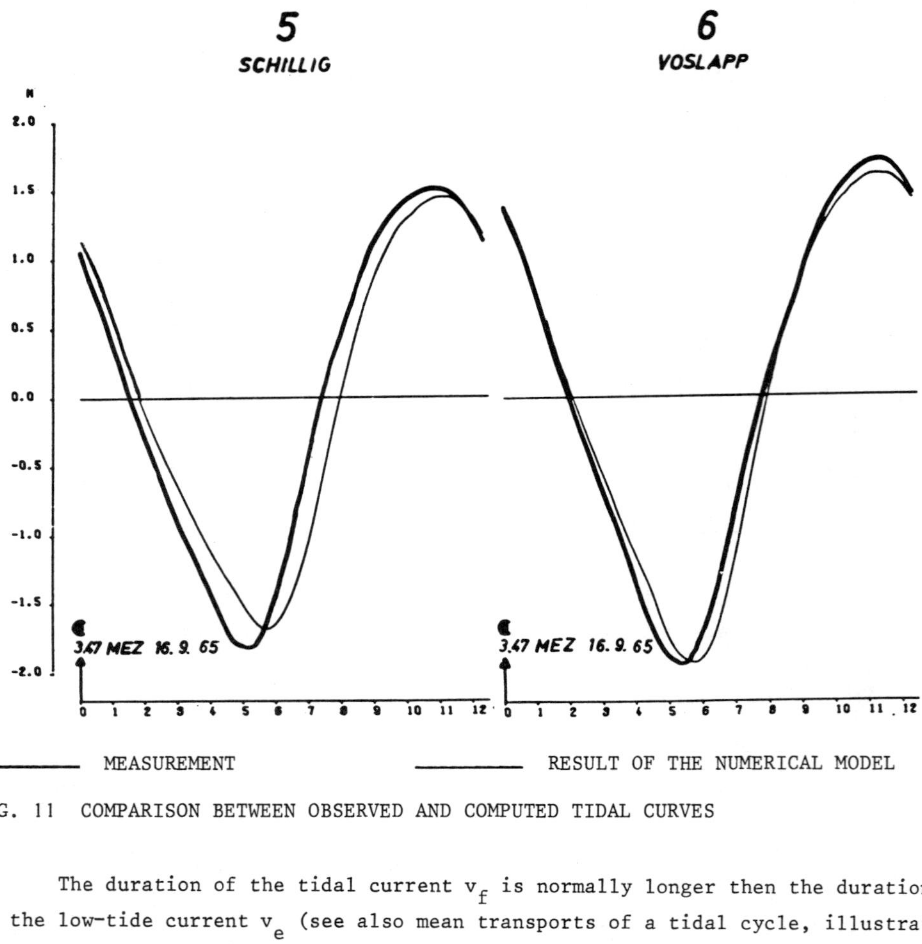

——— MEASUREMENT ——— RESULT OF THE NUMERICAL MODEL

FIG. 11 COMPARISON BETWEEN OBSERVED AND COMPUTED TIDAL CURVES

The duration of the tidal current v_f is normally longer then the duration of the low-tide current v_e (see also mean transports of a tidal cycle, illustrated in table 2). The high tidal current velocities with v_{max} = approx. 1.8 m/s in the southern part of the Harle are noteworthy.

Model II

The velocities at the breakwater head are greater and retain this order of magnitude also over a longer period in comparison to model I. The flow-in time is also longer. Both of the afore mentioned processes are of importance for the interpretation of the numerically determined average transports of a tidal cycle - and they will find here their confirmation.

Extremely high tidal current velocities (1.5/1.75/2.1 m/s) occur in the Harle at the south of the breakwater, whereas the low-tide velocities will be low. These high velocities can - when the water is running up on the flat, i.e. with

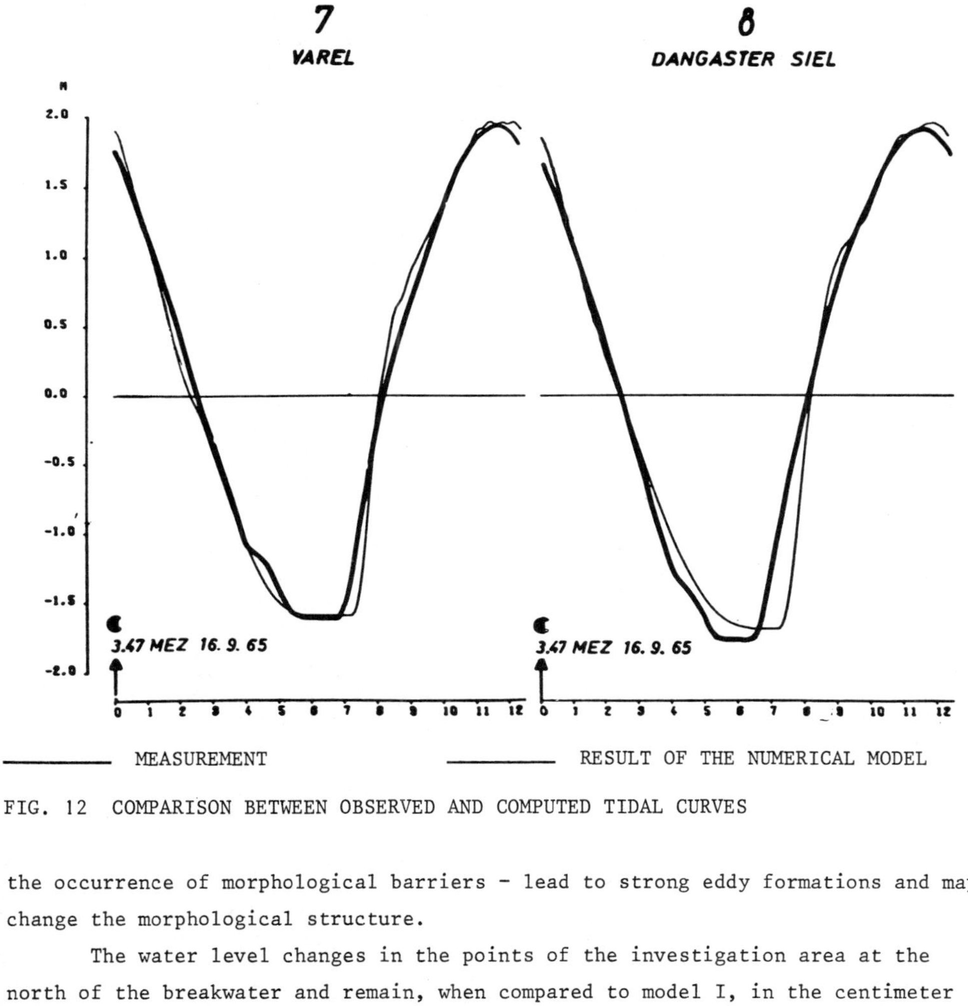

FIG. 12 COMPARISON BETWEEN OBSERVED AND COMPUTED TIDAL CURVES

the occurrence of morphological barriers - lead to strong eddy formations and may change the morphological structure.

The water level changes in the points of the investigation area at the north of the breakwater and remain, when compared to model I, in the centimeter range. The high water, however, lies in the Harle and south of the breakwater at a phase difference of approximately 30 minutes clearly lower - about 25 to 30 cm. This influence is effective up to 3 100 m south of the breakwater.

Model III

Apart from the changes which results solely on account of conditions - shortening of the breakwater by 50 % - there is a strong similarity between the results of the model I and the model III. This local change is obviously of no special importance for the total process - with the exception of a phenomen which is of a considerable importance in my opinion.

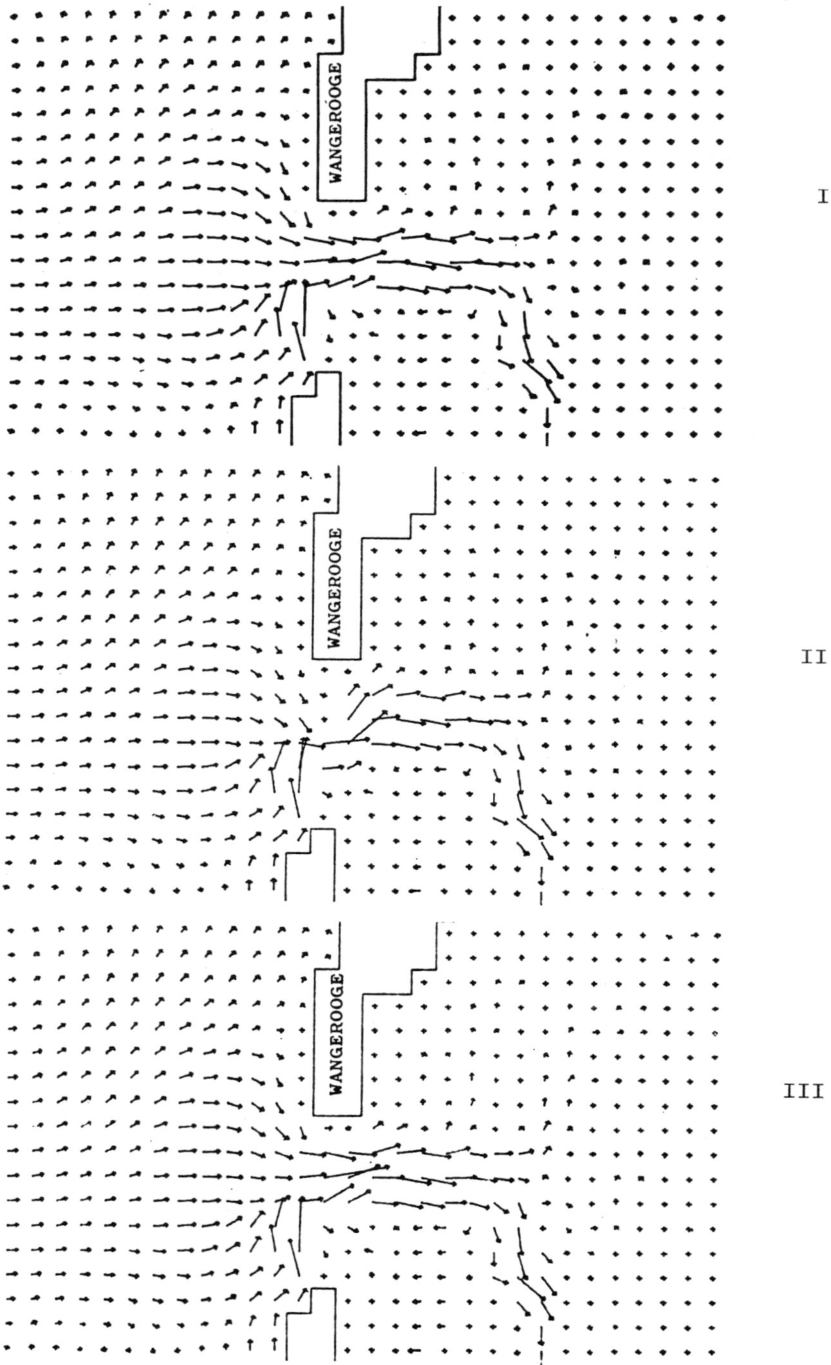

SCALE — 1 m/s

FIG. 13 AREA A - DISTRIBUTION OF VELOCITIES 6 HOURS AFTER MOON'S TRANSIT THROUGH THE MERIDIAN OF GREENWICH (20.6.69 - 11.02 h)

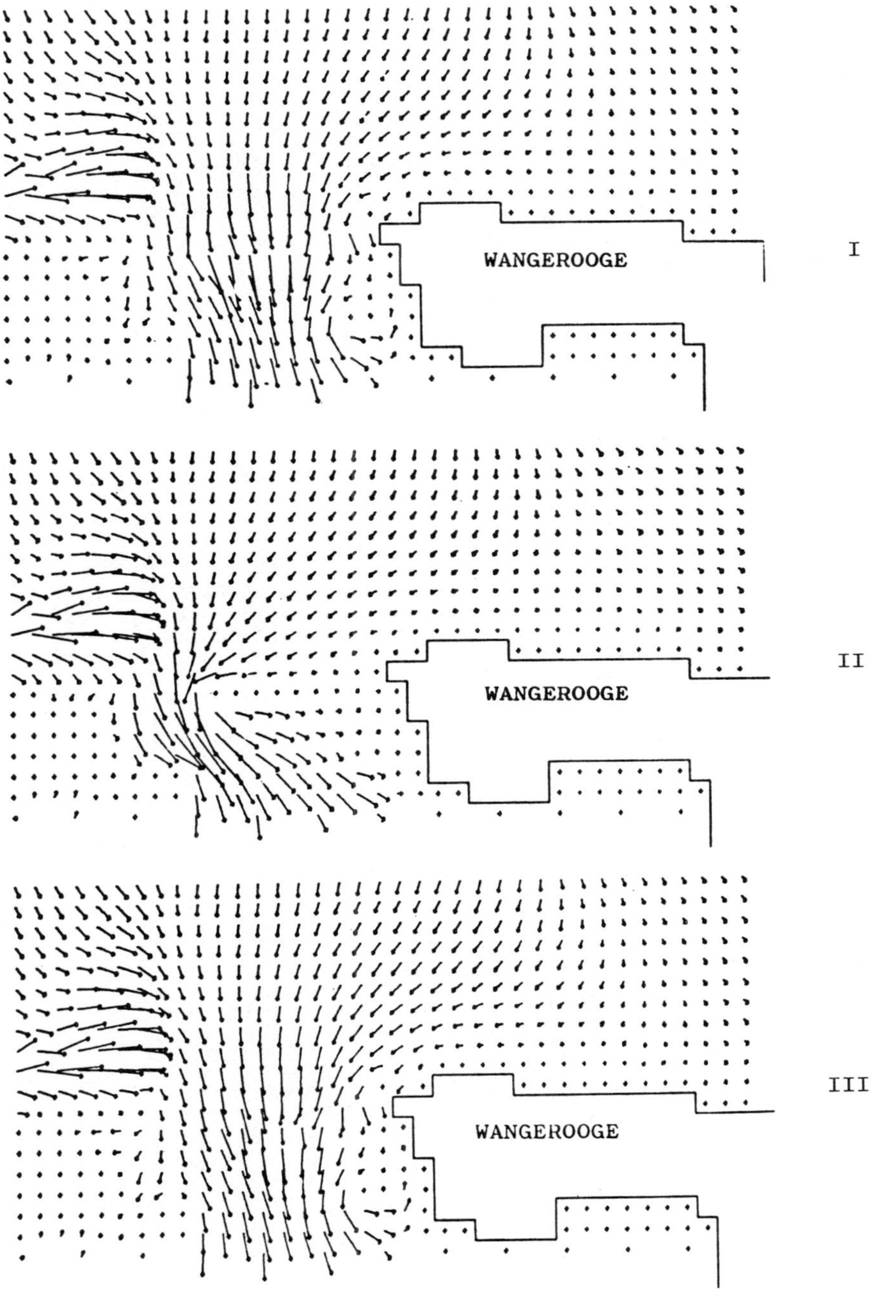

SCALE — 1 m/s

FIG. 14 AREA B - DISTRIBUTION OF VELOCITIES 7 HOURS AFTER MOON'S TRANSIT THROUGH THE MERIDIAN OF GREENWICH (20.6.69 - 12.02 h)

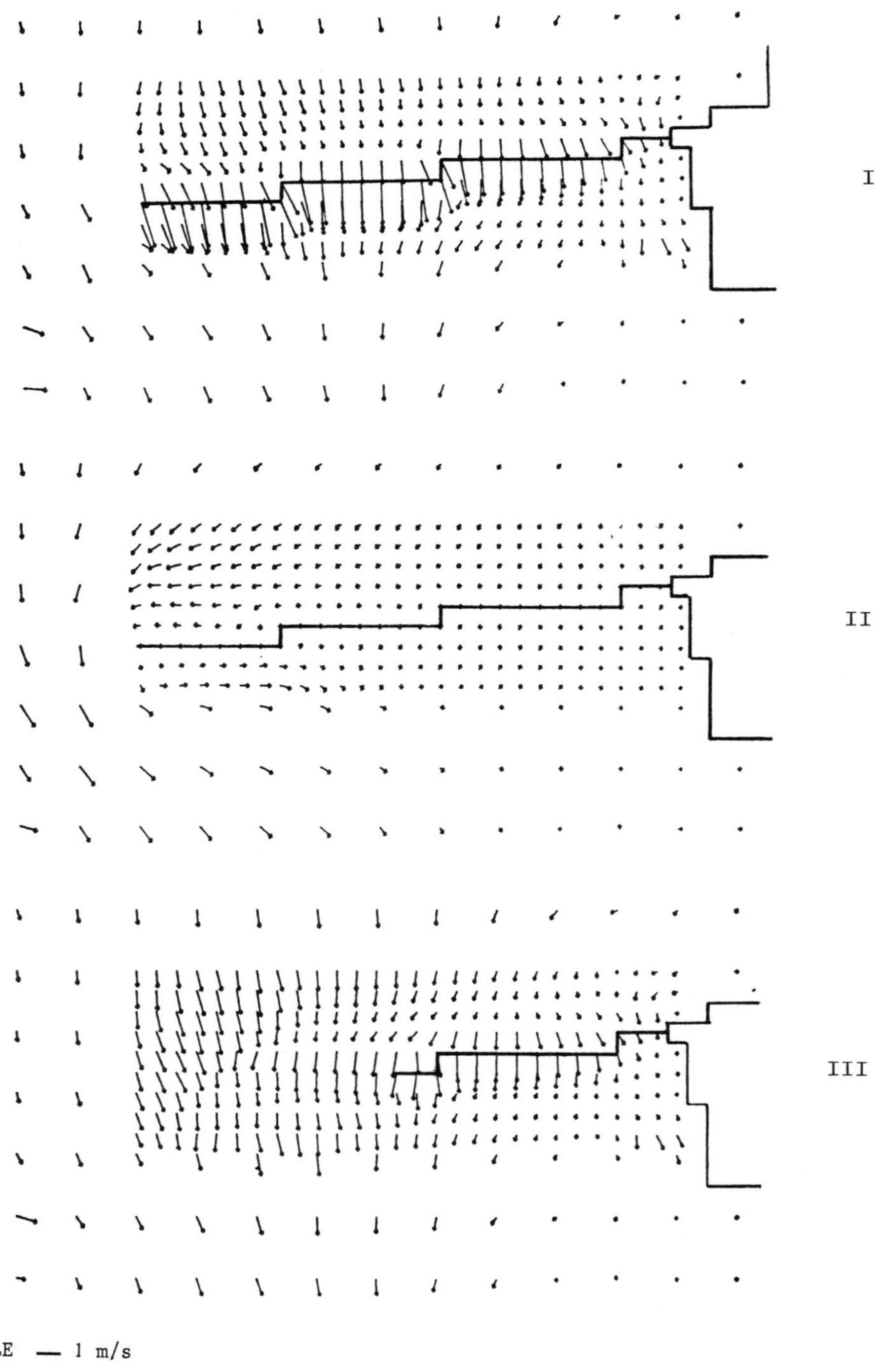

FIG. 15 AREA C - DISTRIBUTION OF VELOCITIES 6 HOURS AFTER MOON'S TRANSIT THROUGH THE MERIDIAN OF GREENWICH (20.6.69 - 11.02 h)

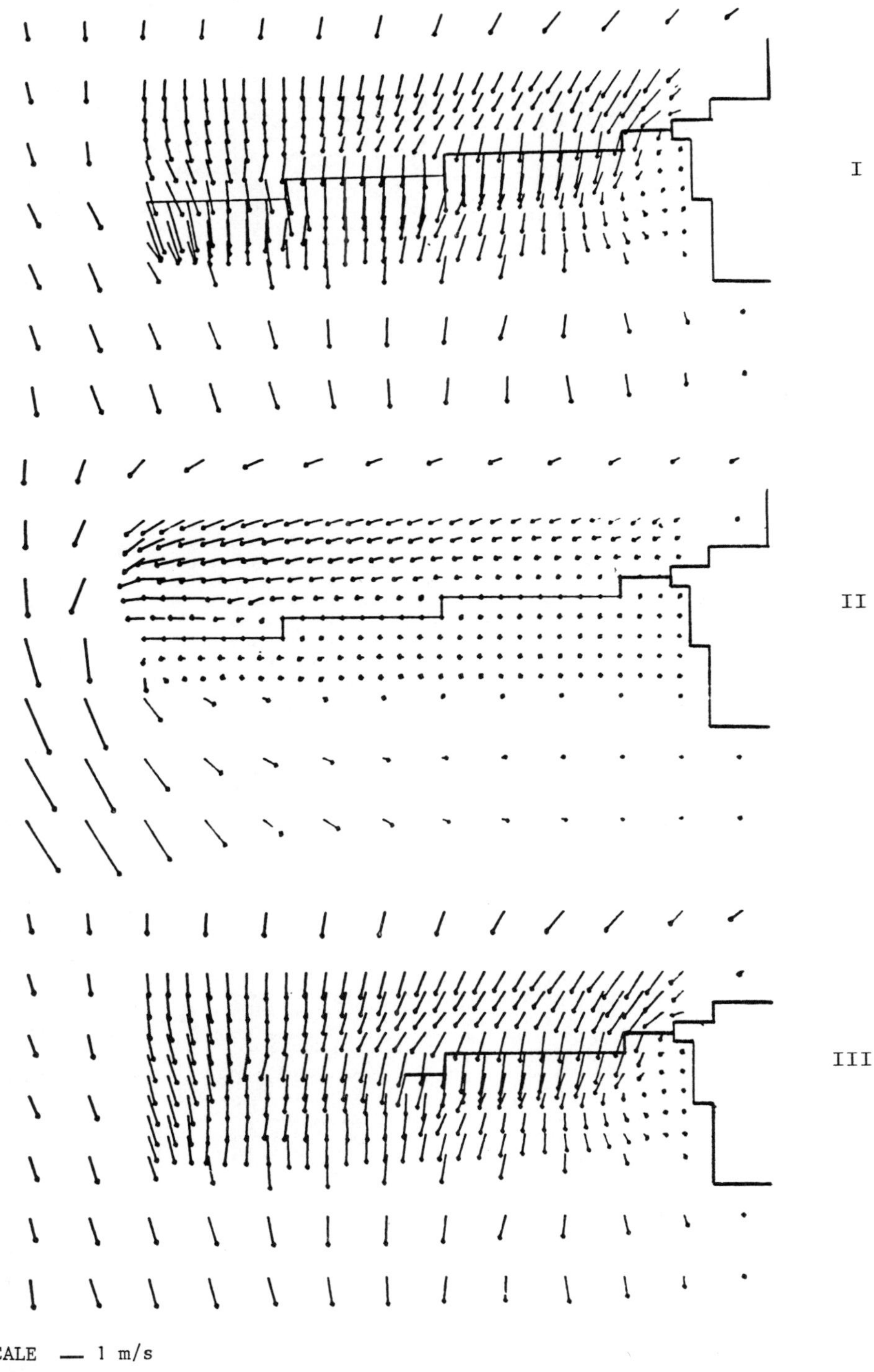

FIG. 16 AREA C - DISTRIBUTION OF VELOCITIES 9 HOURS AFTER MOON'S TRANSIT THROUGH THE MERIDIAN OF GREENWICH (20.6.69 - 14.02 h)

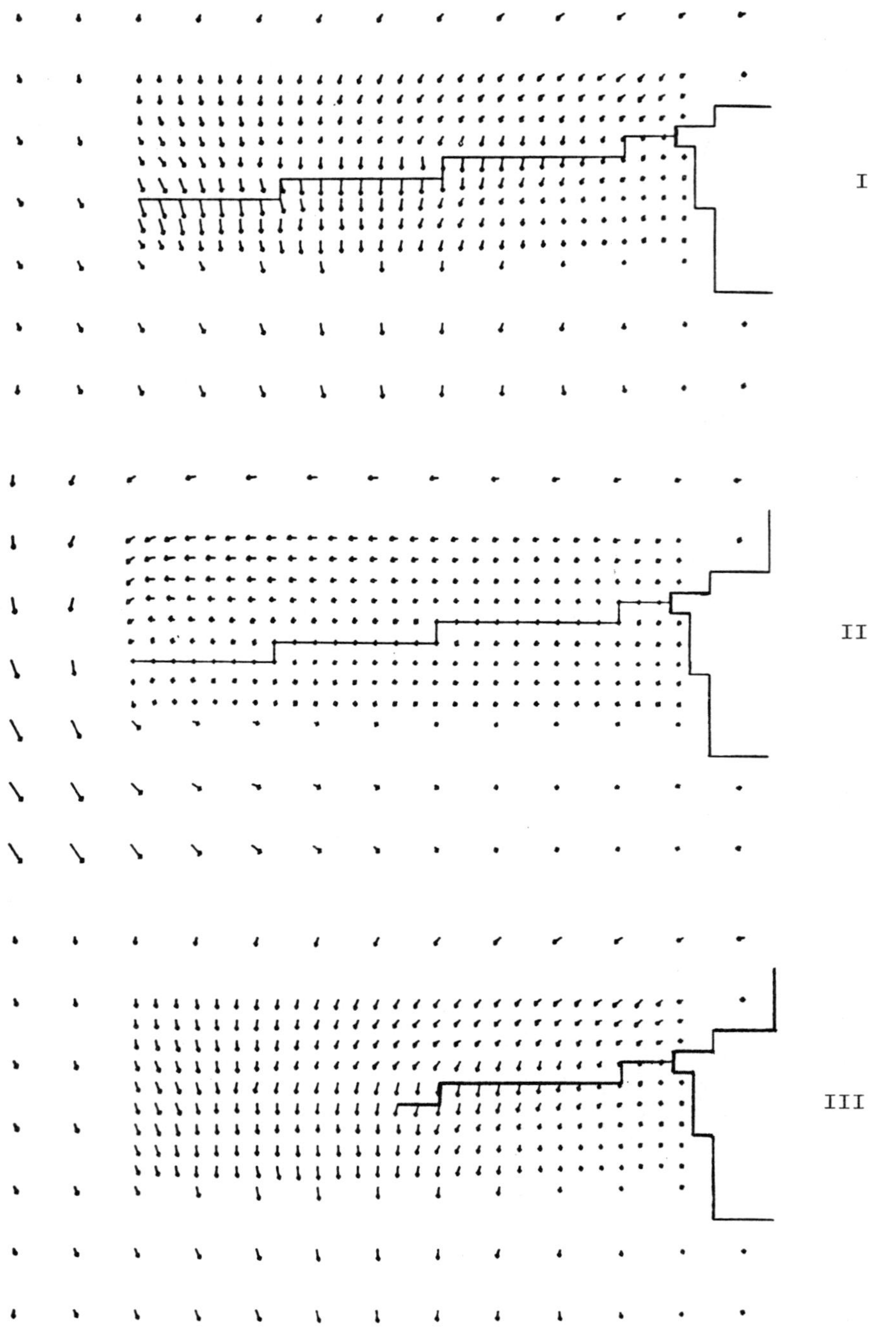

SCALE — 1 m/s

FIG. 17 AREA A - DISTRIBUTION OF VELOCITIES 12 HOURS AFTER MOON'S TRANSIT THROUGH THE MERIDIAN OF GREENWICH (20.6.69 - 17.02 h)

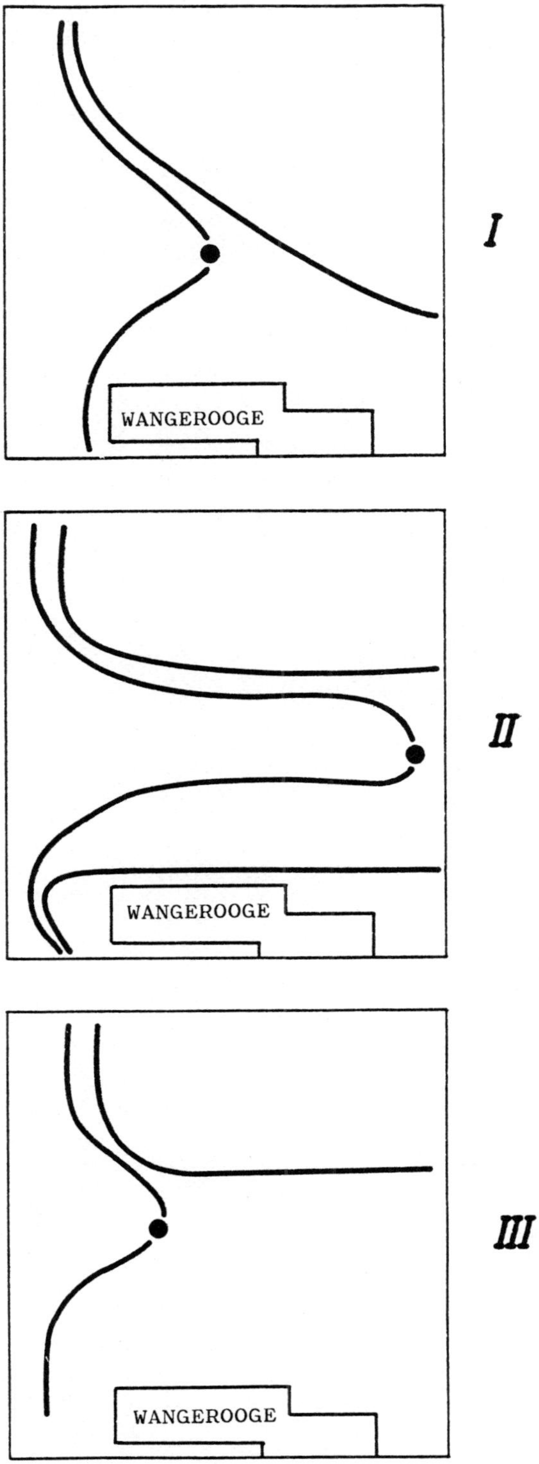

FIG. 18
DISPLACEMENT
OF DIVERGENCE

The through-current flows more uniformly as has been expected but the turn-in of the current takes place farther west in comparison to the present condition of the breakwater.

<u>The current splitting area</u>

The effect of the alteration of the breakwater towards the position of the current dividing area is of special interest. From the lines of the directions of maximum velocity per each model investigation, the point can be found in which in the near vicinity before the island, the turning point can be located of the current path. All the further this point is wandering to the east, all the greater are the strand-parallel running tide and low-tide velocitis which cause simultaneously the transport of solid particles. The following result is on hand:

By a rise of the breakwater to NN + 1.50 m, the turning point is displaced by 3 500 m to the east and by a shortening of the breakwater by 50 % at the present height of NN - 1.40 m it will be displaced by 1 000 m to the west.

<u>Water level comparisons</u>

In connection to the evaluation of the results, the water levels at the north and the south of the breakwater were middled locally at hourly intervals. Thereby, it was determined in the present condition as well as in case of a shortening of the breakwater by 50 % - with deviations in the centimeter range - the water levels at the north and south of the structure will be equal. If the breakwater is raised, however to NN + 1.50 m - i.e. free of tidal high water - striking differences will occur at low water.

The average water levels at the south of the breakwater are as follows:

10.02 h	7 cm
11.02 h	9 cm
12.02 h	38 cm
13.02 h	42 cm
14.02 h	41 cm
15.02 h	34 cm
16.02 h	21 cm
17.02 h	7 cm lower.

As stated in table 1, only the area C was drawn in, i.e. the listed values indicate only the local effect.

If the water level values are middled over the tidal period, then it can be clearly seen that with a rise of the breakwater to NN + 1.50 m in all three areas A, B and C, i.e. in the coastal as well as in the remote range, the mean water level of a tidal period south of the breakwater will be 13 cm lower than north of the structure.

TABLE 1

Mean water levels every hour in the Northern and Southern area of the breakwater (in cm)

Time	Breakwater Actual State		Breakwater Height NN + 1.50 m		Breakwater Half Lenght	
June 20, 1969	North	South	North	South	North	South
05.02	32	33	31	31	30	27
06.02	- 17	- 17	- 18	- 17	- 18	- 18
07.02	- 60	- 60	- 61	- 59	- 60	- 60
08.02	- 94	- 94	- 94	- 89	- 93	- 93
09.02	-112	-112	-112	-110	-112	-112
10.02	-112	-114	-111	-118	-113	-113
11.02	- 82	- 83	- 80	- 99	- 83	- 83
12.02	- 15	- 24	- 12	- 50	- 17	- 17
13.02	47	45	53	11	46	45
14.02	90	87	93	52	88	87
15.02	104	102	106	72	103	102
16.02	93	93	94	73	93	92
17.02	54	54	54	47	54	54

TABLE 2

Mean water level of one tidal cycle (June 2o, 1969) in the northern and southern area of the breakwater (in cm)

	Breakwater Actual State	Breakwater Height NN + 1.50 m	Breakwater Half Length
Area A			
North	- 5	- 5	- 5
South	- 22	- 35	- 22
Area B			
North	- 6	- 6	- 6
South	- 8	- 22	- 8
Area C			
North	- 7	- 6	- 7
South	- 8	- 22	- 9

Average transports of a tidal period

If the expressions $\frac{1}{T}\int_0^T Hu\,dt$ and $\frac{1}{T}\int_0^T Hv\,dt$ with $H = h + \zeta$ and T the period time during an investigation are formed one will obtain at the end of a period the average mass transports at the discrete u- and v-points with $\overline{Hu}$ and $\overline{Hv}$. These values are middled to the enclosed ζ-points

```
      v
      •
u    ↓ζ    u
x →  +  ← x
      ↑
      •
      v
```

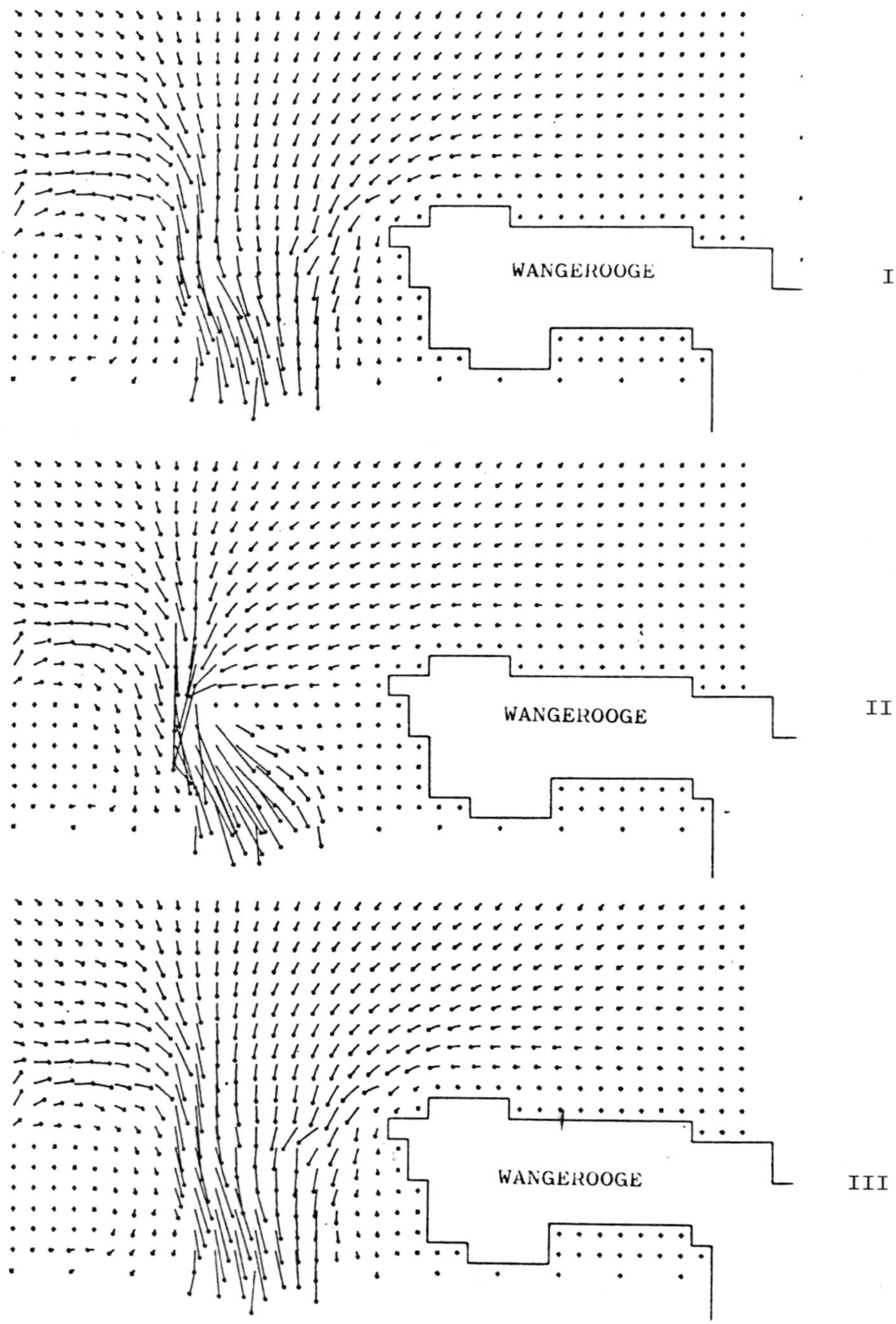

SCALE — 1 m^2/s

FIG. 19 AREA B - MEAN MASS TRANSPORT OF ONE TIDAL CYCLE

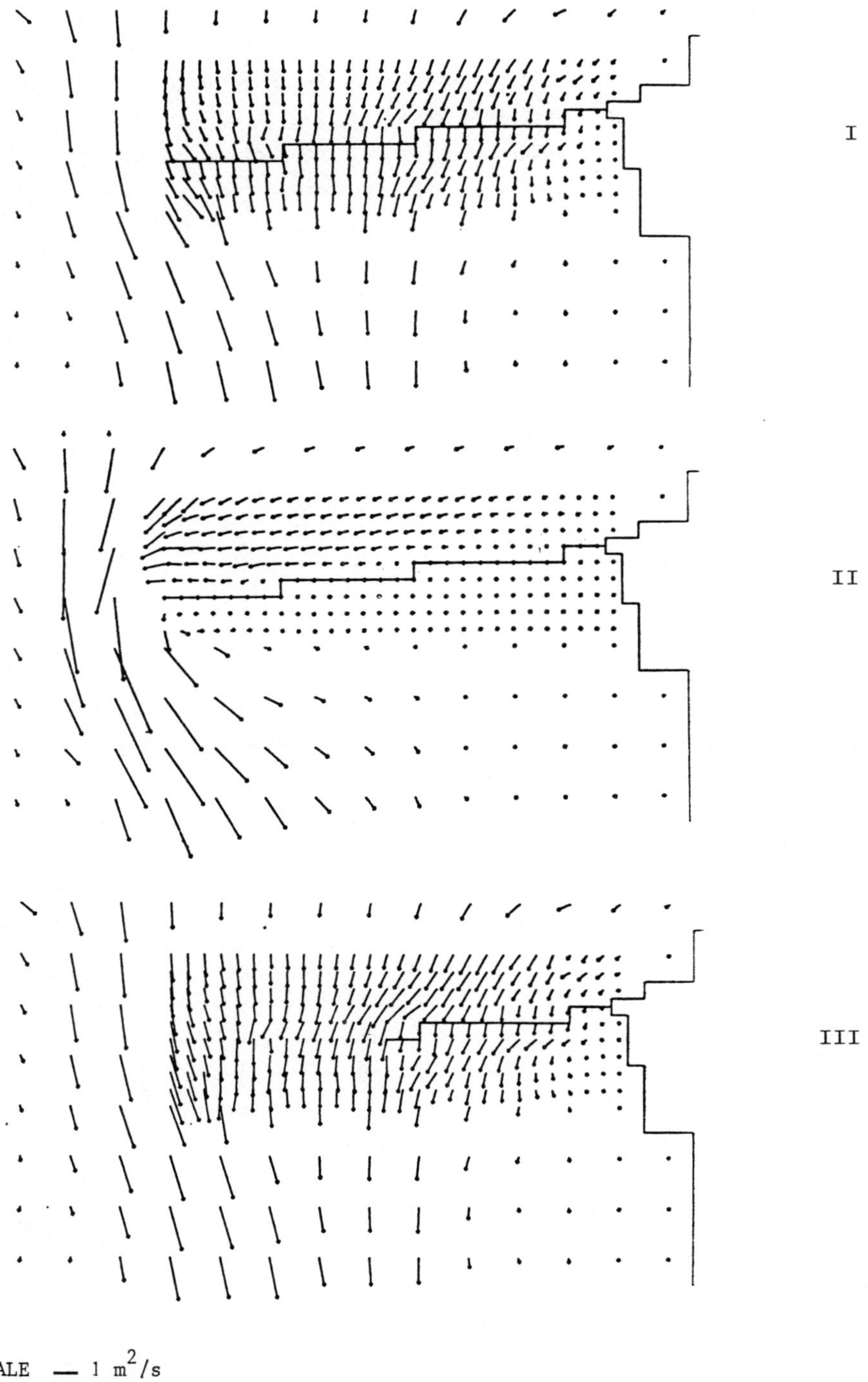

SCALE — 1 m^2/s

FIG. 20 AREA C - MEAN MASS TRANSPORT OF ONE TIDAL CYCLE

and by means of $tg\ \varphi = \overline{Hv}/\overline{Hu}$ the direction of the average transport will be gained simultaneously.

The figures 19, 20, 21 will disclose the average trnasport procedure and confirm all together the hitherto described results.

In table 2 are the masses determined for all three tasks and for the three partial sections between the breakwater and Spiekeroog, the western and the eastern half of the breakwater which flow per second into the shoal area in a generally southern direction. The illustration is very informative. It shows in detail:

a) the general conformity of the amount of the numerically determined average transports with the models I and III,
b) an almost doubling of the transport in the Harle at a rise of the breakwater to NN + 1.50,
c) a great increase of the transports in the western half of the breakwater in case of a shortening by 50 %.

Conclusion

a) The screen-grid solution increases the assertion ability of a numerical model on questions from coastal engineering. Apart from this, the numerical model offers the possibility to compute a multiple number of model representations and to determine functional interconnections by comparative observations.
b) The breakwater is a neuralgical point of the investigation area. The existence of this structure, its height and length as well as its direction determine considerably the course of the motive processes at the north of Wangerooge, in the Harle as well as in the wadden area behind of Spiekeroog and Wangerooge.
c) An inclusion of the shoal region in the eastern direction up to the Minsener Oog and further to the west behind of Spiekeroog as well as the total surroundings of the island Wangerooge up to the Blaue Balje seems to be sensible in order to explain suspected influencing of the transport proceedings.
d) The Blaue Balje - particularly after heaping-up the Minsener Oog - seems to win importance for the transport behind Wangerooge since the average transports of a tidal cycle predominate clearly in the direction of the shoal area.
e) In a comparison of the average transports of a tidal cycle in the three cases to the average transports of a tidal cycle in the three cases to be investigated, the amounts should also be checked by measurements. Important for the judgements of the results, however, should be the relations.
f) The distribution of tidal currents with all three investigations permits the determination that in case of a rise of the breakwater to NN + 1.50 m, the water masses before Wangerooge will shift at first very far to the east - further than at the present condition of the breakwater - and the partake in

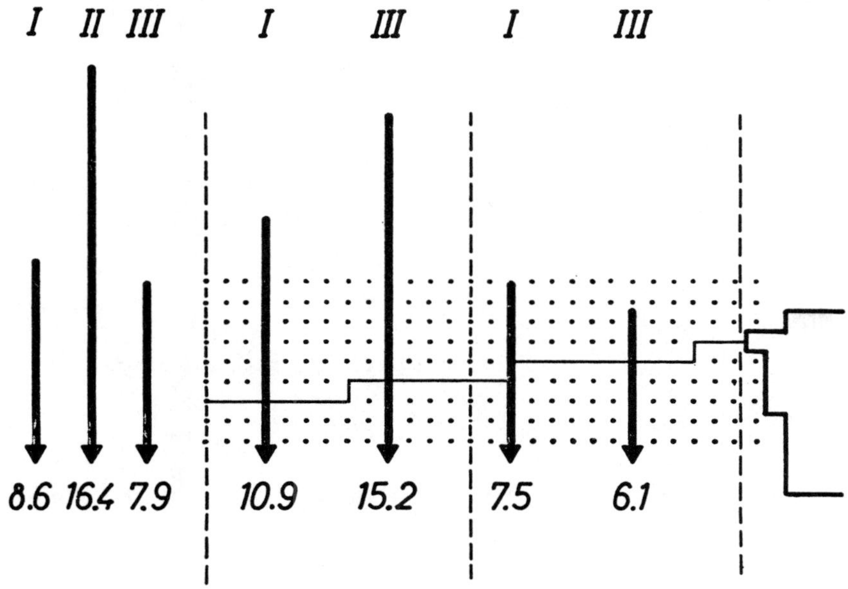

(10^3 m^3 s^{-1})	(10^3 m^3 s^{-1})	(10^3 m^3 s^{-1})
AREA BETWEEN THE BREAKWATER AND THE ISLE OF SPIEKEROOG	WESTERN HALF OF THE BREAKWATER	EASTERN HALF OF THE BREAKWATER

MEAN MASS TRANSPORT IN ONE TIDAL CYCLE IN TOTAL:

I:	8.6 + 10.9 + 7.5	$27.0 \times 10^3\ m^3\ s^{-1}$
II:	16.4 + 0.0 + 0.0	$16.4 \times 10^3\ m^3\ s^{-1}$
III:	7.9 + 15.2 + 6.1	$29.2 \times 10^3\ m^3\ s^{-1}$

FIG. 21 COMPARISON OF THE MEAN MASS TRANSPORT OF ONE TIDAL CYCLE BETWEEN THE THREE DIFFERENT INVESTIGATIONS

the dynamical processes before the breakwater with large streams directed parallel to the strand and in the Harle.

g) A weakening of the strong coat-parallel currents shows the numeric model at a shortening of breakwater to 50 % of the present length.

h) It should be mentioned once more that possible morphological changes were not taken into consideration with these investigations.

Nomenclature

g	acceleration of gravity	$m\ s^{-2}$
h	depth from the undisturbed sealevel to the bottom	m
H	total water depth $(h + \zeta)$	m
ζ	deviation from the undisturbed sealevel	m
H_o, H_1	additional depths in the bottom friction term	m
u, v	components of the vertically integrated velocities	$m\ s^{-1}$
x, y	space coordinates	m
t	time coordinate	s
f	Coriolis parameter	s^{-1}
r	constant bottom friction coefficient with out dimension (0.0025)	
$R^{(x)}$, $R^{(y)}$	components of the bottom friction force	
Δ	Laplace-operator	
A_h	coefficient of horizontal diffusivity	$m^2\ s^{-1}$
Δt	half time step	

REFERENCES

Brettschneider, G. Anwendung des hydrodynamischen Verfahrens zur Ermittlung der M_2-Mitschwingungsgezeit in der Nordsee. Mitteilungen des Instituts für Meereskunde der Universität Hamburg VII, 1967.

Courant, R., Friedrichs, K., Lewy, H. Über die partiellen Differentialgleichungen der mathematischen Physik, Math. Annalen, 100, 1928.

Defant, A. Dynamische Ozeanographie, Berlin 1929.

Hansen, W. Theorie zur Errechnung des Wasserstandes und der Strömungen in Randmeeren nebst Anwendungen, Tellus, VIII, 1956.

Hansen, W. Hydrodynamical Methods Applied to Oceanographic Problems, Mitteilungen des Instituts für Meereskunde der Universität Hamburg I, 1962.

Ramming, H.-G. Shallow Water Tides, Mitteilungen des Instituts für Meereskunde der Universität Hamburg X, 1968.

Ramming, H.-G. Investigations of Motion Processes in Shallow Waters, Report on the Symposium on Coastal Geodesy, München 1970.

Ramming, H.-G. Reproduction of Physical Processes in Coastal Areas. Proc. of the 13th Conference on Coastal Engineering, Vancouver/Canada 1972.

Ramming, H.-G. Reproduktion physikalischer Prozesse in Küstengebieten (in wesentlichen Teilen veränderte Fassung in deutscher Sprache, Die Küste, Heft 22, 1973.

Ramming, H.-G. A Nested North Sea Model with Fine Resolution in Shallow Coastal Areas, Mémoires Société Royale des Sciences de Liège, 6e série, tome X, 1976, pp 9-26.

Sündermann, J. Tidal Waves in Schematic Estuaries, Proc. 12th Conference of Coastal Engineering, Washington/USA, 1971.

Taylor, G. J. Tidal Oscillations in Gulfs and Rectangular Basins, Proc. London Math. Soc., 1920.

APPLICATION OF A FINITE ELEMENT HYDRODYNAMIC MODEL TO THE GREAT BAY ESTUARY SYSTEM, NEW HAMPSHIRE, U.S.A.

RONNAL P. REICHARD and BARBAROS CELIKKOL
Mechanical Engineering Department, University of New Hampshire, Durham, N.H. 03824

ABSTRACT

A vertically integrated, one layer, hydrodynamic model developed by Connor and Wang for Massachusetts Bay has been adapted to the Great Bay Estuary, New Hampshire. This finite element model was found to be well suited to the complexities of the estuary. Initial model development and testing was carried out. The bottom friction coefficient was the dominant parameter in the calibration process. A method for selection of the bottom friction coefficient in the calibration process was developed. Eddy viscosity terms, which primarily act as numerical damping, were included. The model was calibrated using an extensive set of tidal sea level and current data recently collected by the University of New Hampshire and the National Ocean Survey (NOS). The data sets used in the calibration and validation process include sea level data used to specify open boundary conditions in the model. Validation results show good agreement between tidal elevation measurements and model predictions. Predicted currents have a different scale of resolution than the current measurements, making direct comparisons difficult to interpret. A program for processing current data to make more meaningful comparisons is presented briefly.

INTRODUCTION

Increased environmental awareness has created a demand for analysis of estuarine processes and understanding how these processes are affected by development of adjacent coastal land. The foundation of a comprehensive analysis program is understanding the hydrodynamic behavior of the estuary. The work presented here, the implementation of a predictive, computer-based numerical hydrodynamic model, is the first step in establishing a program for analysis of the Great Bay Estuary system.

Hansen (1956) formulated an approach to numerical hydrodynamic modeling used by most modern investigators. Leendertse (1967) made significant contributions in the area of numerical stability and accuracy while developing a two-dimensional vertically averaged numerical hydrodynamic model, utilizing the finite difference method to solve the differential equations. This model has been applied to several estuaries, including Jamaica Bay by Leendertse (1972), and Narragansett Bay by

Hess and White (1974). Connor and Wang (1973) subsequently developed a similar model, using the finite element method, and applied it to Massachusetts Bay. Three-dimensional models have been developed by Leendertse (1973) (multi-layered finite difference) and Connor and Wang (1974) (two-layer finite element), but these models are presently very difficult and expensive to use.

In the work presented here Connor and Wang's two-dimensional finite element model was selected because it is ideally suited to the particular hydrodynamic and topographic conditions found in the Great Bay Estuary system. Initial application has been carried out, including a series of tests used to develop a numerically stabl grid. Field data indicates that the surface slope induced pressure gradient is balanced by bottom friction, which is confirmed by an order of magnitude analysis of the Navier-Stokes equation for one-dimensional channel flow. This balance is used as a quantitative aid in selecting new bottom friction coefficients in the calibration process. Boundary conditions for model calibration and validation were specified directly from sea level data, while both sea level and current data were compared with model predictions. One set of field data was used for calibratio and a different set for validation. Tidal elevation predictions compare favorably with measured values, but scale of resolution differences between measured and predicted currents are apparent in the current comparisons. Plans for obtaining curren measurements on an appropriate scale are presented, which will allow further refinement of the model.

A PHYSICAL DESCRIPTION OF THE ESTUARY

The Great Bay Estuary system (Figure 1), located in southeastern New Hampshir and southwestern Maine, consists of 45 square kilometers of tide water, and its tributaries drain the 2,410 square kilometers of the Piscataqua River Basin. The estuary is a complex combination of tidal flats, and channels which average about 10 meters in depth. The mouth of the estuary is at Portsmouth, which is New Hampshire' only seaport. This segment of the estuary is dominated by a main channel, which is bordered by islands. The surrounding area is heavily developed, with residential, commercial, and industrial uses, including the Portsmouth Naval Shipyard. Farther up the estuary is the lower Piscataqua River, a tidal channel which has been dredged (minimum depth 10 meters) to facilitate shipping. The upper Piscataqua River is a shallow tidal river, formed by the convergence of the Cocheco and Salmon Falls Rivers in Dover. Joining the Piscataqua River at Dover Point is Little Bay, an L-shaped body of water with two tributaries, the Bellamy and Oyster Rivers. The Great Bay is a large, shallow bay characterized by tidal flats with a network of channels. A small channel from the Winnicut River and a larger one from the Squamscott and Lamprey Rivers join to form the main channel, which continues through Furber Strait into Little Bay. Great Bay and Little Bay are undeveloped, and are used primarily for recreation.

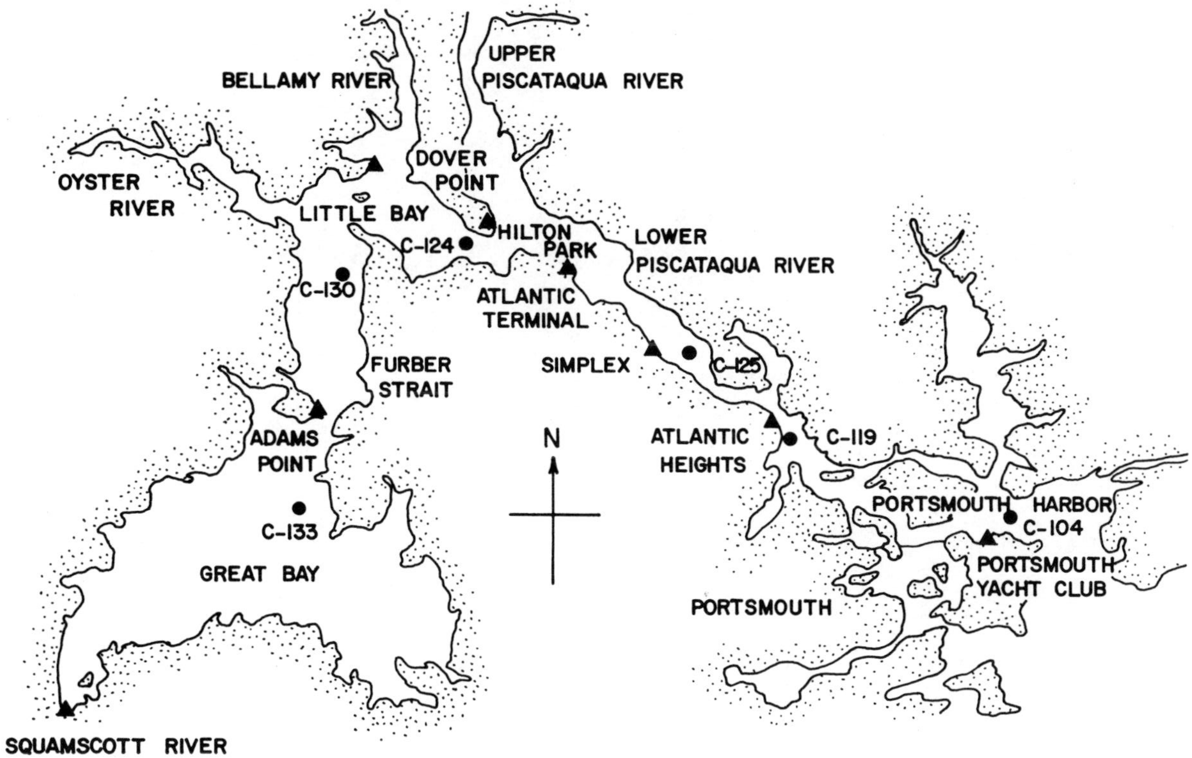

Figure 1. The Great Bay Estuary system, New Hampshire, USA. The location of sea level stations (triangles), and current stations (circles) refered to throughout the text are identified.

The current flow in the estuary is predominantly tidal, as the fresh water component is less than one percent of the tidal prism. The turbulence associated with the tidal flow in the estuary produces a vertically well mixed water column throughout the estuary. To illustrate this, surface and bottom values of temperature and salinity are presented in Figure (2a) and (2b) for periods of high (March) and low (August) fresh water discharge into the estuary. The large deviation in the March salinity curves at km 18 corresponds to a measurement taken at the mouth of the Oyster River. It should be noted that the stratification is local, and gives an indication of the distances required for total mixing.

The tidal range in the estuary is greatest at Portsmouth (2.5 m avg.), decreases to Dover Point (2.0 m avg.), and then increases slightly to the Squamscott River (2.1 m avg.). Tide phase varies similarly, increasing from Portsmouth Harbor to Dover Point, and then increasing only slightly to the Squamscott River. Tide data at six locations in the estuary for a spring tide occurring on July 11, 1975 are presented in several forms to illustrate the tidal characteristics of the estuary. Table (1) presents time and elevation at high and low tides. Figure (3a) is a tidal elevation time series for each of the first four stations in Table (1). The decreasing amplitude and increasing phase from Portsmouth Harbor to Dover Point is readily apparent. Figure (3b) is a tidal elevation time series for Portsmouth Harbor and each of the last three stations in Table (1). Relatively little phase and amplitude variation is seen in the half of the estuary from Dover to the Squamscott River. The tidal elevation data, coupled with the geometry of the estuary, leads quite naturally to characterization of the estuary as a tidal channel (Piscataqua River) connecting a tidal reservoir (Little Bay and Great Bay) to the ocean. The distinction between the channel and reservoir can be seen clearly in Figure (4), which presents hourly tidal elevation along the length of the estuary. From Portsmouth Harbor to Dover Point (0 to 15 km) the surface slopes are relatively large. At Dover Point, extremely large gradients occur, indicating a blockage or choking of the tidal flow. Beyond this area, surface gradients decrease rapidly.

Location	High Tide Elevation (cm)	High Tide Time (hr)	High Tide Phase (hr)	Low Tide Elevation (cm)	Low Tide Time (hr)	Low Tide Phase (hr)	Tidal Range (cm)
Portsmouth Yacht Club	190	-0.03	-	-166	6.32	-	356
Atlantic Heights	165	0.38	+0.41	-140	6.80	0.48	305
Hilton Park	146	1.50	1.53	-111	7.82	1.50	257
Bellamy River	138	2.00	2.03	-101	8.27	1.95	239
Adams Point	140	2.30	2.33	-102	8.98	2.66	242
Squamscott River	150	2.27	2.30	-105	9.00	2.68	255

Table 1 Tidal characteristics of the Great Bay Estuary system from sea level data collected July 11, 1975

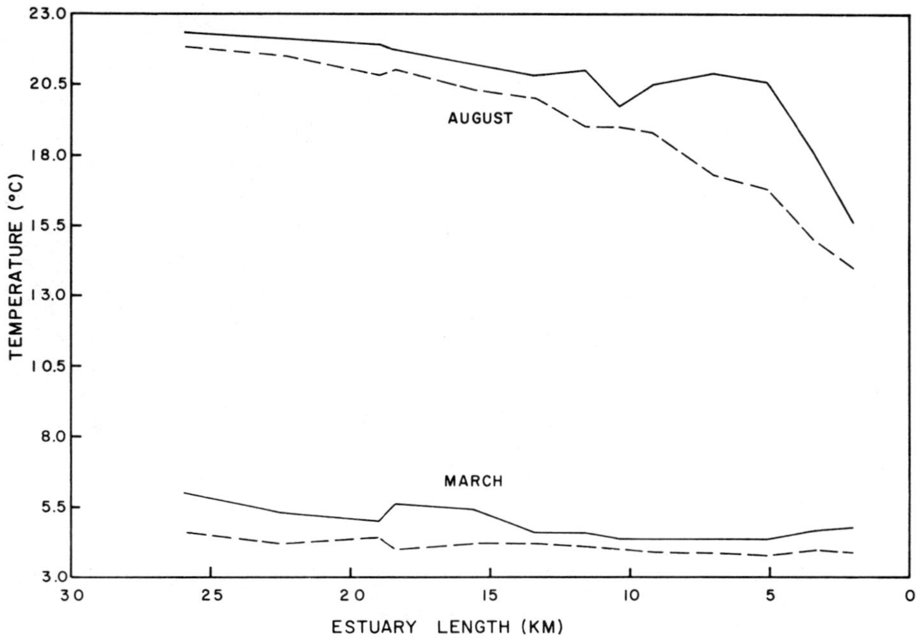

Figure 2a. Surface (———) and bottom (-----) temperature at low tide in the estuary.

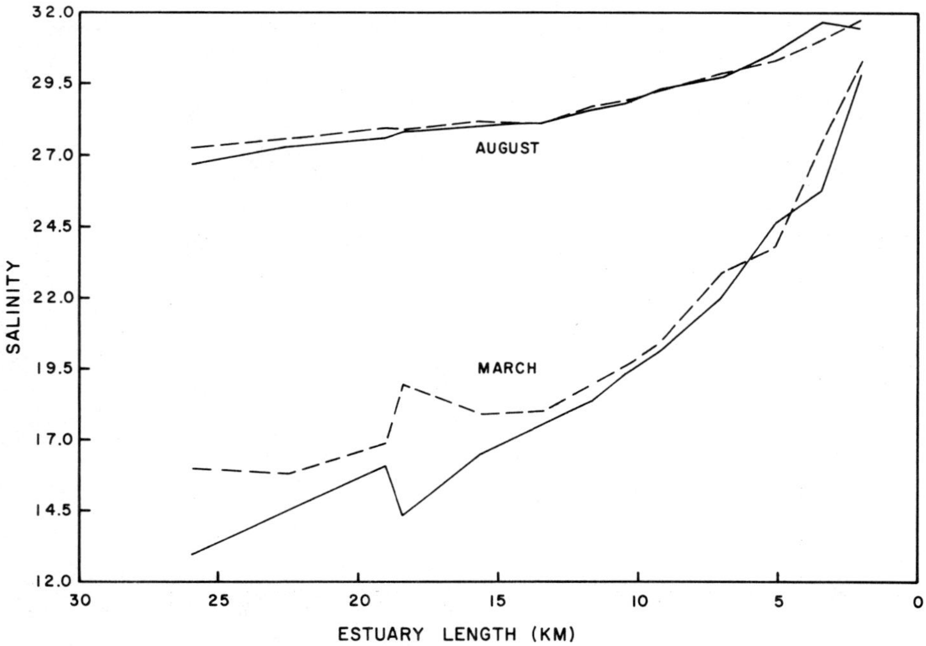

Figure 2b. Surface (———) and bottom (-----) salinity at low tide in the estuary.

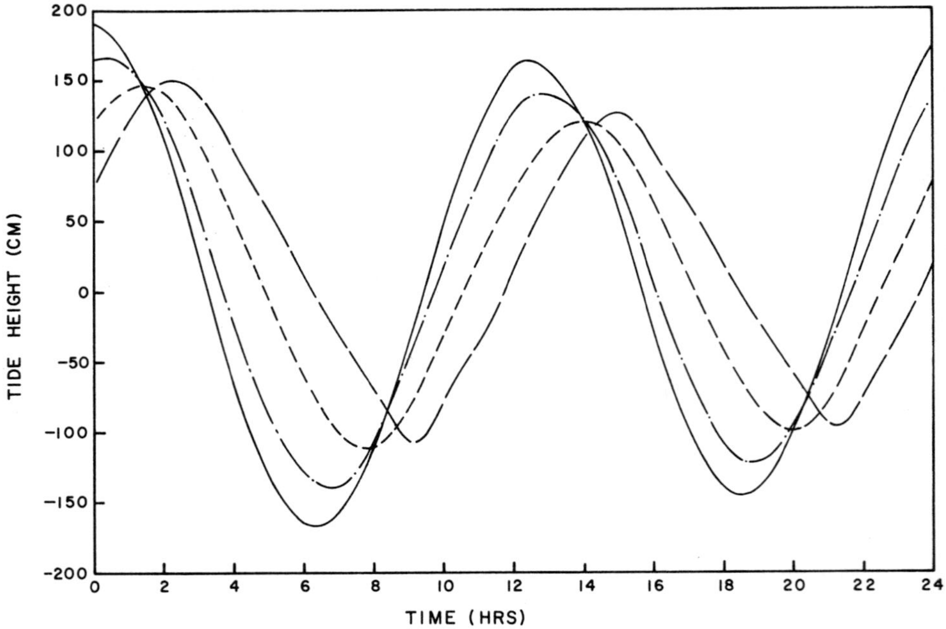

Figure 3a. Tidal elevation for July 11, 1975 at four locations in the estuary: ——— Portsmouth Yacht Club; —·— Atlantic Heights; – – – Hilton Park; — — Squamscott River.

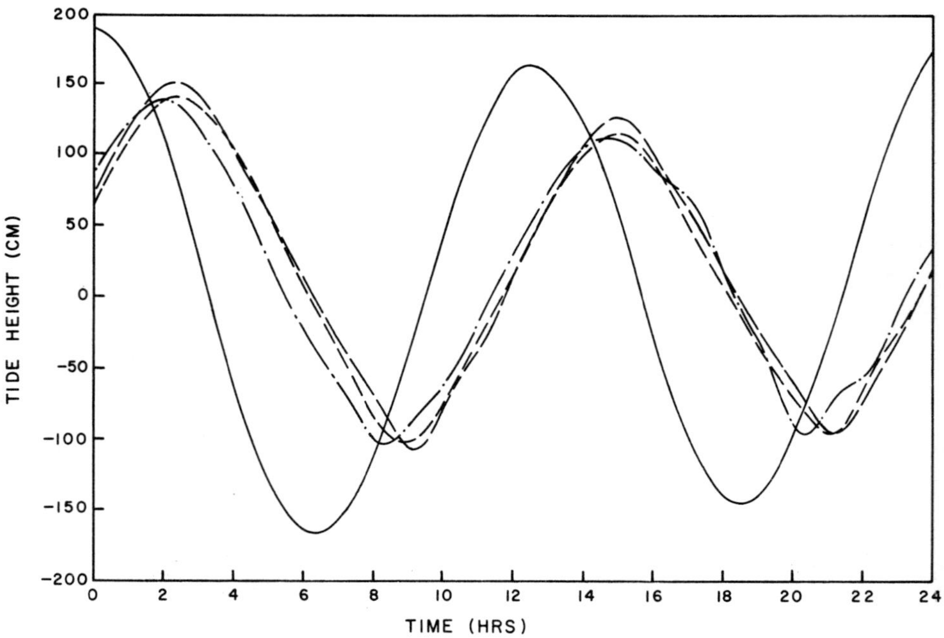

Figure 3b. Tidal elevation for July 11, 1975 at four locations in the estuary: ——— Portsmouth Yacht Club; —·— Bellamy River; – – – Adams Point; — — Squamscott River.

Current velocities vary considerably throughout the estuary, but characteristic velocities can be identified for sections of the estuary.

Location	Velocity (M/S)
Atlantic Heights	2.00
Piscataqua River	1.50
Dover Point	2.00
Little Bay	0.75
Adams Point	1.00
Great Bay	0.50

Current phase changes very little throughout the estuary, with a maximum of approximately one-half hour between Portsmouth Harbor and the Squamscott River.

There is a considerable difference between tide elevation phases and current phases. In this estuary the surface slope is balanced primarily by bottom friction rather than accelerations. Therefore, surface slope is the primary current driving force, and slack water occurs when the surface slope is very small. In Little Bay and Great Bay, slack water coincides with high and low tide, while in Portsmouth Harbor it occurs about two hours after high and low tide. Current velocity, tidal elevation, and surface slope time series are presented for two locations in the estuary. Figure (5a) shows the two hour difference in phase between current velocity and tide height in Portsmouth Harbor. There is about an hour phase differ-

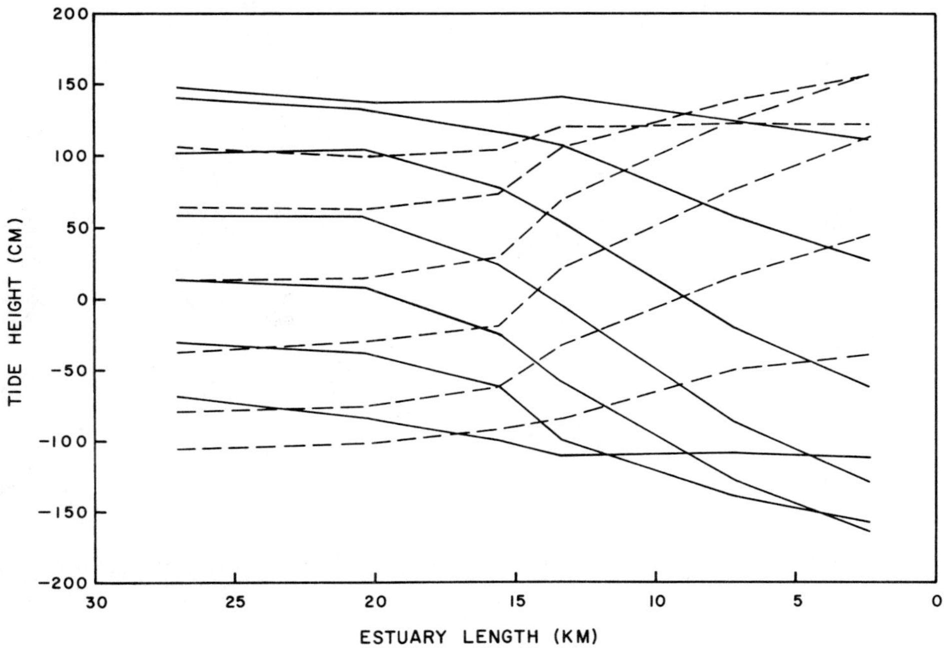

Figure 4. Hourly surface elevation along the length of the estuary for 03:00 to 15:00 July 11, 1975. Solid lines indicate ebb tide, dashed lines flood tide.

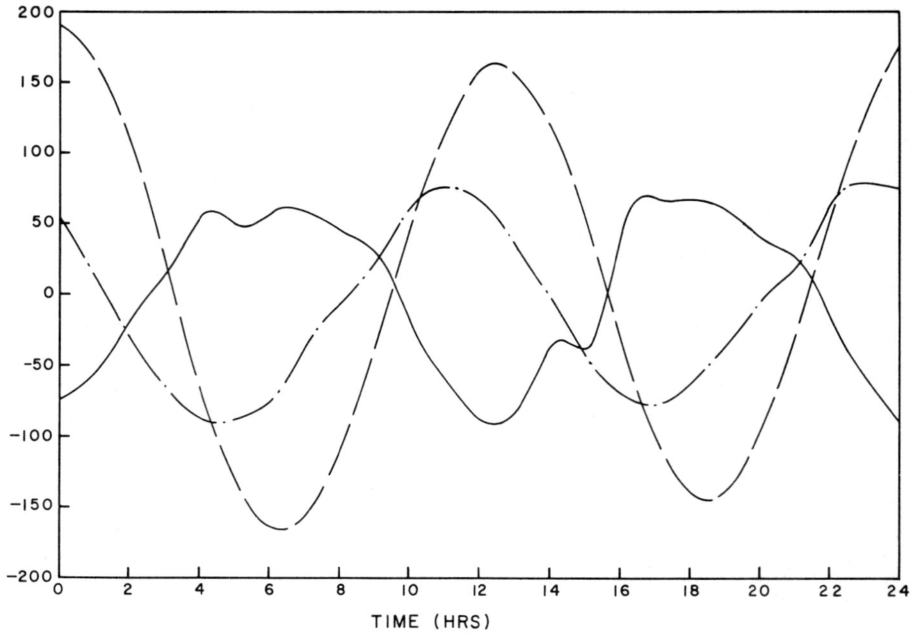

Figure 5a. Current velocity (——— ;cm/s), tidal elevation (— — ;cm), and surface slope (—·— ;x10^{-6}) for July 11, 1975 in Portsmouth Harbor.

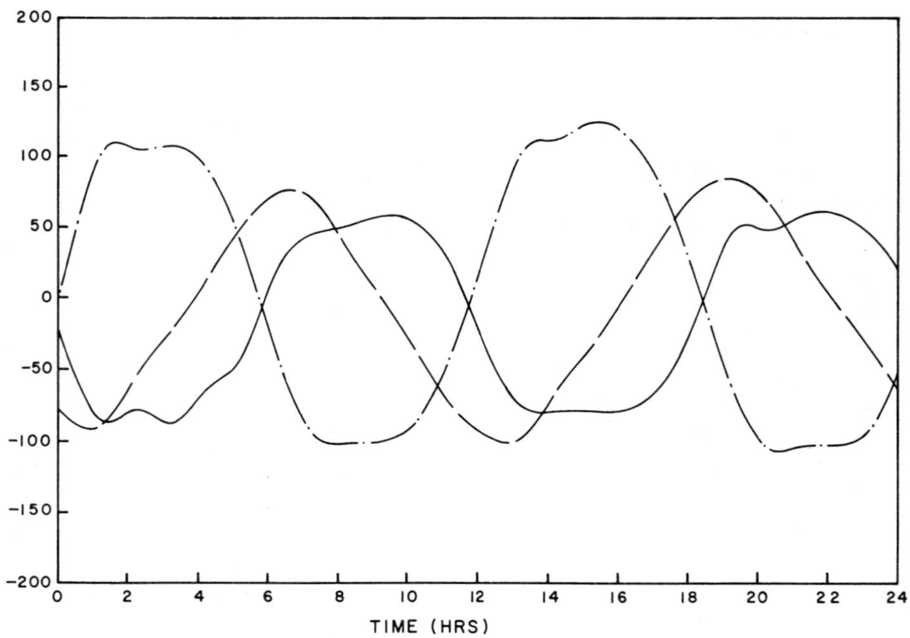

Figure 5b. Current velocity (——— ;cm/s), tidal elevation (— — ;cm), and surface slope (—·— ;x10^{-6}) for July 30, 1975 in Little Bay.

ence between current velocity and surface slope indicating that other forces must be taken into account. Figure (5b) shows slack water coinciding with high and low tide, and little or no phase difference between surface slope and current velocity in Little Bay.

THE HYDRODYNAMIC MODEL

Circulation Analysis by Finite Elements (CAFE) is a two dimensional, vertically averaged, numerical hydrodynamic model developed by Connor and Wang. The model solves a simplified form of the Navier-Stokes and continuity equations using a finite element technique.

The equations governing flow in an estuary are the Navier-Stokes equation:

$$\frac{\partial}{\partial t}(\rho u_i) + \frac{\partial}{\partial x_j}(\rho u_i u_j) = -\frac{\partial P}{\partial x_i} - 2\varepsilon_{ij\ell}\,\Omega_j \rho u_\ell + \rho g \delta_{i3} + \frac{\partial}{\partial x_j}\left[\mu\left(\frac{\partial u_i}{\partial x_j} - \frac{2}{3}\delta_{ij}\frac{\partial u_k}{\partial x_k}\right)\right]$$

$$+ \frac{\partial}{\partial x_i}\left(\xi \frac{\partial u_k}{\partial x_k}\right)$$

and the continuity equation:

$$\frac{\partial \rho}{\partial t} + \frac{\partial}{\partial x_i}(\rho u_i) = 0$$

The equations are simplified by assuming incompressible flow, constant density, constant eddy viscosity, and that the second derivative of velocity with respect to perpendicular coordinates is small. The equations are then ensemble averaged to obtain:

$$\frac{\partial}{\partial t}(\rho \bar{u}_i) + \frac{\partial}{\partial x_j}(\rho \bar{u}_i \bar{u}_j) = -\frac{\partial \bar{P}}{\partial x_i} - 2\varepsilon_{ij\ell}\,\Omega_j \rho \bar{u}_\ell + \rho g \delta_{i3} - \frac{\partial}{\partial x_j}(\overline{\rho u_i' u_j'})$$

$$\frac{\partial}{\partial x_i}(\rho \bar{u}_i) = 0$$

Where the overbar denotes an ensemble average, and the prime a fluctuation about the ensemble average. Assuming vertical variations of the various parameters are small, the equations may be vertically averaged without loss of meaning:

$$\frac{\partial}{\partial t} q_x + \frac{\partial}{\partial x}(U q_x) + \frac{\partial}{\partial y}(U q_y) = -\rho g H \frac{\partial n}{\partial x} + f q_y + \frac{\partial}{\partial x} F_{xx} + \frac{\partial}{\partial y} F_{yx} + \tau_x^b + \tau_x^s$$

$$\frac{\partial}{\partial t} q_y + \frac{\partial}{\partial x}(V q_x) + \frac{\partial}{\partial y}(V q_y) = -\rho g H \frac{\partial n}{\partial y} - f q_x + \frac{\partial}{\partial x} F_{xy} + \frac{\partial}{\partial y} F_{yy} + \tau_y^b + \tau_y^s$$

$$\frac{\partial}{\partial t}(\rho H) + \frac{\partial}{\partial x} q_x + \frac{\partial}{\partial y} q_y = q_I$$

With the following definitions:

$$q_x = \rho \int_{-h}^{n} u\, dz = \rho U H \qquad q_y = \rho \int_{-h}^{n} v\, dz = \rho V H$$

And the constitutive relations:

$$F_{xx} = \int_{-h}^{n} (\tau_{xx} - \overline{\rho u' u'})dz = \varepsilon_{xx}\frac{\partial}{\partial x} q_x$$

$$F_{yy} = \int_{-h}^{\eta} (\tau_{yy} - \overline{\rho v'v'})dz = \varepsilon_{yy} \frac{\partial}{\partial y} q_y$$

$$F_{xy} = \int_{-h}^{\eta} (\tau_{xy} - \overline{\rho u'v'})dz = \varepsilon_{xy} \left(\frac{\partial}{\partial y} q_x + \frac{\partial}{\partial x} q_y\right)$$

$$\tau_x^b = - \frac{C_f}{\rho H^2} q_x (q_x^2 + q_y^2)^{1/2} \qquad \tau_y^b = - \frac{C_f}{\rho H^2} q_y (q_x^2 + q_y^2)^{1/2}$$

Where U and V are vertically averaged horizontal velocity components, η the surface elevation above mean water level, f the Coriolis parameter and C_f the bottom frictio coefficient. The product of velocity fluctuations (with respect to the vertical average velocity) are internal stresses τ_{xx}, τ_{yy}, τ_{xy} which, together with horizonta Reynolds stress, are vertically averaged and then approximated by eddy viscosity terms, with eddy viscosity coefficients ε_{xx}, ε_{yy}, ε_{xy}. The surface and bottom stress, τ^s and τ^b result from vertically averaging the vertical Reynolds shear stress, and are approximated by quadratic functions of wind and current velocity respectively.

Boundary conditions are treated by specifying one component of flow or the surface elevation. Flow normal to the boundary is specified as zero for land boundaries, and set equal to the flow rate at river boundaries. Open or ocean boundaries are treated by specifying surface level elevation. The equations, including the appropriate boundary conditions, are written as variational statements, which serve as the basis for the finite element method.

The finite element method approximates the solution of a boundary value problem with a function of piece-wise continuous polynomials. This involves discretization of the continuum into an equivalent system of finite elements. Connor and Wang selected the simplest configuration, triangles with nodes at the vertices. The values of the variables within the element are assumed to be a linear function of the values at the nodes. The equations are transformed for application to an element using this linear polynominal representation. Treatment of the entire continuum is accomplished through summation of the contributions of each element. Each nodal value influences all of the elements containing that node, and each element value influences the three nodes of the element. Depth is selected at each node point, while bottom friction and eddy viscosity are selected for each element. Solutions for q_x, q_y, and η are obtained at each node.

This model is similar to the finite difference model developed by Leendertse. The principle differences are inclusion of the eddy viscosity terms which Leenderts neglects, and the solution technique. Properly formulated, the method of solution should have little effect on the model results. There are, however, advantages and disadvantages to each method. The finite difference method is more easily understood and applied, and has well developed stability criteria. The principle advant of the finite element method is the flexibility of the grid, making it more appropr ate for situations involving complex geometry, such as the Great Bay Estuary system.

MODEL APPLICATION

The goal of the research program is to model the entire estuary. The present form of CAFE does not treat tidal flats, therefore Great Bay has been excluded from the initial model study. A model of the remainder of the estuary would involve a grid composed of approximately 600 elements and 400 nodes, which would strain the limits of the computational system available to the authors. Therefore, the estuary has been divided into sections, which are modeled separately. The influence of topography on the hydrodynamics of the estuary make the segmentation possible, and determines the location of the segment boundaries. The model of the Portsmouth Harbor and lower Piscataqua River section is presented here. The open boundary in Portsmouth Harbor was located near the mouth of the estuary, where the hydrodynamic characteristics of the estuary are still present. The open boundary in the Piscataqua River was located below the area where Little Bay joins the Piscataqua River. The tidal choking which occurs at Dover Point, and the relative insignificance of the upper Piscataqua River, allow the estuary to be divided at this point.

The grid was selected to coincide with the coastal and bathymetric features of the estuary as presented by United States Coast and Geodetic Survey charts No. 211 and 212. Several factors were considered when developing the grid. Features included in the grid are at least as large as the distance between two nodes (the characteristic length). The size and shape of the elements may vary, but the most stable configuration is equilateral triangles of equal size. To preserve the grid stability, the area of adjacent elements varies less than 20 percent, and the element angles are greater than 30 and less than 90 degress. Another factor considered in selecting the grid is that an element may not have more than two of its three nodes on a land boundary. The final consideration is the number of nodes and elements to be included. The larger the number of nodes, the greater the detail, but computer core and time requirements are also increased. The maximum size of the time step is related to the characteristic length, and in this case the small characteristic lengths constrained the size of the time steps. As a result of these computational restrictions the grid was selected with as large a characteristic length as possible. The grid was modified several times to obtain an acceptable result.

Boundary conditions were set up as specified for the model. Land boundary conditions are handled internally by the model, forcing the flow to be tangential to the land. To accomplish this, normal angles (with respect to the land boundary) must be specified for each node on a land boundary. The open boundaries are initially treated by specifying the tidal amplitude, tidal frequency and phase lag. The tidal frequency was assumed standard, and is the same for all open boundaries. The tidal amplitude was selected for each open boundary, and the phase for each node on each open boundary. The values for these parameters were obtained from the Tide Tables of North America. The bottom friction and eddy viscosity coefficients were assumed constant throughout the grid, and the values for the initial run were selected based on previous applications of the model.

With the data set up, the model was initially run using a constant depth (representative of the depth of the estuary) as it is sensitive to depth variations. This allowed the stability of the grid to be tested and approximate values for the bottom friction and eddy viscosity coefficients to be obtained. After the grid has been evaluated and modified, and approximate values of the coefficients determined, the depth of the estuary at each node was entered. The chart data was supplemented for the Piscataqua River channel with data from a University of New Hampshire bathymetric survey. To preserve the cross sectional areas and volume characteristics of the estuary segment, slight modification of the depths was necessary. The model was run again, with varying depth, to test the stability of the three dimensional grid. Large changes in depth within an element, and very shallow depths, cause numerical instabilities, requiring slight modification of the depths.

To determine the effects of tidal amplitude, bottom friction coefficient, an eddy viscosity coefficient on the model, a series of model runs were executed, systematically varying these parameters (Reichard and Celikkol 1976). The results of this experiment indicates that current speed depends on tidal amplitude and bottom friction coefficient, while current phase is affected by tidal amplitude and eddy viscosity coefficient. Since tidal amplitude is determined by the physical conditi present at the boundary, calibration of the model with respect to current speed is accomplished by changing the bottom friction coefficient, while calibration with respect to phase is carried out through modification of the eddy viscosity coefficient.

MODEL CALIBRATION

The model uses two parameters which must be evaluated for the estuary: the bottom friction coefficient and the eddy viscosity coefficient. Bottom friction plays a major role in estuary dynamics. This frictional effect is expressed in the model in the form CU^2/h, where C is the friction coefficient, U the vertically averaged velocity, and h the water depth. The eddy viscosity term is a combination of Reynolds Stress terms, resulting from ensemble averaging, and internal stress terms, resulting from the vertical averaging. It is expressed in the form $\frac{\partial}{\partial x}(\xi_{xx}\frac{\partial u}{\partial x})$ where ξ_{xx} is the eddy viscosity coefficient and $\partial u/\partial x$ is the horizontal current gradient. Both of the coefficients are assumed constant in time, and constant for an element. They may be specified for each element individually, however allowing spatial changes in the coefficients from element to element. The values of these parameters are selected, in a process called model calibration, to cause model results to compare favorably with field data. A procedure for quantitative selecti of the bottom friction coefficient was developed to assist in the calibration proce

The initial step in the development of such a procedure is a review of the physical processes involved. For a one-dimensional, constant depth channel, neglecting the eddy viscosity term, the conservation of momentum equation can be written:

$$\frac{\partial U}{\partial t} + U\frac{\partial U}{\partial x} = -g\frac{\partial \eta}{\partial x} + \frac{CU^2}{h}$$

Where U is the vertically averaged current velocity, η the surface elevation, h the water depth, and C the friction coefficient. Evaluating the order of magnitude of each term in the equation for the Great Bay Estuary system:

$$\frac{\partial U}{\partial t} + U\frac{\partial U}{\partial x} = -g\frac{\partial \eta}{\partial x} + \frac{CU^2}{h}$$

$$10^{-4} + 10^{-5} = 10^{-3} + 10^{-3}$$

The acceleration terms are small compared to the surface slope and friction terms, and can be neglected. Therefore, the equation can be written as a balance between surface slope and bottom friction:

$$g\frac{\partial \eta}{\partial x} = \frac{CU^2}{h}$$

The current velocity U is related to the bottom friction coefficient:

$$U = K/C^{1/2} \qquad K = (gh\frac{\partial \eta}{\partial x})^{1/2}$$

This expression is used to select new bottom friction coefficients based on model results and field data. The equation is first solved for the constant K using the current velocity predicted by the model for the specified bottom friction coefficient. The equation is then solved for a new bottom friction coefficient using the calculated constant K, and measured current velocity. This procedure was found to work quite well, producing acceptable results with only one or two interations.

COMPARISON OF MODEL RESULTS WITH FIELD DATA

The model predicts currents and tidal elevation, which have been measured in the estuary (Brown et al 1977). To compare these two types of data, the physical interpretation of the prediction and measurement must be clearly understood. The model predictions are arrived at through a numerical solution of the appropriate differential equations. The numerical methods used in the solution impose certain constraints or limitations on the results.

The temporal aspect of the model is obtained by solving the equations at a discrete time, and repeating the solutions at fixed intervals in time. This introduces a minimum temporal scale of resolution approximately the size of the time interval between solutions. In addition, the equations have been ensemble averaged, eliminating stochastic variations from the result. The spatial variation of a parameter in an element is a linear function of the values of the parameter at the three nodes. This imposes a minimum spatial scale of resolution approximately the size of an element. It also means that the parameter predicted by the model represents a spatial average over an area approximately the size of an element.

The cumulative effect of these restrictions on tidal elevation predictions is that a predicted elevation represents a spatial average over an area approximately the size of an element and a temporal average over a time approximately the size of the time step, of the deterministic (ensemble averaged) value of tidal elevation. The differential equations have been vertically averaged, so current predictions represent the average current in the water column. Thus, a predicted current represents a spatial average over a volume of water approximately the size of an element, and a temporal average over a time approximately the size of the time step, of the deterministic value of the current.

The stochastic variations in water surface elevation are of length and time scales small enough that they can be effectively removed from the measurement by mounting the tide gauge in a properly designed well. The surface elevation is measured at one point, while the model predictions represent an average over a large area. However, the deterministic variations in surface elevation are well behaved spatially and temporally, and can be accurately represented by simple mathematical expressions such as those used in the model. As a result, model predictions can be compared directly with tide data, and generally compare quite well.

Water in the estuary exhibits a variety of motion characterized by scales ranging from molecular to some larger than the estuary. Any single method of measurement can deal with only a portion of the wide spectrum of motion, with the remainder appearing as noise. An Eulerian measurement with a length scale of the order of 100 m and a time scale on the order of a minute would be ideally suited for model comparison.

The current data presented in this paper was obtained from conventional mechanical current meters, which can resolve length and time scales of motion on the order of a meter and several seconds respectively. This scale is two orders of magnitude smaller than the resolution of the model, therefore velocity fluctuations and small flow structures which appear in the measurement are not predicted by the model. As a result, there is greater variation present in the current data than in model predictions, and in some cases the differences are quite large. In addition, the model predicts the vertical average velocity, which is generally less than the velocity near mid-depth where the current data was collected.

These problems can be addressed by making measurements at several different points in an area and processing the data appropriately. This has been done for several cross-sections of the estuary. These transects were each conducted for one tidal cycle, but were not done concurrently. This data is presently being processed and is not yet available. We are also presently attempting to measure the flow through a cross-section of the estuary with a geomagnetic measurement. The measurement is marginal and to date no significant results have been obtained.

The model has been run for two distinct tidal cycles, the first occurring on September 19, 1977, and the second on September 21, 1977. The open boundary conditions were specified directly from tide data collected on these two days, thus eliminating errors resulting from approximating the boundary condition, and including meterological as well as astronomical effects. This was accomplished by interpolating from hourly tide height data collected at the boundary for the appropriate day. The model for September 19 was used for calibration, while the model for September 21 was run with the coefficients developed in the calibration process, changing only the open boundary conditions, for model validation.

A surface level isoplot (Figure 6) and current vector plot (Figure 7) at periods of high current velocity are presented as an overall, qualitative view of the model results. The portions of the estuary having the smallest cross-sectional area have the largest surface gradients and current velocities as expected. In addition, the current vector plots indicate the presence of topographically induced eddies in the flow regime. These eddies are at the lower limit of the model's spatial resolution, so the predictions are not quantitatively accurate. The existence of these eddies in the estuary has been confirmed by observations.

Tidal elevation has been compared with model predictions at the Simplex tide station, and is presented in Fig. 8a and 8b. Predicted tidal elevation is presented as a solid line while hourly tide height data is represented by circles. The agreement is quite good with the difference between predicted and measured values generally being less than the accuracy of the measurement.

Current velocities are compared at three locations, stations C-104, C-119, and C-121 in Figs. 9 through 11 respectively, as speed and direction plots. Predicted currents are presented as solid lines, while measured currents are represented as circles (3 m depth) and squares (6 m depth). The speed predictions are generally less than measured values, as expected, since the predicted value is vertically average speed. The current meter moorings were located to one side of the channel to avoid the heavy ship traffic, resulting in asymmetries in the data due to irregularities in the shore. Speed predictions for flood and ebb tide are nearly symmetrical, and predicted directions align with the main channel, as the grid is not fine enough to resolve these local topographical features. An extreme case is Station C-121, which is in the main flow on the ebb tide, but borders on a reverse flow (eddy) on the flood tide.

CONCLUSION

The surface elevations predicted by the model are as accurate as tidal measurements, and provide better spatial resolution. Initial comparisons of predicted and measured currents, within the limitations set forth previously, appear promising. However, the predicted currents should be compared with a data set having a similar scale of resolution. Currents predicted by the model do not provide enough detail to be useful for some of the problems in the estuary presently being addressed, but they are satisfactory for problems concerning large scale mass fluxes. The model is

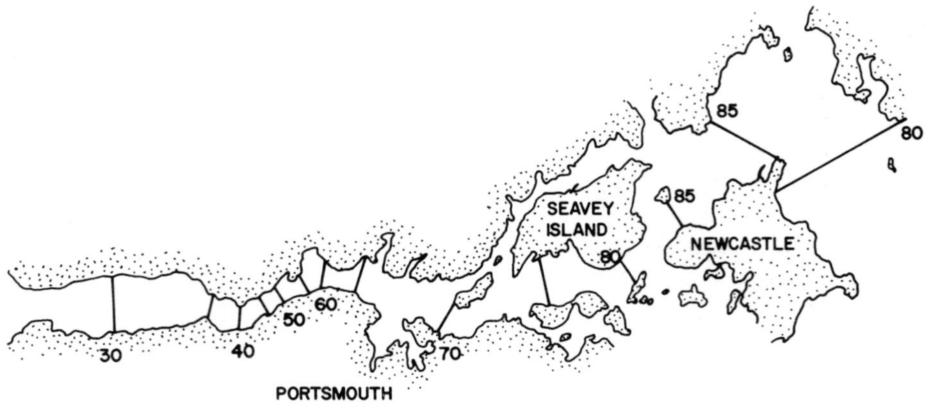

Figure 6. Surface level isoplot (5 cm increments) as predicted by the model one hour after mid-flood.

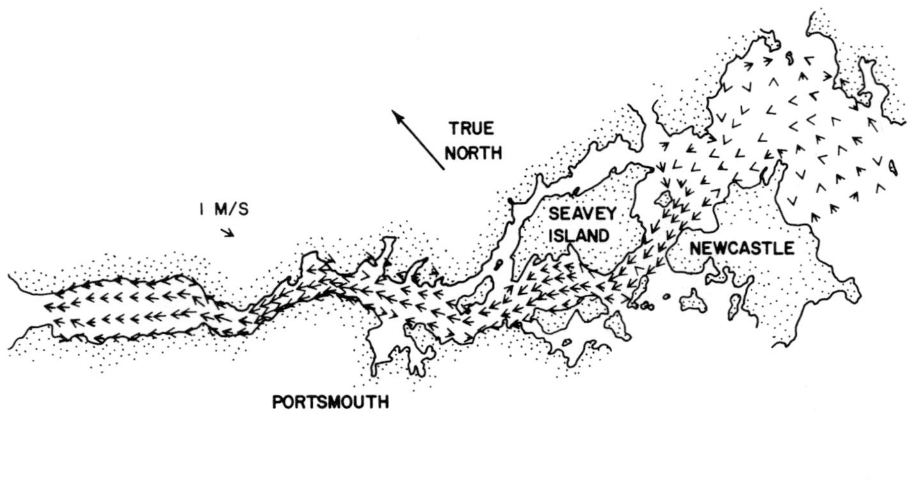

Figure 7. Current vector plot as predicted by the model one hour after mid-flood.

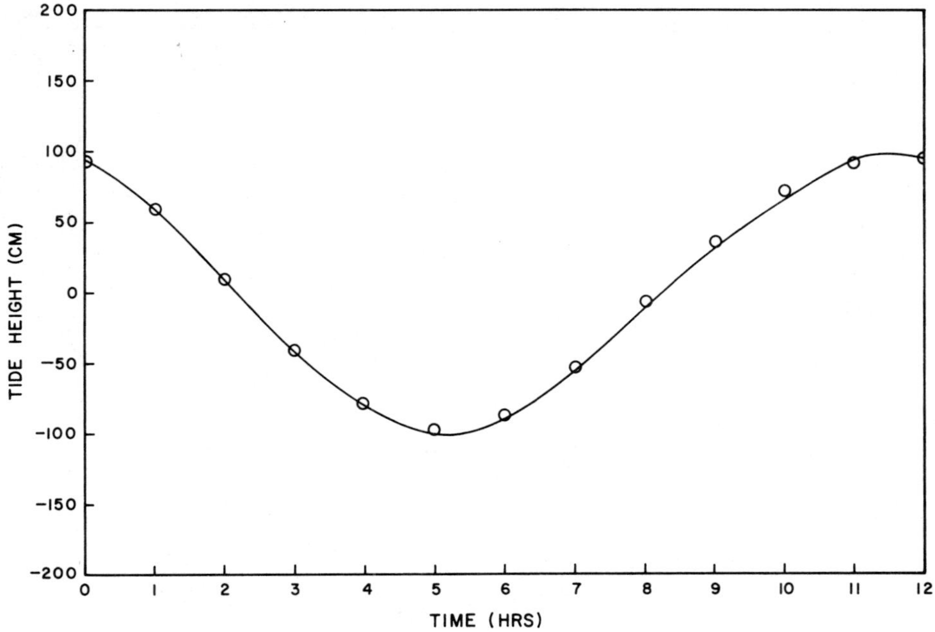

Figure 8a. Comparison of predicted tidal elevation (solid line) and hourly tide height data (circles) at Simplex on September 19 1975.

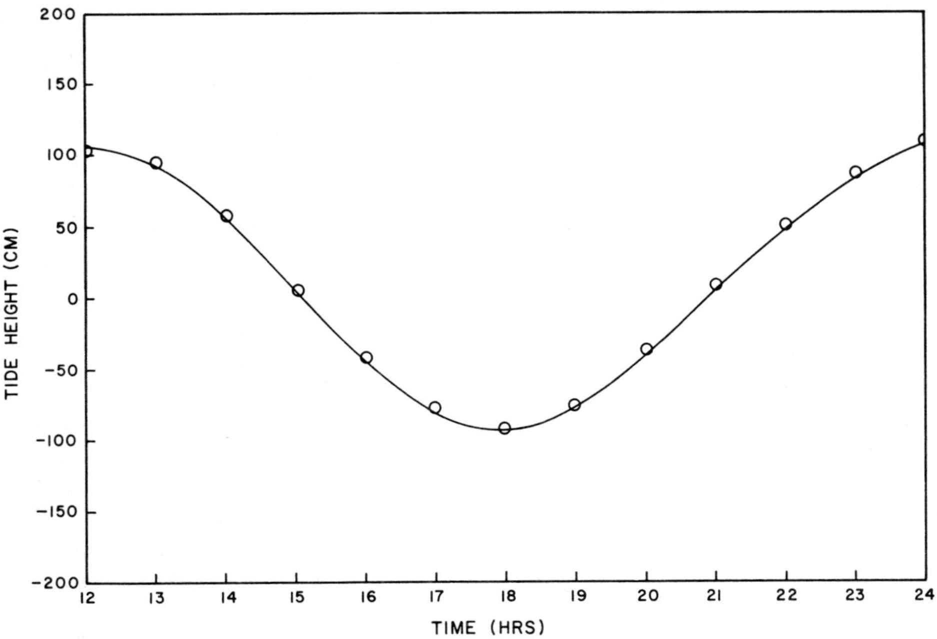

Figure 8b. Comparison of predicted tidal elevation (solid line) and hourly tide height data (circles) at Simplex on September 21 1975.

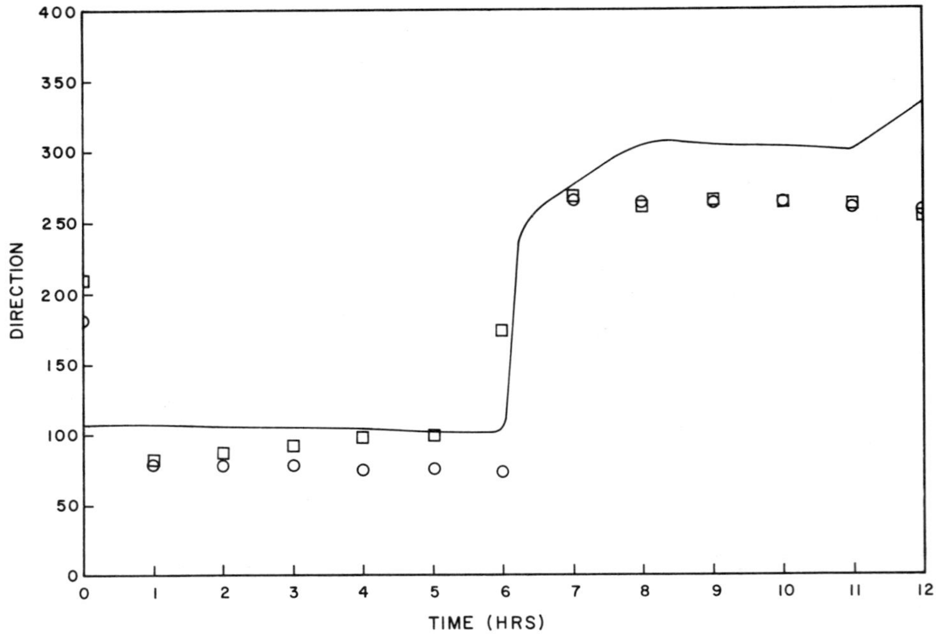

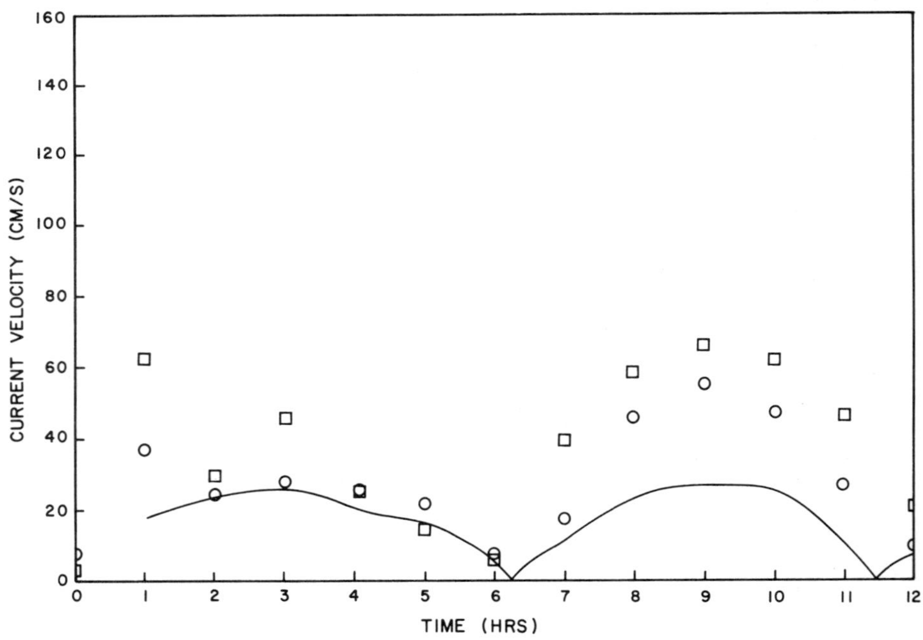

Figure 9a. Comparison of predicted speed and direction (solid lines) with current data (circles and squares) at station C - 104 for September 19, 1977.

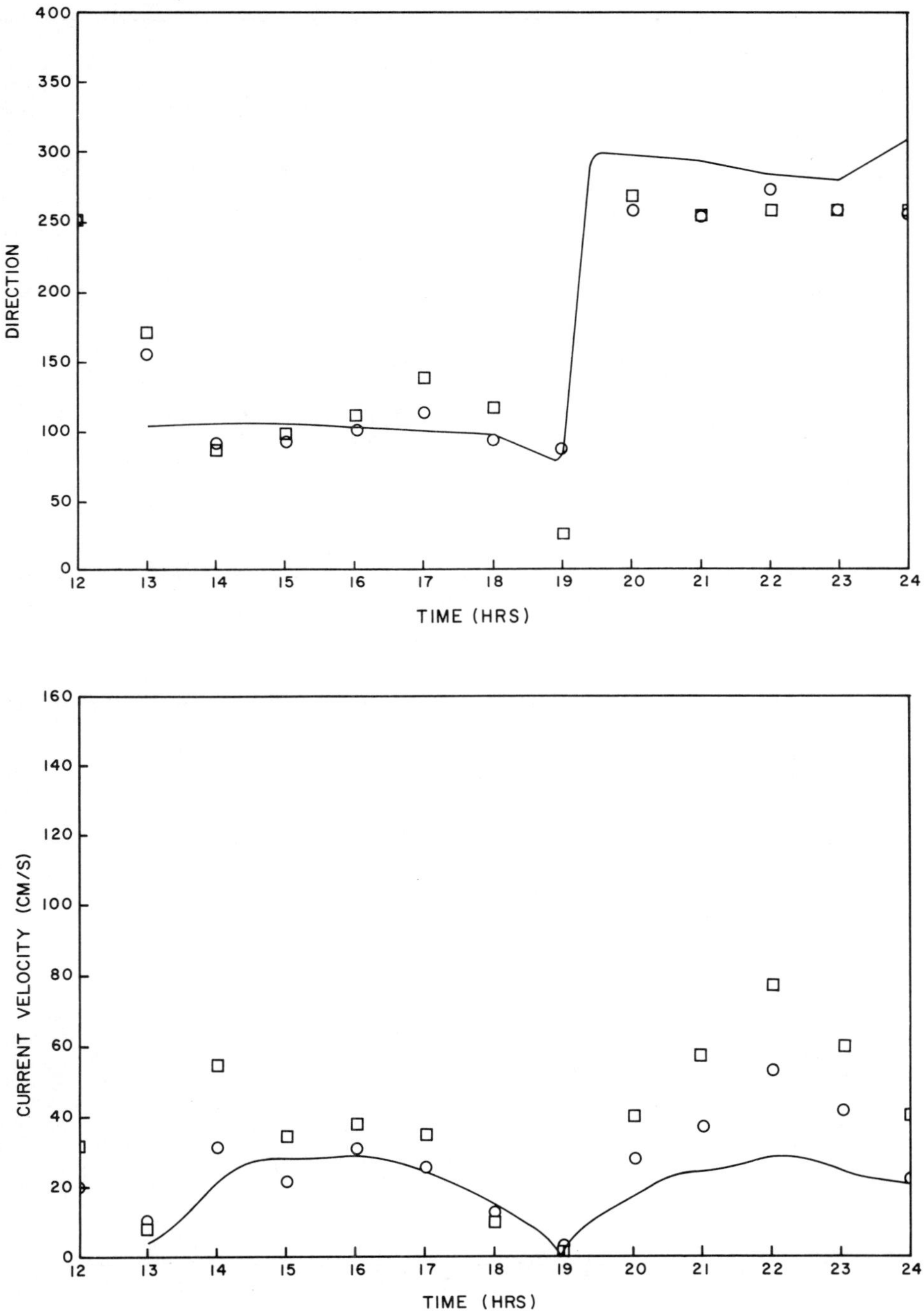

Figure 9b. Comparison of predicted speed and direction (solid lines) with current data (circles and squares) at station C - 104 for September 21, 1977.

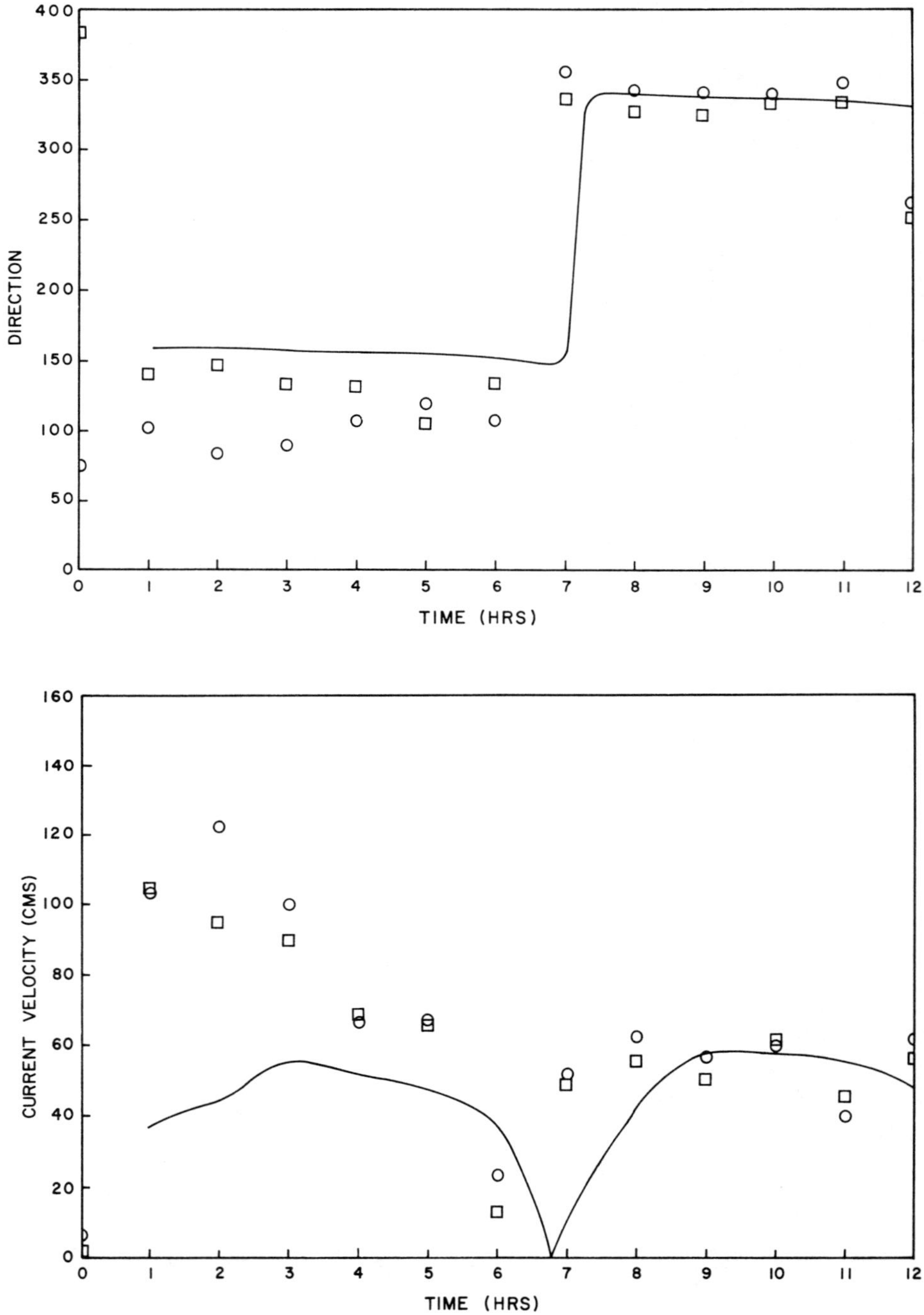

Figure 10a. Comparison of predicted speed and direction (solid lines) with current data (circles and squares) at station C - 119 for September 19, 1977.

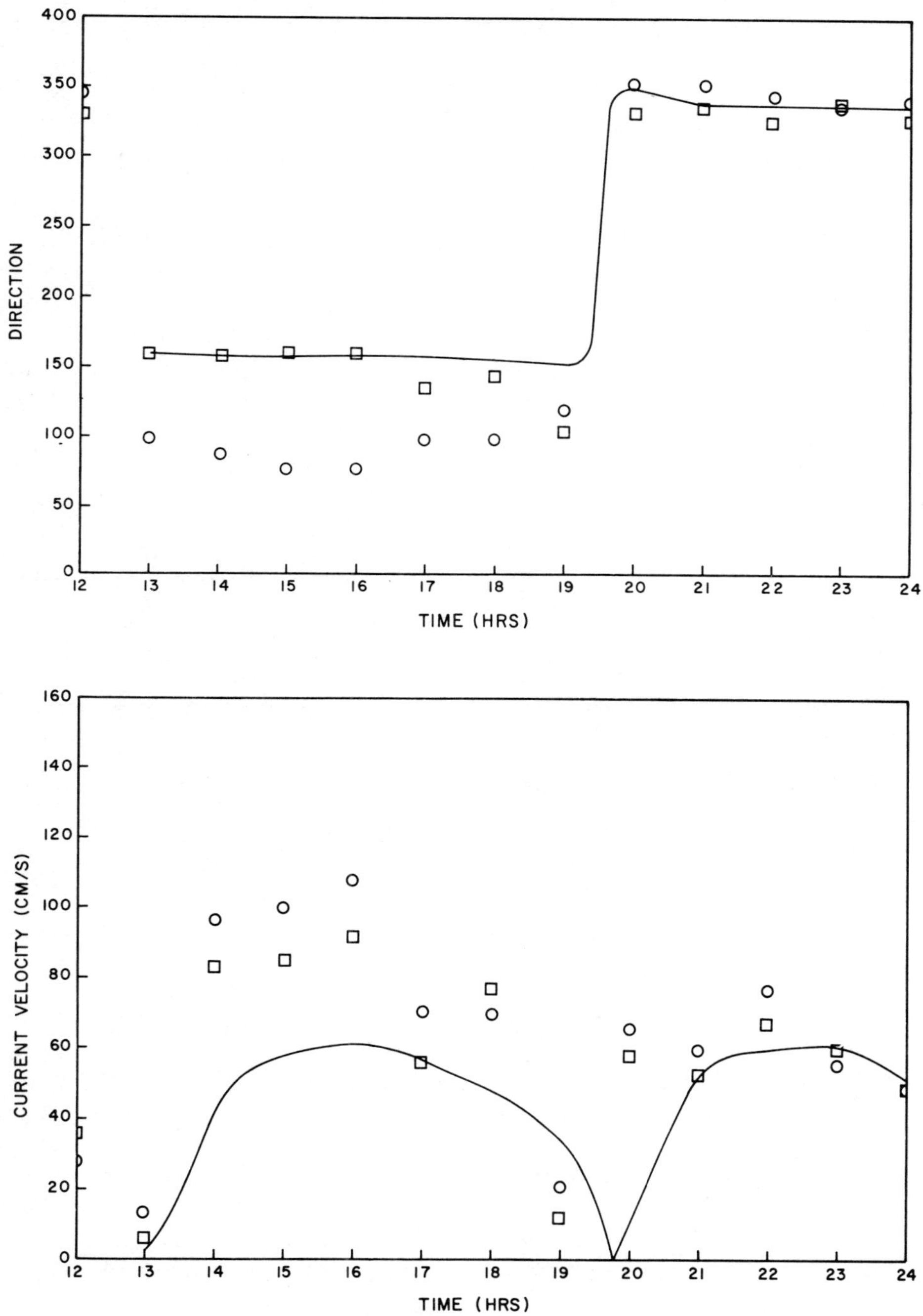

Figure 10b. Comparison of predicted speed and direction (solid lines) with current data (circles and squares) at station C - 119 for September 21, 1977.

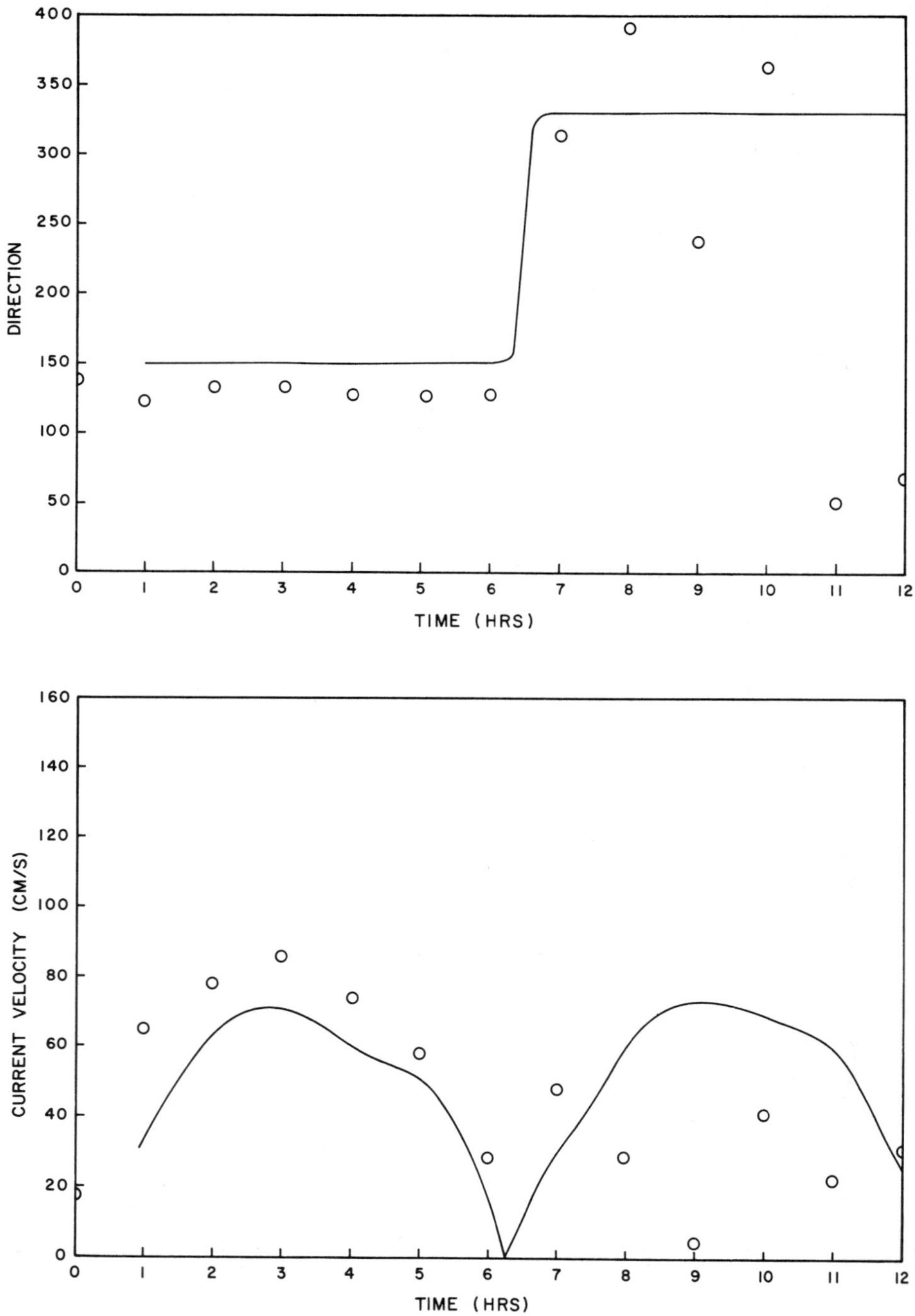

Figure 11a. Comparison of predicted speed and direction (solid lines) with current data (circles) at station C - 121 for September 19, 1977.

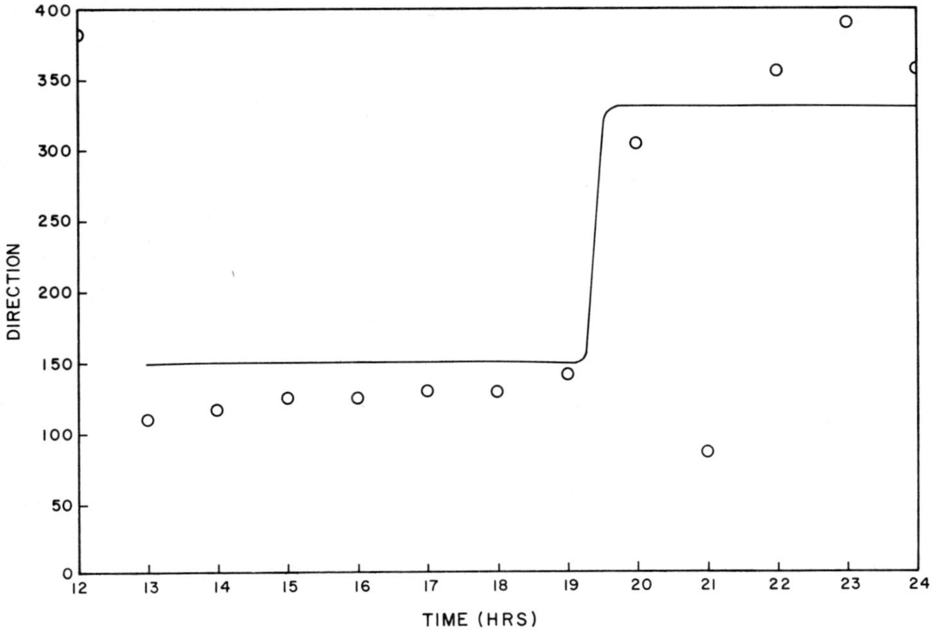

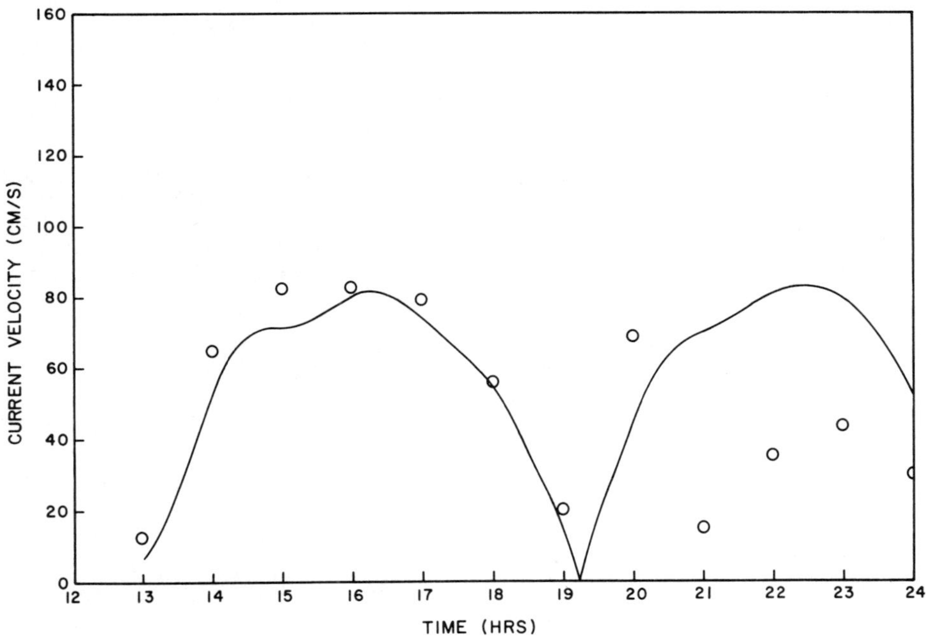

Figure 11b. Comparison of predicted speed and direction (solid lines) with current data (circles) at station C - 121 for September 21, 1977.

presently being used as an aid to understanding the hydrodynamics of the estuary, and to provide a data base for the advective portion of a dispersion model, for which it appears ideally suited.

ACKNOWLEDGEMENTS

We are indebted to the National Ocean Survey and the Public Service Company of New Hampshire for their assistance in collecting field data. This work is a result of research sponsored by NOAA Office of Sea Grant, Department of Commerce, under Grant No. 04-6-158-44056.

REFERENCES

Brown, W., Swenson, E., Trask, R., 1977, The Great Bay Estuarine Field Program 1975 Data Report Part I Currents and Sea Level, UNH Sea Grant Technical Report UNH-SG-157.

Connor, J.J., Wang, J.D., 1973, Mathematical Models of the Massachusetts Bay Part I, MIT Sea Grant Technical Report, MITSG 74-4.

Connor, J.J., Wang, J.D., 1974, Finite Element Model of Two-Layer Coastal Circulation, 14th Coastal Engineering Conference, Copenhagen.

Hansen, W., 1956, Theorie zur Errechnung des Wasserstandes und der Stromungen in Randmeeren Nebst Anwendungen, Tellus, 8-3.

Hess, K.W., White, F.M., 1974, A Numerical Tidal Model of Narragansett Bay, Univ. of Rhode Island, Technical Report MTR No. 20.

Leendertse, J.J., 1967, Aspects of a Computational Model for Long-Period Water Wave Propagation, Rand Corp. RM-5294-PR.

Leendertse, J.J., 1972, A Water Quality Simulation Model for Well-Mixed Estuaries and Coastal Seas: Vol. IV, Jamaica Bay Tidal Flows, Rand Corp. R-1009 NYC.

Leendertse, Jan J., Richard C. Alexander, and Shiao-Kung Liu, 1973, A Three-Dimensional Model for Estuaries and Coastal Seas: Volume I, Principles of Computation, Rand Corp. R-1417-OWRR.

Reichard, R.P., Celikkol, B., 1976, Hydrodynamic Model of the Great Bay Estuarine System, UNH Sea Grant Technical Report UNH-SG-153.

SPREADING AND MIXING OF THE HUDSON RIVER EFFLUENT INTO THE NEW YORK BIGHT

MALCOLM J. BOWMAN
Marine Sciences Research Center, State University of New York, Stony Brook, New York 11794 U.S.A.

ABSTRACT

Results are presented from three Hudson River plume sampling cruises made in the New York Bight, in August 1976.

The data show that the set and shape of the spreading effluent vary widely over time periods ~6 days, and are clearly influenced by local wind stress.

Application of Takano's model of a steady state plume spreading into a stagnant ocean suggests a horizontal eddy viscosity $\sim 10^8$ $cm^2\ sec^{-1}$, and a strong anticyclonic deflection of the plume. This value is considered to be an overestimate, since interfacial shear stress is neglected in the model.

More careful measurements and calculations are needed to separate out the effects of horizontal and vertical viscosities, Coriolis force, advection by a prevailing coastal current and local wind stress, on plume dynamics.

INTRODUCTION

The Hudson River, a major source of freshwater to the coastal ocean of the northeastern United States, has been an important navigable waterway for several centuries and has played an essential role in the commercial development of New York City and State.

In its lower reaches, the River has the characteristics of a classical partially mixed drowned river valley estuary. The effluent from the Hudson estuary flows into that section of the coastal ocean known as the New York Bight Apex, where it spreads as a buoyant plume whose areal extent, in periods of high runoff, may exceed $\sim 500\ km^2$.

The purpose of this paper is to present an analysis of data obtained from three plume sampling cruises made in the Apex on August 13, 16, and 19, 1976, and to compare this with the predictions of Takano's (1954a,b, 1955) steady state analytical theory of a buoyant plume spreading into a rotating stagnant ocean.

PHYSICAL CHARACTERISTICS OF THE LOWER RIVER

The long term mean runoff of the Hudson River is ~550 m^3 sec^{-1} (Giese and Barr, 1967), but its flow is subject to large irregularities. The watershed covers ~3.5×10^4 km^2, and about one half of the annual runoff for a normal year occurs during March, April and May, when the monthly mean may exceed the long term average by a factor of 4 or more.

Salt intrusion up the River may exceed ~130 km under low runoff conditions, but during mean discharge, penetrates ~70 km upstream. Further information on the Hudson River and estuary may be found in Giese and Barr (1967), Jay and Bowman (1975), and Bowman and Wunderlich (1977).

Ketchum et al (1951) made the first extensive survey of the Hudson effluent during six cruises between February,1948 and January, 1950. Further studies have been published by Pearce (1970), National Marine Fisheries Service (1972), Charnell and Hansen (1974) and Bowman and Wunderlich (1976). A large collection of papers on the oceanography and water quality (rather the lack of it) of the Apex is Gross (1976).

The total discharge of the Hudson during a mean year (~1.7×10^{10} m^3) is comparable to the volume of water contained in the region surveyed during the three 1976 cruises (see Fig. 1). However, the volume of fresh water in the plume (~2.5×10^8 m^3, equivalent to ~8 days' discharge) during the survey represented only ~1.5% of the total volume of sea water in the Apex.

The Apex water column is relatively unstratified in winter, but sustains a strong, almost linear stratification during quiescent summer conditions due to the presence of a halocline, the summer thermocline and weak surface wind and bottom tidal stirring.

Ketchum calculated the residence time of water in the plume to be ~6 to 10 days, in spite of ninefold variations in river discharge, and found a rapid reestablishment of the basic circulation pattern after only 2 days following a major winter storm disruption.

Under conditions of high discharge, the effluent often flows southward as a slope current along the New Jersey shore, while under low summer discharge, a much more variable pattern can exist with a weak plume often spreading over most of the Apex.

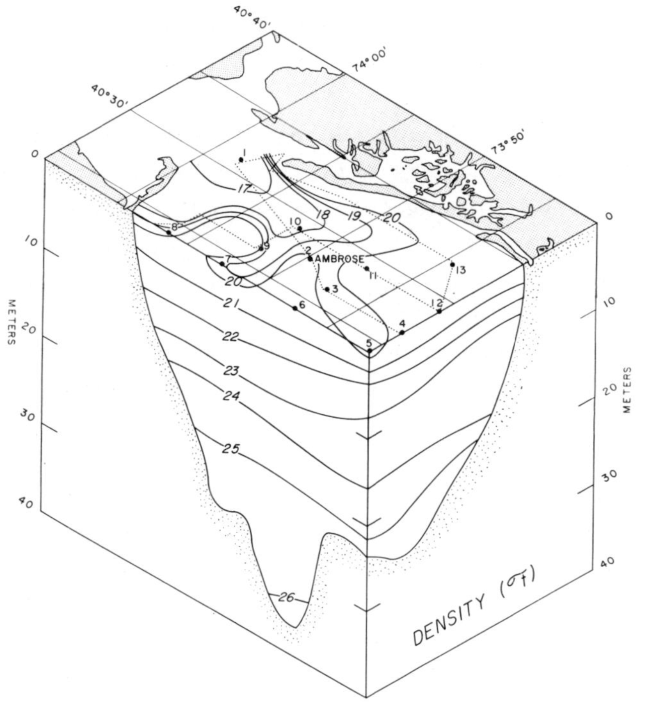

Fig. 1. Isometric perspective of density (σ_T) of the Hudson effluent and the New York Bight Apex.

DETAILS OF THE CRUISES

Three cruises (Aug. 13, 16, 19, 1976) were executed to map continuous surface temperature, salinity and chlorophyll *a*, along cruise tracks chosen in real time to delineate the plume; in addition, during cruise 1 only, profiles were taken at selected stations to determine the vertical structure of the above variables plus inorganic nutrients and suspended particulate matter as well. Only the physical aspects are discussed in this paper; chemical and biological properties will be published elsewhere.

Temperature and salinity at 1 m continuously were measured with a Plessey 6600T Thermosalinograph. Water column samples were gathered with an in situ pumping system (Hulse, 1975) also connected to the Thermosalinograph.

Each cruise lasted about 12-18 hours. Some errors in contouring due to tidal advection are unavoidable; however, away from the mouth of the estuary, the tidal excursion is small relative to the scale of the plume. Contours near the mouth were adjusted to reflect conditions at local slack after estuarine ebb.

An isometric perspective of density is shown in Fig. 1, viewed from space to the northwest (other diagrams illustrating hydrographic, chemical and biological properties are found in Bowman and Iverson, 1977).

The river discharge during the period was steady at ~380 m^3 sec^{-1}. The plume spreads as a very thin (~5 m) lens over the coastal receiving waters, exhibiting complex interdigitations (Fig. 2). The prevailing wind (stress ~2 dyne cm^{-2}) during the preceding 3 days was southwesterly; wind stress clearly influenced the set of the plume and it drifted to the north and the east.

Strong fronts formed along the 29°/oo isohaline (σ_T ~20) on Hudson estuary ebb tide particularly near the bases of the plume lobes, since the surface slopes of the effluent were presumably greatest there.

Surface convergence currents inside the plume are very effective in sweeping flotsam into these fronts, and vast amounts of floating garbage can be traced at times for 30 km or more as filaments stretched along the front. Such debris includes oil, grease and tar balls, bottles, plastic devices and containers of all descriptions, paper, contraceptives, hospital refuse, abandoned boats, entrails, timber, fruit, toys, even cadavers.

Figures 3 and 4 illustrate the surface salinity characteristics of the plume on August 16th and 19th (cruises 2 and 3). Characteristics measured during cruise 2 represented a transition state: the winds during this period were rotating from southerly through westerly to northwesterly, and the plume was drifting back to a southerly set. By cruise 3, the plume had reverted to a long ribbon along the New Jersey Coast. The winds during this period were northwesterly (wind stress during cruises 2 and 3 remained steady at ~2 dyne cm^{-2}).

A vertical cross section of the plume is shown in Fig. 5. A sharp front at station 4 represented its outer limit. This front formed just before the section was taken and coincided with the beginning of ebb tide in the mouth of the estuary. The isohalines (S=25-29°/oo) of Fig. 4 collapsed into singular front which was subsequently followed after taking the transect, unbroken, northwards into the mouth of the estuary.

Discontinuities in properties were dramatic. Plume water was yellow-brown in color, with an oily smooth appearance. Oceanic water outside the front was coastal blue-green, with ~1 m surface gravity waves approaching the front, piling up, and breaking. Again numerous filaments of flotsam delineated the front. It is interesting

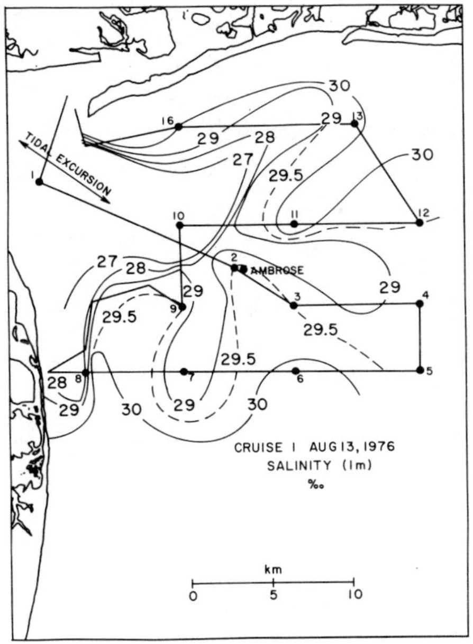

Fig. 2. Surface (1 m) salinity of Hudson effluent during cruise 1 (Aug. 13, 1976).

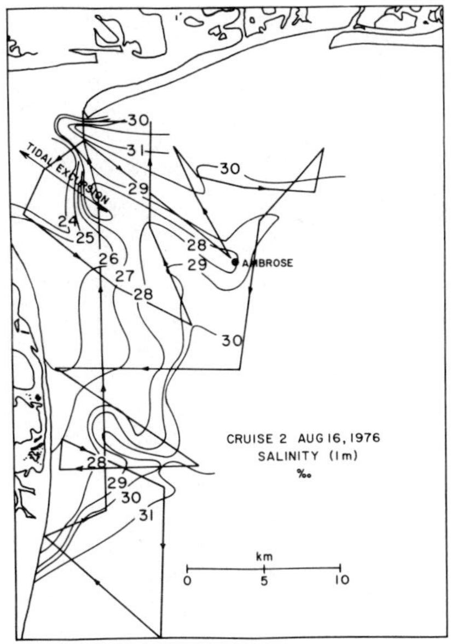

Fig. 3. Surface (1 m) salinity of Hudson effluent during cruise 2 (Aug. 16, 1976).

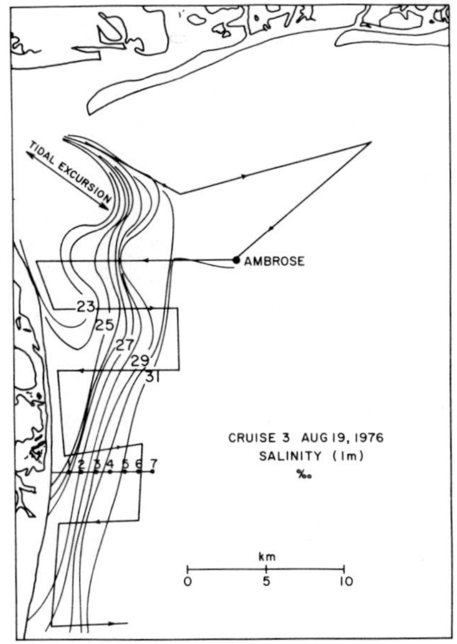

Fig. 4. Surface (1 m) salinity of Hudson effluent during cruise 3 (Aug. 19, 1976).

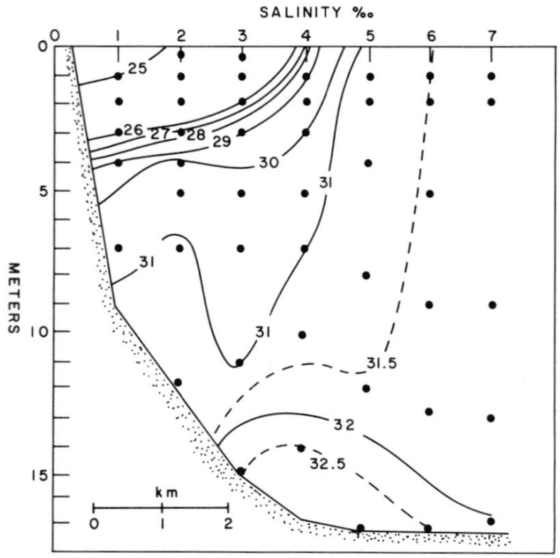

Fig. 5. Vertical salinity cross section across plume during cruise 3. Station positions are shown in Fig. 4.

to note that the areal extent of the plume during the time period of the surveys remained constant (190 km^2) within experimental error.

FRESH WATER CONTENT OF THE PLUME

In order to compare the dynamics of the plume with the steady state vertically integrated model of Takano (1954a,b, 1955), salinity data from cruise 1 were used to separate out the fresh water content of the effluent. The plume consists of water of salinity $S(z) < S_0$, where S_0 is some background oceanic salinity, and z is the vertical ordinate (positive downwards).

The effluent can then be separated into a fresh water component of depth h, overlying a salt water component of salinity S_0, of thickness z_1. Thus, $h = z_0 - z_1$

$$= z_0 - \int_0^{z_0} \rho S(z)\,dz \Big/ \rho S_0$$

where z_0 is the depth at which $S(z) \rightarrow S_0$, i.e., the under surface of the plume, and $\rho(z)$ is density. The results are illustrated in Fig. 6, where the x axis is taken perpendicular to the line across the mouth of the estuary (y axis).

DYNAMICS

The coordinate system is illustrated in Fig. 7. The linearized, steady state equations of motion and continuity are:

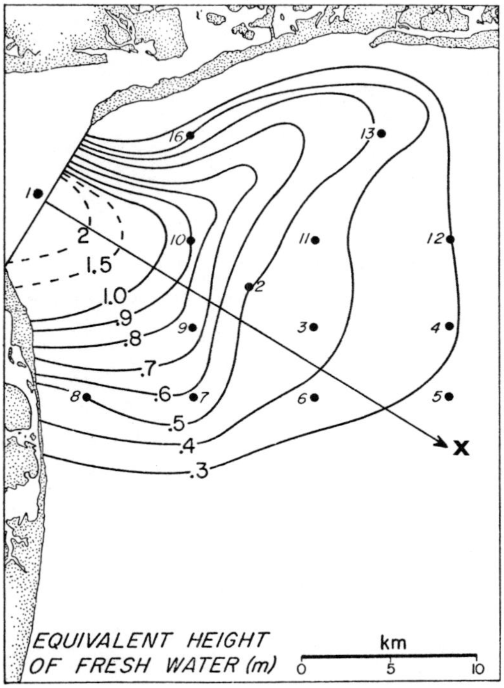

Fig. 6. Freshwater content of the plume during cruise 1.

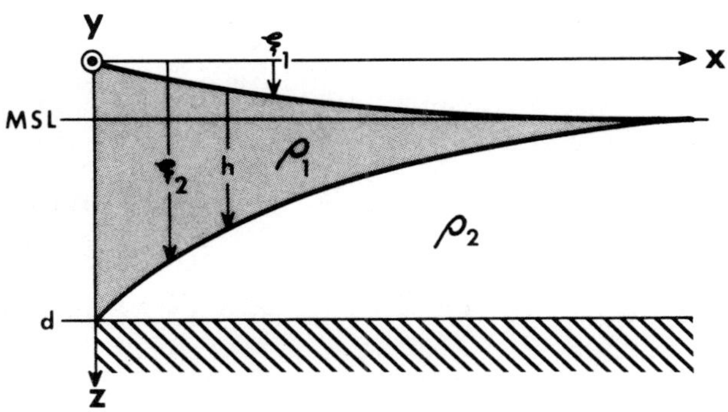

Fig. 7. Coordinate system for plume dynamical calculations.

$$0 = -\frac{1}{p}\frac{\partial p}{\partial x} - fv + N_h\nabla^2 u + \frac{\partial}{\partial z}(N_z \frac{\partial u}{\partial z})$$

$$0 = -\frac{1}{p}\frac{\partial p}{\partial y} + fu + N_h\nabla^2 v + \frac{\partial}{\partial z}(N_z \frac{\partial v}{\partial z})$$

$$0 = \frac{\partial u}{\partial x} + \frac{\partial v}{\partial y}$$

where p = pressure

f = Coriolis parameter $\sim 10^{-4}\sec^{-1}$.

N_h, N_z = horizontal and vertical eddy viscosities

u,v = horizontal current velocities

The boundary conditions are:

$$\frac{\partial u}{\partial z} = \frac{\partial v}{\partial z} = 0 \ , \quad z = \xi_1, d$$

(no stress at surface, bottom)

$$u = v = 0 \ , \quad z \geq \xi_2$$

(no flow in underlying fluid).

Takano defines longitudinal and transverse mass transports per unit width as:

$$M_x = \int_{\xi_1}^{d} \rho u dz = \partial\psi/\partial y$$

$$M_y = \int_{\xi_1}^{d} \rho v dz = -\partial\psi/\partial x$$

where ψ is the stream function.

The integrated pressure P is:

$$P = \int_{\xi_1}^{d} p dz.$$

River flow is constant and uniform across the mouth of width 2ℓ:

$$M_x = M_o \ , \quad x = 0 \ , \quad -\ell < y < \ell$$

$$M_x = 0 \ , \quad x = 0 \ , \quad -\ell > y > \ell$$

$$M_y = 0 \ , \quad x = 0 \ , \quad \forall y$$

The solution for ψ is:

$$\psi = \frac{M_o}{\pi}\left[(y+\ell)\tan^{-1}(\frac{y+\ell}{x}) - (y-\ell)\tan^{-1}(\frac{y-\ell}{x})\right]$$

Note ψ is independent of f.

The solution for P is:

$$P = \frac{M_o}{\pi} \{f \left[(y+\ell)\tan^{-1}(\frac{y+\ell}{x}) - (y-\ell)\tan^{-1}(\frac{y-\ell}{x})\right]$$

(Coriolis contribution)

$$+ 2N_h \left[\frac{y+\ell}{x^2+(y+\ell)^2} - \frac{y-\ell}{x^2+(y-\ell)^2}\right]\}$$

(eddy viscosity contribution)

The thickness of the (freshwater) plume, h, is given by:

$$h(x,y) = (\frac{2P}{\Delta\rho g})^{\frac{1}{2}}$$

$\Delta\rho$ = density contrast across plume ~.02, and

g = acceleration due to gravity.

The thickness h(x,o) along the x axis is:

$$h(x,o) = (\frac{8 M_o \ell N_h}{\Delta \rho \pi g})^{\frac{1}{2}} \frac{1}{(x^2+\ell^2)^{\frac{1}{2}}}$$

The anticyclonic deflection of the effluent as it spreads at sea is a function of the horizontal Ekman number E:

$E = N_h/f\ell^2$.

COMPARISON WITH DATA FROM CRUISE 1

Fig. 8 is a plot of h(x,o) from the above equation and experimental data points taken from Fig. 6. The curve, visually fitted by adjusting N_h and ℓ, gave values of

$N_h = 1 \times 10^8 \ cm^2 \ sec^{-1}$, and

ℓ = 1.95 km

This leads to a value of E ~27.

Figure 9 shows the expected deflection of the plume for various values of E, including that deduced from the data.

DISCUSSION

The deduced value of N_h is some two orders of magnitude larger than values often taken for open ocean viscosities. This is a direct consequence of the large surface slopes found within the effluent (much larger than those found in, for example, the Columbia River plume (Barnes et al, 1972).

The zero shear stress boundary conditions at the surface and plume interfaces are tantamount to neglecting vertical eddy viscosity, and hence all frictional retardation is ascribed to lateral shear stresses. Thus the computed value of N_h is overestimated, and under a more realistic boundary condition of finite interfacial stress, would be accordingly reduced.

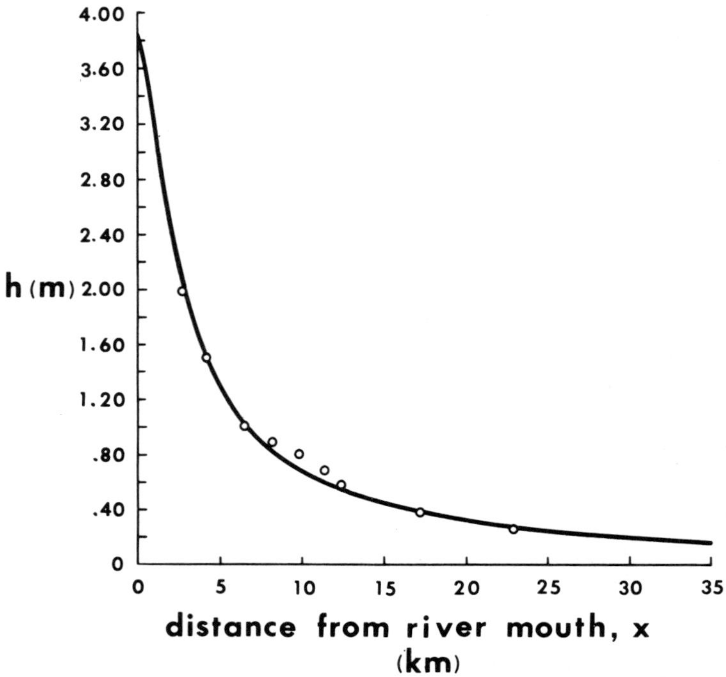

Fig. 8. Plot of thickness of fresh water content of plume versus distance from mouth obtained from the model. The experimental points are obtained from Fig. 6.

As seen in Fig. 10, the estuary supports a well developed upstream non-tidal flow at depth. This counterflow will increase, perhaps quadruple, the interfacial stress between the two layers above that for a stagnant lower layer as assumed in the model.

Beardsley and Hartman (1978) have extended Takano's work by considering the effects of bottom friction and topography on a single layer discharge, and in addition, interfacial friction for a double layered system.

However, they go to the opposite extreme and assume that lateral friction is dynamically irrelevant. For the double layer model, the surface flow deflects to the *left*, whereas the inflow into the estuary at depth approaches from the right, looking out to sea. This opposite deflection for the surface layer is ascribed to the direction of the interfacial shear, and tilting of the interface.

It is obvious that further careful measurements and calculations are needed to adequately model the effects of lateral and vertical viscosities in such plumes.

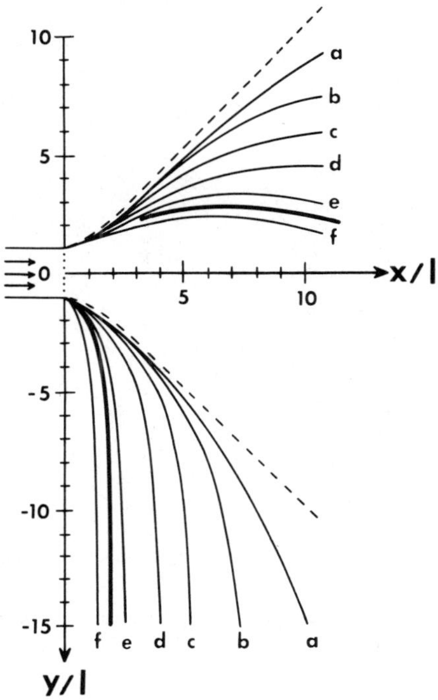

Fig. 9. Expected deflection of the plume as a function of the horizontal Ekman number E. The straight dashed lines represent the edges of the plume for f = 0. The heavy curved lines represent the expected edges of the plume for conditions observed during cruise 1 (neglecting wind stress and coastal currents. Adapted from Officer, 1976). a) E = 500; b) E = 250; c) E = 125; d) E = 62.5; e) E = 31.2; f) E = 15.6.

The value of $2\ell \sim 4$ km is also much less than the width of the estuary mouth ($\sim$9 km). However, as can be seen in Fig. 10, the seaward surface flow is constricted to a width of $\sim$7.5 km, and if the plume is decomposed into fresh and oceanic water, the width reduces even further, so the computed value of ℓ seems reasonable.

CONCLUSIONS

The application of Takano's classical model has provided insight into the dynamic balances of a spreading effluent at sea, but obviously further careful studies are needed to separate out the various influences of horizontal and vertical viscosities, Coriolis effect, advection by a prevailing shelf current, and local wind stress. Shallow shelf waters in New York Bight can be reversed within six hours by a sudden change in wind stress (G. Hann,

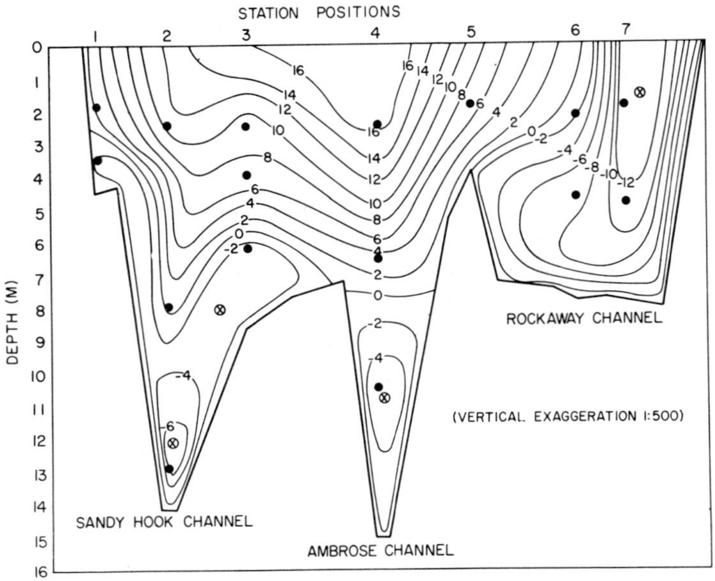

Fig. 10. Averaged non-tidal velocities across the mouth of the Hudson estuary, June 2-7, 1952 (from Kao, 1975).

private communication); the plume itself would be expected to respond even more quickly.

And finally, we know little about how the actual mixing of the plume into the ambient shelf water occurs. Is it primarily by upward and/or downward entrainment between plume and underlying oceanic water, or is the mixing primarily a consequence of horizontal exchange? And what role do the ubiquitous plume fronts play in effecting cross plume transport and mixing? The answer to these questions is not only important to dynamical insight but to environmental concerns as well.

ACKNOWLEDGEMENTS.

This work was supported by the Marine Ecosystems Analysis (MESA) Program of NOAA, and by the University Awards Committee of the Research Foundation and the State University of New York. Contribution 205 of the Marine Sciences Research Center (MSRC) of the State University of New York at Stony Brook.

REFERENCES

Barnes, C.A., Duxbury, A.C. and Morse B.-A., 1972. Circulation and Selected Properties of the Columbia River Effluent at Sea. IN: The Columbia River Estuary and Adjacent Ocean Waters. Bioenvironmental Studies. Eds. A.J. Pruter and D.L. Alversen University of Washington Press, Seattle, 1972.

Beardsley, R.C. and Hart, J., 1978. A simple theoretical model for the flow of an estuary onto a continental shelf. To appear in J. Geophys. Res.

Bowman, M.J. and Wunderlich, L.D., 1976. Am. Soc. Lim. Oceanogr. Special symposia, Vol. 2., pp. 58-68.

Bowman, M.J. and Wunderlich, L.D., 1977. New York Bight Atlas monograph #1, New York State Sea Grant Institute, Albany, N.Y.

Bowman, M. and Iverson, R.L., 1977. Estuarine and plume fronts. In: Oceanic fronts in coastal processes: proceedings of MSRC workshop, May 25-27, 1977. MSRC, State Univ. of N.Y., Stony Brook.

Charnell, R. L. and Hansen, D.V., 1974. Summary and analysis of physical oceanography data collected in the New York Bight Apex during 1969-1970. MESA Rep. 74-3. Boulder, CO:NOAA.

Giese, G.L. and Barr, J.W., 1967. NYS Conserv. Dep. Water Resources Comm. Bull. 61, Albany, N.Y.

Gross, M.G., ed., 1976. Middle Atlantic Continental Shelf and New York Bight. Am. Soc. Lim. Oceanogr. Special symposia, Vol. 2.

Hulse, G.L., 1975. The Plunket: a shipboard water quality monitoring system. Tech. Rep. 22, Marine Sci. Res. Cent., State Univ. of New York, Stony Brook.

Jay, D.A. and Bowman, M.J., 1975. Physical oceanography and water quality of New York Harbor and western Long Island Sound. Tech. Rep. 23. Marine Sci. Res. Cent., State Univ. of New York, Stony Brook.

Kao, A.F., 1975. A study of the current structure in the Sandy Hook-Rockaway Point transect. Unpubl. MS res. paper. Stony Brook: Marine Sci. Res. Cent., State Univ. of New York.

Ketchum, B.H., Redfield, A.C. and Ayers, J.C., 1951. Papers in Phys. Oceanogr. and Meterol. 12(1)1-46.

National Marine Fisheries Service, 1972. The effects of waste disposal in the New York Bight, Summary final rep. NTIS AD 743936.

Pearce, J.B., 1970. The effects of waste disposal in New York Bight. Interim rep. for 1 January 1970. Sandy Hook, NJ: Nat. Marine Fish. Serv.

Officer, C.B., 1976. Physical oceanography of estuaries (and associated coastal waters) Ch. 4, John Wiley, N.Y.

Takano, K., 1954a. Journal of the Oceanographical Society of Japan, 10, 60-64.

Takano, K., 1954b. Journal of the Oceanographical Society of Japan, 10, 92-98.

Takano, K., 1955. Journal of the Oceanographical Society of Japan, 11, 147-149.

A THREE-DIMENSIONAL TURBULENT ENERGY MODEL FOR NONHOMOGENEOUS ESTUARIES AND COASTAL SEA SYSTEMS

J. J. Leendertse and S. K. Liu
The Rand Corporation, Santa Monica, California (U.S.A.)

ABSTRACT

Three-dimensional flows in water bodies with nonhomogeneous density can be computed effectively by use of a finite difference model which contains an equation of continuity, equations describing conservation of momentum, salinity, temperature, subgridscale energy, and an equation of state. In the model, vertical accelerations are neglected, but not the vertical velocities. The vertical exchange coefficients are computed from the subgridscale energy intensity. The introduction of turbulence closure by computing vertical exchange coefficients in the transport equations eliminates the necessity of evaluating exchange coefficients in the whole computation field, but rather requires the evaluation of only a few characteristic parameters.

Experiments made with the model produced velocity distributions which typically occur in coastal areas.

BASIC CONCEPTS OF TURBULENT EXCHANGE PROCESSES

In mathematical models of geophysical problems the approach is taken of using discrete representations of the differential equations which describe the main physical processes.

Generally mass balance equations are used, together with momentum equations. This approach is taken here also, as we will describe later in more detail. The main difficulty encountered, however, is that the discrete representation forces us to work with values for the variables which are characteristic for the region described by the finite representation. If a horizontal grid representation of 1 mile is used, the model can only represent the average velocity, pressure, etc. of the grid distance of 1 mile. Naturally, significant variations do occur which, for example, cause mass and momentum exchanges which can be considerably larger than the advective transports expressed by average velocities and concentrations.

The exchanges are traditionally expressed as functions of the local mass and momentum gradients. In the horizontal motions it is generally assumed that these functions are linear with the gradients; thus we used horizontal momentum and mass exchange coefficients.

In the vertical this approach was even far less satisfactory, and generally the exchange coefficient is taken as a function of the square of the vertical velocity gradient. This assumption seemed justified, as with larger gradients more turbulence is generated, thus more vertical exchange.

In addition to the above-mentioned relationship between the mean flow field and the vertical exchange coefficient, an additional relationship is generally used between the exchange coefficient and the Richardson number. The Richardson number is an expression for the ratio of the density gradient and the local turbulent energy.

In the model presented here the vertical exchange coefficients are related to the subgridscale turbulent energy. The movements of the water on a small scale cannot be described, but we are able to compute the small-scale energy by considering the energy transfer from the larger scales to the smaller scales and the decay of these small-scale motions.

The vertical exchanges are typically induced by the turbulence generated by the flow moving over the bottom, by the turbulence generated by the air moving over the water, and by the turbulence generated by differences in the mean velocities at different locations in the vertical. If the turbulent energy is used as an indicator for the intensity of turbulence, then exchange coefficients can be taken as a function of the subgridscale energy and a length scale. This approach requires that the subgridscale energy be computed. In the model presented here the computed subgridscale energy is transported like a constituent, and is dissipated at a rate dependent on the local energy intensity. The subgridscale energy in our model is produced as a function of the local mean flow gradients and, as stated above, at the upper and lower boundaries of the water body. With this concept we generate the subgridscale energy predominantly near the bottom, and if wind is present, also in the fluid near the surface.

The vertical momentum and mass exchange concept presented here also has a length scale which has to be determined. Two approaches are available, namely, use of an algebraic expression or by computing the length scale also from mean flow information. Presently the most simple approach is used, and we made the length scale a function of the depth.

The vertical mass and momentum exchanges are suppressed if vertical density gradients are generated. In the model the vertical exchange rates are also taken as a function of the Richardson number computed from the local intensities of turbulent energy.

The horizontal exchanges are mainly due to horizontal eddies which are considerably larger than the eddies in the turbulence mentioned earlier (Fig. 1). These large eddies are two-dimensional, since they will be limited by the water depth in the vertical scale, and also by density differences if these are present. The horizontal exchanges should actually also include an exchange

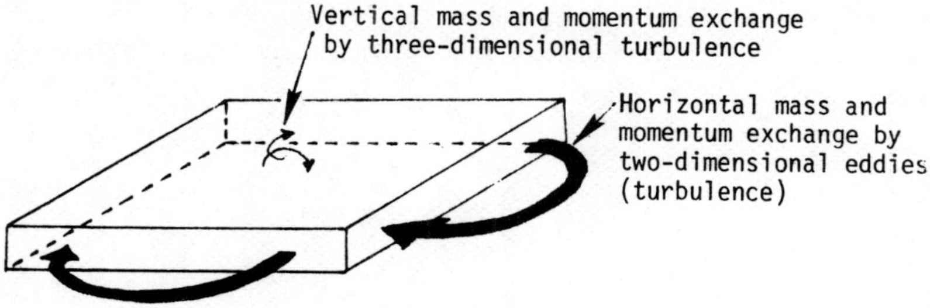

Fig. 1. Concept of mass and momentum exchange in the model.

generated by turbulence from the bottom and the wind-driven water surface, but in our model we neglected this for the present time. Thus the horizontal mass and momentum exchanges were taken only due to turbulence with lower frequencies (smaller wave numbers) than are generated by the bottom and surface momentum transfers.

The simplest and most primitive model for the horizontal momentum exchange is to take it as a function of the local gradient with fixed constant horizontal exchange coefficients for the momentum and mass transfer. This was the approach taken previously by us [1,2]. It seems more appropriate, however, to take this coefficient as a function of the local velocity deformation calculated in the finite difference grid.

Even though the larger subgridscale motions are predominantly two-dimensional rather than three-dimensional, upon which this hypothesis is based, this assumption has produced good results in similar three-dimensional computations with predominant two-dimensional horizontal motions. We will discuss the computation of the horizontal exchange terms in more detail.

MATHEMATICAL DESCRIPTION OF THE PROCESSES

Flows in an estuary or coastal sea are mainly horizontal and primarily turbulent. The equations of horizontal motion for an incompressible, internally source-free fluid on a rotating earth in Cartesian coordinates with the z-axis positive upward are:

$$\frac{\partial u}{\partial t} + \frac{\partial (uu)}{\partial x} + \frac{\partial (uv)}{\partial y} + \frac{\partial (uw)}{\partial z} - f\,v + \frac{1}{\rho}\frac{\partial p}{\partial x} - \frac{1}{\rho}\left(\frac{\partial \tau_{xx}}{\partial x} + \frac{\partial \tau_{xy}}{\partial y} + \frac{\partial \tau_{xz}}{\partial z}\right) = 0 \qquad (1)$$

$$\frac{\partial v}{\partial t} + \frac{\partial (vu)}{\partial x} + \frac{\partial (vv)}{\partial y} + \frac{\partial (vw)}{\partial z} + f\,u + \frac{1}{\rho}\frac{\partial p}{\partial y} - \frac{1}{\rho}\left(\frac{\partial \tau_{yx}}{\partial x} + \frac{\partial \tau_{yy}}{\partial y} + \frac{\partial \tau_{yz}}{\partial z}\right) = 0 \qquad (2)$$

The vertical acceleration of fluid motion associated with the predominant hydrodynamic processes such as tidal and wind-induced circulations are extremely small in comparison with the gravitational acceleration. Therefore we can neglect the vertical acceleration and advection, and the equation of motion becomes the hydrostatic equation:

$$\frac{\partial p}{\partial z} + \rho g = 0 \tag{3}$$

The equation of continuity is:

$$\frac{\partial u}{\partial x} + \frac{\partial v}{\partial y} + \frac{\partial w}{\partial z} = 0 \tag{4}$$

The equations of salt and heat balance are:

$$\frac{\partial s}{\partial t} + \frac{\partial (us)}{\partial x} + \frac{\partial (vs)}{\partial y} + \frac{\partial (ws)}{\partial z} - \frac{\partial (D_x \frac{\partial s}{\partial x})}{\partial x} - \frac{\partial (D_y \frac{\partial s}{\partial y})}{\partial y} - \frac{\partial (\kappa \frac{\partial s}{\partial z})}{\partial z} = 0 \tag{5}$$

$$\frac{\partial T}{\partial t} + \frac{\partial (uT)}{\partial x} + \frac{\partial (vT)}{\partial y} + \frac{\partial (wT)}{\partial z} - \frac{\partial (D_x \frac{\partial T}{\partial x})}{\partial x} - \frac{\partial (D_y \frac{\partial T}{\partial y})}{\partial y} - \frac{\partial (\kappa' \frac{\partial T}{\partial z})}{\partial z} = 0 \tag{6}$$

and the equation of state

$$\rho = \bar{\rho} + \rho'(s,T) \tag{7}$$

Similar balance equations can be written for turbulent densities and dissolved pollutant constituents:

$$\frac{\partial e}{\partial t} + \frac{\partial (ue)}{\partial x} + \frac{\partial (ve)}{\partial y} + \frac{\partial (we)}{\partial z} - \frac{(\partial D_x \frac{\partial e}{\partial x})}{\partial x} - \frac{\partial (D_y \frac{\partial e}{\partial y})}{\partial y} - \frac{\partial (\kappa \frac{\partial e}{\partial z})}{\partial z} + hDe \pm S_e = 0 \tag{8}$$

and

$$\frac{\partial P}{\partial t} + \frac{\partial (uP)}{\partial x} + \frac{\partial (vP)}{\partial y} + \frac{\partial (wP)}{\partial z} - \frac{(\partial D_x \frac{\partial P}{\partial x})}{\partial x} - \frac{\partial (D_y \frac{\partial P}{\partial y})}{\partial y} - \frac{\partial (\kappa \frac{\partial P}{\partial z})}{\partial z} + hK'_P \pm S_P = 0 \tag{9}$$

where u,v,w are respective components of velocity; s and T are salinity and temperature; e and P are turbulent (subgridscale) energy density per unit mass and pollutant concentration, respectively; D is the dissipation rate for turbulent energy; K' is the decay rate for the pollutant concentration; and S_e and S_p are the source and sink terms for the turbulent energy and pollutant concentration, respectively.

The model described in this paper uses a grid system with equidistant points in the horizontal direction. In the vertical direction, it is possible to use an unequal grid distance. The top layer, bounded by the free surface, has a time-variable thickness. In the computation, the origin of the vertical coordinates is taken at the mean sea level, whereas the water surface $z = \zeta(x,y,t)$ is the upper boundary of the system.

The finite difference approximation of the differential equations (1) through (6), and (7) and (9) are accomplished in the following manner: First, the equations for the layers were divided by vertically integrating the variables over the layer thickness, and subsequently, finite difference approximations for the layer equation were established [1,2]. Figure 2 shows the location of variables on the vertical grid. A space-staggered grid (Fig. 3) was selected. Figure 4 shows the location of velocities and other variables on the horizontal grid. The position of the variables in the grid is indicated by indices i,j, indicating a position $i\Delta x$ and $j\Delta y$ from the origin of the coordinates, with $i,j = 0, \pm 1, \pm 3/2, \ldots$ The vertical position is determined as to its location in the center of the layer numbered from the top with integer $k = 1,2,3 \ldots$ or at horizontal interfaces with half-integer values $k = 1/2, 3/2, 5/2 \ldots$ etc. Time is discretized into number of time steps ($n\Delta t$) from the reference time, with n an integer value. We adopted compact sum, difference and time-level notations for x, y, z and t. In the x direction,

$$\overline{F}^x = \tfrac{1}{2}\left\{F[(i + \tfrac{1}{2})\,\Delta x, j\Delta y, k\Delta z, n\Delta t] + F[(i - \tfrac{1}{2})\,\Delta x, j\Delta y, k\Delta z, n\Delta t]\right\} \tag{10}$$

$$\delta_x F = \frac{1}{\Delta x}\left\{F[(i + \tfrac{1}{2})\,\Delta x, j\Delta y, k\Delta z, n\Delta t] - F[(i - \tfrac{1}{2})\,\Delta x, j\Delta y, k\Delta z, n\Delta t]\right\} \tag{11}$$

$$F_+ = F[i\Delta x, j\Delta y, k\Delta z, (n + 1)\Delta t] \tag{12}$$

$$F_- = F[i\Delta x, j\Delta y, k\Delta z, (n - 1)\Delta t] \tag{13}$$

With these notations, Eqs. (1) through (6), and (8) and (9) for the level k with layer-averaged values take the following form:

$$\overline{\delta_t \zeta}^t = - \sum_k \left\{\delta_x(\overline{h}^x u) + \delta_y(\overline{h}^y v)\right\} \qquad \text{at } i,\ j,\ n \tag{14}$$

$$\overline{\delta_t(\overline{h}^x u)}^t = - \delta_x(\overline{\overline{h}^x u}^x\, \overline{u}^x) - \delta_y(\overline{\overline{h}^y v}^x\, \overline{u}^y) - \overline{h}^x \delta_z(\overline{u}^z \overline{w}^x) + f\overline{h}^x \overline{v}^{xy} - \frac{1}{\overline{\rho}^x}\, \overline{h}^x \delta_x p$$

$$+ \frac{1}{\overline{\rho}^x}\left[h\delta_z E_{x} \delta_z \overline{u}^{2t} + \delta_x\left\{hA_{x}\delta_x u\right\}_- + \delta_y\left\{\overline{\overline{h}^x \overline{A_{y}}^x}^y \delta_y u\right\}_-\right] \qquad \text{at } i + \tfrac{1}{2},\ j,\ k,\ n \tag{15}$$

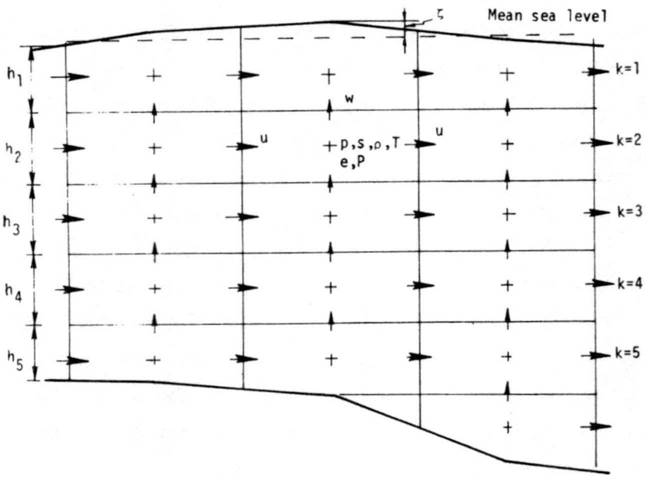

Fig. 2. Location of variables on the vertical grid.

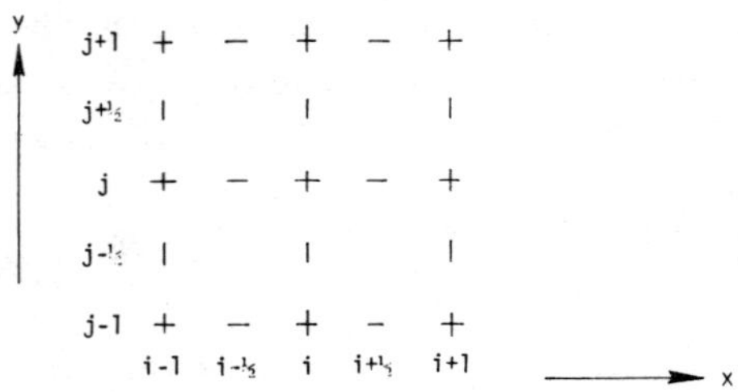

Fig. 3. The location of u (-) and v (|) and other parameters (+) in the space-staggered grid.

$$\overline{\delta_t(\overline{h}^y v)}^t = -\delta_x(\overline{\overline{h}^x u}^y\,\overline{v}^x) - \delta_y(\overline{\overline{h}^y u}^y\,\overline{v}^y) - \overline{h}^y\delta_z(\overline{v}^z\,\overline{w}^y) - f\overline{h}^y\,\overline{u}^{xy} - \frac{1}{\overline{\rho}^y}\,\overline{h}^y\delta_y p$$

$$+\frac{1}{\overline{\rho}^y}\left[h\delta_z E_y \delta_z \overline{v}^{2t} + \delta_x\left\{\overline{\overline{h}^y\,\overline{A_x}^y}^x\,\delta_x v\right\}_- + \delta_y\left\{hA_y\,\delta_y v\right\}_-\right] \quad \text{at } i,\ j+½,\ k,\ n \tag{16}$$

$$\overline{\delta_t(hs)}^t = -\delta_x(\overline{h}^x u\,\overline{s}^x) - \delta_y(\overline{h}^y v\,\overline{s}^y) - h\delta_z(w\overline{s}^z)$$

$$+\delta_x\left\{\overline{h}^x D_x \delta_x s\right\}_- + \delta_y\left\{\overline{h}^y D_y \delta_y s\right\}_- - h\delta_z\left\{\kappa\delta_z \overline{s}^{2t}\right\}$$

$$\text{at } i,\ j,\ k,\ n \tag{17}$$

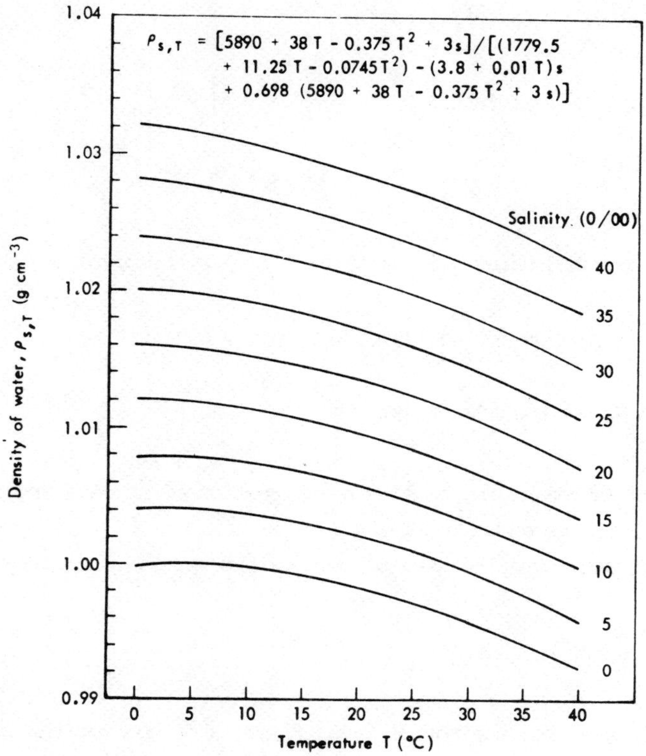

Fig. 4. Equation of state and graphical representations of several selected ranges of S and T values.

$$\overline{\delta_t(hT)}^t = - \delta_x(\overline{h}^x u\overline{T}^x) - \delta_y(\overline{h}^y v\overline{T}^y) - h\delta_z(w\overline{T}^z)$$

$$+ \delta_x\left\{\overline{h}^x D_x \delta_x T\right\}_- + \delta_y\left\{\overline{h}^y D_y \delta_y T\right\}_- + h\delta_z\left\{\kappa' \delta_z \overline{T}^{2t}\right\}$$

at i, j, k, n (18)

$$\overline{\delta_t(he)}^t = - \delta_x(\overline{h}^x u\overline{e}^x) - \delta_y(\overline{h}^y v\overline{e}^y) - h\delta_z(w\overline{e}^z)$$

$$+ \delta_x\left\{\overline{h}^x D_x \delta_x e\right\}_- + \delta_y\left\{\overline{h}^y D_y \delta_y e\right\}_- + h\delta_z\left\{E_e \delta_z \overline{e}^{2t}\right\}$$

$$+ S - Dh$$ at i, j, k, n (19)

$$\overline{\delta_t(hP)}^t = - \delta_x(\overline{h}^x u\overline{P}^x) - \delta_y(\overline{h}^y v\overline{P}^y) - h\delta_z(w\overline{P}^z)$$

$$+ \delta_x\left\{\overline{h}^x D_x \delta_x P\right\}_- + \delta_y\left\{\overline{h}^y D_y \delta_y P\right\}_- + h\delta_z\left\{\kappa\delta_z \overline{P}^{2t}\right\} + S_P$$

$$\text{at } i,\ j,\ k,\ n \tag{20}$$

The density is computed using s(i,j,k,t) and T(i,j,k,t) according to the equation of state

$$\rho = \left[5890 + 38T - 0.375T^2 + 3s\right] \Big/ \left[\left(1779.5 + 11.25T - 0.0745T^2\right) - \left(3.8 + 0.01T\right)s + 0.698\left(5890 + 38T - 0.375T^2 + 3s\right)\right] \tag{21}$$

$$\text{at } i,\ j,\ k,\ n + 1$$

A graphical representation of the density as a function of temperature for water with different salinity is shown in Fig. 4.

The finite difference equation used to compute the vertical velocity component w:

$$\delta_z w = - \delta_x\left(\overline{h}^x u\right) - \delta_y\left(\overline{h}^y v\right) \qquad \text{at } i,\ j,\ k,\ n + 1 \tag{22}$$

This equation is used for the bottom layer first, and then for the layer above, etc. The horizontal pressure gradients are computed from the top layer downward with increasing k by use of

$$\delta_x p = g\overline{\rho}^x \delta_x \zeta + \tfrac{1}{2}\,\overline{h}^x \delta_x \rho \qquad \text{at } i + \tfrac{1}{2},\ j,\ 1,\ n + 1 \tag{23}$$

$$\delta_y p = g\overline{\rho}^y \delta_y \zeta + \tfrac{1}{2}\,\overline{h}^y \delta_y \rho \qquad \text{at } i,\ j + \tfrac{1}{2},\ 1,\ n + 1 \tag{24}$$

$$\delta_z(\delta_x p) = g\delta_x \overline{\rho}^z \qquad \text{at } i + \tfrac{1}{2},\ j,\ k + p,\ n + 1 \tag{25}$$

$$\delta_z(\delta_y p) = g\delta_y \overline{\rho}^z \qquad \text{at } i,\ j + \tfrac{1}{2},\ k + \tfrac{1}{2},\ n + 1 \tag{26}$$

Once the horizontal pressure gradients are known, the water level and velocities can be computed again by the sequence of finite difference equations (14) through (26). Special procedures are required for the computation of variables on or near the boundaries, particularly the seaward boundaries. They are discussed, together with the specification of the initial conditions in Ref. 2.

VERTICAL EXCHANGE OF MOMENTUM AND CONSTITUENTS

In the computational model the turbulent exchanges of momentum and constituents in the vertical which cannot be represented by the computational grid are accounted for by means of exchange coefficients. The magnitude of these exchange coefficients varies not only with space and time, but it is also a function of local turbulence level. Their values are also influenced by the local vertical stability induced by the density stratification. In our computational model, the vertical exchange coefficients are taken according to the basic theoretical considerations of Kolmogorof (1941) and Prandtl (1945) as a function of local turbulent energy level.

$$\varepsilon = L\sqrt{e} \tag{27}$$

where L is a length scale, evaluated as a function of distance to the bottom and free surface according to

$$L = k' z(1 - z/d)^{\frac{1}{2}} \tag{28}$$

where k' is the von Karman constant, z represents the vertical distance from the bottom to the point considered, and d is the local water depth.

In fluids with vertically stable density gradients, the turbulent intensity is suppressed. Therefore, lower values of exchange coefficient would result. The criteria for the onset of the turbulence-suppressing process in a non-homogeneous flow system can be obtained by multiplying by a proportionality coefficient which is a function of the local Richardson number, defined as

$$Ri = -\frac{g}{\rho}\frac{\partial\rho/\partial z}{e}L^2 \tag{29}$$

Mamayav (1958) indicates that the reduction of vertical exchange due to stratification can be expressed in an exponential manner. For momentum exchange in the x-direction,

$$E_x = \overline{\rho}^{xz}\,\overline{L\sqrt{\underline{e}}}^{xz}\,\exp\left[m\,\frac{g}{\overline{\rho}^{xz}}\,\frac{(\overline{L}^z)^2\delta_z(\overline{\rho}^x)}{\overline{\underline{e}}^{xz}}\right] \tag{30}$$

The exponential term in this equation describes the Richardson number dependency, and m represents a constant.

The mass exchange coefficients are computed at a different location, namely, at the layer interface between the points where the concentrations are computed, as shown in Fig. 2.

Consequently, the expression for the mass-exchange coefficient is somewhat different than the momentum exchange coefficients. In the model we are using:

$$\kappa = a_4 \overline{L\sqrt{e_-}}^z \exp\left[r \frac{g}{\overline{\rho}^z} (\overline{L}^z)^2 \frac{\delta_z \rho}{\overline{e_-}^z} \right] \tag{31}$$

where r = a constant.

A factor a_4 appears in this formula, as the mass exchange is not the same as the momentum exchange.

The subgridscale energy is transported in a similar manner as the transport of constituents, thus the energy exchange coefficient can be written in the same form as the mass exchange.

$$E_e = a_1 \overline{L\sqrt{e_-}}^z \exp\left[m \frac{g}{\overline{\rho}^z} (\overline{L}^z)^2 \frac{\delta_z \rho}{\overline{e_-}^z} \right] \tag{32}$$

In the horizontal direction we have assumed that the exchange processes are predominantly governed by two-dimensional turbulence, as the horizontal dimensions are much larger than the depth, and thus no isotropic three-dimensional turbulence can be generated. The two-dimensional turbulence has the property that the enstrophy (one-half vorticity squared) cascades from smaller to larger scales, while in three-dimensional turbulence, it cascades from larger scales to smaller scales. In the model the local enstrophy has to be dissipated, as we have limits to the vorticity which can be expressed on the grid. This dissipation is accomplished by introduction of nonlinear horizontal eddy viscosity coefficients:

$$A = \gamma \left|(\delta_x \overline{\omega}^y + \delta_y \overline{\omega}^x)\right| (\Delta \ell)^3 \tag{33}$$

where ω is vorticity, γ is a coefficient and $\Delta\ell$ is grid dimension.

GENERATION AND DISSIPATION OF SUBGRIDSCALE ENERGY

In the interior of the fluid in our model, it is assumed that the interlayer shear generates the subgridscale energy. This source can be expressed as:

$$S = \varepsilon \left(\frac{\delta \overline{u}}{\partial z}\right)^2 \cdot f(Ri) \tag{34}$$

where $\overline{u}$ = mean velocity

$\varepsilon = L\sqrt{e}$

In the model this source is determined at the interface between the layers at i, j, k + 1/2.

$$S = a_3 \overline{L\sqrt{e_-}}^z \left\{ \left(\delta_z \overline{u}^x\right)^2 + \left(\delta_z \overline{v}^y\right)^2 \right\} f(Ri) \qquad (35)$$

It will be noted that only the energy is computed at the lower time level. This was necessary for stability of the computation.

The energy generated at this location is assumed to be distributed equally into the adjacent layers.

In the bottom layer another source exists. It is assumed that energy which is taken out of the mean flow through the bottom stress immediately enters the subgridscale energy system.

The stress term in the momentum equation in the direction of the mean flow is

$$\frac{\tau}{\rho} = gU^2/C^2 \qquad (36)$$

where U = velocity in bottom layer in the direction of flow.

If this term is multiplied by U, we obtain the energy which is taken out of the mean flow system and the local source (S) for the subgridscale energy.

$$S = gU^3/C^2 \qquad (37)$$

In the model, the subgridscale energy generation is computed at the layer interfaces and the local finite difference source term becomes

$$S = g\left[(\overline{u}^x)^2 + (\overline{v}^y)^2\right]^{3/2} \Big/ C^2 \quad \text{at } i,\ j,\ K + \tfrac{1}{2},\ n \qquad (38)$$

The energy is completely introduced in the layer K.

At the water surface the generation of the subgridscale energy is different. Here the energy source is the wind which generates surface waves, and through these waves, turbulence. Wave and swell conditions depend on wind intensity, duration of the wind and the fetch. In the test cases a fully-developed sea under moderate wind speed was used as input. Under these conditions the waves are so-called deep water waves, and the total wave energy can be found from the Pierson-Moskowitz spectral sets (Neumann and Pierson [6]). Per unit area, the total wave energy is

$$E_t = 5.6 \times 10^{-9} u_w^4 \qquad (39)$$

where u_w = wind speed in cm/sec at 19.5 m above mean sea surface

E = wave energy

Half of this energy is kinetic energy. If we assume that all this kinetic energy is in the top layer (h_1) of the model, then the vertically-average subgridscale energy intensity in this layer is

$$e = 2.8 \times 10^{-9} u_w^4 / h_1 \tag{40}$$

at i, j, 1, n

As the wave theory presents an energy intensity for a given wind condition, we are not concerned with influx of the subgridscale energy into the system, but with maintaining this energy level during the duration of the wind condition in the simulation.

We have assumed that all the kinetic wave energy is in the top layer. From deep water wave theory it is known that the wave-induced water motions are effectively zero at a depth which is half the wave length. This puts an upper limit upon the wind speed which we were able to allow in the simulation. This wind speed can be estimated from the average wave period belonging to the wind speed (Neumann and Pierson [6]).

$$\tilde{\bar{T}} = .81 \times 2\pi u_w / g \tag{41}$$

and from the wave-length-wave-period relation

$$\tilde{\bar{L}} = 2\pi \; g\tilde{\bar{T}}^2 \tag{42}$$

The maximum wind that is allowed in a model with an upper layer thickness h for use in Eq. (40) can then be found from Eqs. (41) and (42):

$$u_w = \left[\frac{2g\sqrt{\pi}}{.81\sqrt{g} \cdot 2\pi}\right]\sqrt{\tfrac{1}{2}\tilde{\bar{L}}} < 21.8\sqrt{h_1} \tag{43}$$

Higher wind velocities would also involve the computation of subgridscale energy in lower layers. The model at present does not include inputs other than in the surface layer.

For the dissipation of energy, use is made of the now classical concepts developed by Kolmogorov [3] and Prandtl [4] that the dissipation rate depends on the transfer process from larger eddies to smaller eddies according to

$$D = a_2 \; \underline{e}^{3/2} / L \tag{44}$$

To demonstrate the generation and decay of turbulent energy in a transient hydrodynamic system, the computational scheme has been tested on a seiche oscillation with an initial amplitude of 25 cm in a rectangular basin. This basin was 46 km long, 14 km wide and 12 meters deep. The interior computational field contained 23 × 7 × 6 grid points and an equal layer thickness for each layer of 2 meters. Constants in the computation are

$$a_1 = 0.68$$

$$a_2 = 0.1$$

$$a_3 = 1.0$$

$$a_4 = 1.0$$

$$K' = 0.4$$

The initial value of SGS energy density was set to 0.4 ergs per unit mass throughout the system. At the starting time, all the energy resolvable in the mean flow is stored in the form of potential energy. Immediately after the start of the computation, a system of current is set up. The stresses at the bottom introduce vertical gradients for the horizontal velocity, which causes subgridscale energy generation. The gradual establishment of the vertical velocity distribution is illustrated in Fig. 5. The velocities obtained from an analytical solution of the long wave equation in the basin for ideal fluid is also shown in the graph.

The subgridscale energy maxima occur shortly after the occurrence of the maximum velocities in the system. The subgridscale energy fluctuation for the first two oscillations is shown in Fig. 6. Note that immediately after the start of the simulation the energy decays, as the velocities are still small, and consequently the generation is small.

The simulation was carried out to 1500 time steps (18.75 hrs real time). At that time the oscillation had decayed significantly. In Fig. 7 the maxima and minima of the total SGS content are shown. Both the maxima and minima decay exponentially.

Figure 8 shows the vertical distributions of the u velocity components at several time intervals. The distribution appears to be near logarithmic only if the flow is well established and approaches a steady state condition.

In a model of Long Island Sound, U.S.A., [7], turbulent energy and the total kinetic energy at each level appeared to increase gradually from three and one-half percent at the surface to sixteen percent near the bottom, where both local production and dissipation are most intensive.

During ebb tide, the velocity magnitude and the subgridscale energy are plotted in Fig. 9. The velocity appears to have a nearly logarithmic profile. Since in this simulation the wind speed was set to zero, the local production of subgridscale energy is small in the top layer.

The inputs of subgridscale energy in that layer are through vertical advection and diffusion and generation at the interface with the second layer. This is

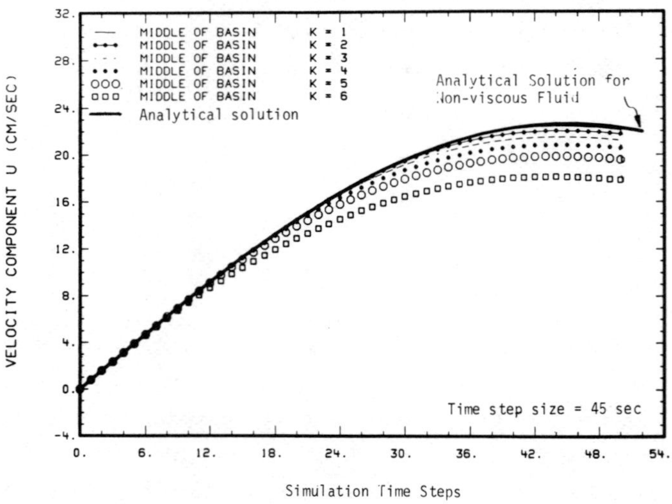

Fig. 5. Comparison of computed mean flow velocity component u by numerical method to non-viscous fluid by analytical solution.

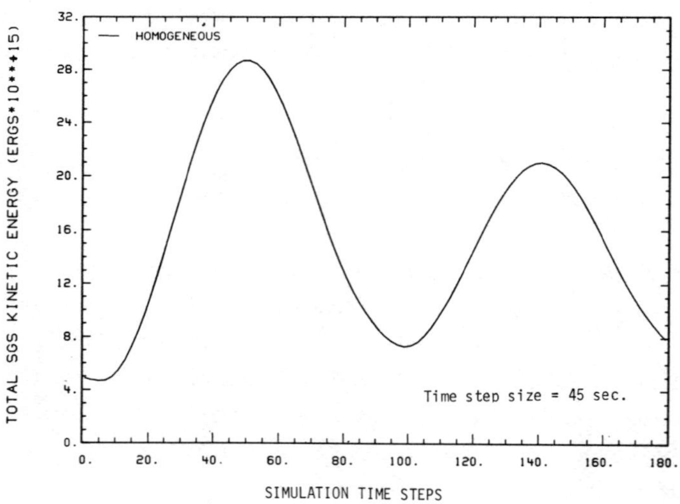

Fig. 6. Generation and decay of the total subgridscale kinetic energy in the oscillating basin.

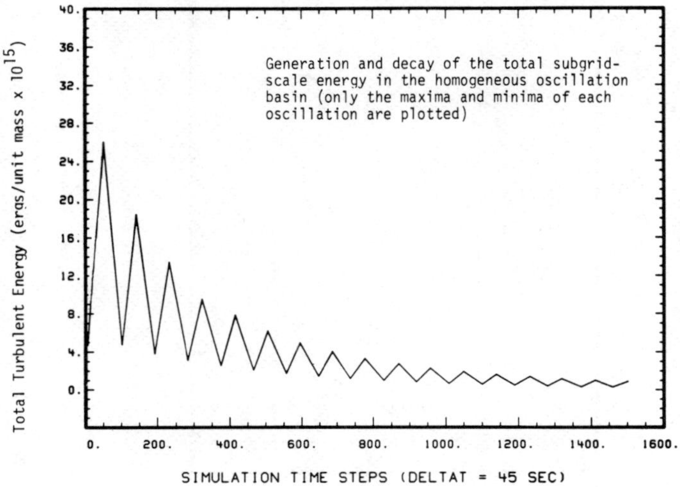

Fig. 7. The generation and decay of turbulent energy in the homogeneous rectangular oscillating basin.

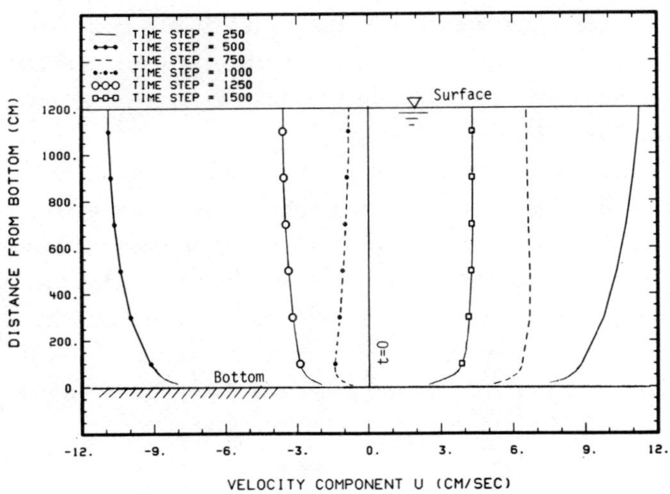

Fig. 8. Vertical distribution of velocity component u in the middle of homogeneous oscillating basin (linear vertical scale).

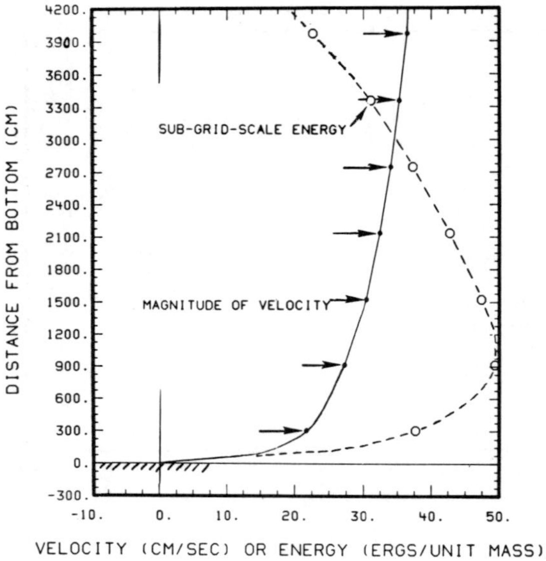

Fig. 9. Computed vertical distributions of horizontal velocity component u and turbulent energy densities at location m=69, n=11, during ebb tide.

balanced by the local dissipation. It appears that the maximum energy is near the bottom, but not in the lowest layer.

In Fig. 10 the vertical distribution of the energy in the mean flow and the energy in the subgridscale range are shown. The computed water surface elevation and the computed horizontal circulation in the surface layer at a particular time are shown in Figs. 11 and 12.

From the simulation we also abstracted the total subgridscale energy in the system, which varies considerably during the tidal cycles. As shown in Fig. 13, in the beginning of the simulation the total subgridscale energy drops, as the tidal flow is only effective in the most eastern part of the bay. From the graph it will be noted that after the second tidal cycle the flow is completely established. The fourth peak is slightly lower than the third peak, as the tide-induced velocities are slightly lower and less subgridscale energy is generated.

DISCUSSION

The introduction of the subgridscale energy in numerical simulations opens many possibilities for the investigation of the behavior of time-dependent hydraulic systems and for the investigation of many of the turbulence hypotheses and theories. Introduction of computations would provide an excellent check on validity of these theories, which otherwise can only be inferred from measurements. We do then expect to see many suggestions for improvement, particularly

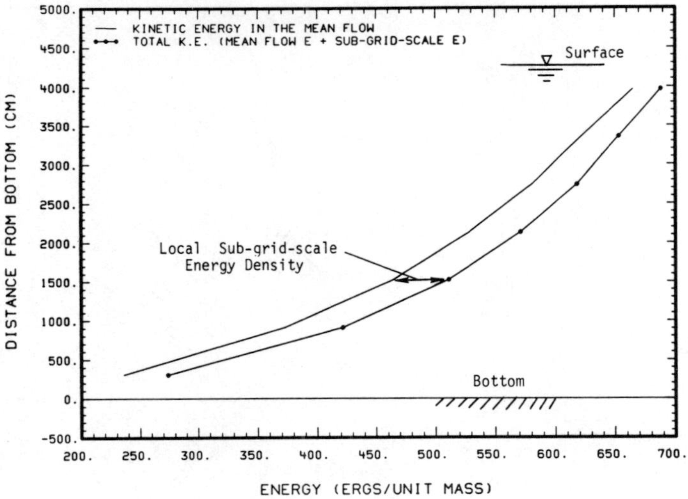

Fig. 10. Computed vertical distribution of tide-induced total kinetic energy at a particular location during ebb tide in a model of Long Island Sound (Leendertse and Liu, 1977).

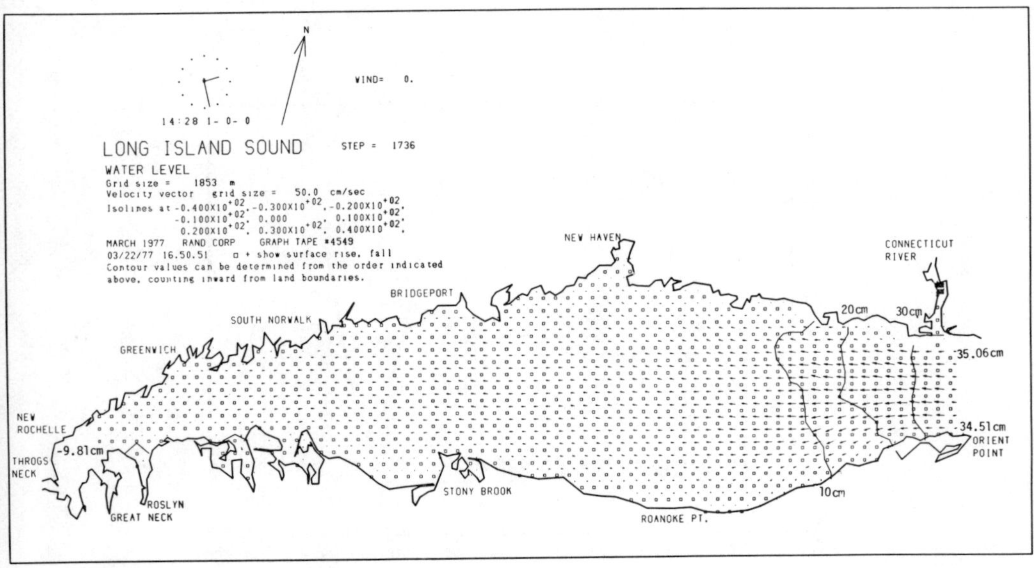

Fig. 11. Computed water surface elevation 14 hours after beginning of simulation (time step 1736).

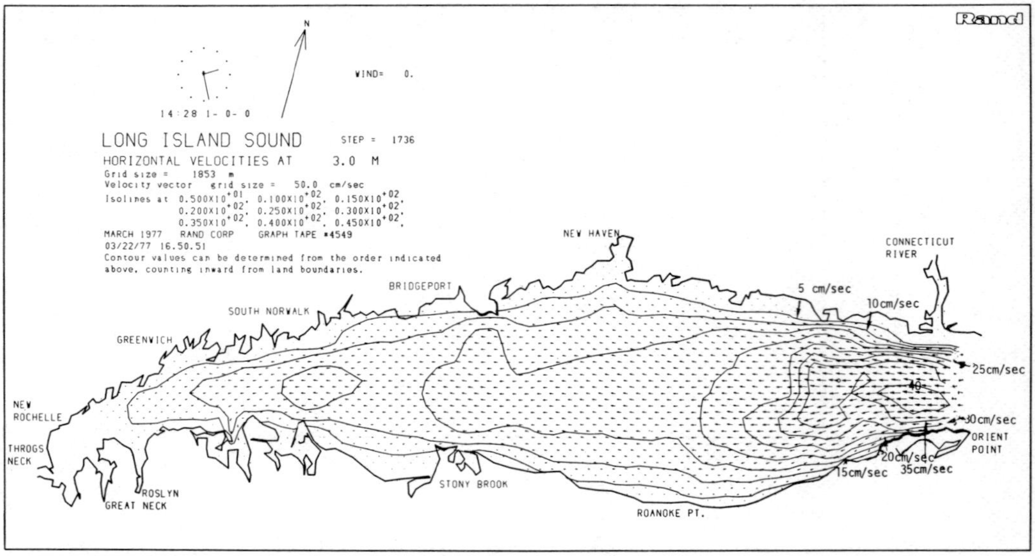

Fig. 12. Computed horizontal circulation in the surface layer 14 hours after beginning of simulation (time step 1736).

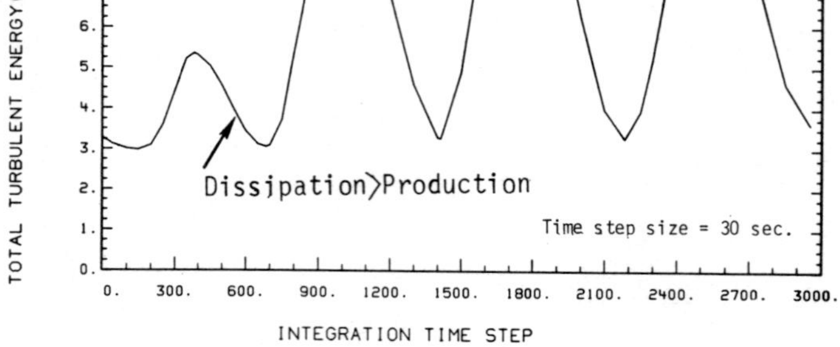

Fig. 13. Computed time history of total turbulent subgridscale energy in the Long Island Sound model [7].

since we are publishing the computer program [8]. With the program listing, a sample case is provided which can be used directly for experimentation.

In the model particular emphasis was placed on the vertical exchanges. Naturally, a similar approach could have been taken for the horizontal exchanges. If energy is determined in the two-dimensional turbulence, then the influence of the two-dimensional turbulence on the subgridscale energy intensity can likely also be computed. Unfortunately, very limited information is available on energy intensities with the length scales of the horizontal grid. Most basic studies directed toward finding rates of horizontal exchanges are based on the spread of dissolved substances, which are not very useful in determining the basic statistical properties of the fluid motions we will be needing for further development of numerical models.

CONCLUSION

Three-dimensional flows in estuaries, bays and coastal seas with nonhomogeneous density distributions can effectively be computed by use of the model described in this paper. The introduction of turbulence closure by computing vertical exchange coefficients in the transport equations for mass, momentum and small-scale turbulent energy from the computed turbulence intensity eliminates the necessity of evaluating exchange coefficients in the whole computational field, but rather requires the evaluation of only a few characteristic parameters.

ACKNOWLEDGMENT

The work presented in this paper was supported wholly by funds provided by the United States Department of the Interior, as authorized under the Water Resources Research Act of 1964, Public Law 88-379, as amended.

REFERENCES

1 Jan J. Leendertse, Richard C. Alexander, and Shiao-Kung Liu, *A Three-Dimensional Model for Estuaries and Coastal Seas: Volume I, Principles of Computation*, The Rand Corporation, R-1417-OWRR, December 1973.
2 Jan J. Leendertse, and Shiao-Kung Liu, *A Three-Dimensional Model for Estuaries and Coastal Seas: Volume II, Aspects of Computation*, The Rand Corporation, R-1764-OWRT, June 1975.
3 A. N. Kolmogoroff, *Compt. rend. acad. sci. USSR*, 30(1941)301 and 32(1941)16.
4 L. Prandtl, "Uber ein neues Formelsystem fur die ausgebildete Turbulenz," *Nachr. Akad. Wiss. Gottingen*, 6(1945)19.
5 O. I. Mamayev, "The Influence of Stratification on Vertical Turbulent Mixing in the Sea," *Izv. Geophys. Ser.*, (1958)870-875. Tr. Victor A. Salkind.
6 G. Neumann, and W. J. Pierson, *Principles of Physical Oceanography*, Prentice-Hall, New Jersey, 1966.
7 Jan J. Leendertse, and Shiao-Kung Liu, *A Three-Dimensional Model for Estuaries and Coastal Seas: Volume IV, Turbulent Energy Computation*, The Rand Corporation, R-2187-OWRT, May 1977.
8 Shiao-Kung Liu, and Alfred B. Nelson, *A Three-Dimensional Model for Estuaries and Coastal Seas: Volume V, Turbulent Energy Program*, The Rand Corporation, R-2188-OWRT, May 1977.

A BRIEF REVIEW OF PRESENT THEORIES OF FJORD DYNAMICS

Fl. Bo Pedersen
Institute of Hydrodynamics and Hydraulic Engineering
Technical University of Denmark.

LIST OF CONTENTS

I. INTRODUCTION

As it is impossible to treat peculiarities in a brief review, we shall confine ourselves to the dynamics of "good-natured" fjords. Even with this restriction a great variability is observed in fjord dynamics from one place to another and from one time to another, see Carstens (1970), Dyer (1973), Gade (1970), Officer (1976), Pickard (1961, 1967, 1971), Pickard et al. (1959), Pickard et al. (1956), Saelen (1967), Tabata et al. (1957), Tully (1949).

In order to define the frame of the review, we shall first look at the main parameters affecting the dynamics and then at their influence. It then turns out to be quite natural to treat the lower and the upper part of the fjords separately.

Finally, we shall try to outline the possible improvements in future models of fjords.

II. THE PARAMETERS AFFECTING THE DYNAMICS OF FJORDS

The observed differences in fjord dynamics originate mainly from the variation in (see fig. 1):

1. The geometry of fjords
2. The hydrology of the adjacent watersheds
3. The oceanographic conditions outside the fjords
4. The wind fields.

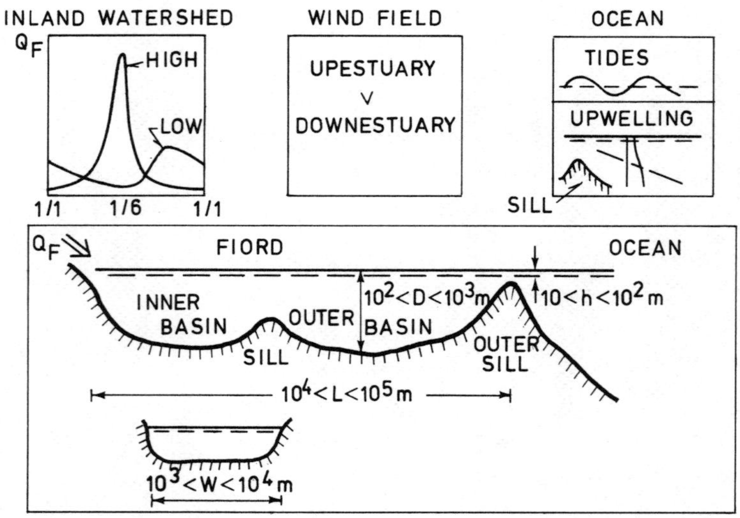

Fig. 1. The main parameters affecting the dynamics of fjords.

II.1. The geometry of fjords

reflects their geomorphologic origins as glacial-cut structures. There are some

common features $\begin{cases} \text{steep sides} \\ \text{deep basin} \end{cases}$

and some

different features $\begin{cases} 0,\ 1,\ 2,\ \ldots.\ \text{sills} \\ \text{side constrictions at the mouth.} \end{cases}$

The order of magnitudes of the sill depth, the basin depth, the width and the length of fjords are shown in fig. 1.

II.2. The hydrology of the adjacent watershed

The yearly distribution of the fresh water discharge Q_F to the fjord shows a pronounced variation from one locality to the other. Let us illustrate this fact by two extreme situations. A high level inland watershed has a great storing effect on the precipitation during winter, which results in a low run-off during winter and a very high runoff during summer from glacial and snow field melting, see fig. 1. A low inland watershed exhibits normally just the opposite variation during the year, as the runoff is nearly in phase with the precipitation due to a weak storing effect of the watershed, see fig. 1. Between these extreme variations we find runoff variations which exhibit a combination of the outlined behaviour.

II.3. The oceanographic conditions outside the fjord

The stratification in the ocean is weak compared with the stratification in estuaries, but changes in the meteorological conditions create variations in the density of the salt water which is in free connection with the bottom water of the fjord. Another characteristic feature of the ocean which influences the dynamics of fjords is the tide, see fig. 1.

II.4. The wind field

The special geometric shapes of fjords have the effect that the wind is normally either up-estuary or down-estuary.

III. THE DEEP BASIN

Oceanographic observations show clearly that the dynamics of the deep basin and of the upper strata are practically uncorrelated, and can thus be treated separately.

As far as the conditions in the deep basin are concerned, there is a remarkable difference between fjords with and without sill(s).

III.1. Fjords without sill

In the fjord without sills, the density distribution in the basin reflects the density distribution in the adjacent ocean. Hence, continuous replenishment of the basin water takes place, which means that the DO-content is normally constantly high.

III.2. Fjords with sill(s)

The combination of one or more sills and a stable density gradient create a blocking effect with a nearly stagnant basin below the sill level. The effect is a decreasing salinity and DO-content until the boundary conditions over the sill level become extraordinary salty, see fig. 2. Then an inflow to the basin takes place, and part of the bottom water is replenished. As the inflowing water deposits at a level corresponding to its density, the frequency of replenishment is decreasing with depth (and distance from the mouth in the case when more than one sill is present).

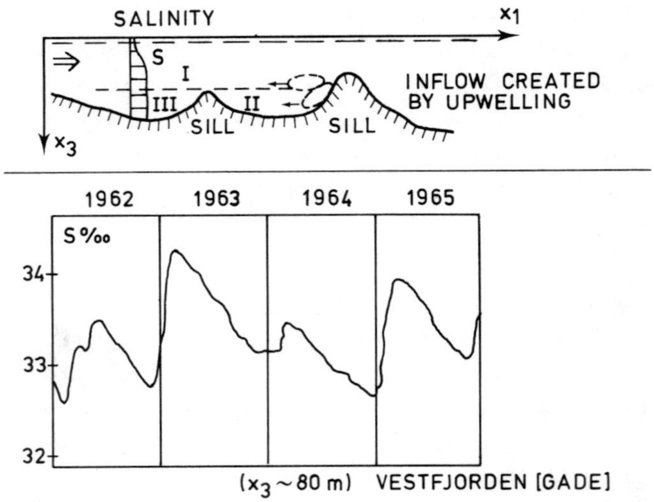

Fig. 2. Replenishment of part of the bottom water in a fjord with sill. The observations shown are from Vestfjorden, Norway, see Gade (1971).

The influence on the deep water from the wind field is very week. This can be verified by looking at the measurements from the Rørholt fjord in Bamble, Norway, see fig. 3 from Carstens (1971). Some 6000 years ago, it was disconnected from the ocean, but the lower 13 m still contain this 6000 years old salt water.

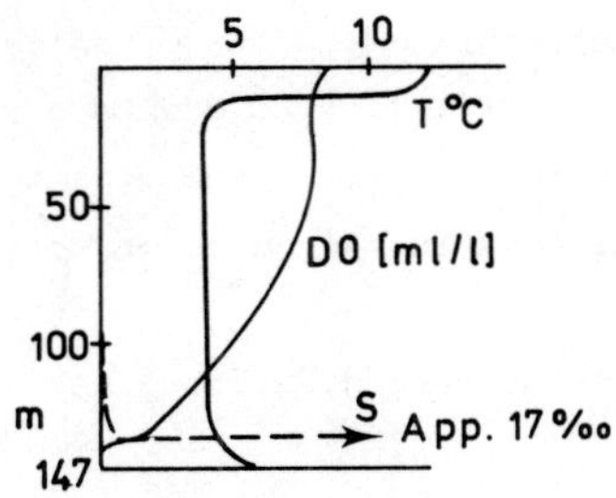

Fig. 3. Wind effect on the bottom water. Measurements from the Rørholt fjord, Carstens (1971).

The mechanism of inflow should create a step-layered bottom water which becomes more and more salty, because the wind effect is confined to the upper, say 20 m. But this is not the case and a plausible explanation is that the tide generates internal waves, which by breaking against the boundaries create turbulence and hence mixing. The mixed water spreads to the interior of the fjord causing an effective vertical mixing, see Stigebrandt (1976).

In order to be able to calculate the bottom water replenishment, the boundary conditions outside the fjord must be known, but this knowledge is seldom available. Therefore, instead of making dynamic models, stochastic models are more fruitful in describing the conditions in the deep basin, see Gade (1971). A pseudo-dynamic model could be established by correlating the inflow with some relevant meteorologic quantities.

IV. THE UPPER LAYERS IN FJORDS

IV.1. Types of fjords

As the combinations of the different factors affecting fjord dynamics are numerous and rather complex, it is more convenient to look at some typical kinematic and dynamic situations found in the upper layer of fjords and then give a general description of the conditions causing this behaviour. In order to cover the range of conditions, three different types are suggested, see fig. 4 (Pickard (1961)).

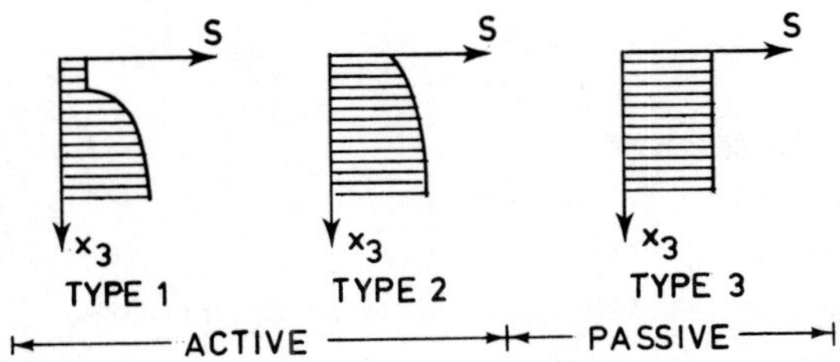

Fig. 4. Types of fjords. After Pickard (1961).

Types 1 and 2 can be characterized as the dynamically active fjords in the sense that fresh water supply is sufficient to create a circulation mode in the upper strata. Type 3 fjords are in the same sense dynamically passive, as they just reflect the boundary conditions in the adjacent ocean - they are merely a bay of the ocean.

All other parameters kept unchanged, a fjord has a tendency to change from type 1 to 2 and 3 when:

Geometry : Length/outlet depth ratio increases (i.e. increases tidal-generated velocity and hence mixing).

Width of the outlet section increases (i.e. decreases critical depth and hence decreases the layer depth).

Distance from head increases (i.e. decreases the density difference and hence increases the mixing).

Hydrology : Freshwater discharge decreases (i.e. decreases the critical depth and hence decreases the layer depth).

Ocean cond.: Tidal amplitude increases (i.e. increases tidal-generated velocities and hence increases mixing).

Wind field : Wind velocity increases (i.e. increases mixing).

In the following, we are dealing only with the dynamics of the upper layers of dynamically active fjords.

V. DYNAMICALLY ACTIVE FJORDS

V.1. Common features of present fjord-theories

All the present theories of fjord dynamics treat the steady state, i.e. the conditions averaged over a tidal period. This is a reasonable approach to the fjord estuaries, as the tidal influence is normally weak, especially in type 1 fjords. For the sake of simplicity, we furtheron confine ourselves to consider a fjord with constant width.

Another feature common to the present theories of fjord dynamics is that they all assume similarity in the velocity profiles as well as in density (or salinity) profiles. Hence - with reference to fig. 4 - the present theories of fjords cover either type 1 or type 2 fjords.

Nevertheless, independently of the type of fjord, the following conditions apply:

V.1.1. The flow characteristica

The main part of the fresh water supply Q_F is normally supplied at the head of the fjord. During the flow from the head to the mouth, the upper down-estuary flowing water is continuously increased by entrainment of salty water from the lower up-estuary counterflow, see fig. 5.

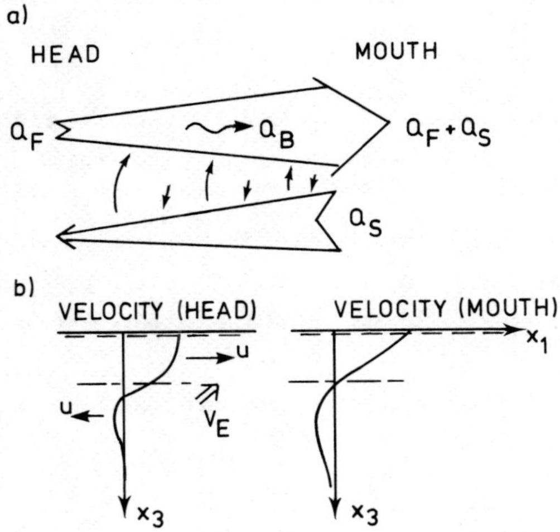

Fig. 5. a. Longitudinal variation of the discharge.
b. Typical velocity distributions.

By considering volume and salt continuity under steady-state conditions, the following two conditions must be fulfilled: *)

Condition 1: $\int_0^D u \, dx_3 = q_F$

which states that the total cross sectional down-estuary discharge per unit width in any cross section is equal to the fresh water supply upstream to the actual cross section (per unit width)

and

Condition 2: $\int_0^D \Delta \, u \, dx_3 = \Delta_F q_F$

which is the analog mass-deficit flux condition. The mass-deficit is defined by

$$\Delta = \frac{\rho_s - \rho}{\rho_s}$$

where ρ_s is a reference density.

*) Footnote: The first person to apply these conditions to oceanographic data was Martin Knudsen in 1899 when he treated the observations from the Danish estuaries (particularly Kattegat - an enormous salt-water wedge). Known as Knudsen's hydrographical theorem.

V.1.2. The dynamic characteristica

The entrainment of salty water has two effects on the dynamics of fjords, viz.

1. Influence on the longitudinal pressure gradients, and hence on the shear stress due to the longitudinal density gradients, and
2. influence on the momentum created by the upward flow across the main longitudinally directed flow, which creates shear stresses as well.

As type 1 fjords are associated with low densimetric Froude numbers $\mathbb{F}_\Delta^2$ and type 2 fjords with high $\mathbb{F}_\Delta^2$ (but still < 1), the pressure gradient effect is most pronounced for type 1 fjords, while the momentum effect is most important for type 2 fjords.

Independently of the type of fjord, two conditions must apply in the steady state, namely

Condition 3: $\left\{\dfrac{p_0}{\rho_s g}\right\}_{x_3=D} = 0$

which states that the pressure measured against a constant reference pressure (the pressure in the adjacent ocean) must vanish at the bottom, and

Condition 4: $\tau_{x_3=D} \simeq 0$

which states that no shear stress is present at the bottom, as the deep water basin is stagnant.

If we apply condition 3, an expression for the superelevation η of the water surface in the fjord can be established. The hydrostatic pressure in the fjord is

$$p_F = \int_{-\eta}^{x_3} \rho g \, dx_3 = \rho_s g(\eta + x_3) - \rho_s g \int_{-\eta}^{x_3} \Delta \, dx_3$$

If we subtract the constant reference pressure $p_R = \rho_s \, g \, x_3$, the overpressure becomes

$$\frac{p_0}{\rho_s g} = \eta - \int_{-\eta}^{x_3} \Delta \, dx_3 = \int_{x_3}^{D} \Delta \, dx_3$$

where we have made use of condition 3:

$$\eta = \int_{-\eta}^{D} \Delta \, dx_3$$

Conditions 1, 2, and 3 are directly or indirectly incorporated in all the present theories of fjords.

With regard to condition 4, the various authors use it in different ways. The problem is not to apply the condition, but to make physically plausible assumptions to which the condition can be applied. As all the present theories assume similarity in velocity as well as in density profiles, it is necessary to treat type 1 and 2 fjords separately in the following. This does not mean that it will always be so.

V.2. Different features of present fjord theories

With a view to the historic development, type 1 fjords will be treated first.

V.2.1. Fjords of type 1 - general >> facts <<

The most characteristic feature of type 1 fjords is the abrupt salinity (and hence the density) jump across the interface, see fig. 6. Above the interface, the density deficit is nearly uniformly distributed, while it decreases gradually below the interface. Accordingly, the velocity is nearly uniform in the upper layer with a weak compensation flow in the lower layer. These features are characteristic for low densimetric Froude numbers, stratified flows, i.e.

$$\mathbb{F}_\Delta^{\,2} = \frac{V^2}{\bar{\Delta}\, g\, y} \ll 1$$

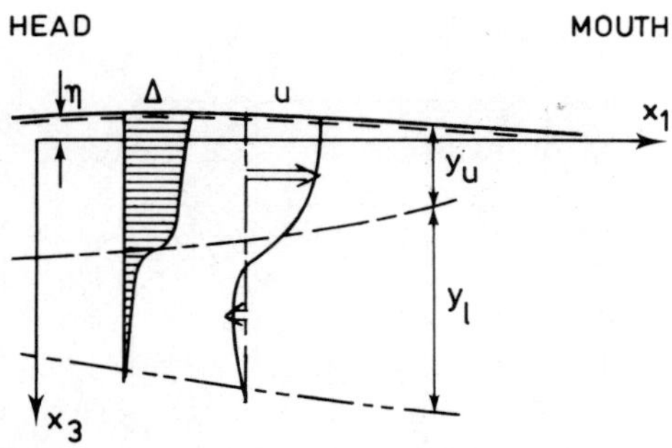

Fig. 6. Typical density and velocity distribution in type 1 fjords.

The fact that $\mathbb{F}_\Delta^2$ is small can be utilized in an illustration of the dynamics of type 1 fjords, as it means that the shear stress in a first order analysis is balanced by the longitudinal pressure gradients

$$\frac{\tau}{\rho_s g} = - \int_{-\eta}^{x_3} \frac{\partial (p_0/\rho_s g)}{\partial x_1}\, dx_3$$

By taking a plausible density-deficit distribution and making use of conditions 3 and 4, the general shape of the shear stress distribution can be found.

Example:

In order to demonstrate the procedure, we take

$\Delta = \Delta_u$ above the interface, and

$\Delta = \Delta_l \exp\left\{- \dfrac{x_3 - y_u}{y_l}\right\}$ below the interface

where y_u = the upper layer depth, y_l = a length scale for the lower layer.

The hydrostatic overpressure distribution is then

$$\left\{\frac{p_0}{\rho_s g}\right\} = \eta - \int_{-\eta}^{x_3} \Delta \, dx_3$$

or

$$0 < x_3 < y_u : \left\{\frac{p_0}{\rho_s g}\right\} = \eta - \Delta_u x_3$$

$$y_u < x_3 < D : \left\{\frac{p_0}{\rho_s g}\right\} = \eta - \Delta_u y_u - \Delta_1 y_1\left(1 - \exp\left\{-\frac{x_3 - y_u}{y_1}\right\}\right)$$

By applying condition 3 we get the surface elevation

$$\eta = \Delta_u y_u + \Delta_1 y_1$$

By differentiating the pressure with respect to x_1 and integrating the pressure gradient, the shear stress is then obtained. As this example is given only for illustration, no further calculations have been performed, but the general shape of τ is shown in fig. 7.

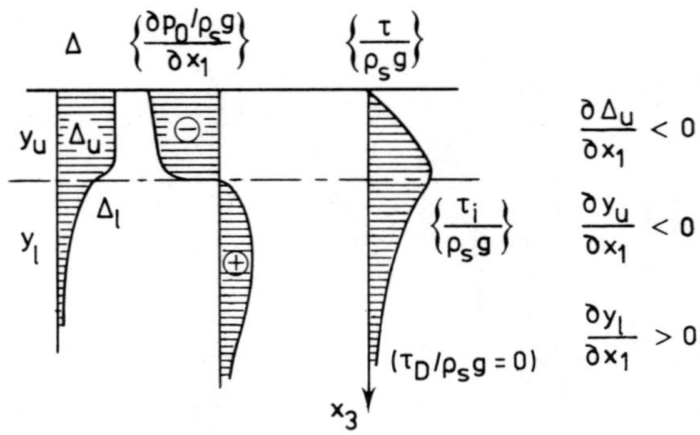

Fig. 7. The general shape of the distribution of the longitudinal pressure gradients and the shear-stress in type 1 fjords.

It is important to realize that the interfacial shear stress is balanced by pressure gradients in the lower layer counterflow. These pressure gradients can be established only by longitudinal density gradients originating from the weak transport of brackish water downwards, combined with the up-estuary convective transport. This gives a recirculation of part of the brackish water, a fact of great importance for the distribution of pollutants. None of the present theories of type 1 fjords describes this mechanism!

V.2.2. Fjords of type 1 - present theories

Instead of giving a detailed description of the theories of type 1 fjords, we shall concentrate on the weakness of the theories, see fig. 8:

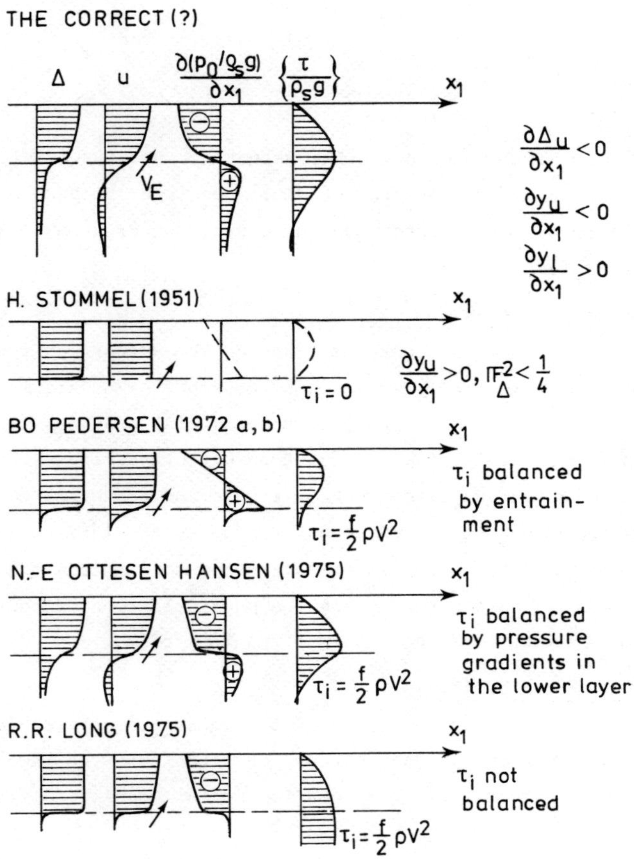

Fig. 8. The distribution of the mass deficit Δ, the velocity u, the longitudinal pressure gradient $\partial(p_0/p_s g)/\partial x_1$, and the shear stress $\tau/p_s g$ in the present theories of type 1 fjords.

H. Stommel (1951)

Stommel's theory is consistent in the sense that it includes all the four mentioned conditions, but by putting the interfacial shear stress equal to zero, the theory fails, for instance by predicting

$$\frac{\partial y_u}{\partial x_1} > 0 \qquad \text{for} \qquad \mathbb{F}_\Delta^2 = \frac{V^2}{\overline{\Delta}\, g y_u} < \frac{1}{4}$$

Fl. Bo Pedersen (1972a, b)

The main progress in Bo Pedersen's theory was to focus on the problem of balancing the interfacial shear stress

$$\tau_i/\rho = \frac{f}{2}\, V^2$$

in order to fulfil condition 4 ($\tau_{x_3=D} \simeq 0$).

By using the work energy equation for the mean flow, a relation between the friction factor f and the entrainment velocity V_E was established

$$f = \frac{V_E}{V}$$

The physical meaning of this result is that the interfacial shear stress is balanced by the entrainment. This has later been shown by Ottesen Hansen (1975) to be contradictory to the assumption of a small $\mathbb{F}_\Delta^2$ and hence the theory failed.

N.-E. Ottesen Hansen (1975)

Ottesen Hansen's theory is based on the assumption that the interfacial shear stress is mainly balanced by pressure gradients in the compensation flow although the mechanism for transferring brackish water downwards was not incorporated in the theory. Even by introducing probable assumptions concerning the interfacial shear stress and the entrainment, it turned out to be difficult to solve the equations. As the introduced simplifications do not seem convincing, it is sufficient to conclude that the basic ideas in Ottesen Hansen's theory are the physically most acceptable presented so far (March, 1977) concerning type 1 fjords.

R.R. Long (1975, 1976)

The fjord theory presented by Long contains an entrainment function and the interfacial shear stress. Nevertheless, the theory is inconsistent, as the interfacial shear stress is not balanced (condition 4 not fulfilled).

V.2.3. Fjords of type 1 - general remarks

It is interesting to notice that the interfacial shear stress, when incorporated in the theories, is represented by

$$\frac{\tau_i}{\rho} = \frac{f}{2} V^2$$

and, accordingly, the mass flux through the interface as

$$\frac{V_E}{V} = g(\mathbb{F}_\Delta^2)$$

But this is not sufficient for a type 1 fjord theory, because the downward mass flux is necessary for creating the pressure gradients in the counterflow. This can be facilitated by using an approach which incorporates an eddy diffusivity $K_{\Delta V}$, defined by

$$-\overline{\Delta' V_3'} = K_{\Delta V} \frac{\partial \overline{\Delta}}{\partial x_3}$$

and accordingly an eddy viscosity K_τ defined by

$$-\overline{V_1' V_3'} = \tau/\rho = K_\tau \frac{\partial u}{\partial x_3}$$

It is evident that a K-theory with known values is superior to the presented theories - and needed. So far, lack of knowledge of the K-values has stopped to this approach. We shall revert to this problem later.

By chance, we can demonstrate the K-theory approach in this review as the fjords of type 2 have been treated in this way.

V.2.4. Fjords of type 2 - general >> facts <<

In fjords of type 2, the two-layer characteristic is absent. The velocity as well as the mass-deficit decrease continuously with the distance from the water surface, see fig. 9. These features are characteristic of stratified flows with higher densimetric Froude numbers and with a more intense vertical mixing. For the dynamics of the fjord, this implies that the shear stress is not balanced solely by the pressure gradients as the momentum is important as well. Hence, a simple illustration as that presented for type 1 fjords is not possible.

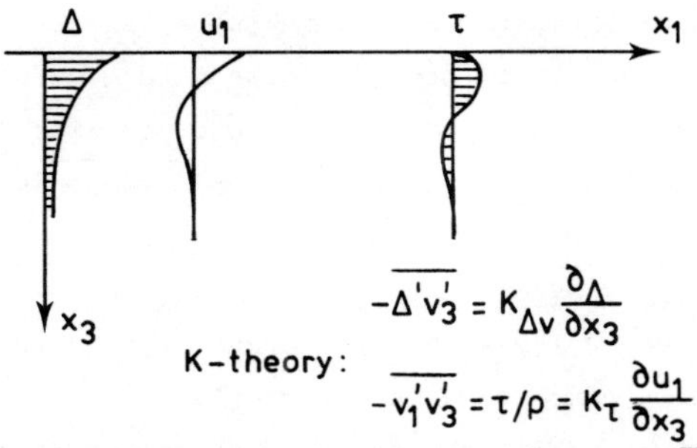

Fig. 9. The general shape of the distribution of the mass deficit, the velocity and the shear stress in type 2 fjords.

W.M. Cameron (1951)

Unfortunately, the doctoral dissertation by Cameron has not been at the author's disposal.

M. Rattray, jr. (1967)

By applying the well-known procedure used by Rattray et al. (1965) to coastal plain estuaries, Rattray transformed the conservation equations for mass, volume, and momentum to a set of two coupled, non-linear, ordinary differential equations*) and a definite integral. These equations were then solved to the lowest order of approximation - i.e. with focus on the upper part of the circulation.

In order to obtain similarity, Rattray was forced to restrict the theory to fjords, in which the fresh-water discharge is a small fraction of the total circulation. This condition is likely to occur in the down-estuary part of type 2 fjords, as demonstrated by the measurements of various authors, see the introduction.

---

*) Footnote: There are two rather misleading misprints in Rattray's eq. 26: $\gamma\phi\phi'' + \nu\phi\phi''$ should be $\gamma\phi\phi''' + \nu\phi'\phi''$.

In order to compare the theory with actual measurements, Rattray estimated the K-values as exponentially damped linear functions, but any attempt to correlate the absolute values with the dynamics was not made. Hence the theory can be used in a descriptive model but not in a predictive model.

D.F. Winther (1972, 1973)

The approach used by Winther is quite similar to that of Rattray. The main progress in Winther's model was the introduction of a distorted scale, in which the characteristic horizontal and vertical length scales were related to the physical behaviour of the flow. This improvement makes it easier to ascribe values to the involved parameters, when the theory is compared with actual field measurements. As an example can be mentioned that the K-values are proportional to the characteristic vertical length and velocity scale, and hence probably of a more general nature than the K-values used by Rattray. As shown by examples in Winther's papers the theory reflects the behaviour of type 2 fjords reasonably well.

V.2.6. The wind effect on fjords

The mixing of the upper layer in a fjord due to wind is normally very strong, but as the problems concerning the wind effect are not specific for fjords they will not be treated in this review. The reader is referred to the papers by Farmer (1972), Ottesen Hansen (1975), and Bo Pedersen (1977b), where further reference lists are given.

VI. CONCLUSIONS AND POSSIBLE IMPROVEMENTS

Measurements in fjords (see the introduction) and in laboratories (see Bo Pedersen et al. (1974)) as well as theories (see Rattray (1967), Winther (1972, 1973), and Ottesen Hansen (1975)) show clearly that the basic assumption in all the present theories concerning similarity is not fulfilled. Furtheron, it has been demonstrated that an approach using an interfacial shear stress and an entrainment function is insufficient. This gain in knowledge could not have been obtained without the present theories, and in that respect, they are all of great value.

The next step in developing a fjord theory is therefore to cancel the similarity assumptions concerning the velocity and density profiles. This demands that quite new ideas are introduced in the fjord theories. One practicable way is to extend the set of equations by the conservation equations for the turbulent kinetic energy, the Reynolds' shear stress, the mass deficit flux, etc. In this way it may be possible to set up a general fjord theory applicable to all types of fjords. The only problem is how to simulate the conservation equations for the turbulent properties.

A first step in this direction has been made by the author (1976, 1977a). If we confine ourselves to the interfacial equilibrium layer for type 1 fjords, the extended sets of equations can be solved analytically. In this way the following quantities have been evaluated: The eddy viscosity for momentum and salt, the velocity and the density gradient, the Monin-Ubokhov length scale, the dissipation length scale, the Brunt-Väiäsäla frequency, the gradient and the flux Richardson number.

It is probably impossible to go much further analytically, but by means of computers the way is prepared for further progress in fjord models.

VII. REFERENCES

Cameron, W.M., 1951. On the dynamics of inlet circulations. Doctoral dissertation. Univ. of California, Los Angeles, California.

Carstens, T., 1970. Turbulent diffusion and entrainment in two-layer flow. Proc. ASCE, WW1, Feb. 1970.

Carstens, T, 1971. Transport processes in water courses and fjords (in Norwegian). Institutt for Vassbygging.

Dyer, K.R., 1973. Estuaries. Wiley and Sons, 140 pp.

Farmer, D.M., 1972. The influence of wind on the surface waters of ALBERNI INLET. Pacific Marine Science Rep. No. 72-16. July 1972.

Gade, H.G., 1970. Hydrographic investigations in the Oslo fjord, a study of water circulation and exchange processes. Geofysisk Institutt, Rep. 24.

Gade, H.G., 1971. Deep water renewals in a sill fjord. Geofysisk Institutt, Rep. 25, 1971/72.

Long, R.R., 1975. Circulations and density distributions in a deep, strongly stratified, two-layer estuary. J. Fluid Mech., 71.

Long, R.R., 1976. Estuarine circulations and mass distributions. Tech. Rep. No. 9 (Series C). John Hopkins University. Dep. of Earth and Planetary Sciences and Mechanics and Materials Science. Baltimore, Maryland, Dec. 1976.

Officer, C.B., 1976. Physical oceanography of estuaries (and associated coastal waters). New York, Wiley, 465 pp.

Ottesen Hansen, N.-E., 1975. Entrainment in two-layered flows. Series Paper 7. Institute of Hydrodynamics and Hydraulic Engineering, Tech. Univ. of Denmark, 99 pp.

Pedersen, Fl. Bo, 1972a. Gradually varying two-layer stratified flow. ASCE, Journal of Hydraulics Division. Vol. 98, No. HY1, Jan. 1972.

Pedersen, Fl. Bo, 1972b. Gradually varying two-layer stratified flow in fjords. International Symposium on Stratified Flows. Paper - 19. Novosibirsk, pp 413-429.

Pedersen, Fl. Bo and Ottesen Hansen, N.-E., 1974. Entrainment in two-layered flows, part 1. Prog. Rep. 33, Aug. 1974. Institute of Hydrodynamics and Hydraulic Engineering, Tech. Univ. of Denmark.

Pedersen, Fl. Bo, 1976. The flux Richardson number and the entrainment function. Prog. Rep. 39, Aug. 1976. Institute of Hydrodynamics and Hydraulic Engineering, Tech. Univ. of Denmark.

Pedersen, Fl. Bo, 1977a. Part 1: The entrainment function for gradually varying two-layered stratified flow. Part 2: A turbulence model and its application to fjord estuaries.
Copies of preprint available from the Institute of Hydrodynamics and Hydraulic Engineering, Tech. Univ. of Denmark, March 1977.

Pedersen, Fl. Bo, 1977b. The bulk flux Richardson number applied on turbulent entrainment at a stable density interface. Prog. Rep. 43, in print. Institute of Hydrodynamics and Hydraulic Engineering, Tech. Univ. of Denmark.

Pickard, G.L., 1961. Oceanographic features of inlets in the British Columbia Mainland Coast. J. Fish. Res. Bd. Canada, 18(6): 907-999.

Pickard, G.L., 1967. Some oceanographic characteristics of the larger inlets of southeast Alaska. J. Fish. Res. Bd. Canada, 24(7): 1475-1506.

Pickard, G.L., 1971. Some physical oceanographic features of inlets of Chile. J. Fish. Res. Bod. Canada, 28: 1077-1106.

Pickard, G.L. and Rodgers, K., 1959. Current measurements in Knight Inlet, British Columbia. J. Fish. Res. Bd. Canada, 16(5): 635-678.

Pickard, G.L. and Trites, R.W., 1956. Fresh water transport determination from heat budget with applications to British Columbia Inlets. J. Fish. Res. Bd. Canada, 14(4): 605-616.

Rattray, M., jr., 1967. Some aspects of the dynamics of circulation in fjords. Estuaries, edit. G.H. Lauff, pp 52-62. AAAC, Washington, D.C.

Rattray, M., jr. and Hansen, D.V., 1965. Gravitational circulation in straits and estuaries. J. Marine Res. 23(2): 104-122.

Saelen, O.H., 1967. Some features of the hydrography of Norwegian fjords. Estuaries, edit. G.H. Lauff, pp 63-70. AAAC, Washinton, D.C.

Stigebrandt, A., 1976. Vertical diffusion driven by internal waves in a sill fjord. Jour. Phys. Ocean. 6(4): 486-495.

Stommel, H., 1951. Recent developments in the study of tidal estuaries, Ref. No. 51-33, Woods Hole Oceanographic Institution, Woods Hole, Mass.

Tabata, S. and Pickard, G.L., 1957. The physical oceanography of Bute Inlet, British Columbia. J. Fish. Res. Bd. Canada, 14(4): 487-520.

Tully, J.P., 1949. Oceanography and prediction of pulp mill pollution in Alberni Inlet. Bull. Fish. Res. Bd. Canada, 83, 169 pp.

Winther, D.F., 1972. A similarity solution for circulation in stratified fjords. Intern. Symp. on Stratified Flows, Novosibirsk, pp 715-724.

Winther, D.F., 1973. A similarity solution for steady state gravitational circulation in fjords. Estuary and coastal marine science, 1: 387-400.

PROPERTIES OF THE ROBERT R. LONG MODEL OF ESTUARINE CIRCULATION IN FJORDS

H.G. GADE and E. SVENDSEN
Geophysical Institute, University of Bergen (Norway)

ABSTRACT

It is shown that the circulation model proposed by Long (1975) implies that the brackish water thickness is completely determined by the interfacial Froude number, the drag ratio and the density difference across the interface. Near the head this relationship approaches $h^2 x \Delta\rho$=constant. Numerical solutions to the set of differential equations are established to show the behaviour of the flow of the brackish layer. The solutions are illustrated in terms of integral curves referring to selected values of layer thickness, initial interfacial Froude number, drag ratio and growth factor of the width.

INTRODUCTION

The problem of estuarine circulation in fjords has been dealt with in the literature along two lines, one pertaining to continuous density distributions, the other requiring vertically homogeneous layers. Both approaches have in principle involved induced horizontal advection as a consequence of mixing, mainly vertical, and the results have depended heavily upon the choice of assumptions made with regard to the magnitude of the mixing involved.

The model presented by Long (1975) is of the layered kind. It involves two vertically homogeneous layers corresponding to a brackish top layer and a sea water body below. The dynamics for the two-dimensional case has been studied by Stommel (1951) and Stommel and Farmer (1952), but their results are incomplete as little was known at that time of the exchange mechanisms to be encountered in the system.

In his model Long makes use of results obtained from laboratory experiments of mixing across an interface (Kato and Phillips 1969, Moore and Long 1971). From these experiments it was possible to specify an entrainment velocity proportional to the cube of the rms turbulent velocity and inversely proportional to the thickness of the upper layer and the density jump across the interface. In the integrations the rms turbulent velocity applicable to the brackish layer is assumed constant in time and space.

Departing from essentially the same conditions as used by Stommel and Farmer (1952), and with the inclusion of the entrainment condition Long proceeds to establish analytical solutions for the case of a fjord of constant width. Similarly, for the case of varying width, the system is reduced to a set of two partial differential equations from which the properties of the flow can be obtained by numerical methods.

The solutions depend on the choice of an outer boundary conditic which is taken to be at the mouth of the fjord. Here, the flow is assumed to become critical, a condition consistent with the results from the tank experiments by Stommel and Farmer (1953). By this condition the solution in the interior is completely determined.

In deriving the analytical solutions Long is forced to make rather complicated substitutions which render the discussion difficul to follow. Also, some of the very simple properties of the model are not immediately apparent. In the present paper an effort is being made to elucidate some of the very basic properties of the model and furthermore give examples of solutions which have been obtained by numerical integration. In doing so it is hoped to make the model more readily available as a means to the study of the behaviour of two-layered fjords.

THEORY

In the following a brief outline of the theory will be given. For details of the derivations the reader is referred to the papers by Stommel and Farmer (1952), Long (1975) and Stigebrandt (1976). However, a few additional remarks of the model are included.

The fjord is treated as a deep channel in which the circulation is essentially two-dimensional. The channel is, however, allowed to vary in width. The flow in the upper layer is assumed to be sufficient turbulent to erode the interface with the sea water below. The sea water thus entrained into the brackish layer is assumed to be instantly mixed vertically so that the upper layer remains vertically homogeneous.

Continuity is maintained by a return flow in the sea water layer. As the mixing is assumed to be unidirectional, the lower layer is unaffected by the brackish flow above and remains completely homogeneous. The lower layer is furthermore assumed to be infinitely deep so that the barotropic flow becomes infinitesimal. Velocities and accelerations in the lower layer are accordingly neglected.

The dynamical equations derive from vertical integration of the momentum equations for each layer. In addition to the resistance encountered in the entrainment, interfacial boundary friction is expected to be of some importance and is retained in the model. Shear stress from wind action at the surface is not a specific feature of the model, although wind is considered of prime importance in maintaining the turbulence of the upper layer.

Tides and effects from rotation of the earth are neglected.

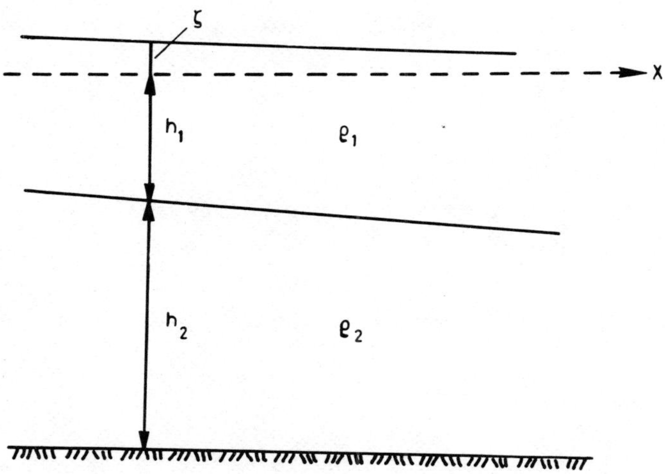

Fig. 1. Schematic diagram of model conditions and notation employed. In the model the lower layer is assumed to be infinitely deep.

With reference to the notation in Fig. 1 the vertically and laterally integrated momentum equations for stationary flow are

$$\frac{\gamma}{W}\frac{\partial u^2 hW}{\partial x} = -\frac{gh}{\rho}\left(\rho\frac{\partial \zeta}{\partial x} + \frac{h}{2}\frac{\partial \rho}{\partial x}\right) - C_2 u^2 \tag{1}$$

$$h\frac{\partial \rho}{\partial x} + \rho\frac{\partial \zeta}{\partial x} = \Delta\rho\frac{\partial h}{\partial x}, \qquad \Delta\rho = \rho_2 - \rho_1 \tag{2}$$

where u is the mean velocity of the upper layer and W the width of the channel. Whenever possible, indexes referring to the upper layer have been left out. Here, a coefficient γ appears as a result of replacing the vertical average energy by the kinetic energy of the mean flow. The quantity γ depends upon the distribution of velocity in the layer. According to Long the value of γ lies in the range of 1 to 2. In the subsequent computations the value 1.3 has been used consistently.

Similarly, the vertically integrated continuity equation becomes

$$Wuh \cdot \frac{\partial \rho}{\partial x} \doteq \Delta\rho \frac{\partial uhW}{\partial x} \tag{3}$$

Furthermore, an entrainment relation of the form

$$w_e = \frac{1}{W}\frac{\partial uhW}{\partial x} = \frac{k \cdot \sigma_n^3}{hg^{\Delta\rho}/\rho} \tag{4}$$

is used, where σ_n is the rms turbulent velocity of the upper layer, and k a constant. The problem is thus described by the four partial differential equations above.

The slope of the free surface can easily be eliminated from Eqn.(1) by means of Eqn. (2), leading to the following form

$$\frac{\gamma}{W}\frac{\partial u^2 hW}{\partial x} = -gh\left(\frac{\Delta\rho}{\rho}\frac{\partial h}{\partial x} - \frac{h}{2\rho}\frac{\partial \rho}{\partial x}\right) - C_2 u^2 \tag{5}$$

Eqn. (3), (4) and (5) constitute the reduced set of equations we shall use.

PROPERTIES OF THE SOLUTIONS

As Long (1975) has shown Eqns. (3), (4) and (5) can be integrat[ed] by a suitable change of variables. It is, however, tempting to derive some of the properties of the solutions directly from the basic equations.

Consider first the case of a fjord with constant width and with no shear stress being transmitted to the lower layer. Near the head of the fjord the velocities are often quite small, making it possible to neglect the left hand side of Eqn. (5). In this case the equation can be integrated directly to yield the fundamental relationship

$$h^2 \cdot \Delta\rho = C = \text{constant}, \tag{6}$$

governing the initial, or inner form of the solutions. Thus, as the density difference gradually diminishes as the brackish water becomes more salty, the thickness of the brackish layer grows accordingly.

The corresponding form of the interfacial Froude number

$$F_-^2 = \frac{\gamma u^2}{gh^{\Delta\rho}/\rho} = \frac{\gamma u^2 h}{g \cdot C} \qquad (7)$$

indicates that at least initially the Froude number will grow as u^2h. If, however, at some location in the fjord $dh/dx=0$ Eqn. (5) will reduce to

$$\gamma 2u\frac{\partial uh}{\partial x} = \frac{gh^2}{2\rho}\frac{\partial \rho}{\partial x}$$

which together with the equation of continuity

$$\Delta\rho \frac{\partial uh}{\partial x} = uh \frac{\partial \rho}{\partial x}$$

permit the elimination of d uh/dx, leaving

$$(\frac{gh^2\Delta\rho}{2\rho} - \gamma 2u^2h)\frac{\partial \rho}{\partial x} = 0$$

which is equivalent to the condition that the interfacial Froude number be 1/4.

It is consistent with this result that for values of F^2 smaller than 1/4, dh/dx is positive, and for values of F^2 larger than 1/4, dh/dx becomes negative. Actual computations of the complete solutions show that for a given inner boundary condition the interfacial Froude number will grow monotonically until the value 1/4 is reached after which it very rapidly will reach the value 1. At this point the flow becomes critical.

In his discussion of the solutions Long assumed the critical flow condition to be associated with the opening of the fjord towards the sea. In this way the sudden widening at the mouth would serve as a hydraulic control. By applying this condition the solutions would be completely determined throughout the fjord.

It is not always possible to identify such a sudden widening of a fjord as it opens up towards the sea. Neither is it always possible to recognize two well defined layers throughout the entire length of the fjord. Most often the transition between the layers widens toward the sea, gradually eroding into the lower layer and also eventually extending all the way to the surface. In this case the concept of a two-layer system may well be applicable to the inner parts of the fjord, but will fail in the outer reaches. For this reason it may be more rewarding to disregard the outer boundary condition and rather develope the solutions from the inner end.

Long found analytical solutions to the problem for the case of uniform width. In his analysis it is shown that these solutions depend upon two non-dimensional parameters m and s, both involving the rms turbulent velocity. It is possible to show that although inplicitly, the solutions are more readily discussed in terms of two other related parameters, F_o and C_2/C_1, where the former stands for the interfacial Froude number at the head of the fjord, and the latter being the ratio of the drag coefficient (C_2) to the relative entrainment velocity $C_1=2w_e/u_o$ at the head of the fjord.

The solutions do also depend upon the value of the entrainment parameter C_1, but in such a way that variation of this parameter causes the solutions to contract or expand along the x-axis. Thus, if C_1x is used as horizontal axis, all solutions can be expressed in terms of the two parameters F_o and C_2/C_1.

The entrainment parameter C_1 does not enter at all into relationships such as that between the thickness and the density of the upper layer. For the case of constant width the solutions are particularly simple. By a suitable substitution Long's Eqn.(21) (Long 1975) transforms to

$$\frac{h}{h_0} = \left(\frac{h}{h_0}\frac{\Delta\rho}{\Delta\rho_0}\right)^3 \left(\frac{2F_0^2+s-1}{2F_0^2+(s-1)\left(\frac{h}{h_0}\frac{\Delta\rho}{\Delta\rho_0}\right)^3}\right)^{\frac{s+1}{2}} \tag{8}$$

where $s=1+(1+2C_2/C_1)^{-1}$. The general behaviour of these solutions is shown in Figs. 2 and 3. In Fig. 2 the density versus thickness relati ship for the case of no boundary friction ($C_2=0$. $s=2$) is shown for selected Froude numbers at the head of the fjord (F_o), all within the range of naturally occurring values. The results are drawn in a double logarithmic diagram which is particularly suited for comparison with observed features in fjords.

The diagram exhibits two striking features. Firstly, it is seen that all lines depart from origin with the slope $-\frac{1}{2}$ in accordanc with the result of Eqn. 6. The other feature is that the thickness reaches a maximum at some point beyond which the thickness of the layer drops drastically. There is also a tendency for the lines to curve backward implying loss of density. This cannot be attained by the physical mechanisms involved so that the mathematical solution cannot be carried behond the point of maximum density which correspon to the point where the interfacial Froude number reaches one.

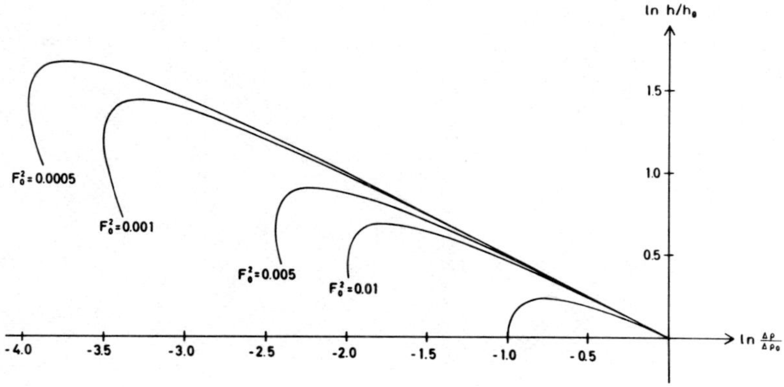

Fig. 2. Logarithmic presentation of thickness of upper layer relative to head conditions versus relative density difference between the layers. Integral curves for the constant width, non-frictional case for indicated initial (inner end) interfacial Froude number squared.

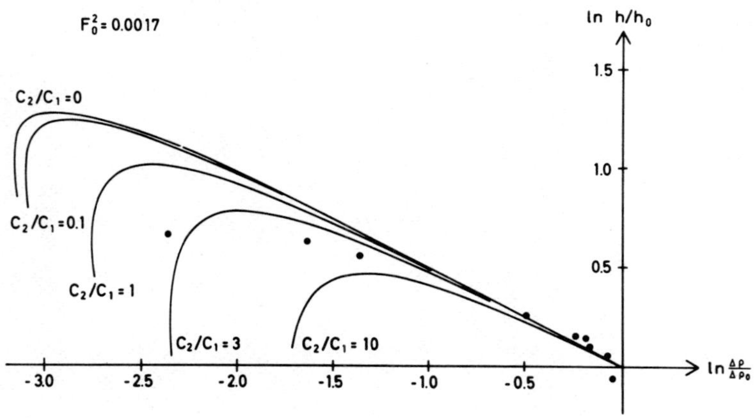

Fig. 3. Logarithmic presentation of thickness of upper layer relative to head conditions versus relative density difference between the layers. Constant width integral curves for indicated interfacial Froude number squared and drag ratios.

The influence of boundary friction on the thickness versus density relationship is illustrated in Fig. 3. For this purpose an initial Froude number is chosen to be typical of one of the major Norwegian west coast fjords, the Sognefjord, where F_0^2 was estimated at .0017. The family of curves corresponds to the various vàlues of the drag ratio C_2/C_1, the uppermost curve representing the frictionless case.

It follows from Fig. 3 that interfacial boundary friction tends to limit the variation of both thickness and density growth of the upper layer. For extreme large values of boundary friction the upper layer may not grow at all, but reduce steadily.

NUMERICAL INTEGRATION

The actual form of the solutions throughout the fjord is specified by a couple of partial first order equations which cannot readily be integrated except by numerical methods. This renders the discussion somewhat cumbersome and one is obliged to go through a series of cases with selected parameter values.

The complete equations are given by Long (1975, Eqns. (35) and (36)). In concordance with the notation used here these equations can be written

$$\frac{\partial h}{\partial C_1 x}=\frac{1}{1-F^2}\left\{\frac{W}{W_0}\left(\frac{1}{4}-\left(1+\frac{C_2}{C_1}\frac{W_0}{W}\right)F^2\right)+\frac{hF^2}{W}\frac{\partial W}{\partial C_1 x}\right\} \tag{9}$$

$$\frac{\partial F^2}{\partial C_1 x}=\frac{3F^2}{2(1-F^2)h}\left\{\frac{1}{2}+\left(1+2\frac{C_2}{C_1}\frac{W_0}{W}\right)F^2-\frac{F^2(F^2+2)}{(1-F^2)W}\frac{\partial W}{\partial C_1 x}\right\} \tag{10}$$

where W is the width of the fjord, W_o being the value at the head.

For the case of constant width C_1x can be eliminated from Eqns. (9) and (10) making it possible to integrate the set directly to yield

$$h=K\cdot\frac{F^{2^{\frac{1}{3}}}}{\left(\frac{1}{2}+1+2\frac{C_2}{C_1}\right)F^2)^{\alpha}} \tag{11}$$

where

$$\alpha=\frac{1+\frac{4}{3}\frac{C_2}{C_1}}{1+2\frac{C_2}{C_1}}$$

and K is a constant of integration.

Eqns. (9) and (10) have been integrated by numerical methods for a range of parameter values. Beginning from the inner end the thickness, density and mean velocity of the upper layer are determined as functions of location in the fjord, i.e. the horizontal coordinate. The solutions are developed for selected values of initial Froude number, inner end brackish water thickness, drag ratio C_2/C_1 and the rate of change of width of the fjord.

Corresponding to a reference value of the inner end Froude number squared $F^2 = .002$ and a drag ratio $C_2/C_1=10$, the solutions for constant width are presented in Figs. 4a, b and c for indicated inner end thicknesses. As also indicated by Eqn. (9) the family of curves shown in Fig. 4a all begin with the slope $(4\gamma)^{-1}$.

Similarly, for the case of given inner end thickness of the brackish layer $h_0=4$m, the solutions for indicated range of F^2_0 are shown in Figs. 5a, b and c, the other parameters being the same.

In Figs. 6a, b and c is shown the influence of varying the drag ratio C_2/C_1. The other parameters are the same as in the previous computations.

Finally, the case of a fjord of varying width is illustrated in Figs. 7a, b and c. Solutions are shown for selected values of the constant gradient of width $b=dW/dC_1x$.

APPLICATION TO OBSERVATIONS

Reliable observations from fjords suitable for testing the model are rather scarce. As a general rule, the fjord should have a "nice" geometry, the width being near constant or changing only gradually with distance, and the longitudinal axis free of strong curvatures.

Immediately prior to and at the time of observation the winds in the area should be weakto allow for a dynamic balance free of surface stresses. Furthermore, a long term near steady state condition should prevail.

A single set of data is available from the Sognefjord, one of the larger Norwegian west coast fjords. The main fjord, some 200 km long, is fairly uniform in width (a= .02, b=1.6x10^4), but has a number of minor branches. Almost all of the fresh water discharge into the fjord occurs in the inner reaches of the system, outside which the fjord can be treated as a single channel.

Corresponding to an estimated value of $F^2_0 = .0017$ the observed average densities and layer thicknesses from 20-21 June 1971 are plotted on Fig. 3. Although the fit is not particularly good, the general form of the solutions is acceptable with the best fit being obtained with a drag ratio near 3.

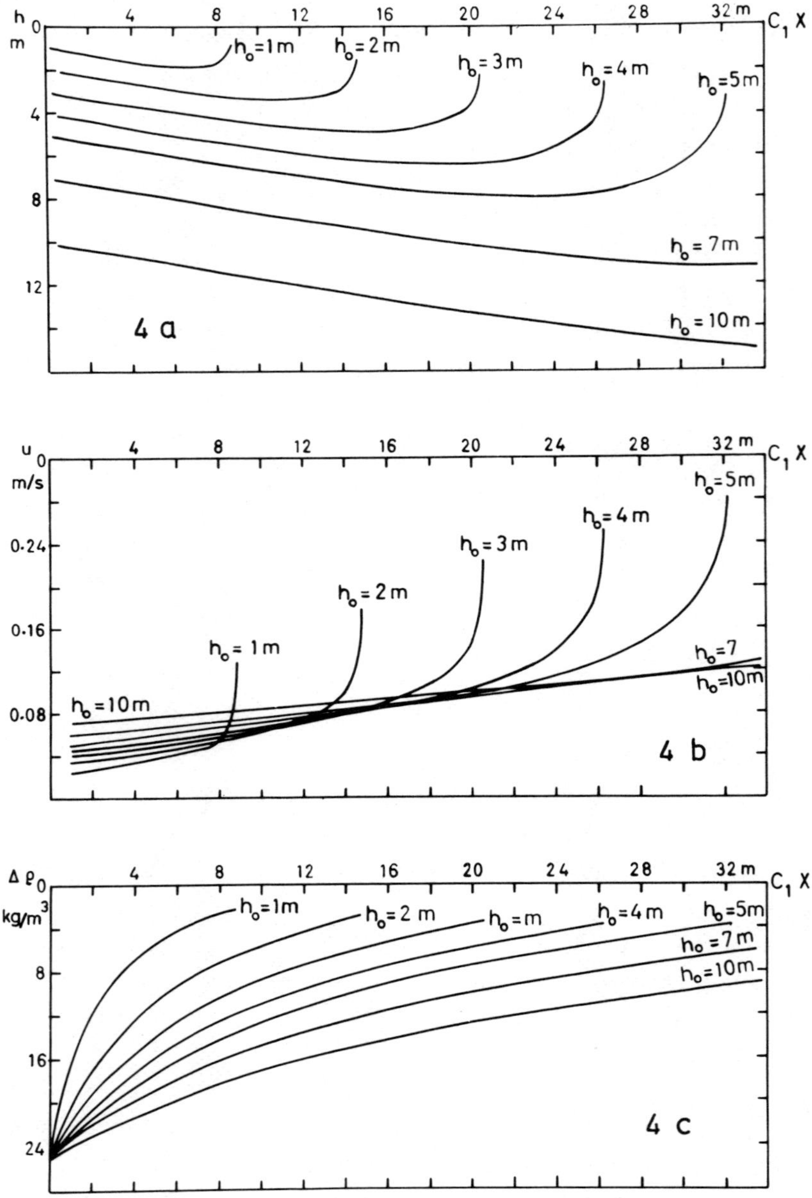

Fig. 4. Constant width integral curves for $C_2/C_1=10$ and $F_0^2=$ .0C2. Solutions of thickness (Fig. 4a), mean velocity (Fig. 4b), and density (Fig. 4c) versus longitudinal distance.

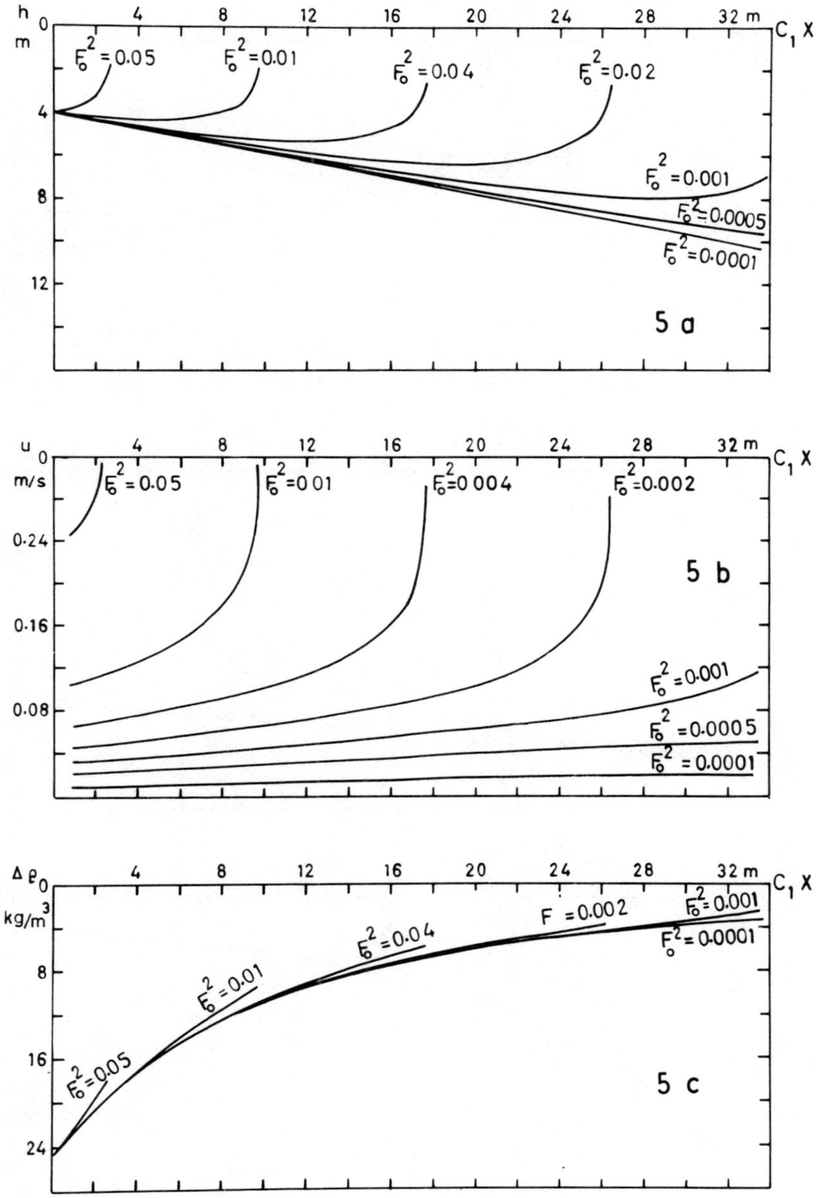

Fig. 5. Constant width integral curves for $C_2/C_1=10$ and $h_0=4$m. Solutions of thickness (Fig. 5a), mean velocity (Fig. 5b), and density (Fig. 5c) versus longitudinal distance for indicated inner end interfacial Froude number squared.

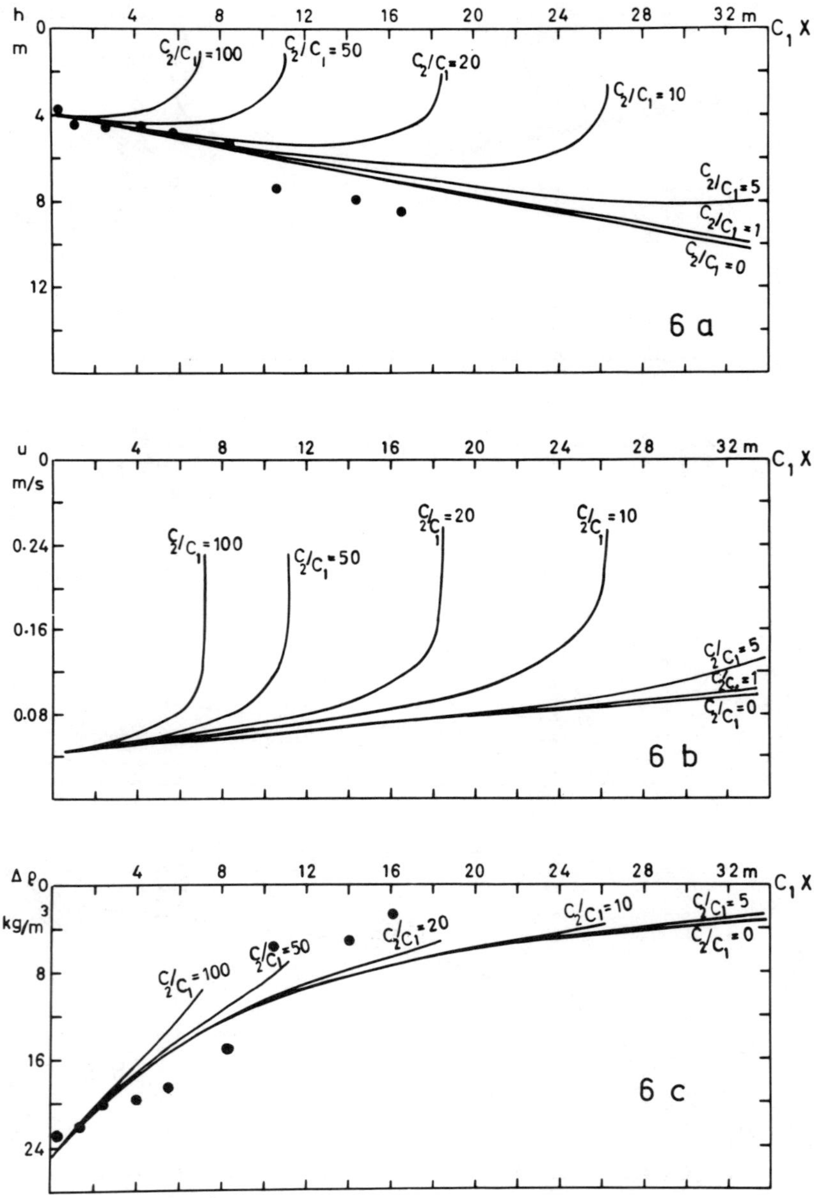

Fig. 6. Constant width integral curves for h_0=4m and F_0^2= .002. Solutic of thickness (Fig. 6a), mean velocity (Fig. 6b), and density (Fig. 6c) versus longitudinal distance for indicated drag ratios.

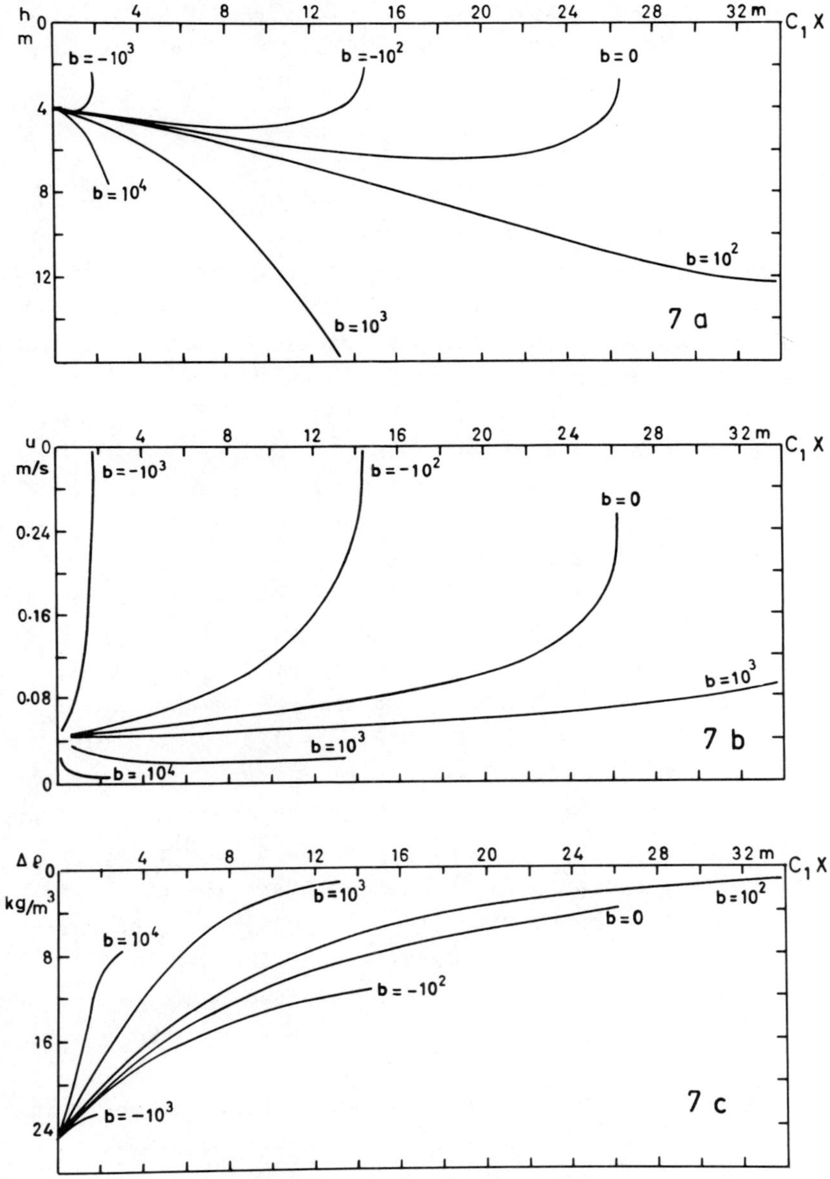

Fig. 7. Effect of a constant width gradient b on the variation of thickness (Fig. 7a), mean velocity (Fig. 7b), and density (Fig. 7c) versus longitudinal distance. Integral curves for $C_2/C_1=10$, $h_0=4$m, and $F_0^2 = .002$.

Similarly, with the observations plotted in the thickness distance relationship in Fig. 6a, it is possible to obtain a reasonable fit with an entrainment parameter $C_1=8x10^{-5}$, again indicating a drag ratio in the range of 1-10, the higher values pertaining to the middle and inner parts of the fjord.

DISCUSSION

At the outset of this study it was not obvious to the author that interfacial boundary friction was essential to the problem. In his discussion Long (1975) had suggested a drag coefficient of a magnitude of the order of one, a value which would lead to an overwhelming influence of the boundary friction. Such a high value would also be inconsistent with that no account was taken of effects of shear stresses being transmitted to the lower layer (Bo Pedersen, personal communication).

The data from Sognefjord presented in this paper indicate an entrainment parameter of the order of magnitude of 10^{-4}, whereas the interfacial boundary friction appears to be determined by a drag coefficient of the order 10^{-3}. These values may, however, be marginal and thus not typical for the majority of fjords.

Although quite simple to obtain, and usually with a fair degree of accuracy, measurements of upper layer thicknesses are nevertheless poorly suited for testing the validity of a circulation model. As typical for Norwegian west coast fjords it appears that the entrainment of sea water into the brackish layer depends largely on the action of wind. For this reason it would seem rather meaningless to seek out periods of calm for carrying out observational programs. On the other hand, winds tend to converge or diverge the water masses, thereby seriously affecting the depth of the pycnocline. The conflict can possibly be avoided by focussing the attention on the change of density of the upper layer along the fjord.

An effort in this direction is shown in Fig. 6c, where the observed densities from the Sognefjord are plotted against the longitudinal axis C_1x. Here, C_1 is maintained $8x10^{-5}$, a value which appears to be too high in the inner reaches, and too small further out in the fjord. This is actually no surprise, as the inner reaches of the fjord are characterized by generally much ligther winds than further out. The same argument holds for tidal and other externally induced disturbances. The effects of the latter in this respect are, however, not ascertained. The application in Fig. 6c gives no indication of the value of the C_2/C_1 ratio.

REFERENCES

Kato, H. and Phillips, O.M., 1969. On the penetration of a turbulent layer into a stratified fluid. J. Fluid Mech., 37:643-655.

Long, R.R., 1975. Circulations and density distributions in a deep, strongly stratified, two-layer estuary. J. Fluid Mech., 71: 529-540.

Moore, M.J. and Long, R.R., 1971. An experimental investigation of turbulent stratified shearing flow. J. Fluid Mech., 49:635-655.

Stigebrandt, A., 1976. Stasjonær tvålagerstrømning i estuarier. (In Swedish). River and Harbour Laboratory, Trondheim.

Stommel, H., 1951. Recent developments in the study of tidal estuaries. WHOI Rep. No. 51-33.

Stommel, H. and Farmer, H.G., 1952. On the nature of estuarine circulation. WHOI Rep. No. 52-88.

Stommel, H. and Farmer, H.G., 1952. Abrupt change in width in two-layer open flow. J. Mar. Res., 11(2):205-214.

Stommel, H. and Farmer, H.G., 1953. Control of salinity in an estuary by a transition. J.Mar.Res., 12(1):13-20.

WIND-DRIVEN CIRCULATION IN A FJORD

H. Svendsen

Geophysical Institute, University of Bergen (Norway)

and

R.O.R.Y. Thompson

CSIRO Divison of Atmospheric Physics, Aspendale (Australia)

ABSTRACT

Currents, temperature, salinity, wind, runoff and water level were observed for a month in the Jøsenfjord of Southern Norway. Tide gauges and currents show little semi-diurnal tide. There is a strong diurnal signal in the upper 20 m, which a linear model shows to be caused by the wind. There is a week-long event in which the entire water-mass above the sill is flushed out; this is interpreted to be caused by down-welling outside the fjord. The strong stratification near the surface of the fjord greatly modifies the diurnal response of the fjord, but any density-driven mean circulation is at least an order of magnitude smaller than the wind-driven currents.

(Manuscript submitted to Journal of Physical Oceanography, October 1977.)

SUMMER REPLACEMENT OF DEEP WATER IN BYFJORD, WESTERN NORWAY: MASS EXCHANGE ACROSS THE SILL INDUCED BY COASTAL UPWELLING.

Hans B. Helle
Geophysical Institute, University of Bergen (Norway)

ABSTRACT

From previous published data on replacement of deep water in Byfjord (Linde, 1970) and obvious relation is found to exist between northerly winds along the west coast of Norway during summer and the volume of deep water being replaced. Northerly summer winds induce upwelling of denser water to shallower depths, which then can spill over the sill and intrude into the fjord basin. The magnitude of replacement appears to be independent of variations in freshwater discharge and climate.

Recent observations of current and water properties reveal that the exchange flow across the sill is basicly two-layered: a landward flow in the lower and a seaward flow in the upper layer, with the level of no-net motion located at mid-depth. Short period (2-7 days) variations in the long-shore wind component are well reflected in the exchange flow. Wind speed extremes are followed one to two days later by corresponding extremes in the exchange flow, indicating the approximate response time for the upwelling process. In periods of sustained northerly winds along the coast the inflow attains a saturation speed of about 30 cm/s, which is slightly less than the critical speed obtained from the mean stratification by assuming an ideal two -layer flow across the sill.

INTRODUCTION

The fjord under consideration is a larger estuarine system contiguous to the town of Bergen in the western Norway (Fig. 1). Major subdivisions are Byfjord, Herdlafjord, Osterfjord and Sørfjord with basin depths ranging from 350 m in Byfjord to a maximum of 650 m in Osterfjord. The interconnections between the different basins are quite open, except for the inner part of Sørfjord where a 400 m deep basin is secluded behind a 160 m deep sill.

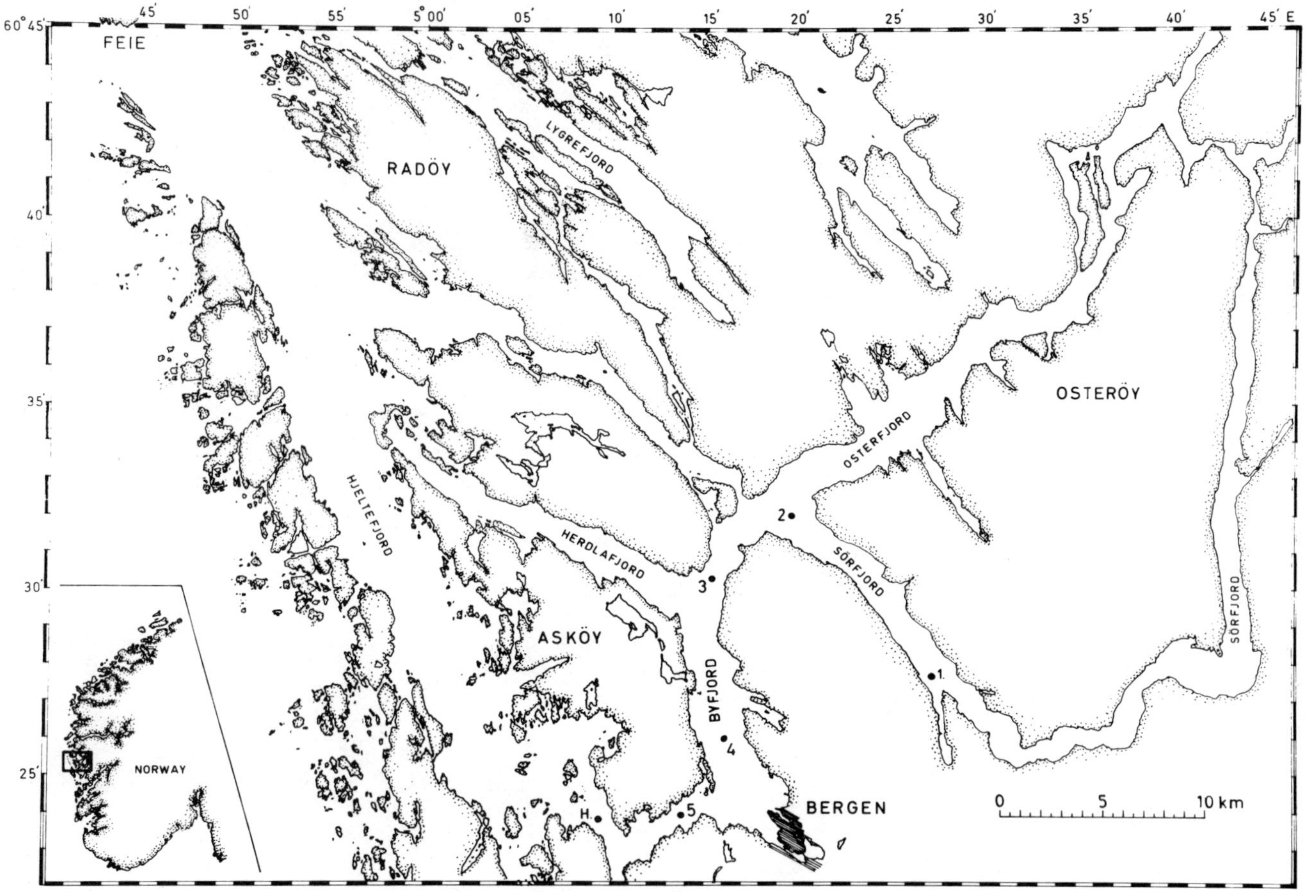

Figure 1 . Map of the area with locations of current meter station (H) and hydrographic stations (1 - 5)

The estuarine system, denoted by Byfjord for abbrevation in the following text, is connected to the adjacent coastal water through a complex network of narrow sounds formed by an archipelago to the west and the north. The sounds are all constricted by sills. The deepest passage is through Hjeltesund at the south-western entrance to the estuarine system through Hjeltefjord. This narrow passage coincides with the location of a 170 m deep sill. No other sound provides a free passage deeper than 40 m. Hjeltefjord is relative wide, deep (300-500 m) and openly connected to the coastal water at its northmost end where a 300-350 m deep submarine canyon provides a free connection with the deepest portions of the Norwegian Trench.

Despite its rather complex topography Byfjord is considered to be a typical Norwegian sill or treshold fjord. The volume of the fjord is 47 km^3 and its surface covers an area of 55 km^2. The drainage basin for the fjord is an area of 3700 km^2 with extraordinarily high precipitation. Annual fresh-water run-off amounts to about 10 km^3, which is mainly discharged during the summer months by rivers located in the innermost parts of the fjord. Mean tidal range in Bergen is approximately 0.9 m, predominately of semi-diurnal (M_2) type.

Owing to it proximity to oceanographers from Bergen (the Bergen Museum and later the University of Bergen) the fjord has been subject to investigations, for various purposes, since the end of last century. There is no doubt the Byfjord is the place where the first "Nansen cast" was taken. Since that the general hydrographic features of Byfjord have been well known, mainly from the long-term serial observations of water properties as treated by Helland-Hansen (1935) and by Linde (1970). The first paper deals with the general hydrography of Byfjord, with special attention to seasonal and long-term variations of the water temperatures related to climatic changes. Linde's report deals with the deep water exchanges, and using a specially developed method he was able to determine the amount of deep water being replaced over a long series of years. Both papers reveal that the bottom and deep water masses of the fjord are almost completely replaced during an extended period every year, with very few exceptions, some time in the period March-August.

The replacement events are characterized by increasing oxygen concentration in the deep water, usually followed by rising salinity and falling temperature. Hence the new water, which originates from water masses occurring outside the fjord mouth, has a greater density than the resident water of the fjord. The inflow itself thus constitutes a gravity flow following the appearance of denser water outside the mouth, at sill level or above, than that existing within the fjord basin. As pointed out by Gade (1973) and by Sælen (1976), the principal mechanisms governing these exchanges in a sill fjord are linked with the sustained density reduction of the resident water thriugh vertical eddy diffusion and the natural occurring density variations in the adjacent coastal water. Along the west coast of Norway these density variations are clearly linked with the seasonal oscillation of the Norwegian Coastal Current assumably due to the seasonal changes in wind pattern, fresh water discharge from land and possibly influenced by climatic conditions. Which one of these is the governing factor, however, has been subjected to some discussion in the litterature but their relative importance has previously not been properly documented. The main goal of this paper, by using Byfjord as an example of a typical west coast fjord, is to demonstrate the relative importance of the different factors to the deep water exchange, and hence to the seasonal density variation in the near-shore coastal water.

Because deep water replacements previously only have been documented at monthly and bimonthly intervals with water properties, little knowledge exists on the inflow process itself. Particular questions being raised are whether the inflow occurs as a slow and steady flow or as a sudden discharge. In most cases for which data is available, the deep water renewal of the fjord basin is known to be a relatively rapid event, which often is completed within the course of a few weeks. This relatively short time scale of the process, the large quantities of water involved, as apparent from Linde's results, and the small density difference between the new and original basin water points to the possibility that the exchange flow may be subjected to hydraulic control at the sill. Some unique observations of deep water exhange made in Byfjord in 1973 and 1974 will be presented. The result shed some light upon some principal questions related to the exchange process and improve our understanding of the oceanography of the Byfjord and other fjords with similar characteristics.

MEAN SEASONAL VARIATIONS IN THE COASTAL AREA

The normal mean seasonal changes in the wind pattern along the west coast are displayed in the vector plot Figure 2 showing mean monthly wind vectors at three different locations along the coast. The meteorological stations (lighthouses) are situated on islands some distance from the mainland; Utsira and Kinn to the south and the north respectively and well within the edges of the straight north-south running part of the coastline. The meteorological station Hellesøy lies approximately midway between Utsira and Kinn and just outside the fjord area under consideration. The wind vectors are constructed on the base of the standard normals 1931-1960 of monthly wind frequencies and mean wind force (Johannessen and Håland, 1969).

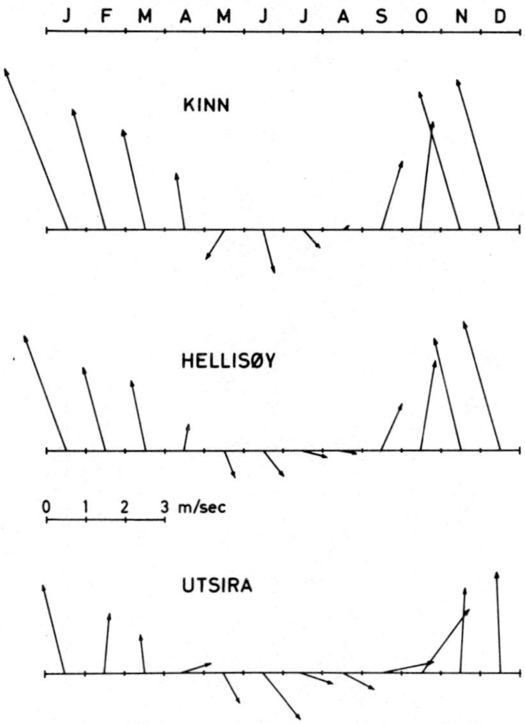

Fig. 2. 30 year mean monthly wind vectors of three different coastal stations. (1931 -60).

Generally, the wind along the west coast is dominated by winds from south in the winter and from north during the summer months. Mean strength and frequency are higher in the winter and the extreme wind situations usually occur in the period December-February. The Norwegian highland has a tendency to influence the wind in a 30-40 km wide belt out from the coast, and the dominant wind directions are parallel to the coast. There is a marked long-shore gradient in the mean wind field, and also a cross-shore gradient due to friction from the rugged landscape.

The water masses occuring outside the coast are characterized by an upper layer of low-salinity water, forming the Norwegian Coastal Current with a mean northward flow, flowing upon a heavier water of Atlantic origin. Usually the 35 o/oo isohaline is taken as the boundary between coastal and oceanic water, but as it will appear from the 35 year mean isopleths in figure 3 from a near-by coastal station the 34.5 o/oo isohaline coincides with the base of the main pycnocline and hence has more physical significance as far as the near-shore zone is concerned. This isohaline follows an almost harmonic varying path, from approximately 100 m depth at the turn of the year to approximately 50 m depth in June through July. The variation of the pycnocline depth quite closely follows the normal mean long-shore wind component.

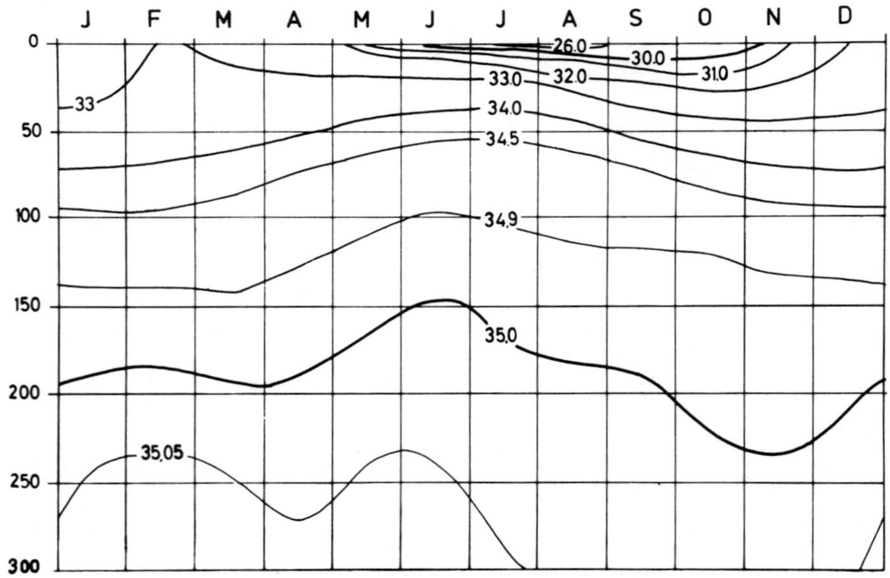

Fig. 3. 35 year mean salinity isopleths from the coastal station Sognesjøen (1935-70).

The typical changes in the distribution of water masses outside the west coast may be best illustrated by the two consecutive salinity sections shown in Figure 10 taken between Feie Island (see Figure 1) and Shetland Islands (U.K.). The first section, obtained during the first days of March 1974, displays a typical winter situation with the highly saline and vertically well-mixed Atlantic surface water occurring on the shelf and the shelf-slope. There is clear evidence for convection of the shelf water down into the deeper portions of the Norwegian Trench, a phenomenon which is known to take place in the some deeper part of the Trench in Skagerrak during cold winters, as described by Ljøen and Svansson (1972), but this more likely occurs every winter in the shallower part of the Trench off the west coast. This convection is of considerable importance to the fjords because it assures high oxygen concentration in the water available for inflow at spring and summer time.

In the winter situation shown, the coastal water itself is confined to a narrow strip of width 60-70 km at the coast and reaching its greatest depth of about 180 m some distance from the coast. The next section was taken about two months later following a period og 3-4 weeks with sustained northerly winds (see Figure 11). The distribution of water masses has entiraly changed; a thin surface layer (20-30 m thick) of coastal water is displaced westward to about 180 km from the coast and upwelling is clearly seen to occur in a 20-25 km wide strip at the coast. The changes apparent from the two consecutive sections are both qualitatively and quantitatively well in agreement with the classical theories of coastal upwelling (see e.g. Yoshida, 1955). The marked dilution of the near-shore coastal water is not necessarily related to increased river run-off, which at the end of April still was relative weak in the western regions of Norway, but is merely an effect of the deep water inflow to the west coast fjords during this period. Even in winter there is a considerable fresh water excess stored in the upper layer fjord water and as a consequence to deep water inflow this low salinity upper fjord water will be poured out into the coastal water.

A TIME SERIES OF DEEP WATER REPLACEMENTS.

The deep water is exchanged or replaced by more dense water which spills or intrudes over the sill. If the intruding water spreads out along the basin bottom, filling the deeper portions of the basin and gently elevating the original water, the amount of water being replaced can be cetermined from the observed elevation of the original water. A convenient method for making such estimates was developed by Linde (1970). He applied his method on the deep water replacements in Byfjord and quantified altogether 18 individual replacements in the two periods of investigation 1927-39 and 1947-56. The last 10 year period constitutes the best dataset from which an unbroken series of replacement estimates can be constructed.

His results are at first utilized for simple statistical treatment as shown in Table 1 and in Figure 4

TABLE 1

	Replacement (km^3)	Initial σ_t	Final σ_t	Incremental $\Delta\sigma_t$
Mean	13.2	27.277	27.337	0.060
St.dev.	3.9	0.019	0.026	0.026
Min	0.0	27.253	—	—
Max	21.0	27.309	27.394	0.127

These statistics are based on all the 18 replacement events. The σ_t values refer to 300 m depth at a fixed station in Byfjord, but because of the high degree of uniformity of the deep water masses in the fjord the given σ_t values are fairly representative throughout the basin. (Although the upper 100 m of the fjord usually is strongly stratified, local density variations in the deep water masses below may be within the accuracy of common determinations). The initial σ_t refers to the situation prior to the inflow event, the final σ_t is the value after the replacement is completed and $\Delta\sigma_t$ is the increment.

In Table 1 is given the range of variation of the replaced volumes and the corresponding changes in the deep water density. Then the mean and the maximum replaced volume corrresponds to 28% and 45% respectively of the total fjord volume. As also apparent from the table is the relatively restricted range of variation of the deep water density. A remarkable point is the very small standard deviation of the initial σ_t - values, which is almost less than the expected standard deviation due to errors in routine determination of salinity by the titration method used.

In Fig. 4 are plotted the volumes replaced versus the incremental densities. The dotted line is a second order least square fit to the data, constrained to pass through the origin. In view of the fact that the basic hydrographic data are collected over a long period of time and the relatively simple method used to make replacement estimates, there is a surprisingly small scatter around the fitted line. This result leaves a feeling of confidence to Linde's and clearly proves the existance of a close relationship between the volume being replaced and the corresponding density change. The least square fit may in fact be utilized with reasonable accuracy to determine the replaced volum directly from density records.

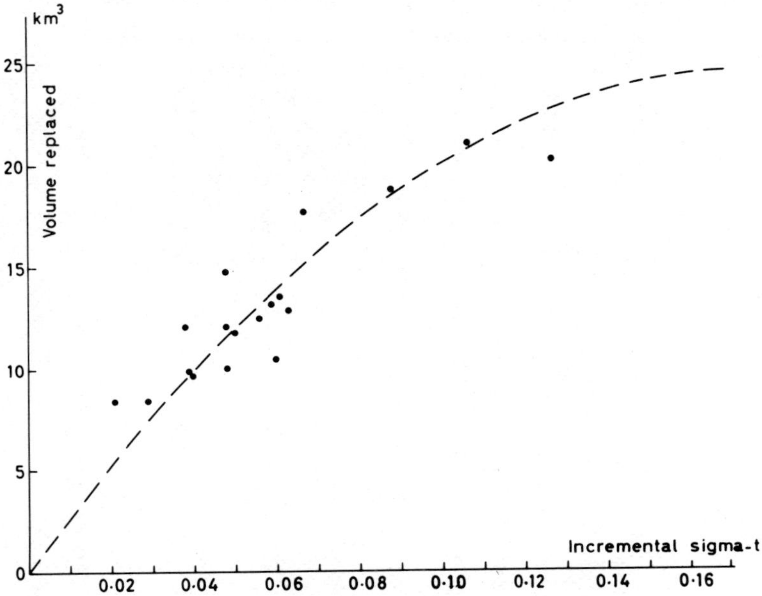

Fig. 4. Replacement - density diagram based on 18 individual replacement events in 1927-39 and 1947-56 according to Linde (1970)

From the statistical treatment it appears that the deep water replacement is a function of the final density rather than of the initial density, and that the initial density attains practically the same contant value prior to each inflow event. This means that the density reduction following a large replacement with high density water must be more rapid than that following a smaller replacement with usually lower density water. This result indicates that the energy which is released, when the above estimated quantities of water change level, is an important contribution to the total energy available for internal mixing and diffusion processes in the sheltered basin.

The assumptions of a slow inflow along the basin bottom and a gently elevation of the original water are not realistic in the general case. Some mixing between new and original water will always occur, particularly at the sill by shear instability of the exchange flow and by stirring induced in the basin. As the inflow proceeds fractions of the new water will escape with the outflowing admixture. When quantifying deep water exchanges it is therefore of some importance to distinguish between the amount which in fact spills over the sill and the replaced volume. From the two basic methods available; Linde's indirect method which determine net replacement and the direct measurement of the exchange flow from which the total input can be determined, the two quantities obtained will probably differ by an appreciable amount and a factor of 2 to 3 seems to be a realistic magnitude (Helle, 1975). Thus, although Linde's estimates represent reasonable values for the replacements, they clearly are underestimates of the total amounts of water which are involved in these exchanges.

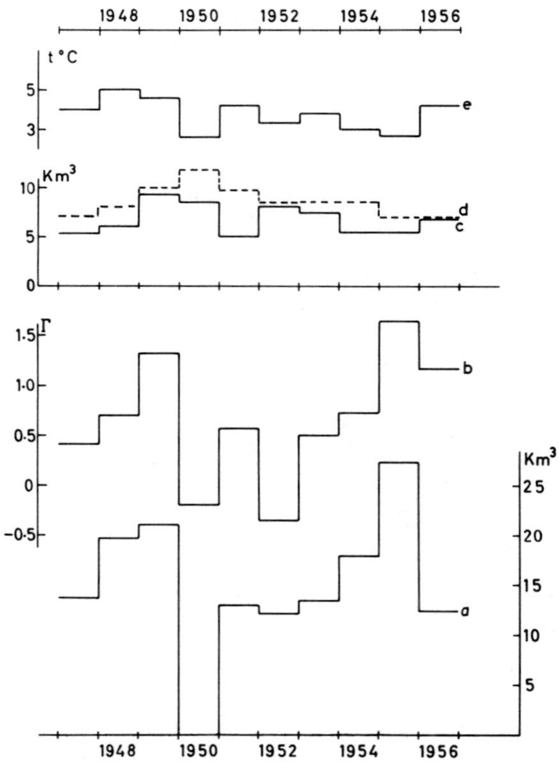

Fig. 5. 10 year series of deep water replacement (a), Northerly wind-flux anomaly (b), local (c) and regional (d: x 0.1) run-off and mean winter air-temperature (e).

The next step in utilizing Linde's result is to contruct a time series of the total annual replacements (which all are confined to the period March-August) which enables us to compare these variations with time series of the different quantities assumed to effect the replacement process. Such an intercomparison is set up in Fig. 5 where the different time series, all covering the 10 year period 1947-56 corresponding to the most comprehensive data set used by Linde, are as foolows:

a) The annual replacement in Byfjord

b) The net northerly wind-flux anomaly ysing Utsira as reference station. The anomaly Γ is defined by:

$$\Gamma = \frac{V - V_{SN}}{V_{SN}}$$

where V is the weighted sum of the monthly mean long-shore wind component given the weight one if southward directed and the weight zero if northward directed, and summed over the six months March-August. The subscript SN refers to the standard normal 1931-60.

c) The summer run-off (March-August) to Byfjord based on discharge records from the main rivers draining to the fjord.

d) The summer run-off (March-August) from the southern Norway is estimated by a procedure equivalent to that used by Tolland (1976) for the normal monthly run-off from the different regions of Norway. The estimates are based on all available discharge records from the rivers in the regions (region I and II according to Tollan).

e) The mean air-temperature in winter (November-April) at Utsira is assumed to reflect the average winter climate in the coastal region.

As stated above, the governing factor for the deep water exchange is the density of the water occurring outside the west coast at spring and summer time, which clearly is affected by the climatic conditions during the previous winter. Cold winters are usually associated with little winter run-off and cooling of the coastal water and the Atlantic surface water occurring on the North Sea shelf. Both the upper coastal water regime and the lower oceanic regime at the coast will hence be more dense than after a mild winter.

The vertical displacement of the isohalines at the coast of the magnitude indicated by the mean isopleth (Fig. 3), however, introduce local density variations which noramlly far exceed the effect from climatic variation. The absence of covariation between air-temperature and replacement in Fig. 5 is thus not unexpected, and it is reasonable to seek for covariations with the factors which are assumed to be major responsible for these vertical oscillations, namely the run-off from land and the coastal wind. A such covariation appears surprisingly clear to exist between replacement and wind-flux, may be with an exception for 1952 when an appreciable amount of deep water was replaced despite weak northerly wind-flux at the coast during the summer.

In the first three years (1947-49) there is also a similar tendency in the run-off diagram, but thereafter the replacement and run-off are completely uncorrelated. No replacement were detected at all in 1950 despite a very cold winter and that maximum run-off for this 10 year period was recorded.

The results of this intercomparison are very convincing and it seems to be no doubt left that the deep water replacements are more or less completely governed by the coastal wind.

The most surprising point, however, is that this process seems to be completely independent of the magnitude of fresh water discharge, both locally and reionally. Though longer time series would be needed in order to make more reliable conclusions about the effect of run-off, this point is of considerable interest in view of the extensive use of hydroelectric resources, with resulting changes in the natural run-off pattern. It appears at least, that although artificial run-off regualtion may cause considerable changes in the upper layer circulation pattern and local ice problems in winter, it will not alter the natural regularity and extent of deep water exchanges.

SOME OBSERVATIONS OF REPLACEMENT EVENTS

An observational program, which included current meassurements at the sill and hydrographic observation at monthly and sometimes bimonthly intervals at selected locations in Byfjord (see Fig. 1), was carried out during 1973 and 1974. The main purpose of this investigation was to determine the capability of the fjord to absorb sewage from the surrounding population centers.

A string of three Aanderaa current meters, equipped with temperature sensors, was moored at the sill (st.H) using a sub-surface float in 25 m depth for buoyancy. The current meters were positioned at depths of 30, 80 and 160 m and operating at a sampling interval of 10 minutes. Two typical replacement events were detected during this period, one in summer 1973 and the other early in spring 1974. Background hydrographic data from the adjacent coastal area were obtained from the Institute of Marine Reasearch in Bergen. Their ships quite regularly occupies stations along the section between Feie and Shetland and fortunately two sections are coincident wiht the 1974 inflow event. Wind data from Utsira lighthouse, sampled every third hour, are assumed representative for the coastal area. Water level records were taken from the tide gauge in Bergen Harbour.

The time series were treated as follows. Current velocities were resolved into components along the axes of maximum and minimum vairiance of total records. Wind velocities were resolved into components along and perpendicular to the coastline. All time series were low-pass filtered by utilizing a 25 hour running mean. The sea-level variation are given relative the mean of total records in each series. Run-off and salinity diagrams from Byfjord are shown for reference in Figure 6 and Figure 7.

The 1973 replacement. This season the replacement took place in two separate events as seen from the two successive peaks occurring in the salinity isopleths Fig. 7. The most pronounced event occurred in March through April when the new and slightly more saline water entered the basin. The original water was lifted to the levels above 100 m where it caused a marked salinity increase. This comparatively high salinity and low oxygen content water occurring in the upper layers gradually escaped from the fjord in the course of May. The inflow itself appears like an inclined plume in the hydrographic section taken along the fjord axis on April 28 (Fig. 8). The salinity in the plume core slightly exceed 35 ‰ at sill level at the entrance and gradually decreased towards the front nose as indicated by the 34.95 ‰ isohaline. Disolved oxygen reveals a similar picture with the highest concentration centred along the core and decreasing concentration towards the front. At this stage the new water had not yet penetrated down into the deepest trough as seen by the comparatively low oxygen consentration near the bottom. Oxygen minimums above the plume indicates the location of displayced original waters.

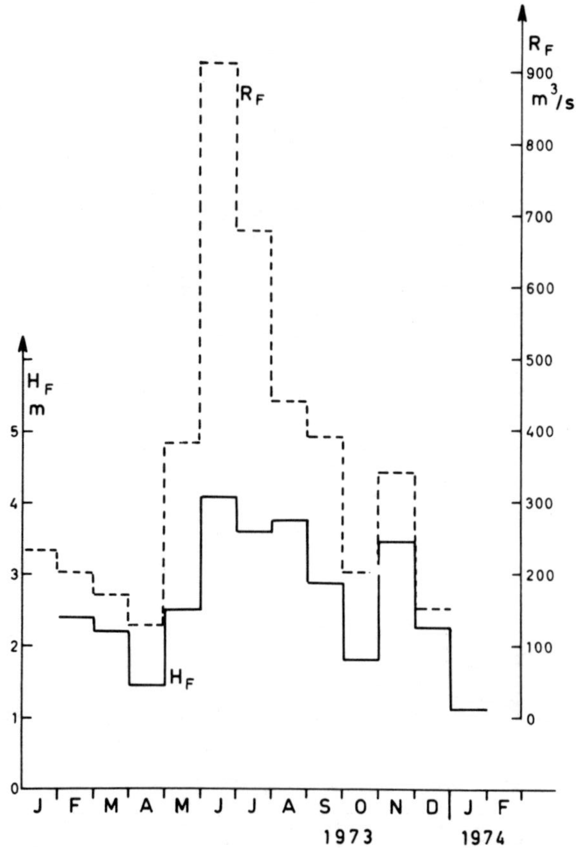

Figure 6. Monthly mean run-off rate R_F and equivalent thickness H_F of the fresh-water layer from monthly observations of salinty profiles.

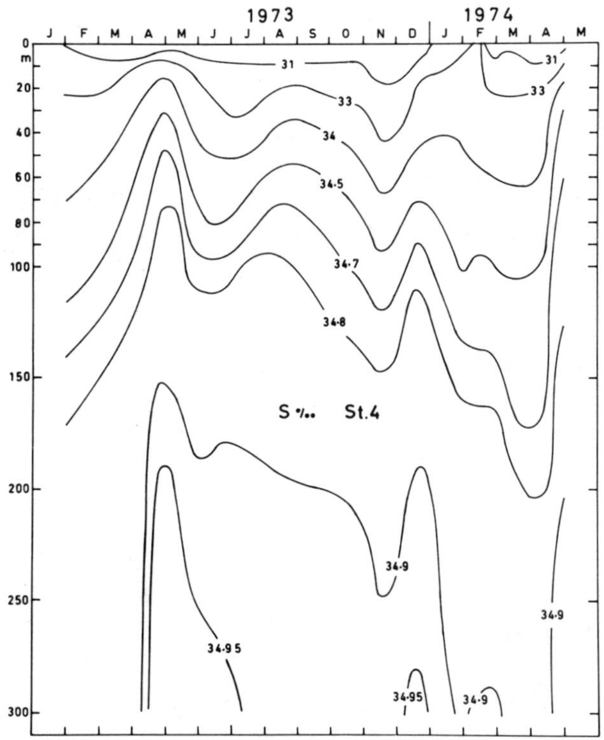

Figure 7. Salinity isopleths from Byfjord St. 4.

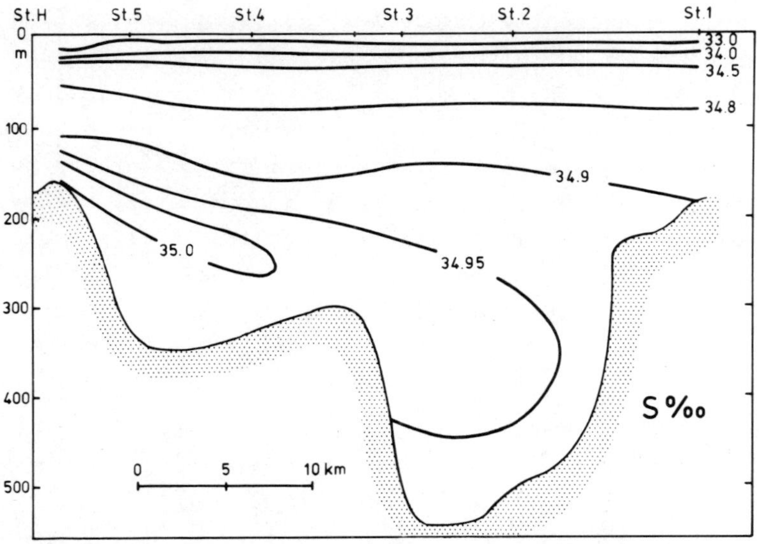

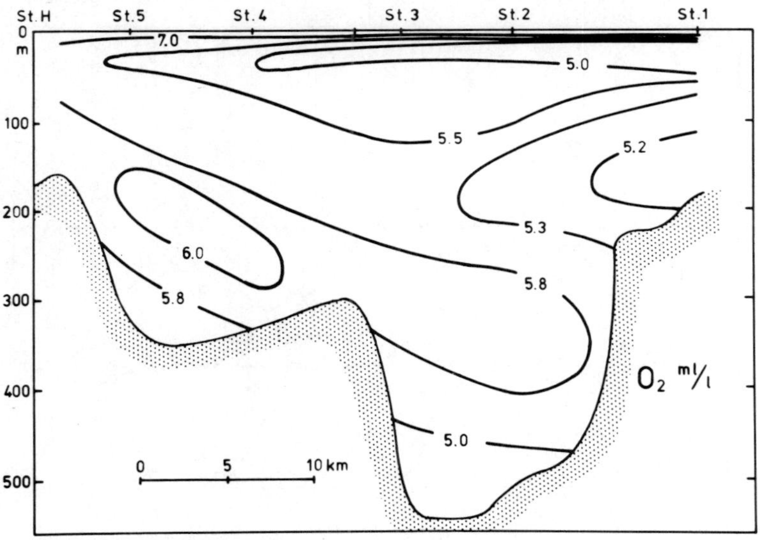

Fig. 8. Sectional diagrams of salinity and oxygen content showing the inflow plume (Byfjord, April 28, 1973).

Unfortunately no current records exist from the early inflow event. The current meter program first started up June 18 and the first record series obtained, lasting until August 3, was coincident with the second 1973 inflow. The low-passed wind vector, current and sea-level records are plotted against a common time axis in Fig. 9. Net current throughout the record series is landward at 160 m, seaward at 30 m and negligeibly small at 80 m. The vertical structure evidently composes a two-layer flow; inflow in the lower and outflow in the upper portion. There is clear evidence for a level of no net motion located at approximately mid-depth. The exchange flow is far from steady. The inflow as detected by the current meter in 160 m forms smooth varying pulses of one day to a week duration arriving at approximately weekly intervals, with speed amplitudes varying between 10 cm/s and 33 cm/s. The variations in the lower flow are well reflected in the upper flow and the speed amplitudes in 30 m and 160 m are of similar magnitude.

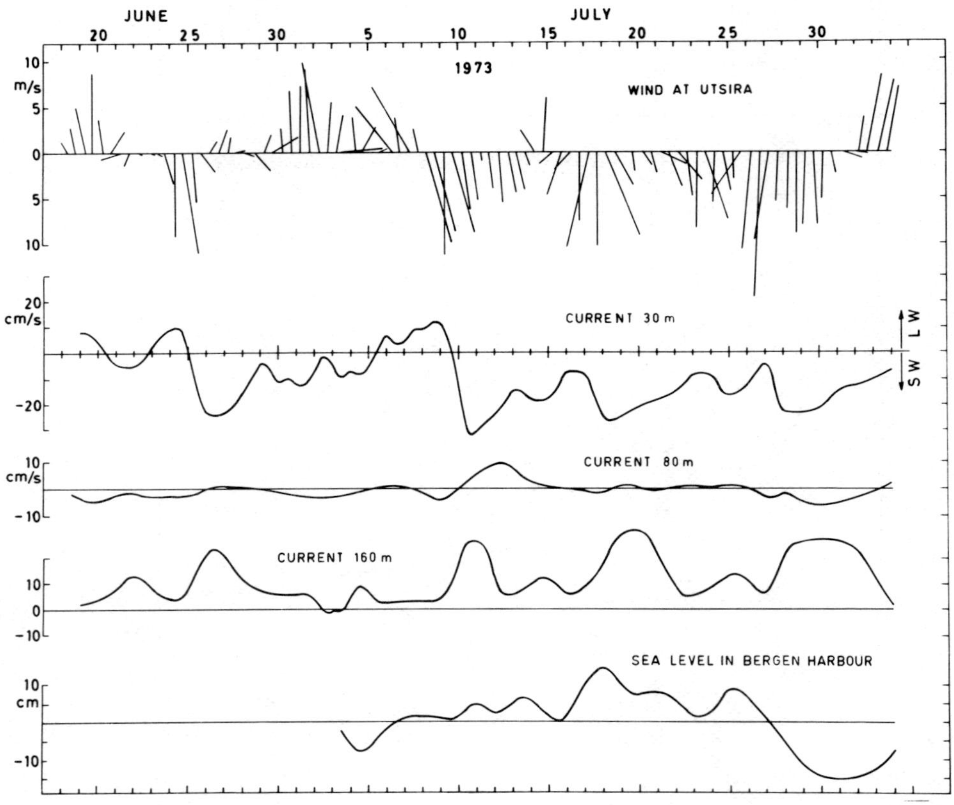

Fig. 9. Low-passed time series of coastal wind, currents at the sill and sea-level in Bergen. June 18 - August 3 1973.

A direct comparision of the current and wind vector plot reveals a similarity where even minor details are reproduced. Northerly winds along the coast are followed by increasing currents

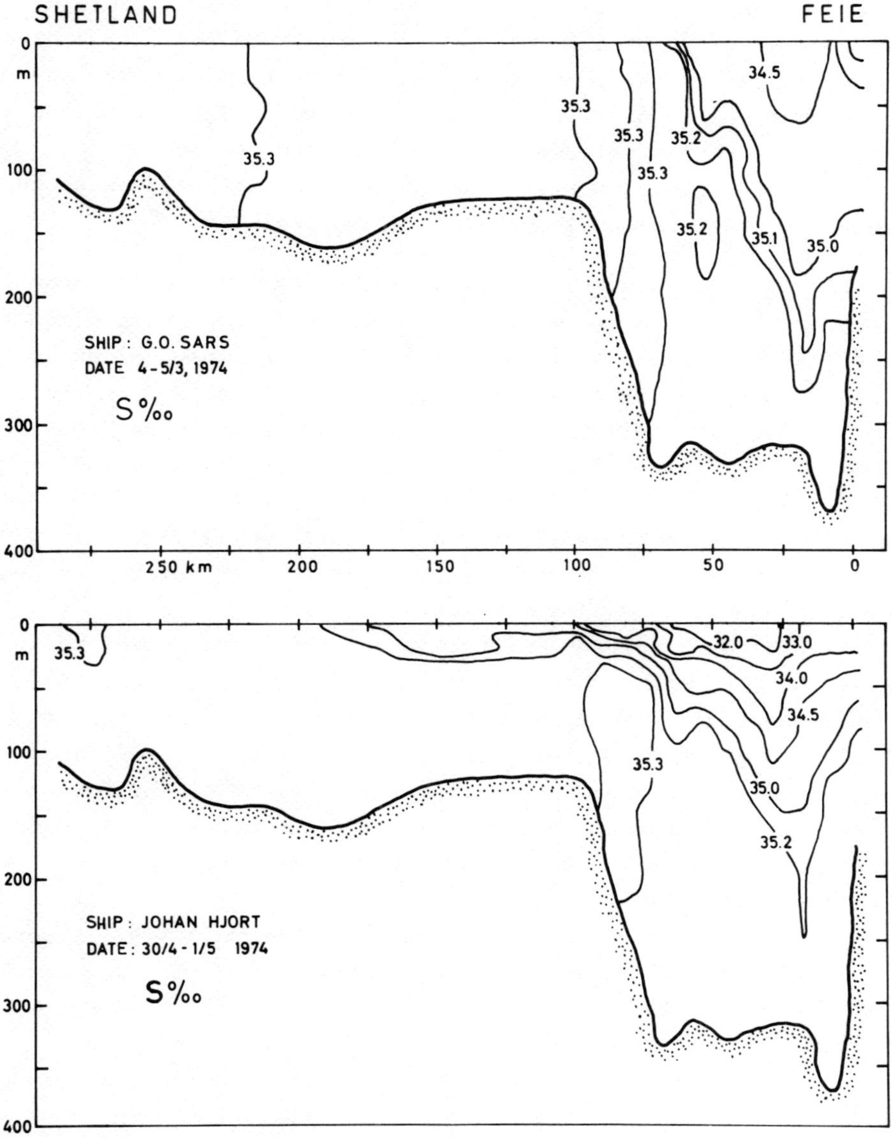

Figure 10. Two distributions of salinity within a vertical section across the Norwegian Trench and the shelf (along the latitude N 60° 45' between Feie and Shetland).

across the sill, southerly winds are followed by decreasing currents and some times by the reverse flow, or possibly a more complex flow structure which cannot be determined by the only three current meters used. Currents and wind are lagged by 1 to 2 days which indicates the approximate response time for the upwelling process involved. Only the sea-level drop occurring at the end of the record series, following a relative long period with northerly winds along the coast, is consistent with the expected behaviour of the sea-level during coastal upwelling.

The 1974 replacement. Following the stagnant period from December to March, during which salt extraction processes dominated in the deep water masses of the fjord (Fig. 7), a new colder and more saline type of water was introduced in the basin from the middle of March. The inflow started with a series of marked inflow pulses similar to what was observed during summer 1973 as described above, but then turned out like a steady flow with current speed within the range 30-35 cm/s at 160 m (Fig. 11). The flow maintained its steady character throughout the rest of the record series, in a period of almost a month.

Mean seaward current at 30 m during this early event was considerably less pronounced (15 cm/s) than during summer 1973, and only about half the mean landward current at 160 m. The two sets of records, however, were obtained under entirely different run-off conditions. Whereas the 1973 event coincided with the seasonal run-off maximum the early 1974 inflow occurred in extraordinarily dry and relatively cold period with insignificant discharge from the rivers.

Two sections of the record series presented in Fig. 11 reveal new features which in several respects differ from what was previously observed. Special attention is attracted towards the first inflow pulse occurring in the records, following a period with relative weak easterly and northerly winds at the coast and associated with a remarkable high surface elevation. The rise of the sea-level of more than 0.5 m in the course of 8 days is related to an atmospheric front arriving at the North Sea from west over Scotland which caused a 60 mbar drop in the air-pressure over the central parts of the North Sea. Whether this inflow, or the associated baroclinic response in the adjacent stratified coastal water (see Fig. 10) was induced by some "coastal trapped wave" (Gill and Clarke, 1974) generated by the atmospheric disturbance, or solely induced by local winds at the coast cannot, however, be clearly verified from the present data. It seems at least that both the barometric surface elevation and the change to southerly winds were responsible for the following decrease and reversing of the flow across the sill.

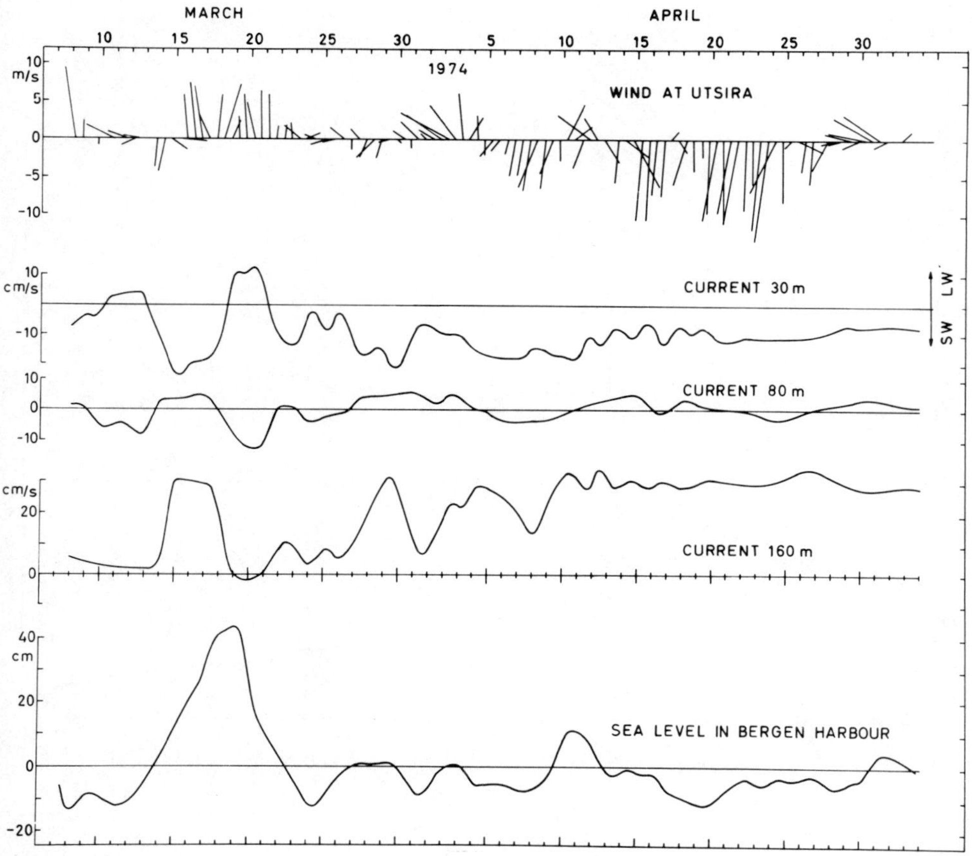

Fig. 11. Low-passed time series of coastal wind, currents at the sill and sea-level in Bergen. March 7 - May 4 1974.

The other section in the records which reveals special features is from about April 10 until the end of the record series May 4. This is a unique example of a major seasonal replacement event with its characteristically marked change of water properties (Fig. 7). Northerly winds dominate at the coast during this period, but despite highly variable strength and even shorter intervals with appreciable southerly wind components the flow across the sill maintains its steady features. In the salinity section taken across the Norwegian Trench (Fig. 10) upwelling in a narrow strip at the coastline is clearly seen and the temperature records at 160 m (Fig. 12) indicates that the intruding water, with temperatures decreasing from 7.6 to 7.2°C, originates from gradually deeper layers in the Trench as the upwelling proceeds.

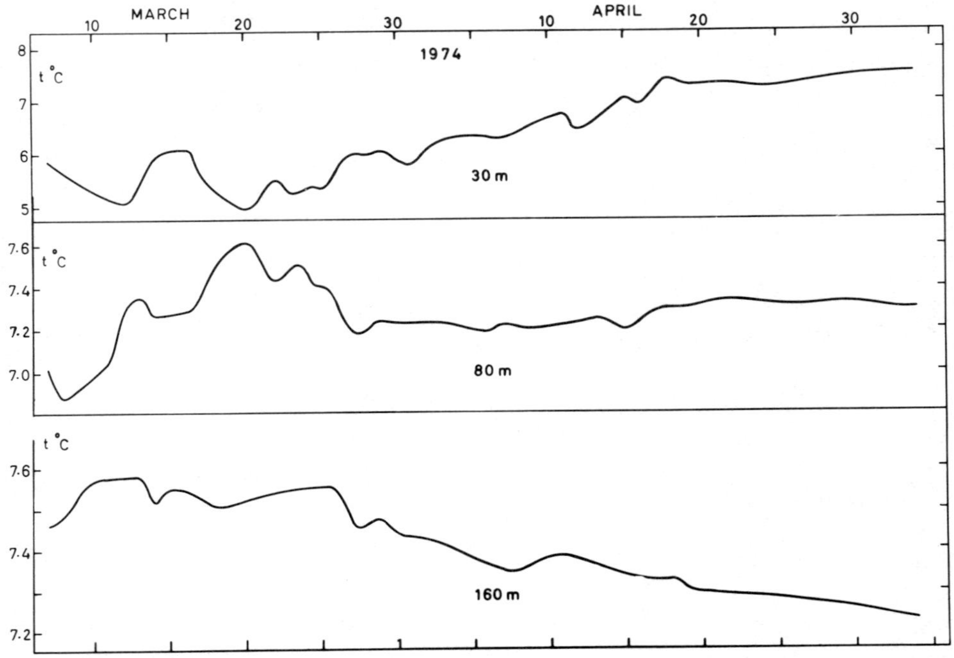

Fig. 12. Low-passed time series of temperature at the sill. March 7 - May 4 1974.

Vertical profiles of water properties and current at the sill itself are shown in Fig. 13 as observed May 3. Ebb and flood current profiles indicates a range of variations which is between 20 and 40 cm/s landward at 160 m, between 7 and 22 cm/s seaward at 30 m and between 5 and 7 cm/s landward and seaward respectivaly at 80 m. A two layer flow structure is also evident from the sigma-t and the oxygen profiles showing the dense and oxygen rich water which intrudes with the deep flow and the original water escaping with the upper flow. A fairly homogenous 20-30 m thick layer contered at mid-depth indicates strong mixing due to shear instability or due to breaking of internal waves.

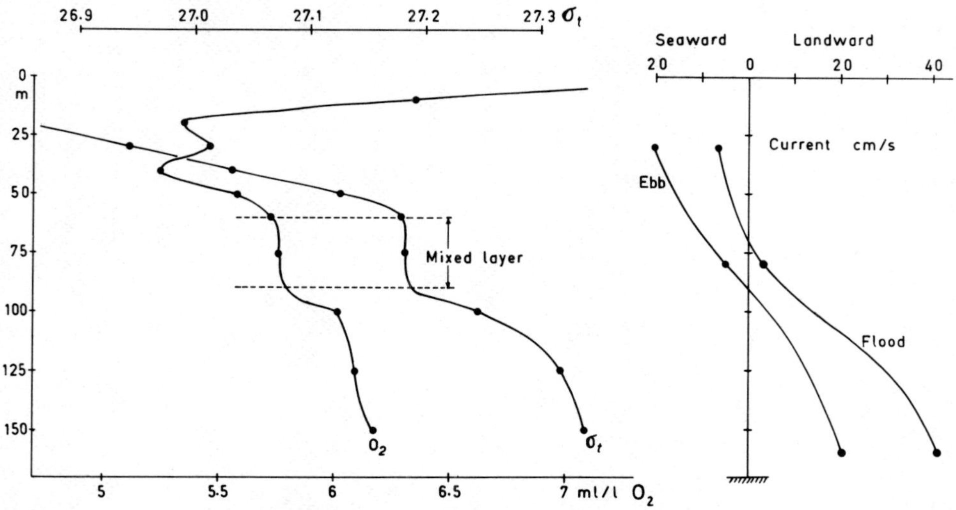

Fig. 13. Vertical profiles of σ_t, 0_2 and current at the sill May 3 1974.

EVIDENCE FOR HYDRAULIC CONTROL OF THE EXCHANGE FLOW.

The passage through Hjeltesund constitutes a constriction of the fjord both in depth and width and hence acts as a control section for the exchange flow.

In order to obtain an estimate of the current velocities neccessary to establish a critical two-layer flow, we adopt the critical conditions of Stommel and Farmer (1953):

1. $F_1^2 + F_2^2 = 1$

where F_1 and F_2 are the internal Froude numbers defined by:

2. $$F_{1,2}^2 = \frac{U_{1,2}^2}{g\frac{\Delta\rho}{\rho} H_{1,2}}$$

Subscript 1 and 2 refer to upper and lower layer respectively, $U_{1,2}$ is the current velocity of the layer averaged over the cross section area $A_{1,2}$. $H_{1,2}$ is the mean thickness, g is the accelleration due to gravity, ρ is sea water density and $\Delta\rho$ is the density jump across the interface.

The condition of no-net transport through the control section is

2. $U_1A_1 + U_2A_2 = 0,$

or for small fresh water discharges as outlined by Long (1976),

4. $H_1 = H_2 = \frac{1}{2}D$

where D is nimimum channel depth (sill depth).
The combination of equation (1), (3) and (4) yields

5. $$U_2 = \sqrt{\frac{gD}{2}\ \frac{\Delta\rho}{\rho}\ \left[1 + \left(\frac{A_2}{A_1}\right)^2\right]^{-1}} \qquad U_1 = -\frac{A_2}{A_1}\ U.$$

The quantities determined by the geometry of the control section are: $A_1 = 128x10^3\ m^2$, $A_2 = 61x10^3\ m^2$ and D = 160 m. The cross section area of the shallower passage ($55x10^3 m^2$ is included in A_1. Representative values of the remaining quantity $\Delta\rho/\rho$ can be obtained from Table 1 using the mean and the maximum $\Delta\sigma_t$. An appropriate value of $\Delta\rho/\rho$ is also obtainable by depth averaging the observed σ_t profile (Fig. 13) over each of the two layer above and below the level of no-net motion. The resulting current velocities are given in Table 2 below.

TABLE 2

$\frac{\Delta\rho}{\rho}x10^{-3}$	U_1 cm/s	U_2 cm/s	
0.060	9.8	20.4	mean $\Delta\sigma_t$ from Table 1.
0.126	14.2	29.5	max $\Delta\sigma_t$ from Table 1.
0.5	28	59	$\Delta\sigma_t$ from actual profile.

In the two first cases there is very good agreement with range of current velocities observed. In the latter case, using the observed density profile, the computed velocities are some too high but still of an acceptable magnitude. The relative great density jump in this case, and hence the high critical velocities are due the contribution from the upper brackish water. This dilution of the outflowing water is not accounted for in the first cases where the actual density measurements are made at 300 m depth in the basin.

A more appropriate value of $\frac{\Delta\rho}{\rho}$ which also is in accordance with the critical flow conditions as the existance of a long non-propagating interfacial wave, may be obteined by taking the difference between the average densities of a say, 10-20 m thick layer on both sides of the mixed layer indicated in Fig. 13. whatever definition of the density jump is adopted, however, the ideal assumption in the hydraulic theory; homogenous layers, steady frictionless flows with no interfacial mixing, are clearly not satisfied. Nevertheless, the estimated velocities in Table2 indicate that critical flow conditions represent quite a realistic situation during deep water replacement in Byfjord. From the current records it is also evident that certain upper limits exist to the exchange rate.

REFERENCES

Braaten, B and Sætre, R., 1973. Oppdrett av laksefisk i norske kystfarvann; Miljø og anleggstyper. Fisken og havet, serie B, 9: 1-95.

Gade, H.G., 1973. Deep water exchange in a sill fjord: A stochastic process. J.Phys.Ocean., 3: 213-219.

Gill, A.E. and Clarke, A.J., 1974. Wind-induced upwelling, coastal currents and sea-level changes. Deep-Sea Res., 21: 325-345.

Helland-Hansen, B., 1935. Temperature variations in a Norwegian fjord. In: S. Hedin (Editor), Geografiska Annaler 1935, Svenska Sällskapet för antropologi och oceanografi, Stockholm, pp. 563-570.

Helle, H.B., 1975. Oseanografisk resepientundersøkelse av fjordene rundt Bergen. Unpublished report. Geophysical Institute, Bergen, 72 pp.

Johannessen, T.W. and Håland, L., 1969. Standard normals 1931-60 of monthly wind summaries for Norway Climatological summaries of Norway. Det norske meteorologiske Institutt, Oslo.

Linde, E., 1970. Hydrography of the Byfjord. Rept. 20, Geophysical institute, Bergen, 38 pp.

Ljøen, R. and Svansson, A., 1972. Long-term variations of subsurface temperatures in the Skagerak. Deep-Sea Res., 19: 277-288.

Long, R., 1976. Mass and salt transfer and halocline depth in an estuary. Tellus, 28,5: 460-472.

Stommel. H. and Farmer, H.G., 1953. Control of salinity in an estuary by transition. J.Mar.Res., 11: 205-214.

Sælen, O.H., 1976. General hydrography of fjords. In: S.Skreslet et al. (Editors), Fresh water on the sea. The Association of Norwegian Oceanographers, Oslo, pp. 43-49.

Tollan, A., 1976. River run-off in Norway. In: S.Skreslet et al. (Editors), Fresh water on the sea. The Association of Norwegian Oceanographers, Oslo, pp. 11-13.

Yoshida, K., 1955. Coastal upwelling off the Californian coast. Records of Ocean.Works in Japan, 2,2: 8-20.

NONLINEAR INTERNAL WAVES IN A FJORD

DAVID FARMER

Institute of Ocean Sciences, Sidney, B.C., Canada

and

J. DUNGAN SMITH

Department of Oceanography, University of Washington, Seattle, Washington, U.S.A.

ABSTRACT

A pilot study in Knight Inlet, British Columbia has yielded measurements of large amplitude, nonlinear internal waves in the strongly stratified surface layer. A train of several waves with amplitudes exceeding 10 meters and periods of a few minutes, passed the vessel on each flood tide. Observations with profiling and fixed depth instruments showed that the waves travelled in a strongly sheared flow. The results are interpreted using Benjamin's (1967) approximate model for nonlinear waves in deep fluids, suitably modified to allow for shear. This model then provides the basis for an examination of conditions leading to shear flow instability. A solution to the eigenvalue problem for exponential stratification and shear is used to examine the interaction of the waves with the background flow; critical amplitudes corresponding to conditions of marginal stability are found to be comparable to those observed in Knight Inlet. The waves are discussed in terms of their likely contribution to vertical exchange processes in fjords.

1. INTRODUCTION

This paper is concerned with one of the many unsteady aspects of flow in fjords. Historically, attempts to study circulation in deep, strongly stratified estuaries have concentrated on a steady state or tidally averaged description. This simplification is understandable and may indeed be justified for the description of circulation in places that have a very small tidal range; it also makes sense as a first step towards understanding what is clearly a most complicated system. Nevertheless, time-dependent effects are often dominant and the assumption of a steady flow in the tidally averaged sense masks much of the underlying physics that must be understood if we are to gain a deeper knowledge of estuarine circulation. As an example that is directly related to the observations described herein, the interaction of tidal flow with topographic features may influence mixing in a very important but nonlinear way. Yet without an understanding of the physics involved, it is impossible to select appropriate entrainment models to use with tidally averaged calculations of the circulation.

Of course tidal flow is not the only time dependent signal with which one must contend. Wind-stress is often more variable and always less predictable than tidal forcing and can play a major role in altering the circulation of deep estuaries. Similarly fluctuations in fresh-water discharge, which can be large during periods of high run-off, will introduce a time dependence into the circulation as will variations in the density structure outside the fjord. However, one of *our* primary concerns is with the response of fjords to tidal oscillations. This interest is motivated partly by the consideration that tidal effects are ubiquitous and tidal ranges are quite large ($\sim$4m) in the inlets of the North East Pacific, and partly by the strong tidally generated effects we have observed in Knight Inlet, the estuary that we have chosen to examine in detail.

Knight Inlet has been the site of several previous oceanographic surveys, so a substantial body of historical data is available; perhaps for this reason it has been a testing ground for recent theoretical models of fjord circulation (Winter, 1973, Long, 1975, Pearson and Winter, 1978). In particular a detailed series of current measurements were taken by Pickard and Rodgers (1959). It is of interest that their observations taken over 20 years ago, are not inconsistent with our own measurements made in 1976 with a modern profiling instrument and internally recording current meters.

Although somewhat longer than most (102 km), Knight Inlet (Figure 1) is fairly typical of British Columbia fjords. It has an average width of 3 km and a maximum depth of 550 m. The inlet has two sills, the innermost of which is 74 km from the head and is located near the middle of a straight reach; this sill has a maximum depth of 63 m. Most of the fresh water enters at the inlet head from the Klinaklini and Franklin Rivers. Mean monthly run-off reaches a maximum close to 800 m^3s^{-1} in June (Pickard and Rodgers, 1959) but the mean annual discharge is estimated to be only 410 m^3s^{-1}. Pickard (1961) classifies the discharge as type Al. A small secondary maximum occurs in the autumn due to rainfall rather than snowmelt but it is followed by relatively low discharges during the winter months.

An air reconnaissance of the inlet was made in August 1976. Flying the length of the channel, several features were observed indicating the great variability that must be expected. Lenses of clear fresh water from streams along the sides of the inlet were floating on the surface of the brackish zone, coloured grey by its suspended load of rock flour, and showed the existence of complex eddy motions. Even more striking were the surface effects of internal waves; these were clearly visible over the sill as a pattern of overlapping curves and further up-inlet as a sequence of straight lines separated by 50-150 m. (See Figure 2). Pickard (1961) has made similar observations and has described trains of internal waves centered on the pycnocline in Knight, as well as other British Columbia inlets. In calm weather the presence of the waves is made apparent at the surface by surface ruffles thought to be caused by convergence effects.

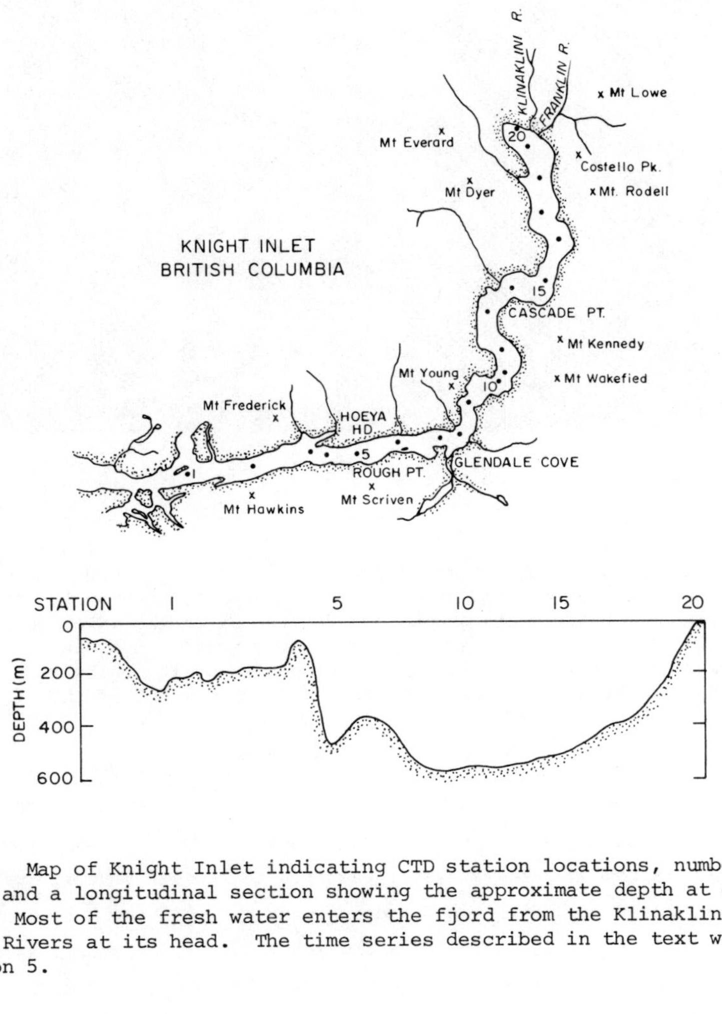

Figure 1. Map of Knight Inlet indicating CTD station locations, numbered from 1 to 20, and a longitudinal section showing the approximate depth at each station. Most of the fresh water enters the fjord from the Klinaklini and Franklin Rivers at its head. The time series described in the text was taken at station 5.

During the week of November 9-14, 1976 we undertook a pilot study in Knight Inlet using the University of Washington Research Vessel ONAR, in order to determine the temporal variability that would have to be considered in planning a more detailed experiment. This study included CTD measurements along the channel with some time series observations at a station approximately 6 km up-inlet of the sill (Figure 1).

Figure 2. Knight Inlet looking seaward. The inner sill extends from Hoeya Head, about 2/3 of the way from the bottom of the photograph on the right hand side, to Prominent Point which is the low feature extending into the inlet 3/4 of the way up on the left hand side. The leading edge of a train of internal waves travelling up inlet, is visible in the center of the photograph. The wave pattern becomes quite complicated in the neighbourhood of the sill.

2. INSTRUMENTATION

Observation of the salient features of tidal, estuarine and wind driven circulations from a single research vessel concurrent with the study of relatively short period internal waves requires the use of several instrument systems, each based on a different sampling strategy. In the November 1976 study the primary physical variables of interest were temperature, salinity and flow velocity and these had to be resolved over the upper 100 meters in a temporal band ranging from a fraction of a minute to several tidal days.

Low frequency information over the entire period and depth range was obtained with a velocity and density profiling system designed for use in the Arctic. This unit cycled continuously between the surface and 140 meters approximately once every five minutes. The Arctic Profiling System (APS) could not be held at one depth to provide an accurate time series, because vertical motion was required to flush the conductivity cell. To satisfy a need for the latter type of information, a Surface Layer Frame (SLF), supporting a horizontally oriented conductivity cell, a temperature sensor and several types of current meters was used. However, even with both of these systems the conductivity and velocity fields due to passing internal waves could not be resolved with sufficient spatial and temporal resolution and a third approach was required. This comprised a string of internally recording (Aanderaa) current meters each fitted with conductivity cells and fast response thermistors and set to sample every 30 seconds. The three systems are sketched in Figure 3 in the manner that they were suspended from R/V ONAR.

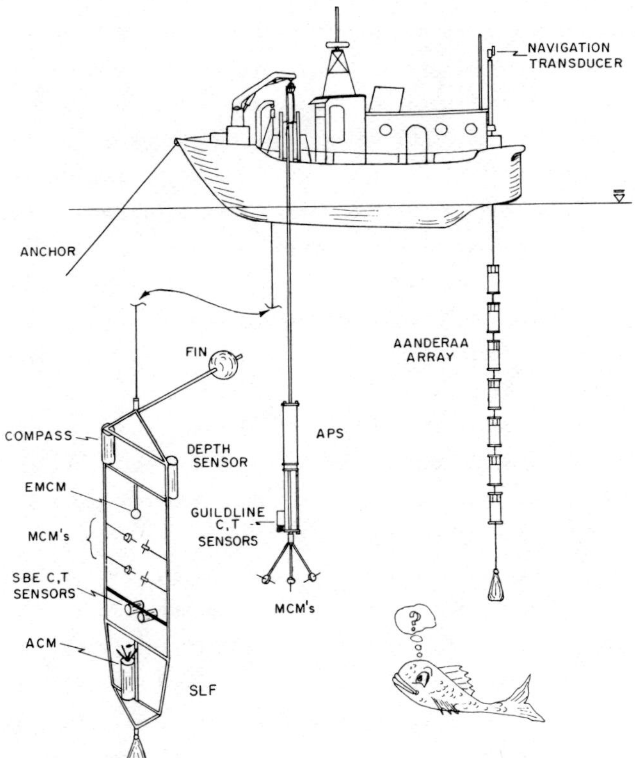

Figure 3. Cartoon showing the configuration of the research vessel and the oceanographic sensors that were suspended from it while anchored at station 5. Three basic systems were employed: a) the Arctic Profiling System (APS) from which velocity and density profile data were obtained; b) the Aanderaa current meter array from which time series of temperature, salinity, and current speed were procured and c) the Surface Layer Frame (SLF) from which time series of near surface salinity and current velocity were measured.

The Arctic Profiling System is constructed around a Guildline Model 8101 CTD. On the nose of the device three precision mechanical current meters are mounted orthogonally and symmetrically, and behind the Guildline pressure case a second stainless steel tube encloses an inertial reference unit capable of resolving bearing to 2° and providing accelerometer and rate gyro data from which the accelerations and tilts experienced by the device can be calculated. The latter is required for removing any high frequency horizontal motion of the profiling fish that are induced by the ship plus all vertical flow components. The current meters deployed with the APS were developed for measuring turbulent fluctuations in rivers, estuaries and ocean boundary layers, and are described by Smith (1974). When used in the profiling mode, their 1 cm/sec threshold velocity is not a problem even in very slow flows and their calibrated accuracy of better than 0.1 cm/sec ensures that the measurements will be obtained with more precision than the stability of the reference frame can be guaranteed.

Pickard and Rodgers (1959) have emphasized the necessity of correcting velocity measurements for ship motion; to this end we used a Mini-Ranger II microwave navigation system with the output recorded on magnetic tape alongside the APS data. After analyzing the APS data, it was discovered that when the velocity profiles were averaged over a period of approximately one hour (10 to 12 profiles) the resulting velocity component profiles were within a fraction of a centimeter per second of each other at all levels whether or not the navigation correction was made. Since ship motion correction certainly should be made to each individual profile when the instrument is near the surface and as such a correction probably should not be made when the APS is at great depth, this important result greatly simplifies the data analysis procedure. It implies also that the accuracy obtained with ship mounted profiling systems may be substantially increased by averaging profiles over a suitable period.

The Surface Layer Frame was designed to head into the mean flow even under relatively rough conditions and it has proven successful in this task on a number of occasions. In the Knight Inlet experiment the frame supported a Seabird Electronics (SBE) conductivity cell, a Seabird Electronics temperature sensor, a Marsh-McBirney two axis electromagnetic current meter, two pairs of precision mechanical current meters, a two axis travel time acoustic current meter made by the Christian Michaelson Institute in Bergen, Norway, a Paro-scientific pressure cell and an Aanderaa compass. The inset in Figure 3 shows the instrument layout on the frame. As the conductivity cell on the SLF was oriented to flush horizontally, this system could be left suspended at a fixed depth or raised and lowered slowly relative to the horizontal current speed. Several of the sensors on this frame including the SBE temperature and conductivity sensors were quite new and their specifications had not yet been checked under field conditions. Results from the latter were compared to profiles from the Guildline System on the APS during quiet periods as was done with the Aanderaa conductivity cells described subsequently. In both cases this approach was sufficient for the scientific goals of the study but it did not comprise a very stringent test of the stability of the SBE sensors. Similarly, the electromagnetic flow sensor outputs were compared to results from the precision mechanical current meters and were found to be reliable to within 2 cm/sec.

The Aanderaa current meters were suspended without vanes from two parallel cables. The array was deployed twice; on the first occasion it remained in the water for about 8 hours after which the tapes were removed from the instrument and processed in the field. Once it was clear that the instruments were set at reasonable levels and running satisfactorily, they were redeployed with new tapes for the rest of the experiment. For the most part the internally recording current meter data discussed in this paper comes from the second deployment during which the instruments were at 7, 12, 13.5, 15 and 20 meters below the surface. Use of such a short sampling period severely limits the resolution of speed data; in the present case resolution is $\pm .041\ ms^{-1}$.

3. OBSERVATIONS

3.1 Background salinity, temperature and velocity field.

Salinity profiles (Figure 4) taken along the channel show the transition from the highly stratified conditions between Cascade Point and the Inlet head, and the weaker stratification between Cascade Point and the sill (Station 3). In Figure 5 this aspect is presented again in the plots of representative salinities and pycnocline depths. This figure also emphasizes the variability of the stratification between stations 3 and 10. A fundamental difficulty in the interpretation of profile sequences such as these arises from aliasing due to internal waves. The structure of these waves may only be resolved by rapid time series taken at one location.

Throughout the time series at Station 5, surface currents of 0.4 to 0.7 ms^{-1} were observed, setting predominately down inlet; this surface flow appeared to be due primarily to the wind which blew down inlet at moderate speed. Beneath the surface the flow velocity fell off rather sharply, especially in the upper 8 meters reaching zero near this level during the flood. Most of the river water was in this surface layer. As the wind driven and estuarine circulations were enhancing one another throughout the entire experimental period it was not possible to separate them with any degree of certainty. Furthermore, the situation was complicated by the tides which at the experimental site caused the flow to reverse below about 8 m, but with a phase lag of two and one half hours relative to the tidal elevation and the tidal current over the sill at Hoeya Head.

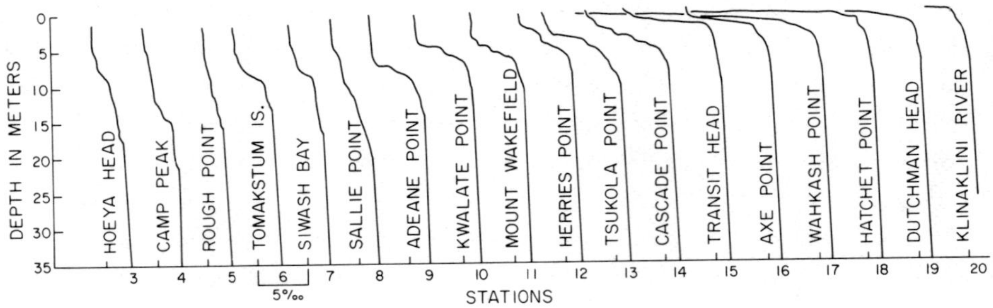

Figure 4. Salinity profiles at 20 stations along the axis of Knight Inlet. From station 17 to the fjord head the surface water was fresh and the surface layer was less than or equal to 1 m; to simplify the diagram the upper parts of these profiles are not shown.

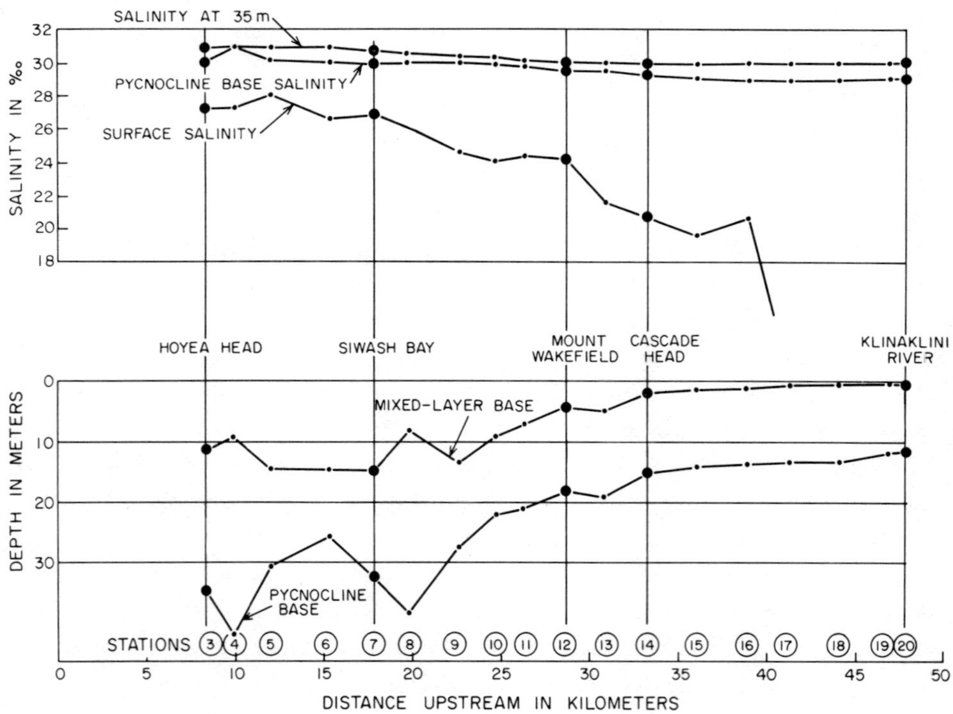

Figure 5. Typical surface layer characteristics between Hoeya Head and the Klinaklini River. The upper graph shows a reference salinity at 35 meters below the surface, the salinity at the base of the pycnocline and the surface salinity as a function of position along the axis of the inlet; the lower one shows the depth of the pycnocline and that of the mixed layer base.

Figure 6 shows approximately one hour average salinity, temperature and velocity profiles. Both the temperature and salinity profiles indicate a relatively poorly mixed surface layer and a clear tidal variation in its thickness. The temperature variation is small and the density profile is dominated by salinity. Although the stratification is quite strong, salinity decreases all the way to the surface at all stages of the tidal cycle and the pycnocline has no well defined base because the rate of salinity increase is more or less uniform in its lower part. As can be seen from Figure 7, this general structure is not an artifact of averaging, but is displayed in each of the individual profiles as well. In fact, the only major difference between the averaged profiles of Figure 6 and the ones presented in the subsequent figure is the removal of high frequency and high wave number structure in the former.

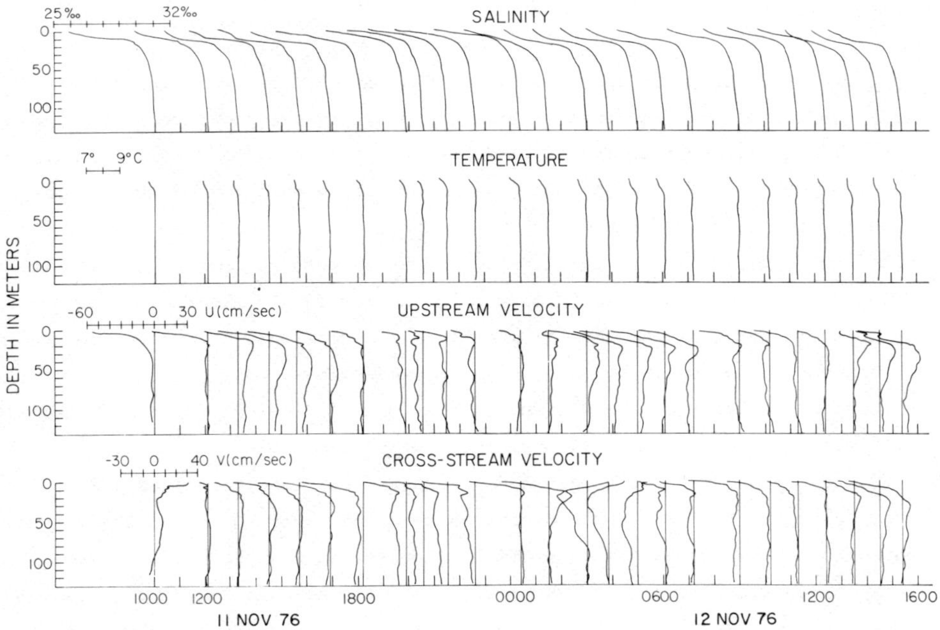

Figure 6. Sequence of one hour average profiles of salinity, temperature and upstream and cross-stream velocity components for the period between 1000 on 11 November and 1600 on 12 November. These profiles were obtained by averaging data from approximately 10 individual lowerings of the APS.

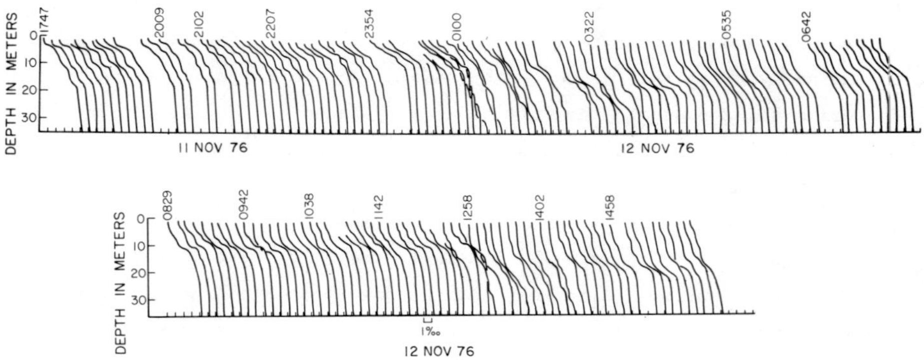

Figure 7. Individual salinity profiles from which the averaged salinity data displayed in Figure 6 were obtained. Note the effects of large internal waves around 0100 and again around 1300 on 12 November. In addition note the layered nature of the pycnocline during the ebb tide especially in the early part of the experiment.

The velocity field is shown in Figure 6c and d; Figure 6c gives the down inlet and Figure 6d the cross-stream component. These indicate a substantial cross-inlet component in the observed velocity field. The profiles in Figure 6c for the period just prior to slack water are representative of the measured flow during passage of the waves. Here the shear is largest in the upper 10 meters of the flow and is fairly small below 30 meters. Moreover the downstream velocity component is close to zero below 20. meters at the time the internal waves arrive at the experimental location. In contrast, the down inlet component of the surface velocity is quite high at these times, particularly at mid-day. At the vessel location the current in the upper 100 meters increases noticeably with the passage of the first nonlinear internal wave.

Figure 8 shows the depth change of characteristic isopycnals as a function of time. Here two main depth zones can be recognized. Below about 30 m the variation is approximately sinusoidal; whereas, in the pycnocline there is a sharp plunge with passage of the large internal waves followed by a gradual rise over the rest of the tidal cycle. These features are clearly tidal in character and appear to be relatively free of wind effects. The data comprising Figure 8 show the overall temporal and vertical structure at the experimental site but also have been averaged over a one hour period to remove high frequency features, such as internal waves.

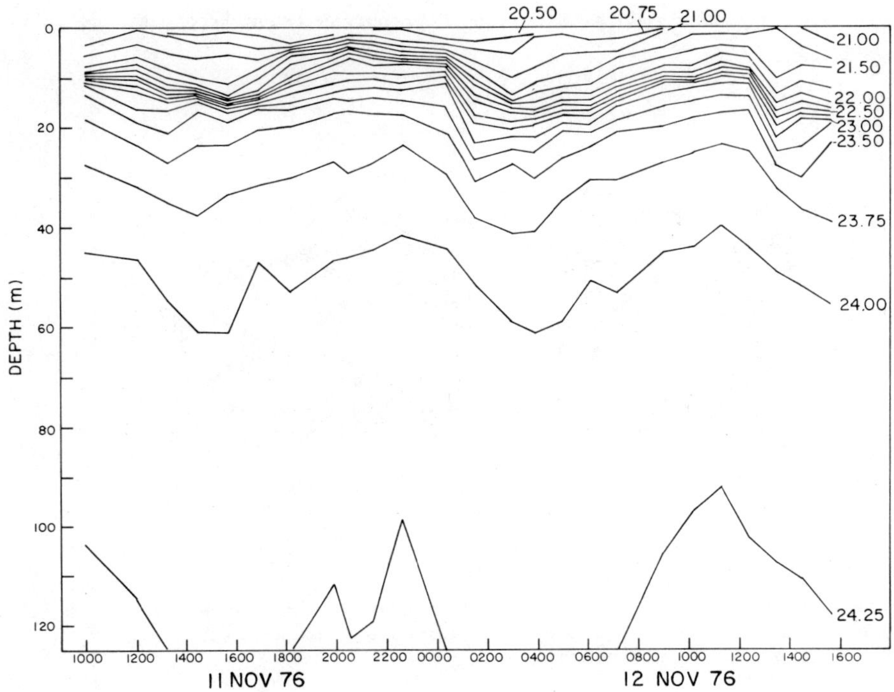

Figure 8. Isopycnal depths as a function of time obtained from the averaged data displayed in Figure 6. Note the gradual rise then rapid descent of the pycnocline depth in concert with the semi-diurnal tide.

3.2 Internal Waves

Unfortunately, even in the salinity profiles from instantaneous APS lowerings, the internal waves are incompletely sampled. Nevertheless, their presence is visible and the nature of the changes in vertical structure caused by their passage is evident. In general, the pycnocline is drastically distorted and deepened with the passage of each wave, but returns almost to its original shape after the wave goes by. Figure 7 shows that the salinity is substantially decreased to a depth of at least 30 meters with the passage of each wave and that under the trough there is an almost linear decrease in salinity between this characteristic depth and the bottom of the surface layer.

Figure 9 presents three time series, the deeper two of which clearly indicate the relationship between the large internal waves and the rapid deepening of the surface layer, followed by a slow rise in pycnocline depth. In Figure 10 the time series from seven instruments are shown on a greatly expanded scale. As each wave passes, the salinity decreases corresponding to a deepening of the isohalines. At no point does the salinity rise above the equilibrium level; thus the waves may be viewed as isohaline depressions travelling past the point of observation.

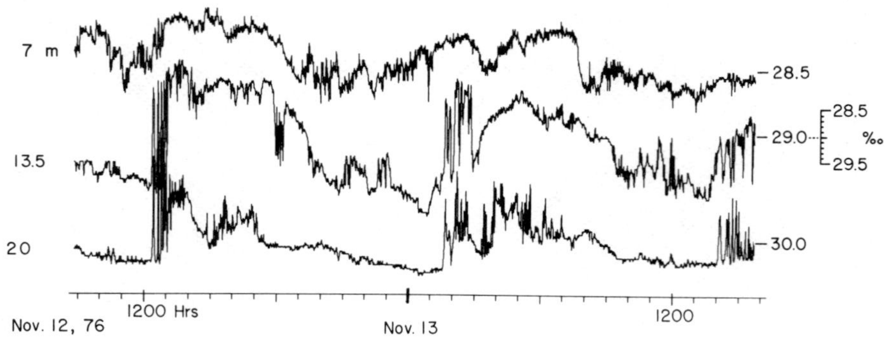

Figure 9. Detailed time series from three Aanderaa instruments for 12 and 13 November. Note the high frequency oscillations associated with the rise in salinity at the two deeper instruments.

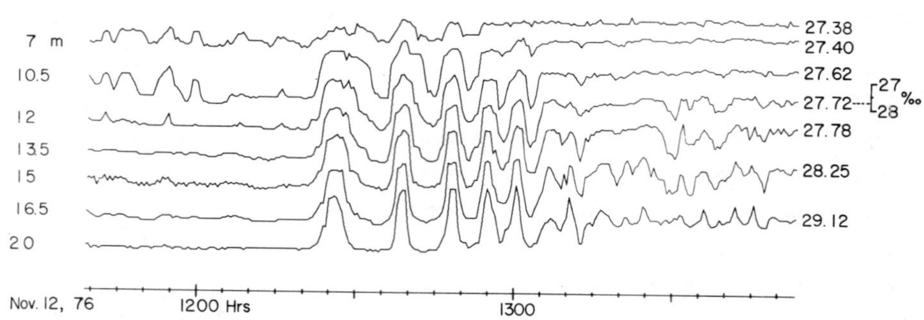

Figure 10. Seven detailed salinity time series for the period of rapid pycnocline depth increase and high frequency salinity oscillations. Here a train of five solitary-like internal waves is evident.

The instrument from which the top recording was obtained in Figures 9 and 10 was situated relatively close to the surface in a layer with a substantially lower salinity gradient and it displays somewhat different features. For example, during the event just after 1200 hours on 12 November, the internal waves in the surface layer produce a signal that is not of significantly larger amplitude than a number of other unidentified events at this level and the former can be found only by its coherence with the clearly defined internal wave induced disturbances in the other two traces. Further, during both the early morning and early afternoon flood tides on 13 November, no evidence of the large internal waves is seen in the surface salinity time series, whereas they certainly are visible in the other two records. All available data show that the abrupt deepening was most pronounced on the afternoon of 12 November, but that it occurred to some degree on all ebb tides during which measurements were being made. From Figure 9 and even more clearly from Figure 10 it can be seen that the change in salinity due to the presence of internal waves is comparable to that due to the abrupt change in pycnocline depth.

Although internal waves were observed during several tidal cycles, our best data are for the mid-day ebb on 12 November. The first wave, which appears to be longer than those following it, arrived just after 1220 Pacific Standard Time. It led a well-defined train of seven waves, the signatures of which are clearly evident in the salinity time series of Figure 10. Data from the Aanderaa meters (Figure 11) indicate the change in velocity due to passage of the first wave was considerably less than that due to subsequent events. Speeds from the Aanderaa instruments were somewhat higher than those obtained from the SLF, probably as a result of ship motion. During most of the experimental period, ship motion was relatively small but during passage of the first wave the ship drifted significantly in the upstream direction causing substantial distortion of the signals. Signatures of the subsequent waves were negligibly affected by ship drift. It is evident from Figure 10 that the period between each of these succeeding waves decreases as they pass.

Conductivity and velocity records from the Surface Layer Frame during passage of the first two internal waves also show the velocity disturbance associated with the first wave to be substantially smaller in magnitude and to lag well behind the salinity response. A first correction using the navigation data and the flow measurements from this frame yields an actual wave period of approximately half that observed. As the ship motion followed the wave motion the wave shape also was distorted resulting in a broader apparent trough.

4. THEORETICAL CONSIDERATIONS

4.1 Theoretical models for internal waves

Estimates based on observed salinity excursions during passage of the internal waves shown in Figures 9 and 10 suggest amplitudes in excess of 10 m. As the strongly stratified part of the water column is only 15-20 meters deep it is evident that nonlinear effects will be important.

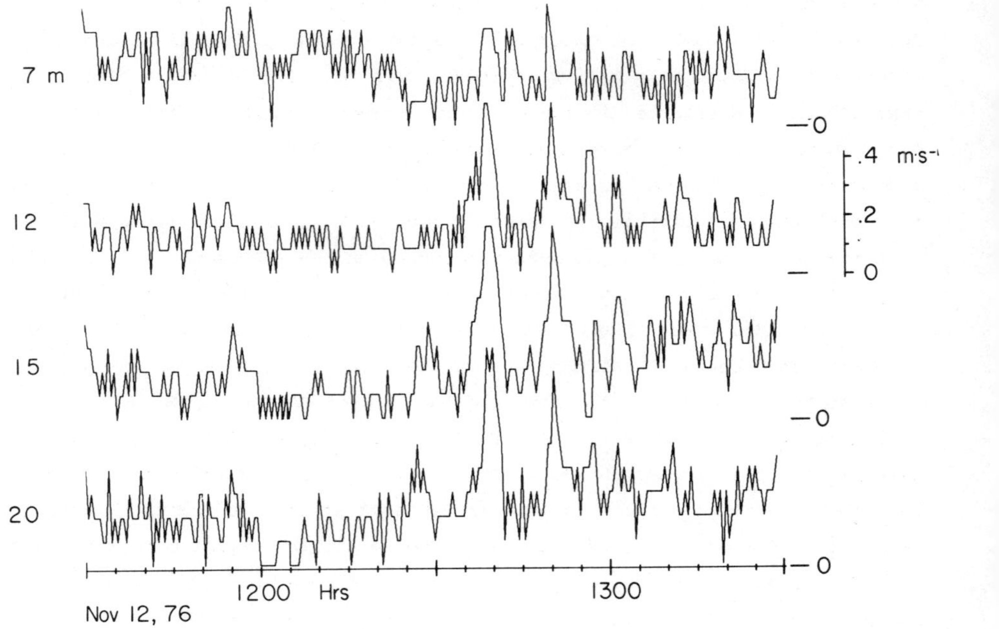

Figure 11. Aanderaa current meter records from four depths for the time period shown in Figure 10. In order to resolve the nonlinear internal waves, a 30 second sampling interval had to be used in the Aanderaa instruments causing the velocity resolution to be reduced to .041 cms $^{-1}$. During passage of the first wave the ship moved with the wave producing a negligible relative velocity whereas this did not occur with the subsequent two waves. The velocity field associated with the smaller internal waves at the end of the train was not well resolved by Aanderaa instruments.

These large amplitude waves may be compared to those photographed during the summer. Figure 2 shows a view looking seaward from approximately 10 km up-inlet of the sill. The waves show up as colour variations presumably associated with the varying thickness of the surface layer, which carries a suspension of rock flour, although the surface ruffles described by Pickard (1961) also help to make the waves visible. This photograph is one of the sequence taken during flood tide. In other photographs internal waves can be seen right over the sill, but in these the pattern is highly complex; the simple banded structure of Figure 2 was found some distance from the sill and appeared to be advancing up-inlet.

From the observed density variation in the surface layer a phase-speed of 0.5 ms $^{-1}$ may be estimated using long wave theory. This may be compared with the much higher speed (1 ms $^{-1}$) reported by Pickard (1961), presumably based mainly on summer measurements when the stratification is more intense. Since the isohaline depressions pass the point of observation in three or four minutes, the horizontal scale of the wave must be of order 70-100 m. Thus, at least on the basis of this calculation the waves appear to be long with respect to the depth of the surface layer. On the other hand the total depth of the fjord in this reach exceeds 400 m, so that the shallow water approximation is not strictly valid beneath the stratified layer.

Trains of large amplitude waves which bear a qualitative resemblance to those shown in Figure 10 have been observed both in lakes (Hunkins and Fliegel, 1973, Thorpe et al, 1972, Farmer 1978) and in the ocean (Halpern, 1971 and Ziegenbein, 1969). Hunkins and Fliegel interpreted the Seneca Lake waves in terms of solutions to the Korteweg-de Vries equation which was derived for long nonlinear surface gravity waves and subsequently found to be applicable to certain classes of long internal waves. The same equation was later extended to second order by Lee and Beardsley (1974) and a numerical integration of the initial value problem was successfully applied to Halpern's Massachusetts Bay observations.

The great depth of Knight Inlet precludes an analysis based on the long wave approximation throughout the fluid. However the existence of internal waves of permanent form in fluids of great depth has been demonstrated by Benjamin (1967) and by Davis and Acrivos (1967). Benjamin considered a density structure composed of two regions: a thin heterogenous layer (Region I) and an infinitely deep layer (Region II) of uniform density. We shall apply a modified version of his model to Knight Inlet and then compare the theoretical results with our observations.

A feature of the waves to be expected in this case is that the relevant dispersion relation has leading terms of the form $\omega = kc_o(1 - \gamma\,|k|)$, $\gamma>0$ where c_o is the speed of an infinitesimal wave. As is also the case for the steady wave solutions of the Korteweg-de Vries equation, the departure of the phase speed from that of an infinitesimal wave increases with amplitude a. However the relationship between the horizontal scale λ and the vertical scale of stratification h, is $a\lambda = O(h^2)$ in contrast to the classical solitary wave relationship $a\lambda^2 = O(h^3)$.

Benjamin postulated the equation:

$$u_t + c_o u_x + Cuu_x = c_o\,\gamma F[u_x] \tag{1}$$

where $F[\xi] = (2\pi)^{-1} \int_{-\infty}^{\infty} |k| e^{-ik\xi} \left[\int_{-\infty}^{\infty} \xi(x') e^{ikx'} dx' \right] d\xi$

and C is a constant, and in effect, he examined solitary and periodic solutions of (1) depending only on the wave co-ordinate x' = x-ct. Although equation (1) was presented by Benjamin, it was not formally derived by him and he did not use it in his subsequent analysis. It is of interest however, in that it constitutes a counterpart to the Korteweg-de Vries equation for the case of an infinitely deep fluid; it was first derived by Gargett (1976) who was able to show by a comparison theorem that at least for harmonic initial conditions, a train of solitary waves evolves, similar to the known solution to the Korteweg-de Vries equation. What has not yet been established is that equation (1) yields a train of solitary waves for an arbitrary initial condition, as does the KdV equation but the evidence is suggestive. The present observations do not permit a detailed analysis of the generation mechanism; however we have a clear picture of the waves themselves as they pass the point of observation, and these observations permit a comparison of the wave properties with Benjamin's theoretical predictions.

While reference may be made to the original paper (Benjamin, 1967) for mathematical details, we note here those aspects directly applicable to an analysis of the Knight Inlet data. Figure 12 shows profiles of density and velocity averaged over approximately one hour prior to the arrival of the waves presented in Figure 10; it also indicates how the fluid was divided into two regions as required by the theory. The strong down-inlet surface flow is likely to be important in the present case and the analysis is modified as suggested in Section 6 of Benjamin's (1967) paper.

Application of the theory for specific density and velocity distributions must begin with the solution of the eigenvalue problem for infinitesimal waves. Let ξ be the vertical displacement perturbation and $\xi(x,\eta) = \phi_o(\eta)e^{ikx}$ with k real, and η the depth below the equilibrium free surface; the resulting Sturm-Liouville problem in Region I then becomes

$$(\rho W_o^2 \phi_{o\eta})_\eta + g\rho_\eta \phi_o = 0 \tag{2}$$

with the free surface condition:

$$\phi_o - W_o^2 g^{-1} \phi_{o\eta} = 0 \qquad \text{at } \eta = 0$$

and (3)

$$\phi_{o\eta} = 0 \qquad \text{at } \eta = h.$$

Alternatively, with little loss of accuracy for most practical applications we may take the fixed upper boundary condition, $\phi_o(0) = 0$. The equivalent problem for Region II, in which there is no shear, remains unchanged. Solutions satisfying these boundary conditions must be found by choosing the appropriate wave speeds c_{o_1}, c_{o_2},... for successive modes. However, in what follows we shall drop the subscript and treat only the lowest internal mode, as only this mode appears to be applicable to the Knight Inlet data. It is also convenient to normalize the eigenfunction by taking $\phi_o(h) = 1$.

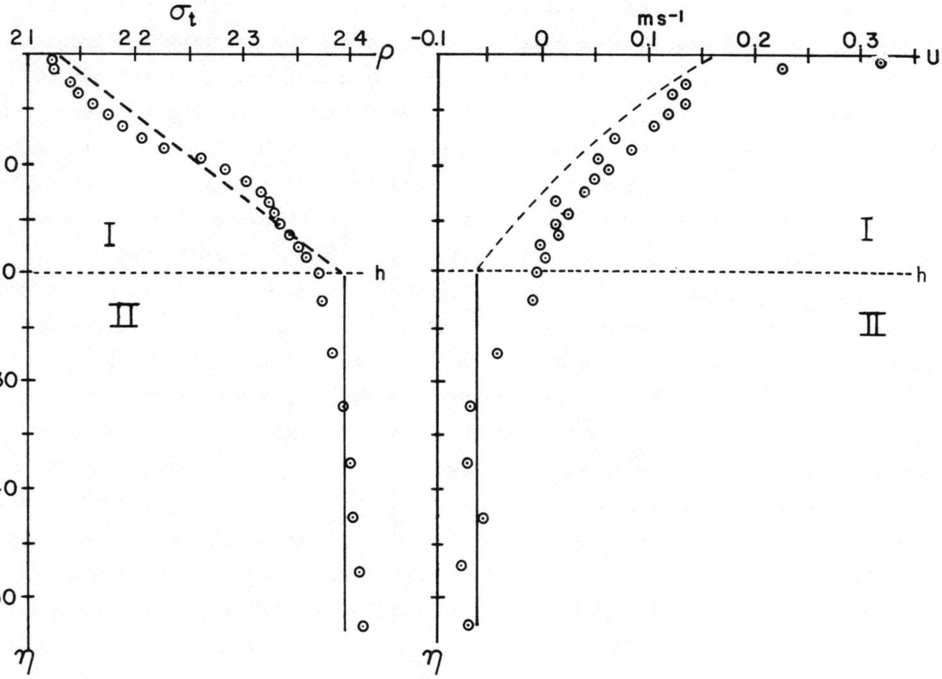

Figure 12. Sigma-t and down-inlet velocity component prior to passage of the first solitary-like internal wave. In the numerical calculations described in section 5.1, the measured σ_t and velocity component data of Region I were used. In the simplified analytical model described in section 4.3, the dashed curve was used. For both models velocity and σ_t are assumed independent of depth in Region II as indicated by solid lines. In the calculations the velocity axis is shifted so that the velocity in Region II can be taken as zero.

In equation (2) W_o refers to the quantity $U \pm c_o$ where $U(\eta)$ denotes the velocity of the undisturbed fluid in the x-direction, calculated with respect to the velocity in Region II (we take $U = 0$ in Region II). Similarly, $W = U \pm c$ where c is the true wave speed, so that W represents the primary velocity as observed in a frame of reference travelling with the wave; W_o is the corresponding function for infinitesimal waves. The ambiguous sign covers the case of propagation with (-) or against (+) the current.

The existence of a non-uniform basic flow introduces some special physical considerations. First, the analysis is based upon the condition of dynamic stability to small disturbances; we may assume stability in this sense provided $4g\,\rho^{-1}\,\rho_z \geqslant U_z^2$. The mean flow and density distributions shown in Figure 12 satisfy this condition, however these profiles are based on an average of several profiles each of which is separated by several minutes. Even if the individual casts showed dynamically stable conditions, we cannot be sure that the condition is instantaneously satisfied; moreover the perturbation induced by the passage of the wave itself contributes to the shear. This interaction of the wave with the basic velocity field must be considered a primary candidate for shear flow instability and resultant generation of turbulence and mixing. Notwithstanding this complication and its oceanographic implications Benjamin's results appear to offer a reasonable description of the waves, and also permit an examination of the conditions likely to lead to such instabilities.

Secondly, the existence of the basic flow has additional implications for the wave properties as discussed by Benjamin (1962). In particular, both the phase-speed and the horizontal scale of wave motion are modified although the functional form of the wave shape remains unchanged. However, a difficulty arises if the basic flow is such that the primary velocity $W(\eta)$ becomes very small or negative at some point of the profile. In this case separation of the flow must be expected, that is, a stagnant region of fluid is carried along by the wave, so that although a steady wave might still be possible, the analysis used here would not be appropriate. Recent work by Redekopp (1977) has shown how such closed streamline problems may be handled and has indicated the complex eddy structures that result.

In the present context, and for a basic flow $U(\eta)$ of the shape shown in Figure 12, separation of flow would be expected at the foot of Region I if the flow opposing a wave was sufficient to cause $W(\eta) = U(\eta) + c$ to vanish or become negative. For a wave travelling *with* the surface flow $W(\eta) = U(\eta) - c$ would be least near the surface so that separation would be expected at this point. Observations taken at one location cannot unambiguously determine the direction of propagation. However, our observations during the summer as well as physical considerations, indicate a train of waves travelling up-inlet, that is to say against the surface flow. Moreover our photographs show the waves to be quite straight so that it seems justified to treat them as plane waves travelling along the axis of the inlet. On this assumption, calculation of $W(\eta)$ shows that this quantity is everywhere greater than zero for lowest mode internal waves, and the problem of flow separation will not be considered further.

Having found the phase speed c_o of infinitesimal waves, as well as the function ϕ_η and $\phi_{\eta\eta}$, we next calculate the integral coefficients $\overline{U}$ and $\overline{V}$ for the nonlinear equation $\overline{U} f - \overline{V} f^2 + F[f] = 0$ where f(x) is now the vertical displacement of the interface $\eta = h$. Again allowing for a basic flow $U(\eta)$, Benjamin's formulae for these coefficients (equation 3.52, 3.53) become

$$\overline{U} = c_o^{-2} \rho(h)^{-1} \int_o^h (W^2 - W_o^2)\rho\phi_\eta \, d\eta \tag{4}$$

$$\overline{V} = 1.5\, c_o^{-2} \rho(h)^{-1} \int_o^h \rho W_o^2 \phi_\eta{}^3 d\eta \tag{5}$$

For solitary waves the relationship between the horizontal scale λ and amplitude a is then given by:

$$a = 2\overline{U}/\overline{V}, \ \lambda = 1/\overline{U}. \tag{6}$$

Solving for the (nonlinear) wave speed c we find

$$c = p + [(c_o - p)^2 + aq]^{1/2}, \tag{7}$$

where $p = -jI_1/I_2$, and where $q = .75\, I_3/I_2$.

The integrals are obtained by writing out the coefficients $\overline{U}$ and $\overline{V}$ in full:

$$I_1 = \rho(h)^{-1} \int_o^h U \rho\phi_\eta{}^2 d\eta$$

$$I_2 = \rho(h)^{-1} \int_o^h \rho\phi_\eta^2 d\eta$$

$$I_3 = \rho(h)^{-1} \int_o^h \rho(U + jc_o)^2 \phi_\eta{}^3 d\eta \tag{8}$$

where j = +1 or -1 for waves travelling against or with the flow respectively. The sign of the root in (7) is established by the requirement that $c = c_o$ for $a = 0$. Knowing the phase-speed, the coefficient $\overline{U}$ may be evaluated and by (6) the horizontal scale λ. Finally, the vertical displacement of a streamline is $f(x) = a\lambda^2(x^2 + \lambda^2)^{-1}$.

4.2 Dynamical stability for a wave in sheared flow

The dynamical stability of the fluid in the neighbourhood of the wave can be examined in terms of changes to the gradient Richardson Number, Ri. In most practical cases horizontal gradients are much less than vertical gradients, so we may take the customary definition

$$R_i = g\,\rho^{-1}\rho_z(u_z)^{-2} \tag{9}$$

The distortion of the flow field due to passage of an internal wave can be found from the streamline displacement relation $z = \eta + af(x)\,\phi(\eta)$. Since $u = (c \pm U)/z_\eta$ (9) becomes

$$R_i = g\,\rho^{-1}\,\rho_\eta\,\Phi^3\,[U_\eta - (c \pm U)\,af\,\phi_{\eta\eta}\,\Phi^{-1}]^{-2} \tag{10}$$

where $\Phi = 1 + af(x)\phi_\eta$.

The dominant term in brackets [ ] in (10) will determine which of two effects alter the Richardson Number as the wave passes. Suppose first that $|U_\eta| \gg |(c \pm U)\,f\phi_{\eta\eta}\Phi^{-1}|$ corresponding to the situation in which the shear of the unperturbed flow was far greater than shear induced by the wave itself. This situation might be expected close to the surface as indicated by the Knight Inlet profiles shown in Figure (12); then

$$Ri \simeq g\,\rho^{-1}\,\rho_\eta\,\Phi^3 U_\eta^{-2} \tag{11}$$

and the Richardson Number is seen to be modulated by the term $(1 + af\,\phi_\eta)^3$. Thus stability *increases* during passage of the wave. Physically we may interpret this as the effect of stretching of the streamlines leading to reduction of the basic shear. Of course isopycnals are also stretched, but the Richardson Number depends upon the square of the velocity gradient so the destabilizing effect of isopycnal stretching is relatively less important.

If the basic shear is small and $|U_\eta| \ll |(c \pm U)\,f\,\phi_{\eta\eta}\,\Phi^{-1}|$, wave induced shear dominates and we find

$$Ri \simeq g\,\rho^{-1}\,\rho_\eta\Phi^5\,(caf\,\phi_{\eta\eta})^{-2} \tag{12}$$

with the x-dependence now governed by the term

$$(1 + af\,\phi_\eta)^5\,(caf)^{-2}. \tag{13}$$

If we neglect for the present the influence of amplitude on wave speed, differentiation of (13) with respect to a shows that the Richardson Number will decrease with increasing amplitude unless the amplitude exceeds a value of $2/(3f\phi_\eta)$. However there will always be a depth at which ϕ_η vanishes, for example at $\eta = h$ in Benjamin's (1967) model, so this constraint on the influence of amplitude on shear flow instability disappears in that part of the water column where ϕ_η is sufficiently small, and the Richardson Number will always drop during passage of the wave. In particular, for $\phi_\eta \ll 1/af$

$$Ri \simeq g\,\rho^{-1}\,\rho_\eta\,(caf\,\phi_{\eta\eta})^{-2} \tag{14}$$

It is in the lower part of the stratified layer, where equation (14) is applicable, that we therefore expect minimum Richardson numbers and thus a greater opportunity for shear flow instability during passage of the wave. Taking the criterion Ri = 1/4 for marginal stability, we can compute a critical amplitude such that this value is just achieved at the wave center x = 0. Thus,

$$g\,\rho^{-1}\,\rho_\eta\,(caf\,\phi_{\eta\eta})^{-2} = 1/4. \tag{15}$$

These results are quite general; a specific application using Benjamin's (1967) model allows comparison with the waves observed in Knight Inlet. In this case we combine equations (15) and (7) to obtain:

$$\Gamma^3 + p\,\Gamma^2 - (c_o - p)^2\,\Gamma - (p\,[c_o - p]^2 + kq) = 0 \tag{16}$$

where

$$\Gamma^2 = (c_o - p)^2 + a_c q$$

and

$$k = 2[g\,\rho^{-1}\,\rho_\eta \phi_{\eta\eta}^{-2}]^{1/2} \quad \text{at} \quad \eta = h.$$

For given coefficients c_o, k, p, q, $\phi_{\eta\eta}$, the critical amplitude a_c may be recovered using Cardan's solution to the cubic (16) or by numerical methods.

4.3 *Solution for exponential stratification and shear*

While the eigenvalue problem (2) in Region I may be solved numerically for observed velocity and density distributions, it is of interest first to consider a simple analytical result. Benjamin (1967) presented an example in which the density changed exponentially; we shall keep the same density distribution but now include a velocity profile that decreases exponentially with depth to zero at the interface h. We shall restrict the analysis to waves travelling upstream, as observed in Knight Inlet and describe the density and velocity structure with three independent parameters β, δ and h.

Let

$$U(\eta) = \alpha\ (e^{-\delta\eta} - e^{-\delta h}) = U\ (o) + \alpha\ (e^{-\delta\eta} - 1) \tag{17}$$

$$\rho(\eta) = \bar{\rho}\ e^{\beta(\eta - h)}$$

The constant α is not known *a priori*, but is determined as a solution to the eigenvalue problem for a particular mode and for given values of β, δ and h. In particular we relate α to the phase speed: $c_o = \alpha - U(0)$. The velocity and density field of (17) may be represented by the dimensionless parameter $R_o = g\beta/\alpha^2\delta^2$ and the eigenvalue problem now consists of determining δh as a function of R_o and thus α and c_o as a function of β, δ and h.

From (17) we have

$$W_o = U + c_o = \alpha\ e^{-\delta\eta}$$

$$\rho\ W_o^2 = \rho\alpha^2\ e^{(\beta - 2\delta)\eta - \beta h}$$

which when substituted into (2) gives

$$\phi'' - (2\delta - \beta)\phi' + g\beta\alpha^{-2}\ e^{-2\delta\eta}\ \phi = 0, \tag{18}$$

$$\phi(0) = \phi'(h) = 0.$$

The general solution to (18) is

$$\phi(\eta) = \psi[A\ J_i(\psi) + B\ Y_i(\psi)]$$

where $\psi = \sqrt{R_o}\ \exp(\delta\eta)$ and where the order of the Bessel functions is $i = 1 + \beta/2\delta$. Even for weak shears $\beta << 2\delta$ so we may set $i = 1$. For the profiles in Figure 12, $\beta/2\delta \simeq 0.5 \times 10^{-3}$.

Application of the boundary conditions then leads to the equation

$$J_1(\psi_o)Y_o(\psi_h) - Y_1(\psi_o)J_o(\psi_h) = 0, \tag{19}$$

where $\psi_o = \sqrt{R_o}$, $\psi_h = \psi_o \exp(\delta h)$. Zeroes of (19) are tabulated (c.f. Abramowitz and Stegun, 1968, page 415); these zeroes now define the relationship between R_o and δh from which α and c_o may be determined as a function of β, δ and h.

The integrals (8) must be numerically evaluated to find the coefficients $\overline{U}$ and $\overline{V}$. Table I gives the relationship between δh, R_o, p/α, q/α^2, $a\lambda/h^2$, a_c/h for tabulated first mode zeroes of (19).

TABLE I

Dimensionless wave parameters for exponential shear and density model

R_o	δh	p/α	q/α	$a\lambda/h^2$	a_c/h
43.00	0.223	-.128	.0201	.685	.796
6.78	0.511	-.217	.0161	.555	.667
2.15	0.811	-.245	.0125	.447	.560
1.54	0.916	-.243	.0114	.413	.528
1.17	1.012	-.238	.0105	.385	.501
.617	1.253	-.215	.0083	.321	.441

As a numerical example, consider the values

$\beta = 1.37 \times 10^{-4}\ m^{-1}$

$\delta = 2.56 \times 10^{-2}\ m^{-1}$

$h = 20$ m

$a = 10$ m

which are an approximate representation of the conditions shown in Figure 12. From Table I we find $R_o = 6.78$ which implies $\alpha = .550\ ms^{-1}$, $U(0) = .220\ ms^{-1}$. In Table II we compare the values for c_o, c and λ with those obtained from Benjamin's model with the same density distribution but without shear.

TABLE II

Comparison of wave properties with and without opposing shear
Wave amplitude 10 m

	With shear	Without shear
c_o	.330 ms^{-1}	.467 ms^{-1} *
c	.420 ms^{-1}	.572 ms^{-1} *
λ	22.2 m	32.4 m

* In a fixed reference frame these speeds are increased by the mean Region II velocity of .062 ms^{-1}.

The presence of an opposing shear has the effect of slowing down the waves and decreasing their horizontal length scale. The shear influences the dynamical stability by reducing the critical amplitude a_c required to produce marginal stability at the foot of Region I. Solving (16) for the numerical values given above we find that in the sheared flow $a_c = 13.34$ m, while in the unsheared case $a_c = 17.5$ m. Calculation of the gradient Richardson Number from (10) throughout the water column, shows that for a wave of 13.34 m the stability at x = 0 is increased above 17 m but sharply decreased below this depth. Figure (13) gives critical amplitudes for other values of R_o, together with the relationship for δh.

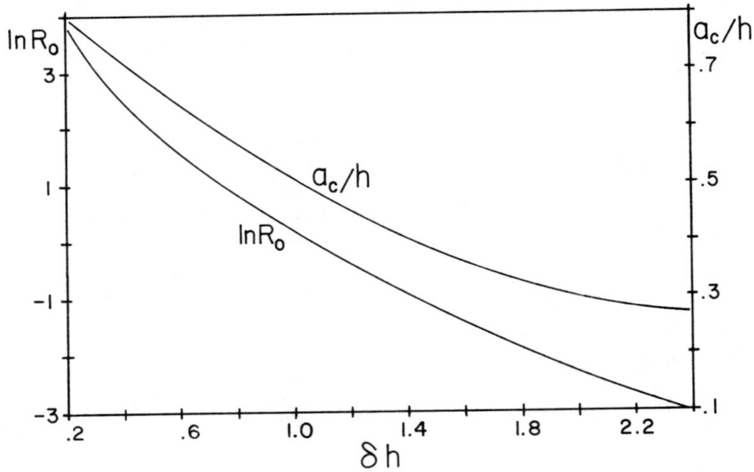

Figure 13. Curves relating R_o to δh and a_c/h to δh. The first curve gives the relationship between the flow parameters α, β, δ and h satisfying the first mode solution to equation (19). The second permits evaluation of the critical amplitude of a wave corresponding to a given value of δh.

Our analysis has been based upon a model which assumes Region II to be unstratified and this has necessarily restricted the discussion of Richardson Numbers to the surface layer. In practice however, there is weak stratification in Region II, as seen in Figure 12 and wave induced shears may be expected to lead to instabilities and mixing on both sides of the interface.

To summarize these results we may say that the waves tend to increase the stability of sheared flow against which they are travelling in the upper part of the stratified layer, while decreasing the stability near its base. For the conditions prevailing in Knight Inlet during November quite large waves are required to reduce Ri below 1/4. Of course the background wave field needs to be considered as well and in the presence of other internal waves turbulence may be generated by smaller solitary waves. It is of interest that the waves observed in our pilot study were comparable in size to those required to produce marginal instability, suggesting that waves of this amplitude constitute an upper limit for these conditions. Larger waves would lose energy rapidly until they approached this critical amplitude.

5. DISCUSSION AND CONCLUSIONS

5.1 *Comparison with numerically evaluated wave properties*

A closer comparison between observations and the theoretical predictions can be made by solving (2) numerically for the measured density and velocity profile of Figure 12. The eigenvalues were obtained by a shooting method. For a trial phase speed c_o, integration by the fourth order Runge-Kutta scheme was begun at $\eta = h$ and the free surface condition checked at $\eta = o$. Convergence was then sought by the method of successive bisection and inverse parabolic interpolation.

Observations at a single point cannot be used to determine the wave speed c directly. However comparison of the measured perturbations with Benjamin's theory may be made using the calculated wave speed to transform the observed temporal variations into spatial variations. For example the time λ_t required for the displacement of an isohaline at depth h, to change from 1/2 a to a, is related to the horizontal scale as $c\lambda_t = \lambda$.

The actual vertical displacements may be inferred from the observed salinity excursion and the measured salinity profile into which the waves are travelling. This seemed a more satisfactory procedure than computing contours of constant salinity based on an interpolation of observations at discrete depths, particularly since the time series measurements only covered the depth range 7 - 20 m whereas the profiles began at the surface. Figure 14 shows the estimated vertical excursion at 20 m for the period during which the waves shown in Figure 10 were observed. It will be noted that the waves represent a depression of streamlines as expected for a density structure of this type. The first order theory indicates that larger amplitude waves would travel faster and also have a smaller horizontal scale.

The amplitude of each wave is such that the validity of an application of Benjamin's analysis is questionable. In particular we are scarcely justified in the assumption made for the expansion in Region I, that $a/h \ll 1$. This also becomes evident upon calculation of the horizontal scale length λ, which for large waves may be comparable to the vertical scale of stratification h. Nevertheless, the results are of some interest and are presented in Figure 14. Calculations were carried out both with and without inclusion of the velocity profile. The corresponding eigenfunctions are compared with observed values in Figure 15. A wave of amplitude 12 m is predicted to have a speed of 0.555 ms^{-1} neglecting the surface flow and 0.467 allowing for the surface flow. Note that the term amplitude as used here refers to the maximum vertical excursion of the streamline at $\eta = h$ rather than one half that value as would be appropriate for a sinusoidal wave.

We have already mentioned that the first wave was distorted by ship motion producing a much broader apparent trough. The theoretical prediction for subsequent waves shows that the shape is similar although the observed wave width is somewhat greater than predicted.

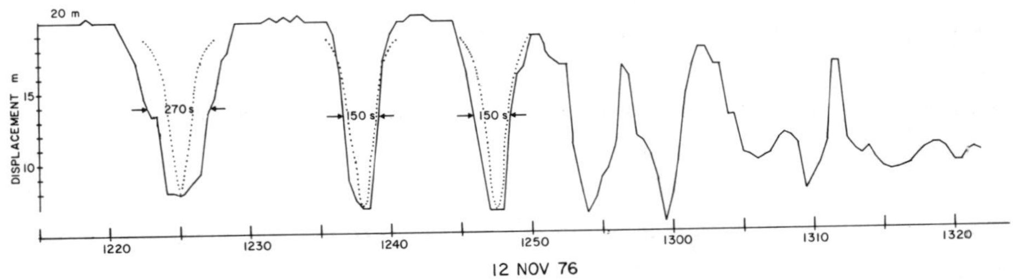

Figure 14. Comparison of measured and calculated pycnocline depths. The solid line in this figure indicates the pycnocline base displacement measured with the Aanderaa instrument string; the dotted line indicates a displacement predicted by the modified theory of Benjamin (1967). The apparent broadening of the crest in the first wave results from ship motion during passage of the internal wave.

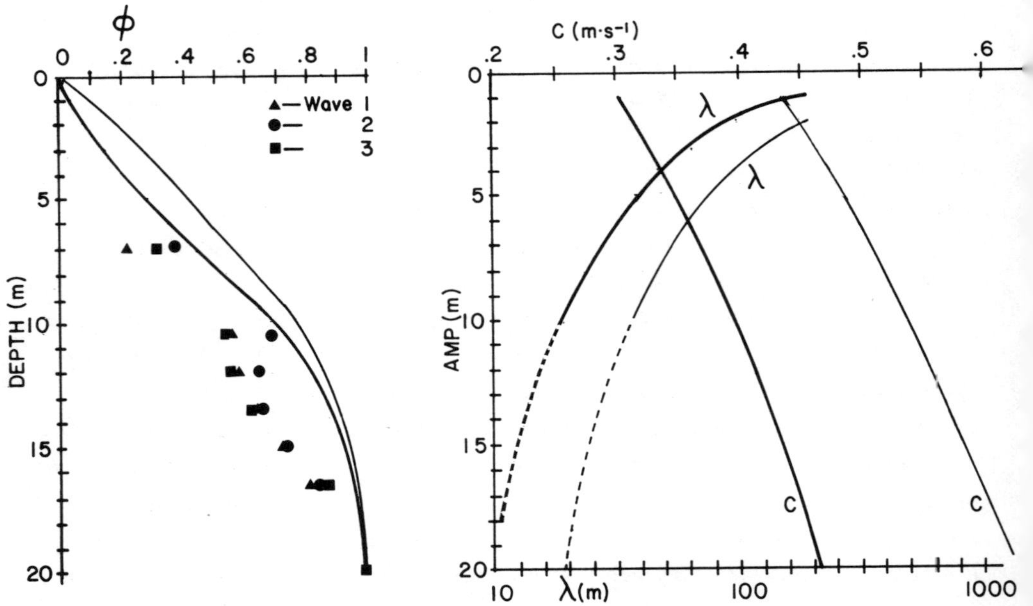

Figure 15. Comparison of calculated internal wave properties with and without shear. The right-hand graph shows the variation of phase speed with wave amplitude and length, and the variation of wave length with wave amplitude and phase speed; the left-hand one indicates the structure of the eigen function. Heavy lines correspond to calculations including shear effects, light ones are for zero shear.

A further test of the theory could be made if the time and position of wave generation were known. Identification of these was the subject of a recently completed experiment, most of the data from which have not yet been analyzed. Nevertheless, from the available results we now know that the waves are generated during the breakdown of an internal hydraulic jump over the down inlet side of the sill an hour or so before low water. Taking the speed calculated for the nonlinear waves in the absence of shear to be applicable throughout the entire distance over which they have propagated and assuming a zero basic flow in the inlet, yields a travel time of 3.9 hours; use of a phase speed lowered by the presence of the opposing shear observed at the experimental site yields a travel time of 5.2 hours. The former is not unreasonable and gives a generation time about 1 1/2 hours before low water. The latter is somewhat too large.

Unfortunately we cannot assume that the waves travel at all times through a flow similar to that observed at the experimental site. Both the vertical shear and tidal advection change with tidal phase and location. Pickard and Rodgers (1959) found relatively less shear near the sill than at other locations; moreover the tidal flow is relatively small during the first two to three hours after the waves are formed. At this stage we might therefore expect the waves to travel at a speed closer to .55 ms^{-1} which is that predicted in the absence of shear. Without simultaneous observations taken at several locations it is hard to estimate the expected wave speed, but it is reasonable to suppose that their average velocity was between these two extremes. If the correct value averages to 0.50 ms^{-1} it would take 4.2 hours for the waves to travel the 7.6 km from their source to the experimental site. This is consistent with a generation time approximately 1.7 hours prior to low water. While these results do not exactly coincide with the summer observations they are close enough to suggest that the generation mechanism is similar, although possibly modified by the weaker stratification during winter.

5.2 Internal waves and fjord circulation

We have discussed observations of large amplitude internal waves in Knight Inlet accompanied by a deepening of the surface layer. This deepening is consistent with the concept of an undular bore, which we believe to be generated by the breakdown of an internal hydraulic jump over the sill. Our exponential shear and density distribution model suggests that critical conditions were approached at the base of the stratified layer for waves of the amplitude observed in Knight Inlet. Further numerical calculations based on the observed profiles and numerically evaluated eigenfunction, yield Richardson Numbers in the wave trough of order unity. This may be compared with values 10 to 100 times greater for the undisturbed flow. Whether the former value is sufficiently low to permit turbulence production with the passage of each wave is an unresolved question. However, there probably are sufficient small scale disturbances available to make this a distinct possibility and we suggest that the passage of the nonlinear internal wave train causes turbulent mixing in the lower part of the pycnocline serving to keep it broad at least under winter conditions. Of course a perturbation model based on small values of the parameter a/h could be inaccurate when applied to observations of large amplitude waves. However, it seems likely that successful comparison of theoretical models with observations will be as dependent on accurate estimates of the three dimensional flow field as on any further refinements of the model.

It appears that large amplitude internal waves of this sort are an effective mechanism for the extraction of energy from tidal flows over sills. If the waves are large enough shear flow instabilities convert part of this energy into turbulence, leading to mixing near the base of the stratified layer. Once below this critical amplitude the waves may proceed relatively unattenuated, probably eventually being destroyed through interactions with the inlet sides, as discussed by Stigebrandt (1976) in connection with internal tides. During the summer of 1977 we have observed waves travelling as far as Kwalate Point, about 12 km beyond the straight reach, but no further.

The wave properties are strongly influenced by shear. Taking Benjamin's (1967) model as a starting point we have shown that when travelling upstream through the type of velocity profile observed in Knight Inlet, the waves are slowed down and their horizontal scale is diminished. However, the shear in the lower part of the stratified zone is increased under these features, thus enhancing the opportunity for shear flow instability and mixing. Numerical calculations for waves travelling with the flow show that in this case the wave speed is greater and the horizontal scale λ is increased. Thus the direction in which the waves travel will have a significant effect on the critical amplitude and hence the opportunity for shear flow instabilities leading to turbulence production and mixing. It is the waves travelling upstream that will have the smallest critical amplitudes and will thus be most susceptible to instability.

Our pilot study was not designed to provide direct evidence of mixing due to internal waves. However it is clear from earlier hydrographic surveys, as well as our own CTD data, that the smoothly changing density structure in the sinuous region of Knight Inlet has a much smaller horizontal gradient than does the straight reach up inlet of the sill. Undoubtedly the waves have caused some aliasing of previously collected hydrographic data, but all the evidence supports the hypothesis of enhanced vertical exchange in this straight reach, and the internal waves that we have described appear to be a most likely source of energy for this process.

References

Abramowitz, M. and Stegun, I.A. (Editors), 1968. Handbook of Mathematical Functions, Dover.

Benjamin, T.B., 1962. The solitary wave on a stream with an arbitrary distribution of vorticity, J. Fluid Mech., 12: 97.

Benjamin, T.B., 1967. Internal waves of permanent form in fluids of great depth, J. Fluid Mech., 29: 559-592.

Davis, R.E. and Acrivos, A., 1967. Solitary internal waves in deep water, J. Fluid Mech., 29: 593-607.

Farmer, D.M., 1978. Observations of long nonlinear internal waves in a lake, J. Phys. Oceanog., 8, *in press*

Gargett, A.E., 1976. Generation of internal waves in the Strait of Georgia, British Columbia, Deep Sea Res., 23: 17-32.

Halpern, D., 1971. Observations of short period internal waves in Massachusetts Bay, J. Mar. Res., 24: 116.

Hunkins, K. and Fliegel, M., 1973. Internal undular surges in Seneca Lake: a natural occurence of solitons, J. Geophys. Res., 78(3): 539-548.

Lee, C. and Beardsley, P., 1974. The generation of long nonlinear internal waves in a weakly stratified shear flow, J. Geophys. Res., 79(3): 453-462.

Long, R.R., 1975. Circulations and density distributions in a deep, strongly stratified, two-layer estuary, J. Fluid Mech. 71(3): 529-540.

Pearson, C.E. and Winter, D.F., 1978. Two-layer analysis of steady circulation in stratified fjords. In: J. Nihoul (Editor), Hydrodynamics of Estuaries and Fjords, Elsevier, Amsterdam.

Pickard, G.L., 1961. Oceanographic features of inlets in the British Columbia coast, J. Fish. Res. Board Can., 18(6): 907-999.

Pickard, G.L. and Rodgers, K., 1959. Current measurements in Knight Inlet, British Columbia, J. Fish. Res. Board Can., 16(5): 635-678.

Redekopp, L.G., 1977. On the theory of solitary Rossby Waves, J. Fluid Mech., 82: 725-746.

Smith, J.D., 1974. Turbulent structure of the surface boundary layer in an ice covered ocean, Rapp.P. -v Reun. Cons. Int. Explor. Mer.

Stigebrandt, A., 1976. Vertical diffusion driven by internal waves in a sill fjord. J. Phys. Oc. 6(4): 486-495.

Thorpe, S., Hall, A. and Crofts, I., 1972. The internal surge on Loch Ness Nature, 237: 96-98.

Winter, D.F., 1973. A similarity solution for steady state gravitational circulations in fjords, Estuarine and Coastal Mer. Sci., 1: 387-400.

Ziegenbein, J., 1969. Short internal waves in the Strait of Gibraltar, Deep Sea Res., 16: 479.

TWO-LAYER ANALYSIS OF STEADY CIRCULATION IN STRATIFIED FJORDS

C.E. PEARSON and D.F. WINTER
Department of Aeronautics and Astronautics and Department of Oceanography, University of Washington, Seattle, WA 98195

ABSTRACT

In deep, narrow, fjord-type inlets with copious runoff near the head, freshwater inflow produces a surface slope and a pressure gradient which drives a brackish, near-surface layer seaward; at greater depths a denser, saline layer derived from oceanic water moves landward. We describe in this paper a self-consistent two-layer representation of this mode of inlet flow, generally referred to as "estuarine circulation." Our approach is different from other layered analyses in that the present model includes important effects of variations in mass density, channel width and depth, and also allows for turbulent and advective exchange between the deep and near-surface layers. The starting point of the analysis is a set of equations expressing incompressibility and conservation of mass and horizontal momentum in each zone. Transfer of mass and momentum across the interface between the layers is parameterized by two interzonal exchange flux rates, F_u and F_d, representing the upward and downward rate of fluid flow per square meter of interfacial area. When the time-averaged mass density variations can be estimated from field data, then the flux rates, F_u and F_d, can be inferred entirely from known or measurable quantities. Two integrals of the motion are immediately available, and the mathematical problem is reduced to solving a pair of nonlinear equations for the layer cross-sectional areas. By way of illustration, the procedure is applied to Knight Inlet, a deep, stratified fjord on the Pacific Northwest coastline.

TWO-LAYER ANALYSIS

Inlets with appreciable fresh water runoff frequently exhibit a two-layer circulation pattern in which a brackish near-surface layer moves seaward while a deeper saline layer moves landward. The near-surface layer salinity generally increases in the seaward

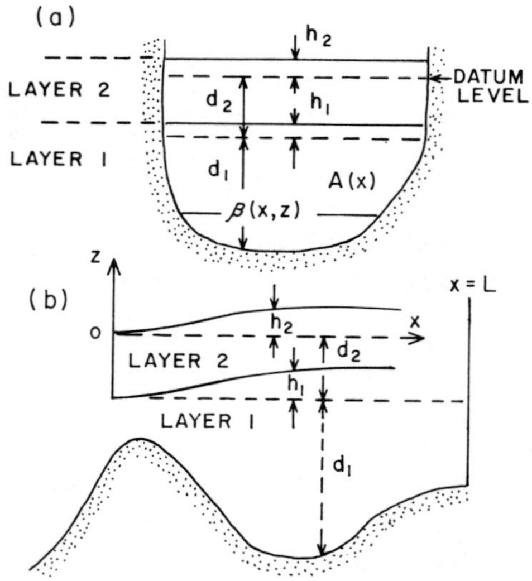

Figure 1. Sketch of an inlet cross-section (a) and longitudinal section (b), depicting the two major flow layers and illustrating geometrical quantities. The horizontal scale is compressed.

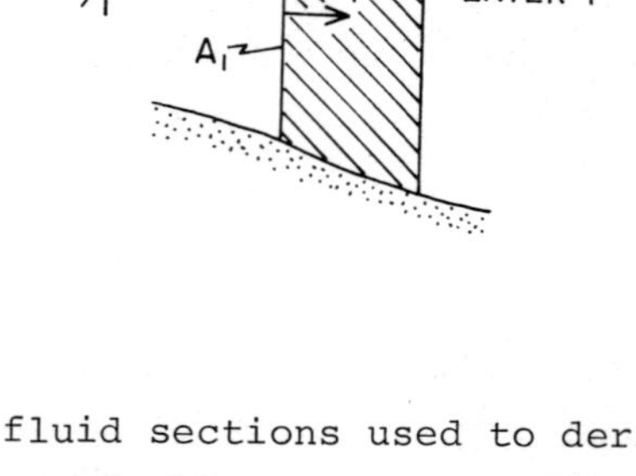

Figure 2. Sketch of fluid sections used to derive equations of conservation of volume, mass, and horizontal momentum.

direction because of entrainment of water from the lower layer. The water in the lower layer is derived largely from sea water external to the inlet, although it may be freshened somewhat during its progress by brackish water from above due to turbulent mixing between the layers. We consider here a simplified model of the estuarine circulation mode, constructed from an hydraulic analysis similar to that developed by Stoker (1957) for river flow and later adapted to flow in stratified estuaries by Vreugdenhil (1970), Long (1975), and others. The present analysis extends previous layer models by including bathymetric effects and by allowing for inter-layer exchange; the discussion is restricted to time-averaged conditions (over a tidal cycle, say).

It is assumed that the inlet is sufficiently narrow that a one-dimensional treatment is appropriate, in the sense that all quantities in each layer depend only on the distance x from the seaward end (Fig. 1). The z-axis is directed vertically upward from a horizontal datum level. Subscripts 1 and 2 are used to denote variables in the lower and upper layers, respectively. It is assumed that the fresh water in the inlet is introduced exclusively into the upper layer with zero horizontal velocity (except at $x = L$) at a cumulative volumetric rate given by $R(x)$; (thus, $R(x)$ denotes the total influx between x and L). Because of the typically small depth of the near-surface layer, its breadth $b(x)$ may be taken as constant over its depth. Let d_2 be an appropriate fixed reference depth, such as the layer thickness at the inlet mouth, and denote by h_1, h_2 the displacements of the lower and upper surfaces, respectively (Fig.1). The cross-sectional area of the near-surface layer, denoted by $A_2(x)$, can be written as

$$A_2(x) = b(x)[d_2 + h_2(x) - h_1(x)] . \qquad (1)$$

Next, suppose that the deep layer has cross-sectional area $A(x)$ relative to the datum level; the breadth of its top surface will be $b(x)$. The actual cross-section of the deep layer can then be represented as

$$A_1(x) = A(x) + b(x)\, h_1(x) . \qquad (2)$$

(It will be convenient subsequently to refer to a maximum depth $d_1(x)$ of the deep layer when the fluid is at rest.) Finally, the horizontal velocities of the deep and near-surface layers will be denoted by $u_1(x)$ and $u_2(x)$, respectively; each is positive when directed landward. Equations expressing incompressibility (or conservation of volume) and conservation of mass and horizontal momentum in each layer can now be written down.

In so doing, it is necessary to allow for (upward) convective motion and for turbulent mixing between the layers. In models where flow parameters are continuous functions of depth, vertical transport is described by vertical flow velocities and turbulent fluxes, the latter involving eddy coefficients of viscosity and diffusion. Since turbulent exchange mechanisms in estuarine flow are not at all well understood, the eddy coefficients are usually assigned values or functional forms which are adjusted until calculated flow and hydrographic patterns are similar to those observed in practice. In layered models of coastal plain and salt wedge estuaries, the approach has not been much different; the traditional procedure is to assign adjustable values to friction coefficients at a layer interface until some correspondence is achieved between computation and observation. In the present work, convective and turbulent transfer between the layers is represented by two interzonal exchange flux rates denoted by F_u and F_d. The symbol F_u represents the upward volume rate of flow of fluid from the deep layer to the near-surface layer per square meter of interfacial area. Likewise, F_d denotes the downward volumetric flux rate from the upper to the lower layer.

Equations expressing incompressibility or volume conservation for each layer are derived from considerations of volume flow rates in and out of sectional "slices" of thickness Δx, illustrated in Fig. 2. For the deep and near-surface layers respectively, one obtains

$$(u_1 A_1)' = b(F_d - F_u) \tag{3}$$

and

$$(u_2 A_2)' = b(F_u - F_d) - R' , \tag{4}$$

where a prime denotes differentiation with respect to x. Similarly, the equations expressing conservation of mass in the lower and upper layers are

$$(\rho_1 u_1 A_1)' = b(\rho_2 F_d - \rho_1 F_u) \tag{5}$$

and

$$(\rho_2 u_2 A_2)' = b(\rho_1 F_u - \rho_2 F_d) - \rho_o R' \qquad (6)$$

where ρ_o denotes the mass density of fresh water. In writing down Eqs. (5) and (6), it has been assumed that horizontal diffusion is negligible compared with advective transport. This is a common assumption in deep inlet studies and is based upon the findings of certain oceanographic field studies of fjord waters (e.g., McAlister et al., 1959; Dyer, 1973). On the other hand, it may not be justified in the vicinity of a long shallow sill where tidally induced longitudinal dispersion of mass can become competitive with advection.

To derive the momentum equations, we use the fact that the net force on each slice of fluid, in Fig. 2, must equal the net rate of efflux of momentum from that slice. Consider first the lower slice. The net force on it is made up of a pressure imbalance across its two faces, the pressure force on the sides due to changes in breadth, the horizontal component of the pressure force acting on the interface, and the frictional stress $\tau_f(x)$ acting on the wetted perimeter C_w:

$$-\frac{d}{dx}\left[\int_{-[d_2+d_1(x)]}^{-d_2+h_1(x)} p\,\beta(x,z)\,dz\right] + \int_{-[d_2+d_1(x)]}^{-d_2+h_1(x)} p\,\beta' dz + g\,\rho_2 b(d_2+h_2-h_1)(h_1)' + \tau_f C_w$$

$$= (\rho_1 u_1^2 A_1)' + b(\rho_1 u_1 F_u - \rho_2 u_2 F_d) \qquad (7)$$

where p is the pressure, g the acceleration of gravity, and $\beta(x,z)$ the breadth of the lower layer at vertical position z. The usual convention of shallow water theory is adopted to the effect that p results from hydrostatic forces only, so that

$$p = g[\rho_2(d_2+h_2-h_1) + \rho_1(-d_2+h_1-z)] \qquad (8)$$

In carrying out the differentiation indicated for the first term of the momentum conservation equation above, there is no contribution from the x-dependence of the lower limit, since $\beta(x,-[d_2+d_1]) = 0$. Furthermore, there is no contribution from this term if channel has a flat bottom since, in that case, the momentum equation contains an additional term of equal magnitude and opposite sign. The final result can be written as

$$u_1(u_1)' = -g\left(\frac{A_1 - A}{b}\right)' - \frac{g}{\rho_1}\left(\frac{\rho_2 A_2}{b}\right)' + g\,\frac{(\rho_1)'}{\rho_1 A_1}\left[\left((1-\gamma)d_2 - \frac{1}{2}\gamma d_1\right)A\right.$$

$$\left. - \frac{A_1^2 - A^2}{2b}\right] + \frac{\rho_2}{\rho_1}\frac{bF_d}{A_1}(u_2 - u_1) + \frac{\tau_f C_w}{\rho_1 A_1} , \qquad (9)$$

where γ is defined by the relation

$$\gamma(-d_2 - \frac{1}{2}d_1)\,A = \int_{-d_2-d_1}^{-d_2} z\beta(x,z)\,dz\ . \qquad (10)$$

Clearly, in the case of a rectangular channel, $\gamma = 1$. A similar line of argument leads to the equation of horizontal momentum applicable to the near-surface layer:

$$u_2(u_2)' = -g\left(\frac{A_1 + A_2 - A}{b}\right)' - \frac{1}{2}g\frac{(b\rho_2)'}{\rho_2}\frac{A_2}{b^2} + \frac{\rho_1}{\rho_2}\frac{bF_u}{A_2}(u_1 - u_2)$$

$$+ \frac{\tau_w b}{\rho_2 A_2} + \rho_o u_2 \frac{R'}{\rho_2 A_2} , \qquad (11)$$

where $\tau_w(x)$ is the stress exerted on the upper surface by wind.

Equations (3) - (6), (9), and (10) can be regarded as a system of equations for the six unknown dependent variables ρ_1, ρ_2, A_1, A_2, u_1, and u_2. Two integrals can be obtained immediately by adding Eqs. (3) and (4), and Eqs. (5) and (6), respectively:

$$u_1 A_1 + u_2 A_2 = -R \qquad (12)$$

and

$$\rho_1 u_1 A_1 + \rho_2 u_2 A_2 = -\rho_o R\ . \qquad (13)$$

These two equations will be used subsequently to eliminate a pair of dependent variables from the system.

We suppose now that ρ_1 and ρ_2 (and therefore also ρ_1' and ρ_2') are known from field measurements as functions of x; R(x) is also supposed known. The quantities (u_1A_1) and (u_2A_2) can then be determined by solving the linear equation system (12) and (13):

$$u_1A_1 = \frac{R(\rho_2 - \rho_o)}{\rho_1 - \rho_2} \tag{14}$$

$$u_2A_2 = \frac{-R(\rho_1 - \rho_o)}{\rho_1 - \rho_2} \tag{15}$$

Using Eq. (14) in Eqs. (3) and (5), we obtain expressions for the exchange flux rates F_u and F_d:

$$F_d = -\frac{\rho_1'(\rho_2 - \rho_o)R}{(\rho_1 - \rho_2)^2 b} \tag{16}$$

$$F_u = -\frac{\rho_2'(\rho_1 - \rho_o)R}{(\rho_1 - \rho_2)^2 b} + \frac{(\rho_2 - \rho_o)R'}{(\rho_1 - \rho_2)b} \tag{17}$$

From Eqs. (14) and (15), u_1 and u_2 can be expressed in terms of A_1 and A_2. Since Eqs. (16) and (17) determine F_d and F_u, it now follows that Eqs. (9) and (11) constitute a coupled pair of differential equations for A_1 and A_2 of the form

$$a_{11}A_1' + a_{12}A_2' = b_1 \tag{18}$$

$$a_{21}A_1' + a_{22}A_2' = b_2 \tag{19}$$

where the a_{ij} and b_i coefficients are certain functions of A_1, A_2, and axial distance x. (The coefficients are readily derived from information provided above, but inasmuch as they are quite lengthy, the expressions are not presented here.) The system (18) and (19) can be readily integrated by, say, a conventional fourth order Runge-Kutta method to obtain A_1 and A_2 as functions of x. Of course, initial conditions at $x = 0$ must be specified; because the choice of the datum level is arbitrary, we can set $h_2 = 0$ at $x = 0$, and the value of h_1 at $x = 0$ can be obtained from the measured salinity (or, preferably, horizontal current) profiles. In turn, these two values determine A_1 and A_2 at $x = 0$. Once $A_1(x)$ and $A_2(x)$ have been calculated, $u_1(x)$ and $u_2(x)$ may be obtained from Eqs. (14) and (15), and the solution is complete. An example is provided in the last section.

STABILITY

The numerical values of the coefficients in Eqs. (18) and (19) are frequently such that the two equations are very similar to one another, and consequently it is advantageous to solve the modified system

$$a_{11}A_1' + a_{12}A_2' = b_1 \tag{20}$$

$$(a_{21} - a_{11})A_1' + (a_{22} - a_{12})A_2' = (b_2 - b_1) , \tag{21}$$

where the second of these equations is obtained by analytically subtracting Eq. (19) from Eq. (18). Even so, the solution of the modified system is fairly sensitive to the quantity $(\rho_1 - \rho_2)$, as might be expected from its appearance in denominators in Eqs. (14) through (17), and from the fact that $\rho_1 - \rho_2$ is generally small compared with either ρ_1 or ρ_2.

It turns out that for sufficiently small values of $(\rho_1 - \rho_2)$ the two equations (20) and (21) can cease to have a solution. This can occur in the neighborhood of a sill, for example, where intense turbulent mixing may take place by means of physical processes whose effects are not adequately represented in the present model. In order to gain further insight into such a situation, we confine our attention to the simple case in which R, B, and A are constant. Then Eqs. (20) and (21) become

$$\left(\frac{g}{b} - \frac{u_1^2}{A_1}\right)A_1' + \frac{g\rho_2}{\rho_1 b}A_2' = -\frac{g}{b}\frac{\rho_2'}{\rho_1}A_2 - \frac{g\rho_1' A_1}{2b\rho_1}$$

$$+ \frac{\rho_2 b F_d}{\rho_1 A_1}(u_2 - u_1) - \frac{bu_1}{A_1}(F_d - F_u) \tag{22}$$

$$\frac{u_1^{\,2}}{A_1} A_1' + \left[\frac{g}{b}\left(1 - \frac{\rho_2}{\rho_1}\right) - \frac{u_2^{\,2}}{A_2}\right] A_2' = -\frac{g\rho_2' A_2}{2\rho_2 b} + \frac{g\rho_2' A_2}{\rho_1 b} + \frac{g\rho_1' A_1}{2\rho_1 b}$$

$$+ (u_1 - u_2)\left[\frac{\rho_1 b F_u}{\rho_2 A_2} + \frac{\rho_2 b F_d}{\rho_1 A_1}\right] - (F_u - F_d)\left(\frac{bu_2}{A_2} + \frac{bu_1}{A_1}\right) \tag{23}$$

Typical numerical values are such that, to a first approximation, Eq. (22) states that $A_1 + A_2$ is approximately constant and equal to (say) K. Then in Eq. (23), A_1' can be replaced by $-A_2'$. Next we write $\rho_1 = \rho_2 + \Delta\rho$ and assume that $\Delta\rho$ and ρ_2' are approximately constant over some range of x-values of interest. In most examples we have considered, A_1 has been much greater than A_2, and the dominant terms on the right-hand side of Eq. (23) have been the first three. It then turns out that Eq. (23) can be approximated by

$$\left[\frac{g}{b}\frac{\Delta\rho}{\rho_o} - \frac{R^2(\rho_1 - \rho_o)^2}{(\Delta\rho)^2 A_2^{\,3}}\right] A_2' \;\tilde{=}\; \frac{g\rho_1' K}{2\rho_o b} \tag{24}$$

If $\rho_1 - \rho_o$ is treated as approximately constant, integration of Eq. (24) yields

$$A_2 + \frac{\alpha}{A_2^{\,2}} = \frac{1}{2}\frac{\rho_1 K}{\Delta\rho} + \text{constant} \tag{25}$$

where

$$\alpha = \frac{bR^2(\rho_1 - \rho_o)^2 \rho_o}{2(\Delta\rho)^3 g} \tag{26}$$

Write $\rho_1 = \rho_{10} - \varepsilon x$, where $-\varepsilon$ is the (constant) value of ρ_1' and approximate α by

$$\alpha \;\tilde{=}\; \frac{bR^2(20)^2(1000)}{2(\Delta\rho)^3(10)}, \tag{27}$$

where we have inserted representative values for $(\rho_1 - \rho_o)$ and for ρ_o (in metric units). Then Eq. (25) may be written as

$$A_2 + \frac{\alpha}{A_2^{\,2}} \cong A_{20} - \frac{K\varepsilon x}{2\Delta\rho} \tag{28}$$

where A_{20} is close to the value of A_2 at $x = 0$. The left-hand side of Eq. (28) has a minimum value when $A_2 = \sqrt[3]{2\alpha}$ and, consequently, from Eq. (28) the maximum value of x must be given by

$$\frac{K\varepsilon x_{max}}{2\Delta\rho} = A_{20} - \frac{3}{2}(2\alpha)^{1/3} \tag{29}$$

If the average depth of the channel is D, then from Eq. (27) it follows that the maximum length over which the quasi-laminar flow model can be valid is given by

$$x_{max} \cong \frac{2(\Delta\rho)h_o - 100(R/b)^{2/3}}{D\varepsilon} \tag{30}$$

where h_o is the thickness of the upper layer at $x = 0$. This equation yields results compatible with numerical solutions of Eqs. (20) and (21) in appropriate cases, in the sense that the thickness of the upper layer, as computed from Eqs. (20) and (21) vanishes near this value of x.

APPLICATION TO KNIGHT INLET, BRITISH COLUMBIA

The method is illustrated by application to Knight Inlet in the southern portion of the mainland British Columbia coastline.

Knight Inlet is a deep fjord located about 320 km northwest of Vancouver, British Columbia. It is approximately 110 km long and has an average width of about 3 km (see map in Fig. 3). Knight Inlet is a positive, high-runoff fjord estuary which is in communication with the Pacific Ocean by way of Johnstone Strait and Queen Charlotte Strait. The inlet is divided into two basins by a threshold of approximately 65 m depth, located at 75 km from the head. The outer basin, which is the shallower of the two, has a second 65 m deep threshold at the confluence of the inlet with marine waters in Johnstone Strait. The outer basin is characterized by depths in the range 150 to 200 m, and has an irregular

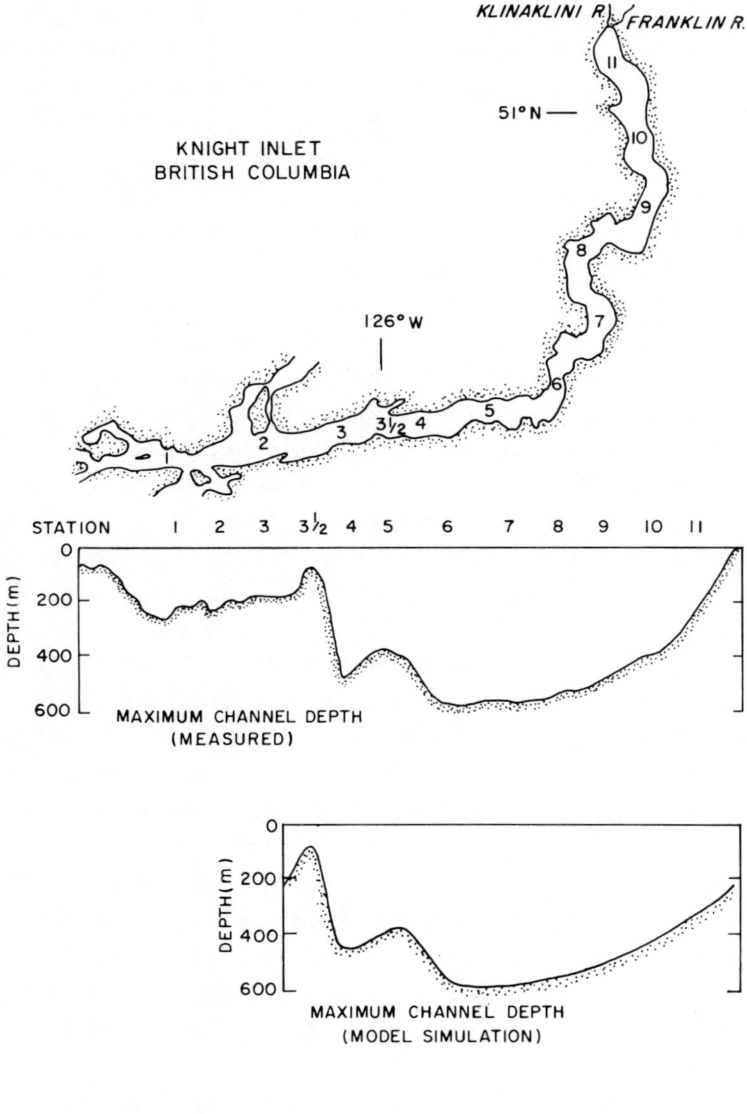

Figure 3. Map of Knight Inlet, British Columbia, together with observed longitudinal section (after Pickard and Rodgers, 1959), and maximum depth variation used in model calculations.

shoreline indented by passes leading to adjacent bodies of water. The inner basin is deep over most of its extent, with the maximum depth in excess of 550 m. A longitudinal section of Knight Inlet is shown in Fig. 3; in the same figure is shown an idealized representation of the inner basin and inner sill depth profile, based on elementary functions (trigonometric and quadratic forms), which was used in the model calculations below.

Most of the fresh water in Knight Inlet is derived from the Klinaklini and Franklin Rivers which discharge into the head. Although runoff is significant throughout most of the year, its intensity exhibits a seasonal variation characterized by a January-March minimum and a May-July maximum, the latter being due to summertime melting of snow which fell on the adjacent mountains during the winter. A secondary runoff maximum may occur during the rainy season in October-November. The Klinaklini and Franklin Rivers are not routinely gauged, but estimates of fresh water discharge for Knight Inlet are available (Trites, 1955; Pickard, 1961), which indicate an annual mean of 410 $m^3\ sec^{-1}$, and high runoff rates in the neighborhood of 600 and 750 $m^3\ sec^{-1}$ during the months of June and July, respectively.

Inlets with the yearly discharge pattern of Knight Inlet have been termed Type A.1 (stored runoff) fjords by Pickard (1961). Typically, the surface salinity is well below 2 o/oo near the head when runoff is intense and it remains at low values over substantial portions of the inlet. Figures 4a and 4b show continuous profiles of salinity obtained during 3-4 June 1951 and 15 July 1953, as presented by Trites (1955). Pronounced stratification over the major portion of the inlet is evident in these traces; the data from Stations 3-5 near the sill can only be regarded as qualitative since it is now known that aliassing is a serious problem throughout the region which includes these stations, (D.M. Farmer and J.D. Smith, personal communication, 1977).

Model calculations were performed for comparison with the Trites' data from June and July, and the computed results for the salinity distributions are exhibited in Figs. 4c and 4d. The depth profile in Fig. 3 was used, together with an assumed constant main channel width of 2 km. The mean monthly cumulative fresh water discharges R for June and July were taken to be independent of x and equal to 600 and 750 $m^3\ sec^{-1}$, respectively. Average upper and lower layer salinities were estimated from the data. The thickness of the near surface layer near the head was assigned a value of 7 m

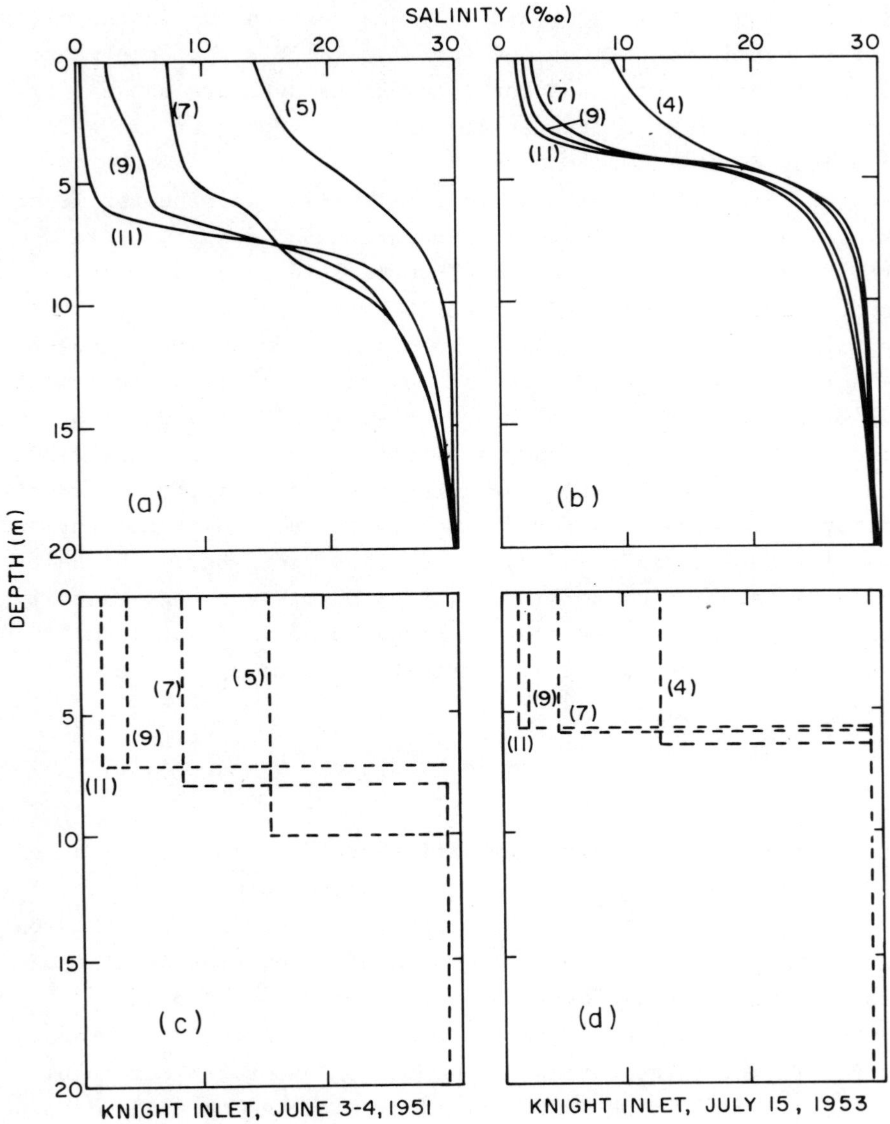

Figure 4. Depth profiles of salinity measured on (a) 3-4 June 1951 and (b) 15 July 1953, as presented by Trites (1955); model salinity profiles for conditions in (c) June 1951 and (d) July 1953.

for June and 6 m for July, as suggested by the measurements, and the integration of Eqs. (20) and (21) was carried out in the direction of decreasing x. It can be seen from Fig. 4 that the general features of the salinity and zonal thickness data are adequately represented over much of the inlet length, except near the sill where aliassing is known to occur.

Figure 5 shows the calculated axial variation of upper layer thickness, zonal velocities, and the interzonal exchange rate F_u for the month of June (F_d was equal to zero in this calculation since ρ_1 was taken to be independent of x). Unfortunately, we lack appropriate field data for making meaningful comparisons between time- and depth-averaged horizontal current profiles and calculated values of u in the upper and lower zones. Field data described by Pickard and Rodgers (1959) suggests that, in the absence of wind stress, near the Knight Inlet sill the average horizontal current speed may be as high as a few decimeters per second at the surface, but it decreases rapidly in magnitude toward zero below the halocline. A consideration by Winter (1973) of distance and velocity scales appropriate to near-surface circulation in stratified fjords suggests that the characteristic transport speed u_o in the upper zone is

$$u_o = R_o/b_o z_o \sigma_o \quad , \tag{31}$$

where R_o is a representative cumulative runoff value, b_o is an effective channel width, z_o is a representative thickness of the upper zone, and σ_o is the fractional salinity excursion from great depths to the surface in the inlet segment under consideration. If we assign representative mid-inlet values of $R_o = 600\ m^3\ sec^{-1}$, $b_o = 2000$ m, $z_o = 8$ m, and $\sigma_o = 2/3$, then the characteristic transport speed for Knight Inlet during the month of June turns out to be $u_o = 0.056\ m\ sec^{-1}$, which is in agreement with the calculated value of $u_2(x)$ in the vicinity of Stations 5 and 7, but lower than might be inferred from Pickard and Rodgers' measurements. If the characteristic horizontal scale x_o is determined by a balance between mixing and advective salt transport, as assumed by Winter (1973), then

$$x_o = u_o z_o^2/K_o \quad , \tag{32}$$

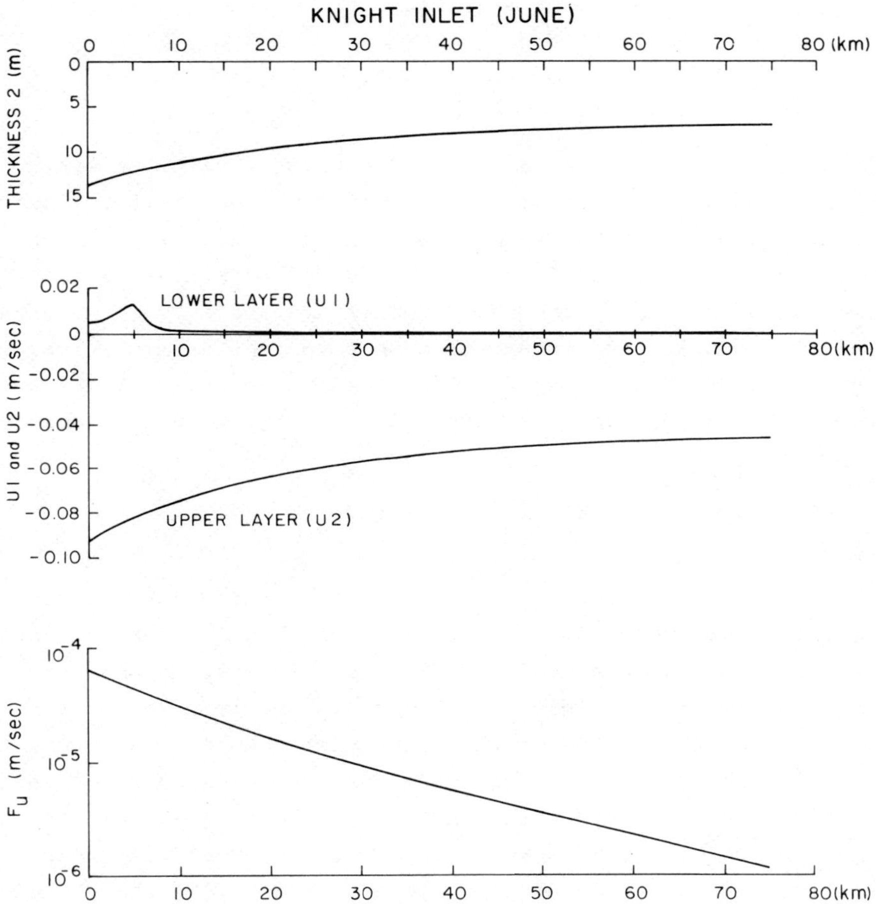

Figure 5. Calculated axial variations of upper layer thickness, average upper and lower layer speeds, and upward flux rate F_u in Knight Inlet, B.C., for June conditions. Coordinate x is measured positive landward from Station 3, seaward of the sill.

where K_o is a representative value of the vertical eddy diffusivity. From the equation of continuity it follows that the characteristic vertical velocity w_o is of the order of

$$w_o = u_o z_o/x_o . \tag{33}$$

Trites (1955) has estimated that K_o in Knight Inlet for early summer conditions is of the order of 2×10^{-4} m^2 sec^{-1}; it follows from the equations above that $x_o = 18000$ m and $w_o = 2.5 \times 10^{-5}$ m sec^{-1}. The latter quantity is of the same order of magnitude as the computed values of the upward interzonal flux rate F_u. In this same connection, it is also interesting to note that from physical model experiments, Keulegan (1949) established an empirical relationship between the entrainment velocity and the longitudinal velocity of the upper layer:

$$F_u = 3.5 \times 10^{-4} u_o , \tag{34}$$

in the present notation. The value 0.056 m sec^{-1} for u_o leads to an estimate of 2×10^{-5} m sec^{-1} for F_u, which is of the same order as calculated mid-inlet values (Fig. 5). It would appear from the figure that in Knight Inlet F_u attains its maximum value at $x = 0$, i.e., at the seaward end of the main inlet segment. However, this feature is an artifact of the functional representation of the upper layer mass density variation $\rho_2(x)$. Field data indicate that in reaches of the inlet seaward of $x = 0$, ρ_2' eventually decreases faster than $(\rho_1 - \rho_2)^2$, and it follows from Eq. (17) that F_u ultimately declines to smaller values.

Figure 6 presents longitudinal variations of surface salinity and salinity at depth as functions of x, where x is measured positive landward from Station 3, located a few kilometers seaward of the inner sill. The points in the figure represent the average of field measurements acquired during the month of June in the years 1951, 1973, and 1974. It would appear from the figure (and from other field data) that mixing processes taking place near the inner threshold cause the salinity of the surface layer to increase substantially from about 30 km upstream from the sill.

Figure 7 shows averages of salinity measurements at selected depths at fixed stations in the inlet during the months of March and June in the three years 1951, 1973, 1974. The runoff during

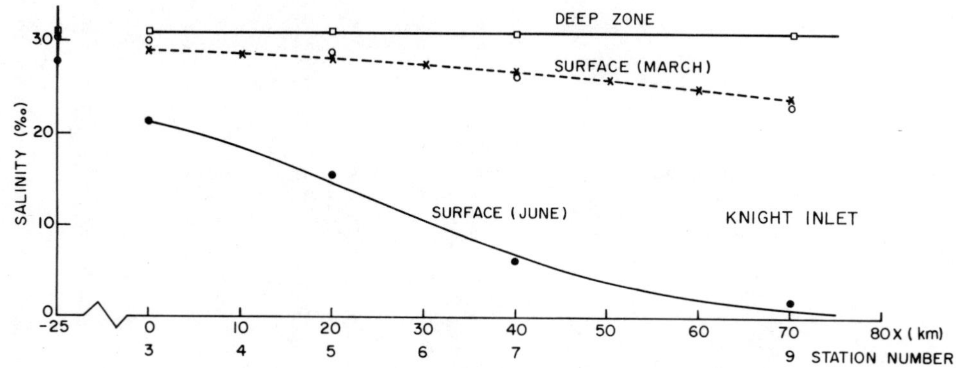

Figure 6. Longitudinal variations of surface salinity and salinity at depth for March and June conditions in Knight Inlet. The data points are averages of field measurements acquired in the years 1951, 1973, and 1974.

March is estimated to be of the order of 250 m^3 sec^{-1}. It may be noted that the scatter in the measurements in the pycnocline leads to an apparent distortion of the variation of salinity with depth, as exemplified by the traces in Fig. 4. Nevertheless, calculations based on the averaged measurements seem to give reasonable representations of the upper layer thickness variations for both runoff regimes in Knight Inlet.

It can be seen from Fig. 5 that, during high runoff, the upper layer thickness increases monotonically in the seaward direction from 6 or 7 m near the head to about 20 meters on the seaward side of the sill. Throughout the greater part of the inlet, however, the calculated interface depth is between 6 and 10 m, in accord with observation. On the basis of a somewhat different analysis (lower layer at rest, critical flow at the mouth, specified entrainment rate), Long (1975) has calculated a halocline depth of 21 m for Knight Inlet, but this appears to be somewhat of an overestimate. Only on the seaward side of the sill does the time-averaged upper layer thickness attain a value of the order of 20 meters. It is an interesting fact that when integrations of Eqs. (20) and (21) are performed with the same longitudinal density profiles ρ_1 and ρ_2,

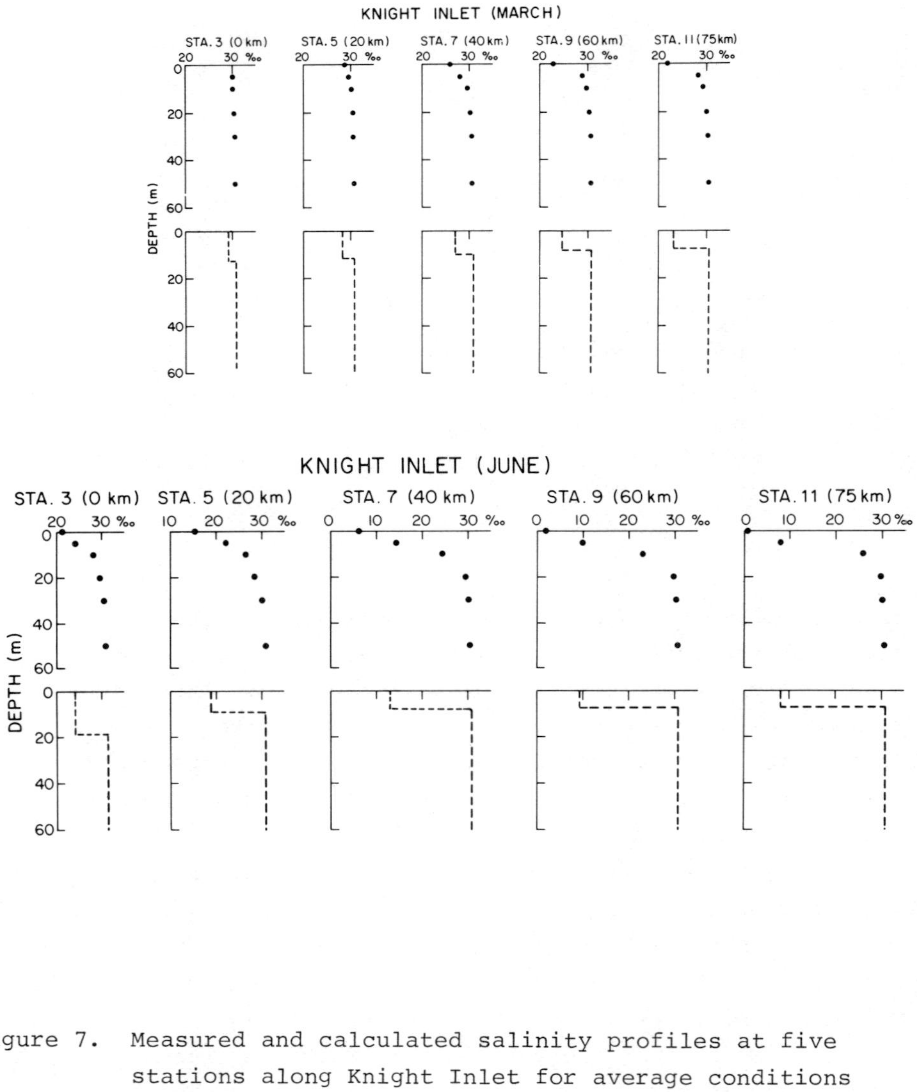

Figure 7. Measured and calculated salinity profiles at five stations along Knight Inlet for average conditions during March and June.

but with somewhat different starting values of upper layer thickness, the calculated upper layer thicknesses from mid-inlet toward the head are still of the order of a decameter or less.

We conclude from the results for Knight Inlet, and from calculations performed for several other fjords along the British Columbia coastline, that the present model can provide reasonable, self-consistent representations of the main features of time-averaged estuarine circulation in stratified inlets.

ACKNOWLEDGMENTS

We are grateful to Dr. David M. Farmer for several useful discussions of the physical oceanography of British Columbia inlets. The research described in this paper was partially supported by the National Coastal Pollution Research Program of the Environmental Protection Agency under Grant No. R-801320, and by the National Science Foundation, Oceanography Section, under Grant No. OCE76-80720. A travel grant from the National Science Foundation made it possible to present this work at the Ninth International Liège Colloquium on Ocean Hydrodynamics. Salinity data for Knight Inlet were obtained from Data Reports prepared by the Institute of Oceanography, University of British Columbia, Vancouver, B.C.

REFERENCES

1. Dyer, K. R., 1973. Estuaries: A Physical Introduction. John Wiley, London, 140 pp.
2. Keulegan, G. H., 1949. Interfacial instability and mixing in stratified flows. Journal of Research, National Bureau of Standards, 43: 487-500.
3. Long, R. R., 1975. Circulations and density distributions in a deep, strongly stratified, two-layer estuary. Journal of Fluid Mechanics, 71: 529-540.
4. McAlister, W. B., Rattray, M., Jr., and Barnes, C. A., 1959. The dynamics of a fjord estuary: Silver Bay, Alaska. Department of Oceanography Technical Report No. 62, University of Washington, Seattle, Washington, 70 pp.
5. Pickard, G. L., 1961. Oceanographic features of inlets in the British Columbia mainland coast. Journal of the Fisheries Research Board of Canada, 18: 907-999.
6. Pickard, G. L. and Rodgers, K., 1959. Current measurements in Knight Inlet, British Columbia. Journal of the Fisheries Research Board of Canada, 16: 635-678.
7. Stokes, J. J., 1957. Water Waves. Interscience, New York, 567 pp.
8. Trites, R. W., 1955. A study of the oceanographic structure in British Columbia inlets and some of the determining factors. Ph.D. Thesis, Institute of Oceanography, University of British Columbia, Vancouver, B.C., 125 pp.

9. Vreugdenhil, C. B., 1970. Computation of gravity currents in estuaries. Delft Hydraulics Laboratory Publication No. 86, 108 pp.
10. Winter, D. F., 1973. A similarity solution for steady state gravitational circulation in fjords. Estuaries and Coastal Marine Science, 1: 387-400.

Contribution No. 1001, Department of Oceanography, University of Washington, Seattle, WA.

VARIABILITY OF CURRENTS AND WATER PROPERTIES FROM YEAR-LONG OBSERVATIONS IN A FJORD ESTUARY

G. A. CANNON and N. P. LAIRD

Pacific Marine Environmental Laboratory, Environmental Research Laboratories, National Oceanic and Atmospheric Administration, Seattle, Washington 98105, U.S.A.

ABSTRACT

Observations were made of velocity, temperature, and conductivity from a vertical array of sensors at a location in Puget Sound for 1 year commencing in September 1975. Winds were measured nearby on land, and water properties were measured along the Sound at 2- to 3-month intervals. Most dense bottom water entered the Sound below sill depth at about fortnightly intervals during early fall near the end of the coastal upwelling season. The fortnightly events were associated with large tides over the entrance sill. Density decreased about linearly until early winter when cold water entered the Sound first at the bottom in a series of step decreases, also at about fortnightly intervals. The largest was 1.6°C in 3-4 days. Density increased during each inflow interval lasting about 5 days, but the overall density continued to decrease until early February when the water column became isothermal but was still stratified. During late winter and early spring, density again increased through a series of fortnightly salinity intrusions, and the initial one was more than 0.5 ‰ in 5 days. Mean daily near-bottom speeds were up to about 20 cm/sec, implying a possible excursion exceeding the length of the Sound. Winds were observed to alter mean daily current profiles to more than half the water depth. These and other winter observations indicated deep-water renewal was much quicker than earlier estimates.

INTRODUCTION

Puget Sound is a fjord-like estuary connecting through Admiralty Inlet and Deception Pass to the Strait of Juan de Fuca and then to the Pacific Ocean (Fig. 1). It is entirely within the State of Washington in the United States and is part of a larger estuarine system contiguous to several major population centers in the United States and Canada. About 98% of the tidal prism flows through Admiralty Inlet over a sill of about 64 m depth near Port Townsend. This sill is about 150 km from the ocean and extends south about 30 km. The main basin has depths exceeding 200 m and extends south about 60 km from the major junction with Admiralty Inlet near Possession Point to The Narrows, a constriction of about 44 m sill depth separating a southern basin. Within the main basin the 183-m contour delineates a deeper section approximately 50 km long and 3-5 km wide. The other main subdivisions include Hood Canal and the Whidbey basin, the latter extending from Possession Point to Deception Pass through Saratoga Passage. Numerous rivers enter the Puget Sound system. However, the Skagit entering in the north supplies more than 60% of the freshwater, about half of which flows southward toward the main basin (Barnes and Ebbesmeyer, 1977).

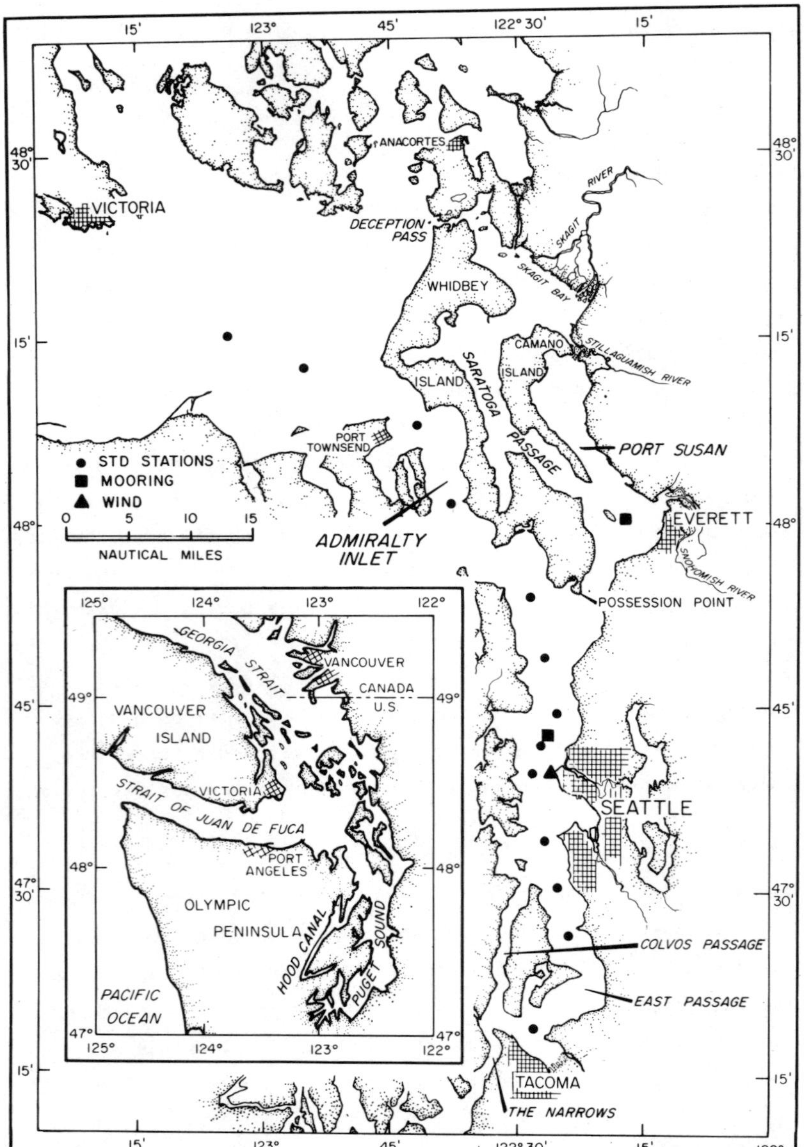

Fig. 1. Puget Sound region showing station locations.

Replacement of Puget Sound bottom water with more saline and denser water has been thought to occur in July-October in response to northerly summer winds along the Pacific coast. Northerly winds induce upwelling of denser oceanic water to shallow depths which then can be transported along the bottom of the Strait of Juan de Fuca into Puget Sound (Barnes and Collias, 1958). Additionally, long-term hydrographic observations show that bottom-water density in the Sound also increases during April-May and decreases during May-June and October-February (Collias *et al.*, 1974). During the winter decrease in density in November-March, mixing and/or replacement results in the deeper water in the main and Whidbey basins becoming relatively isothermal, less saline, and higher in dissolved oxygen (Barnes and Collias, 1958; Cannon, 1975). Because deep-water replacement previously only had been documented at about monthly intervals with water properties, observations of currents and water properties were made at several levels at a central location for 1 month during winter 1973 (Cannon and Ebbesmeyer, 1977).

Our previous work during the winter month showed that water near the bottom was cooled in a series of step decreases of temperature of up to 0.6°C at about 2-week intervals coincident with the onset of about 5-day intervals of net landward flow near the bottom. Between these intervals, bottom flow was tidal with a small seaward component. Temperature sections along the axis of the Sound showed following the first step a vertical temperature front close to the mooring, whereas at the beginning and end of the month there was no evidence of this front. Density sections showed that, following the step, less dense water (colder but less saline) was intruding to near the bottom just north of the mooring and the passing temperature front, and that an overall decrease in density occurred during the month (Fig. 2).

Deep water was shown to be replaced when flood-tide ranges at Seattle exceeded 3.5 m. At this range Strait of Juan de Fuca water had been shown by hydraulic model studies to completely transit the finite-length sill in Admiralty Inlet in one flood period and thus be least mixed (Farmer and Rattray, 1963). In our study the largest temperature decrease occurred following flood tides concurrent with below-freezing air temperatures over Admiralty Inlet. Density differences during the temperature steps were small, and it was not possible to determine whether the new water initially was more dense. Also, an 8-10 day transit time was required for this water to flow from the sill to the mooring site.

Finally, it was speculated that a similar replacement process might occur in late summer when the most dense bottom water can enter the Sound. Confirming current observations on the sill of Port Susan, a secondary basin in Puget Sound, showed flow into the basin at about fortnightly intervals lasting about 5 days each time (Cannon, 1975). It also was speculated that deep water might be replaced quasi-continuously at about 2-week intervals. Thus, a year-long series of observations was planned and made during 1975-76. This paper describes those observations and presents additional ideas concerning deep water replacement which might be applicable to other fjords.

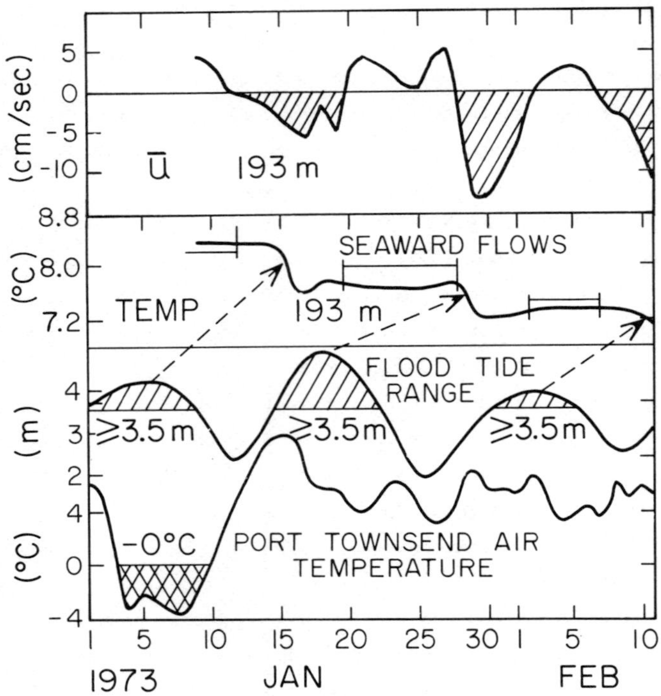

Fig. 2. Daily average longitudinal bottom currents and temperature in Puget Sound; range of greatest daily Seattle flood tide; and daily average Port Townsend air temperature (from winter 1973).

EXPERIMENT DESCRIPTION

The single current-meter mooring with seven Aanderaa meters with temperature and conductivity sensors was first deployed in mid-September 1975 in the main basin of Puget Sound in about 200 m of water (Fig. 1). It was recovered and redeployed in November, February, April, and June, and it was finally recovered in early October 1976. The NOAA ship *McArthur* was used on all but the June cruise which was carried out from the University of Washington's *Onar*. Nominal depths of the instruments were 15, 30, 50, 105, 145, and 195 m (5 m above bottom). The meters recorded every 20 min samples of average speed and instantaneous values of pressure, direction, temperature, and conductivity. The meters had relatively large vanes, making them fairly insensitive to high-frequency direction fluctuations. Buoyancy of about 450 kg was provided by an ORE sphere; the mooring cable was torque-balanced 3 × 19 cable; and the anchor was three railroad wheels weighing about 1000 kg. The mooring was deployed anchor first from the *McArthur* and buoy first from the *Onar*. The buoy or anchor was then lowered with a tension-release trip hook. The mooring was retrieved each time by AMF acoustic release of the anchor. Unfortunately, the mooring was snagged, apparently by a tug and barge, on 20 July (as shown in the data) and moved about 14 km south. Thus the year-long series at the single location was not completed. However, data from a spatial array deployed in the same vicinity on 20-21 July (by chance) and from the relocated mooring showed similar characteristics for the remainder of the year.

A survey of water properties along Puget Sound was made five times during the year, primarily coincident with the deployment cruises (Fig. 1). Observations were made at each station to within a few meters of the bottom using a pinger attached to a salinity-temperature-depth (STD) recorder (Bissett-Berman, model 9006). Nansen bottle samples were obtained to calibrate the STD. The occupied stations (taking about 12 hr) were chosen because of the relatively large historical data base at these sites, and the northern two stations were seaward of the entrance sill. A 25-hr time series made in February at three stations near the mooring showed no tidal signal in the water properties.

Wind speed and direction were measured using a Meteorology Research, Inc. mechanical weather station on a nearby low point of land, where winds seem to be representative of the main basin. The analog records were averaged hourly by hand, resolved into north and east components, and then daily averaged. Gaps in this data set were filled from U.S. Coast Guard observations at the same location.

Aanderaa current data were first reduced to north and east components, 2.86-hr filtered for time series plots and progressive vector diagrams, and 35-hr filtered for stick plots. Instantaneous values of temperature and salinity were plotted, and simple daily averages calculated because of the absence of any tidal signal. Velocities then were resolved into components along axes of maximum and minimum variance of the total record (Cannon, 1971). Primarily due to changing topography, these axes change with depth. The along-channel axes were about 20°-25° in the upper water and about 40°-45° in the deeper water. Tidal periodicities were filtered from the along-axis records utilizing fast-Fourier-transform techniques. Fourier coefficients were calculated for 30-day intervals. Coefficients for periods shorter than 30 hr were set to zero, the resulting series retransformed, and daily averages calculated for conceptual convenience. This technique has been compared to various other filters, and no significant differences have been found.

HYDROGRAPHY

The daily averages of density as sigma-t (where $10^3\ \sigma_t = \rho - 1$), salinity, and temperature from the year-long mooring have been plotted to show the characteristics of the variations (Figs. 3-5). Similar characteristics were observed at spatial array moorings following relocation of the year-long mooring. Density changes predominately were reflected in salinity changes with some important exceptions noted below near the bottom. Note the estuary was always stratified, and the smallest top-to-bottom differences in density (or salinity) probably occurred in late October.

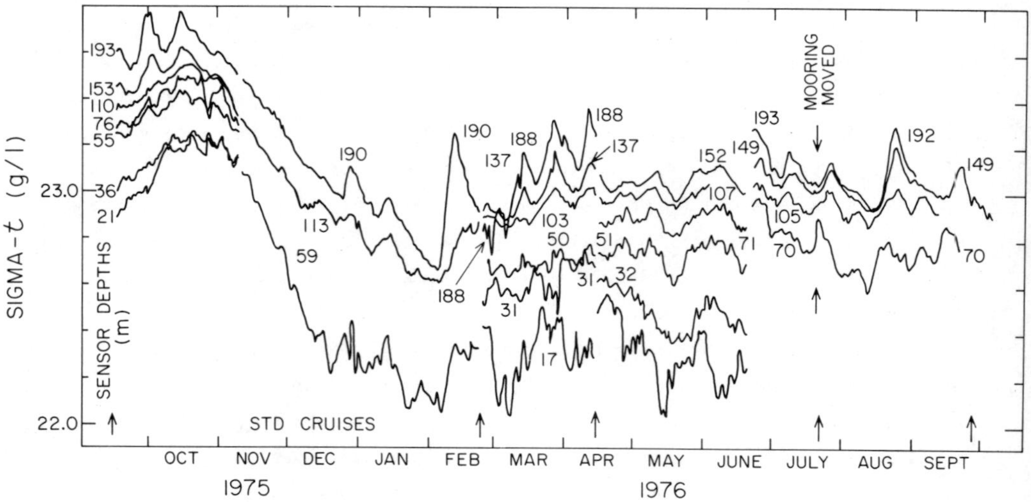

Fig. 3. Daily average sigma-t in Puget Sound main basin.

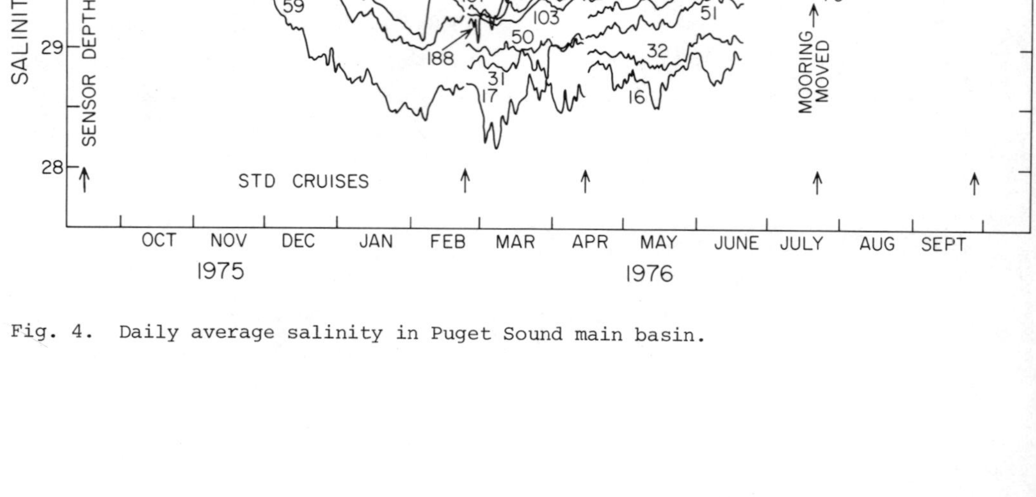

Fig. 4. Daily average salinity in Puget Sound main basin.

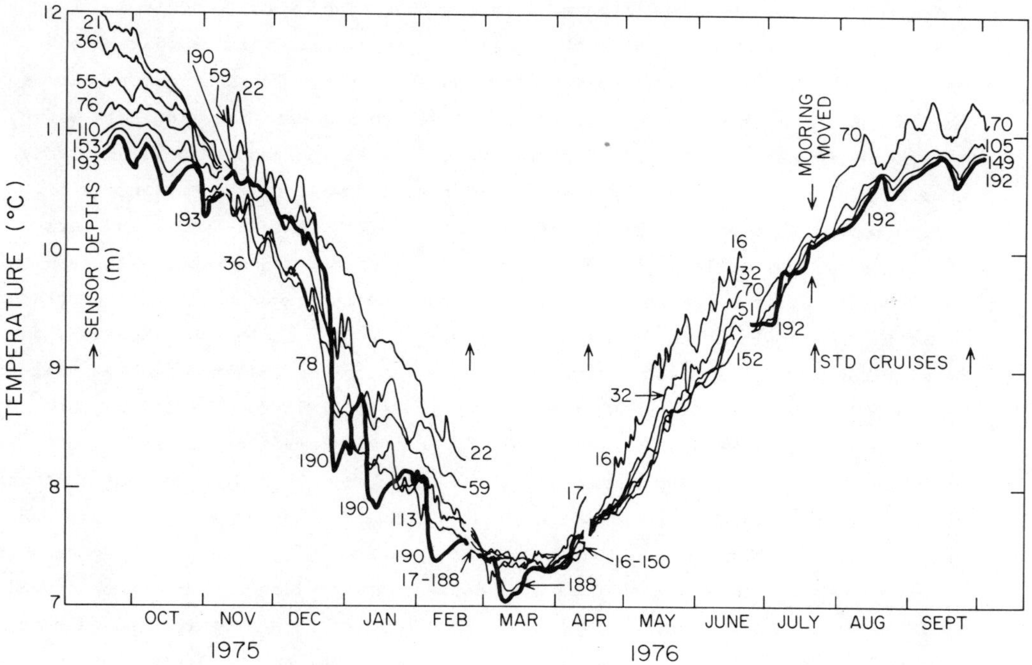

Fig. 5. Daily average temperature in Puget Sound main basin.

The change in bottom sigma-t during the year was relatively large, more than a 1.1 g/ℓ decrease from the maximum in October to the minimum in February. The maximum density appeared to occur later nearer the surface than at depth. The relative increases in density at the deeper levels lasting for a few days to a week are interpreted as new water entering the Sound across Admiralty Inlet sill. Current-meter data shown below verify this concept. Two prominent density increases occurred in September-October. These were followed by a relatively steady decrease in density until late December when two more relative increases occurred. The overall density, however, continued to decrease until the minimum was reached in early February. The single largest sigma-t increase of about 0.6 g/ℓ, more than half the yearly change, then occurred in about 1 week, followed by a series of increases raising the overall bottom density during the next 2 months. During one of these increases in late February, it appeared that the increase occurred at shallower depths before it occurred at the bottom. If this was so, the lower part of the water column may have been unstable temporarily. STD observations just before this event gave no indications of an instability, hence it may have been instrumental error.

Increases in density at the bottom almost always were evident at shallower depths up to about 50 m. Exceptions occurred in early August and in early September when increases observed at 70 and 105 m but not at 149 and 192 m indicated the only intermediate level intrusions. The overall magnitude of density was lower at the end of the yearly observations. This may have been partly the result of unusual summer conditions in 1976. There were more storms than usual, hence more freshwater was available to mix with incoming salt water. Also, the increased abundance of storms would result in lesser developed coastal upwelling, thus less dense water would progress landward to the entrance sill.

The temperature changed from a maximum of $11°-12°C$ in late summer and early fall to a minimum of less than $7.5°C$ in March. The top-to-bottom temperature range was greatest during highest temperatures and least during lowest temperatures. In fact, the top-to-bottom range was only $0.1°-0.2°C$ during March. However, the density range was relatively large, and the water column remained stably stratified. The fall 1976 values were about the same as those of 1975 for similar depths.

There was a remarkable large decrease in bottom temperature of about $1.6°C$ which occurred in 3-4 days commencing on 25 December. This decrease was followed by another of about $0.9°C$ commencing about 8 January. These can be explained by the process outlined above using the winter data from 1973 (Cannon and Ebbesmeyer, 1977). However, these temperature step decreases were larger than any observed in 1973, and, unlike those in 1973, they were evident as a definite increase in bottom density shown by the density peaks in late December and early January. Note, there was no corresponding salinity increase during these two events. All other density increases were accompanied by salinity increases.

There were seven temperature step decreases observed at the bottom from early October until March when the water column became isothermal. All but the one decrease in early November (more similar to those in 1973) was accompanied by a prominent density increase, and even it showed a slight increase. Density increases following the water column becoming isothermal in late March must be due to salinity increases. However, the intrusions then were characterized by relatively warmer water as evidenced in April and twice in July. Commencing in August, the intrusions again were characterized by colder water at all observed depths, and those at 70 m (about sill depth) occurred before those at deeper depths by several days.

STD profiles show the vertical characteristics of water properties at the mooring during the two extreme seasons (Fig. 6). Salinity and sigma-t curves were parallel, and only a slight pycnocline was observed in February because freshwater input was not nearby. Temperatures were warmer at the surface in September and about isothermal in February. Values observed with the moored instruments agreed well with the STD observations.

Sections along the Sound show the horizontal distributions of properties for the same two time periods (Figs. 7-8). The September section was occupied during an intrusion as shown by the temperature and salinity contours. In fact, at the mooring all water below about sill depth (64 m) was flowing southward. The

February section was occupied about 10 days after the largest salinity intrusion and at the onset of a smaller intrusion, but the only evidence of it is the colder temperatures and higher salinities of the deep water (see Figs. 4-5). Prior to the largest intrusion, salinity was less than about 29.0 °/₀₀ at the mooring, and, afterwards, water deeper than 100 m is all greater than 29.2 °/₀₀. Temperatures changed from about 8.0°C to about 7.5°C. The warmest deep water occurred at the southern end of the main basin and is characteristic throughout the cooling cycle.

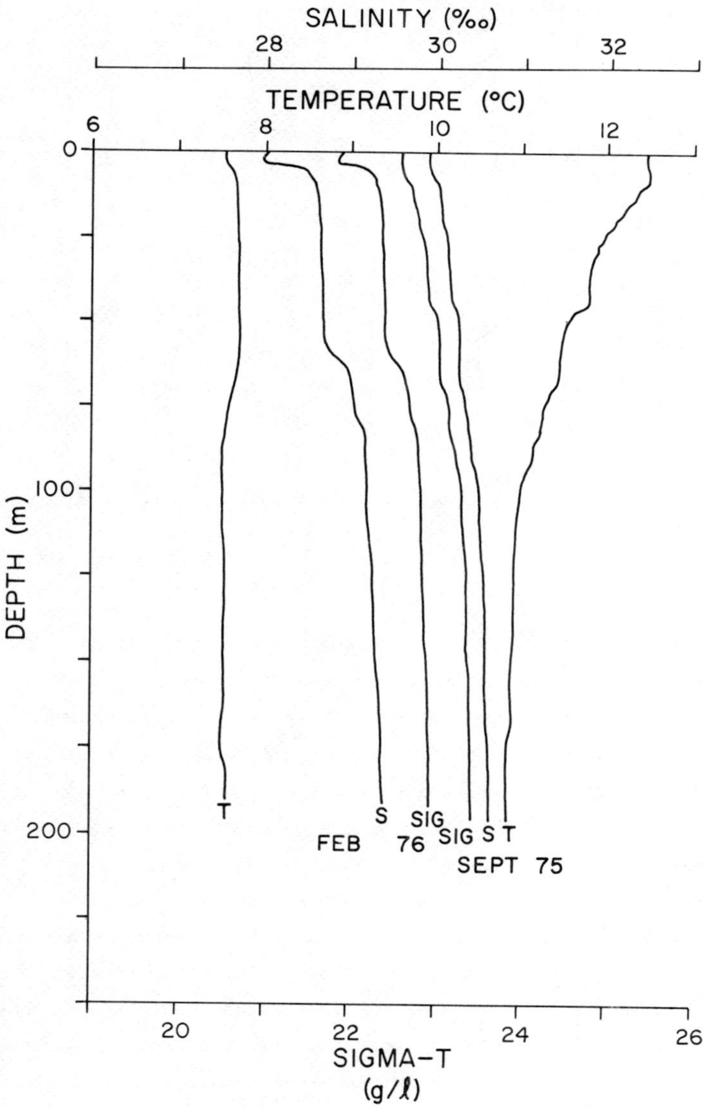

Fig. 6. Salinity, temperature, and sigma-t vertical profiles near the mooring in summer and winter.

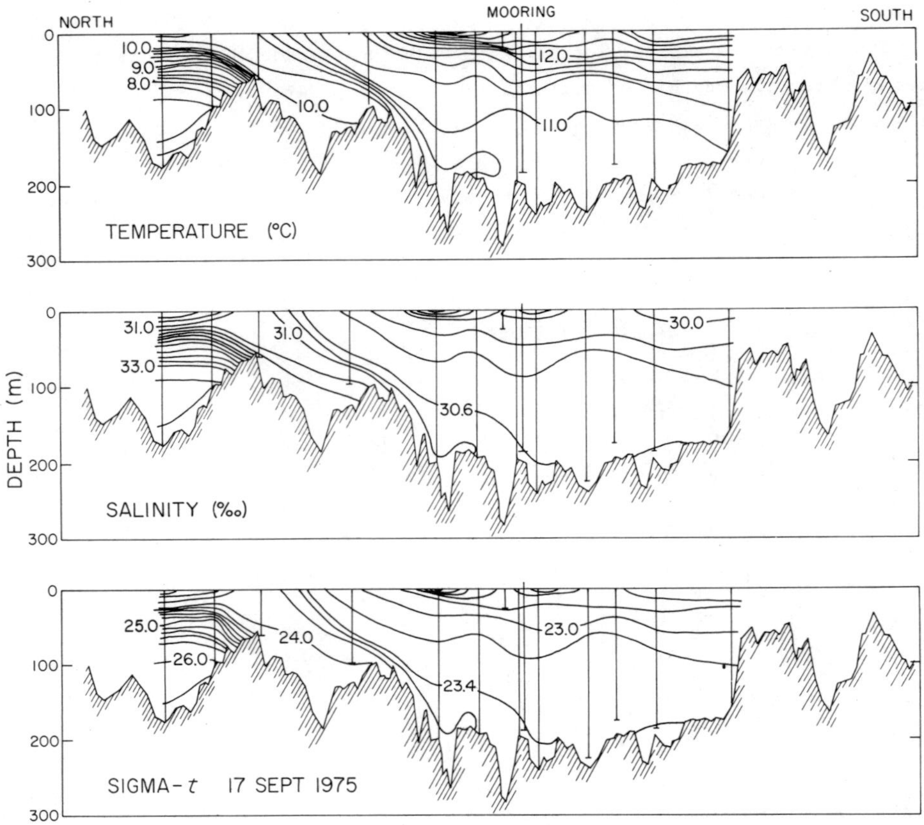

Fig. 7. Temperature, salinity, and sigma-t sections along Puget Sound in late summer 1975.

Salinity sections only show the progression through the remainder of the year (Fig. 9). The April section was immediately following an intrusion; July was during an intrusion; and September was well after an intrusion. Spring-summer conditions usually showed lower surface salinities near the south end of Whidbey Island (two stations north of the mooring) which were the result of the major Skagit River discharge into Puget Sound. There were some common features between the sections. Isopleths that were continuous at depth through the main basin and were more or less horizontal, sloped upward in Admiralty Inlet and intersected the surface. Isopleths from the Strait of Juan de Fuca which crossed the Admiralty Inlet sill sloped downward and usually intersected the bottom in the Inlet. These features were caused by the intense tidal mixing through the Inlet. Although the contours are horizontal, individual STD's indicated slight horizontal gradients in the deep water in the main basin. Saltier (denser) water than in the bottom of the main basin was always present at about sill depth just seaward of the Admiralty Inlet sill.

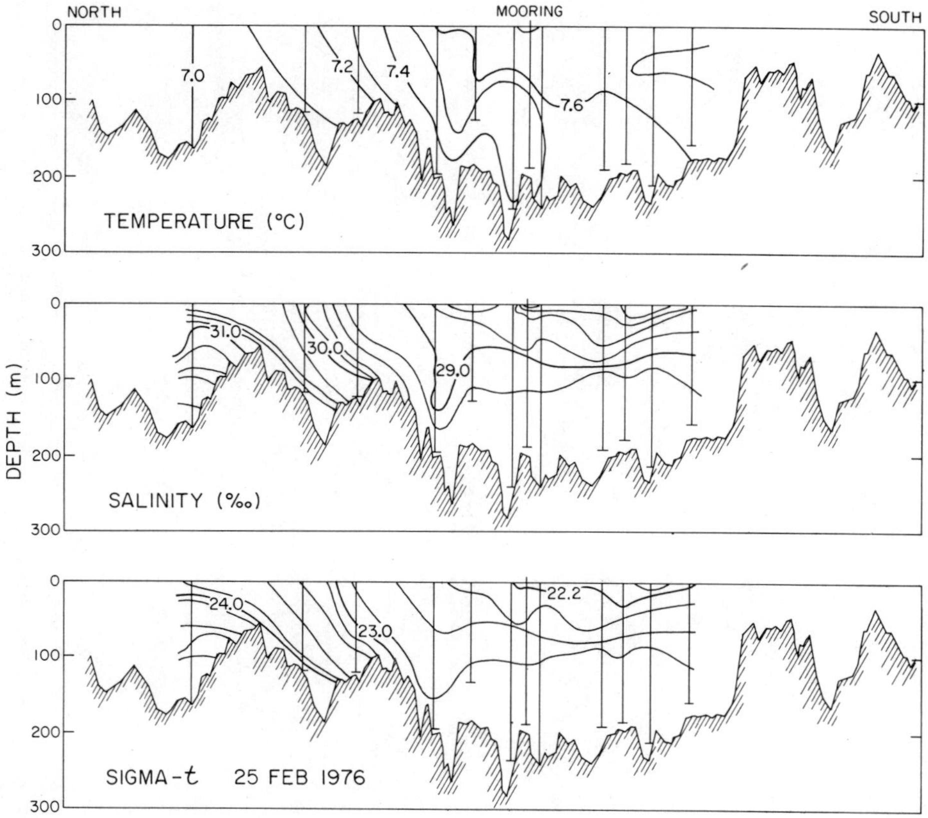

Fig. 8. Temperature, salinity, and sigma-t sections along Puget Sound in winter 1976.

FLOW CHARACTERISTICS

Inflow occurred at the bottom during each of the density intrusions (Fig. 3), and the details of the flow were similar throughout the year. This section describes some of the flow characteristics during the intrusions observed in the November-February deployment. The flow was tidal at all depths and was predominately semidiurnal (Fig. 10). Maximum currents exceeded 30 cm/sec at the surface and were about 20 cm/sec at the bottom. Tidal amplitudes were similar at all depths except right near the bottom which was slightly reduced and was probably the result of frictional influence. Phases also were approximately the same at all depths. Net seaward flow is obvious at the upper three levels, and net landward flow at 113 m. Onset of the largest salinity intrusion occurred midday on 5 February both at the bottom and at 113 m. Tidal signals were still evident in both flows, but the increased average landward flow prevented any reversals at the bottom. Vector plots of the tidally filtered records more clearly depicted the average flows, the influence of topography on the direction of flow, and the onset of the intrusion (Fig. 11).

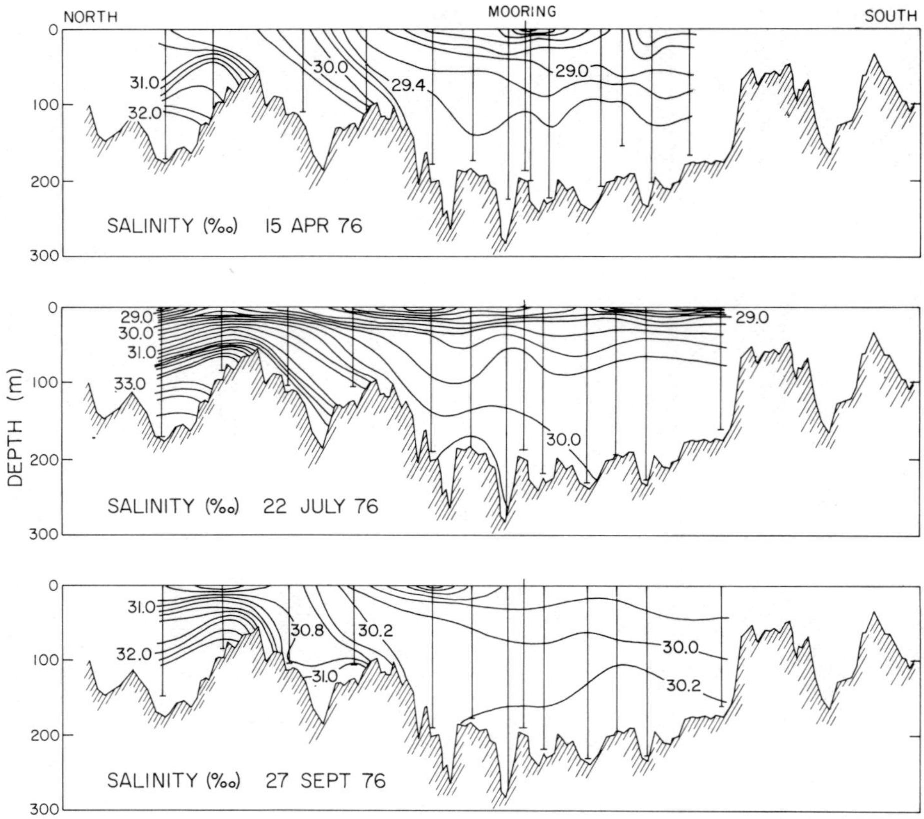

Fig. 9. Salinity sections along Puget Sound for spring, summer, and fall 1976.

The cyclic nature and magnitude of the three bottom inflows during this deployment can be represented using a progressive vector diagram (Fig. 12). If these vectors represented continuous flow along the Sound, instead of the vector summation of flow at a point, then the excursions would have been 35, 35, and 140 km, respectively. These were all relatively large, and the last one, coinciding with the largest salinity intrusion, was remarkably large considering the length of the Sound at these depths is only about 50 km. The implications also are large in that new bottom water could fill the Sound in less than 1 week.

Average daily along-channel currents, calculated as indicated above, are shown as vertical profiles for parts of December and February (Fig. 13). These particular examples are shown because they indicated inflow at the bottom corresponding with two major bottom density increases, one resulting primarily from temperature change and one by salinity change. The increase in December corresponded with the large temperature decrease, and inflow occurred from 26-30 December. The increase in February corresponded with the largest salinity increase which followed the minimum bottom salinity (and density) observed during the year, and inflow occurred from 5-14 February. Note the extraordinarily large average daily currents at the bottom in February. Speeds were about 20 cm/sec for 5 consecutive days. This would correspond to a horizontal excursion of 86 km, which is slightly longer than the main basin. Otherwise, the profiles are characteristic of estuarine flow with landward flow at depth and seaward flow in the upper water. There are large variations, however. The average depth of change from average landward to average seaward flow has been shown to occur in winter at this location at about 50 m and to have significant perturbations related to changing winds (Cannon and Laird, 1976). The average depth-of-change in flow direction is about 25% of the total depth, and this is intermediate between the shallower changeover for most fjords (10-15%) and the deeper changeover for partially mixed estuaries (40-50%).

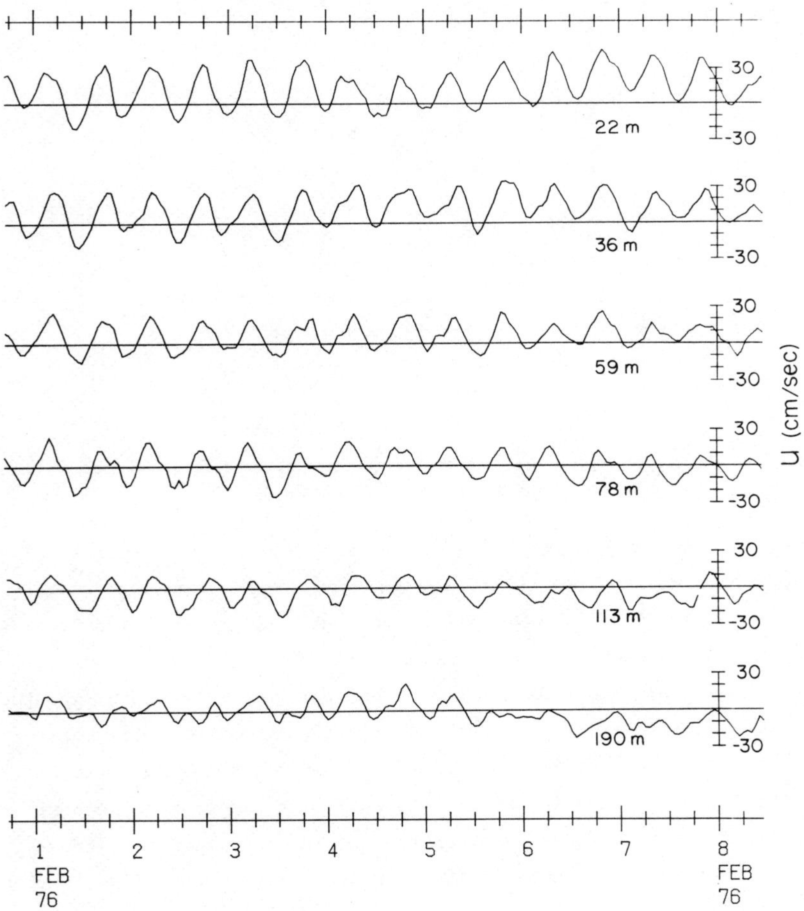

Fig. 10. Along-channel currents for part of the record at the mooring which have been 2.8-hr filtered to remove high-frequency noise. Positive currents are seaward.

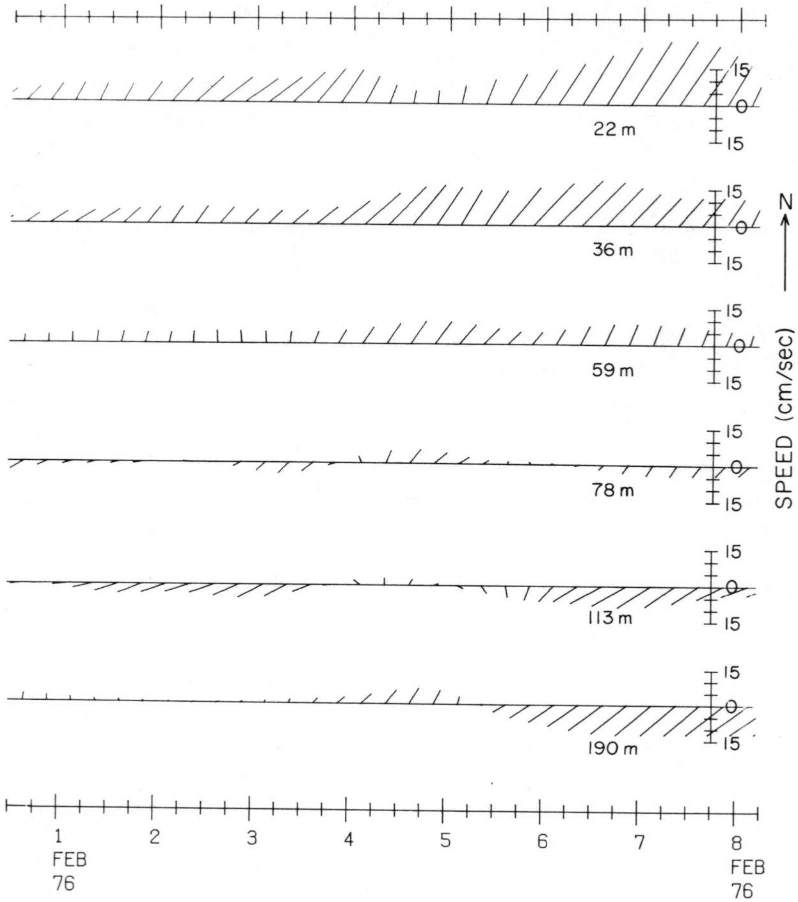

Fig. 11. Vector plots of currents at the mooring constructed from 35-hr filtered data and plotted every 6 hr for the same data as in Fig. 10.

Winds along the Sound for December-February were primarily to the north associated with winter storms, but there were a few reversals associated with cessation of storms (Fig. 14). The 3-month north-south average wind was 3.8 m/sec, while the fall and spring were only about half as large at 1.8 and 1.3 m/sec, respectively. Only in the summer 3 months was the average to the south, and then it was relatively small at 0.2 m/sec. Variance about the mean was comparable in fall, winter, and spring, but it was about half in summer. The wind reversal commencing at the end of January apparently was sufficient to change the horizontal pressure gradient and to significantly alter the current profile on 4 February, the peak of the wind reversal. These shifts were consistent with Rattray's (1967) theoretical study of fjords. Out-estuary winds should raise the inflowing layers to shallow depths and increase the magnitude of the maximum inflow. This was evident here around 23 December, 5 January, and 18 February. Relaxation of the out-estuary winds or reversal to in-estuary winds should result in deepening the outflowing layer and sometimes to development of a third layer flowing in at the surface. This was evident here following the wind reversals around 1 January and 4 February. Cannon and Laird (1976) have discussed other examples of these reversals.

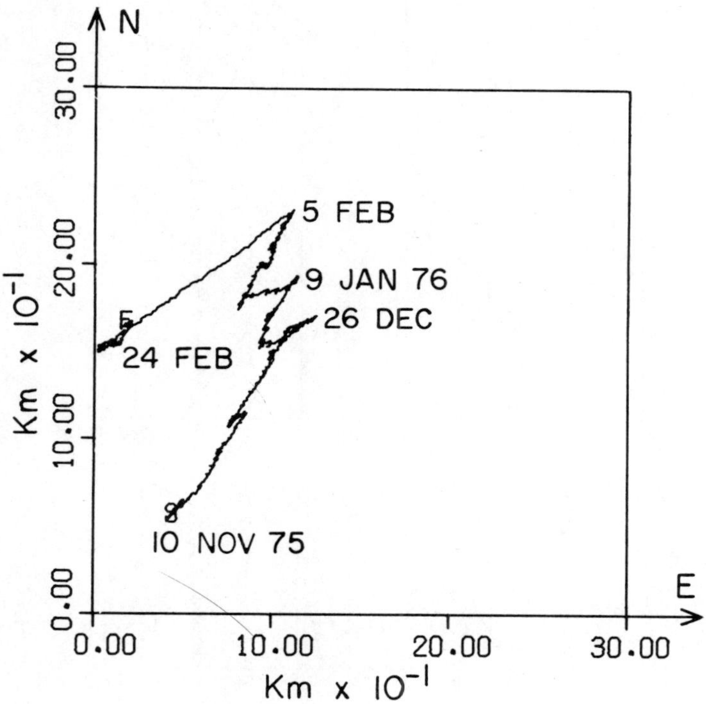

Fig. 12. Progressive vector diagram from the near-bottom current meter for the November 1975-February 1976 deployment.

A 25-hr time series of three STD stations across Puget Sound at the mooring showed that variations in temperature and salinity caused by tidal advection were small. Figure 15 shows the average temperature, salinity, and sigma-t in the section for 24 February and the average daily current profiles for 23 and 25 February. A level of no-net-motion is indicated at about 50 m. The isopycnals sloped down from west to east, and the sigma-t section showed the 22.6 g/ℓ isopycnal at about the bottom of the mixed layer and the depth of no-net-motion. If the isopycnals were adjusted for 22.6 g/ℓ to be flat, then the isopycnal slope would be roughly consistent with outflow in the surface layer and inflow in the deeper layer, perhaps in quasi-geostrophic balance. The temperature section showed a well-mixed layer at middepths (stippled) which extended down to about the depth of no-net-motion, which might be indicative of the water coming from farther south. In fact, it probably is water which has been mixed in The Narrows and is then refluxed into the main basin somewhat south of the mooring (Barnes and Ebbesmeyer, 1977).

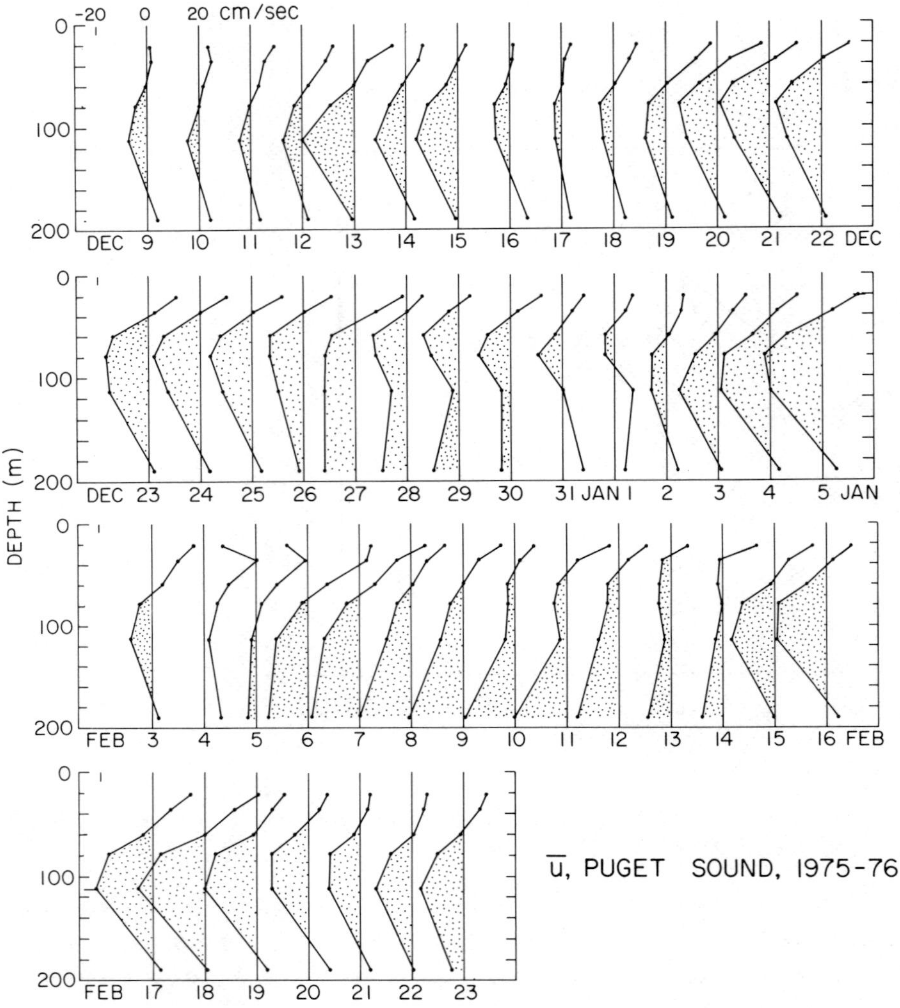

Fig. 13. Vertical profiles of daily average along-channel currents for parts of December 1975-February 1976. Stippled areas designate landward flow. Vertical lines designate zero flow for each profile, and adjacent vertical lines designate ± 20 cm/sec.

REPLACEMENT PROCESS

The replacement of deep water in Puget Sound can be summarized utilizing Figure 16 showing deep-water density, along-channel flow, and the flood-tide range predicted for Seattle. As in our earlier work at the same location (Cannon and Ebbesmeyer, 1977), we observed the winter intrusions (December-February) to occur 8-10 days following periods when the flood-tide range exceeded 3.5 m. This time lag indicated the time required for the intrusions to flow with the tidal currents from the entrance sill to the mooring; the tidal range predicted when the flooding tides would be greatest in Admiralty Inlet over the sill. Similar time lags have been observed for the approximate fortnightly renewal during summer-fall for the Port Susan estuary, a secondary basin within Puget Sound with about a 100-m sill depth (Fig. 1). About a week was required for water to flow from Admiralty Inlet sill to the sill of Port Susan (Cannon, 1975).

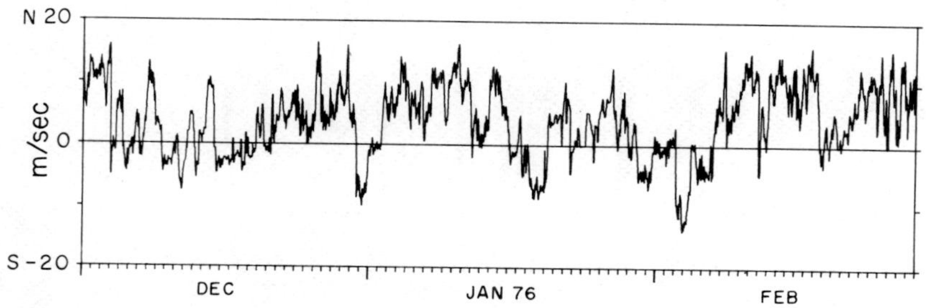

Fig. 14. Hourly average winds measured at West Point for December 1975-February 1976.

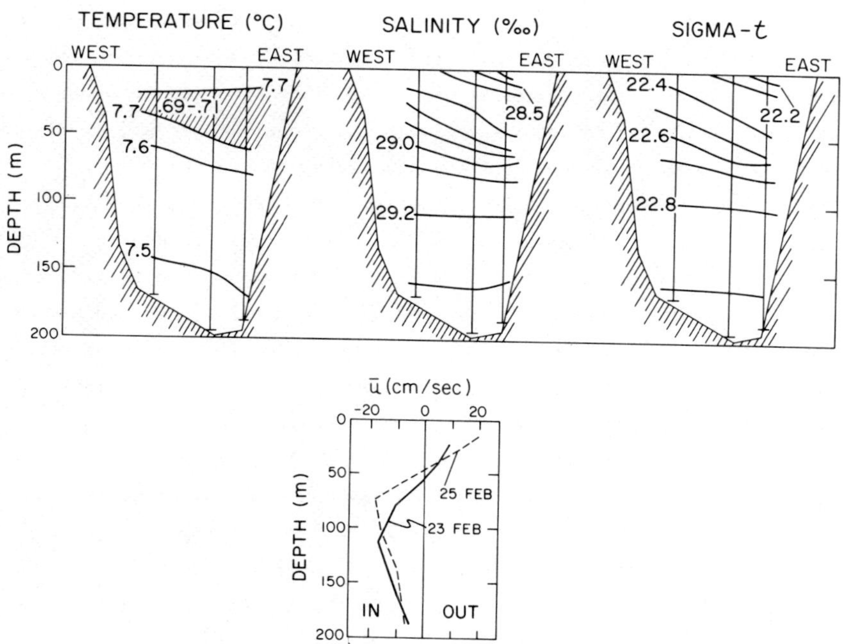

Fig. 15. Daily average water properties in a section across Puget Sound at the mooring for 24 February, and daily average along-channel currents for 23 and 25 February.

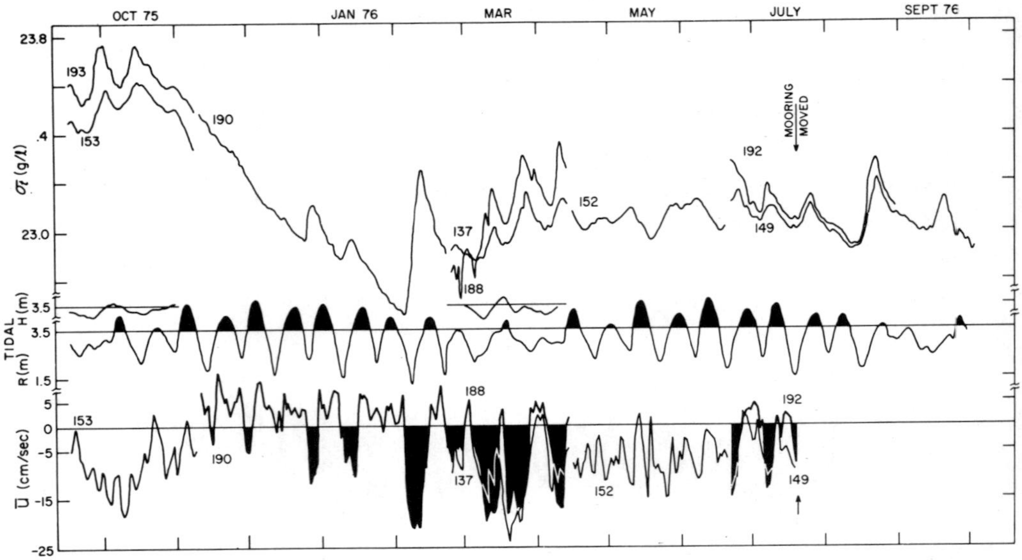

Fig. 16. Daily average sigma-t at 5 and 50 m above the bottom in Puget Sound; range (R) and height (H) of greatest daily Seattle flood tide; and daily average along-channel currents at 5 and 50 m above the bottom for 1975-76.

There are major variations, however, from our winter concepts. First, there were occurrences of large tidal ranges during November-January with almost no or very small inflow and particularly with no apparent changes in water properties. Second, there were occurrences of inflow during low tidal ranges during August-October and during March-April. These latter two periods also had differences in that late summer-fall was when the most dense water could enter the Sound and that winter-spring was about when the least dense water was in the Sound.

Thus, other factors must be important in determining when there will be intrusions of deep water. One apparent factor in discussions concerning the problem (Barnes, personal communication) was to consider the degree of mixing which takes place when crossing the finite-length entrance sill. During summer-fall the water apparently is sufficiently salt (dense) that it does not matter that more than one tide is required to transport it across the sill. Also, there is minimum freshwater entering the Sound to be available for mixing over the sill. On the contrary, the lack of significant inflow during fall when there are large tides probably is influenced by having an abundance of freshwater available for mixing over the sill. The largest temperature step decrease in December occurred following a relative extreme cold spell coincident with the large tidal range. This cold spell also was characterized by clear weather and a lack of freshwater, as was the case with the 1973 data.

Finally, the two large intrusions in March occurred following relatively small tides. They represented excursions as large as the one in February. This case is least well understood. Apparently, the combination of greater density outside and minimum inside is sufficient to cause the intrusion, even though more than one tidal cycle would be required to cross the sill. This period of renewal was observed to be most important in other years and needs additional study. Our earlier work showed it to be the dominant time of renewal of oxygen in the secondary basin of Port Susan (Cannon, 1975).

The overall rate of replacement of intermediate and deep water was approximated for the 1973 data using the net landward transport and assuming negligible effects of entrainment and diffusion (Cannon and Ebbesmeyer, 1977). Computations were made using average daily currents, assumed uniform across channel. The landward transport was divided into the volume beneath the average depth of no-net-motion (about 35 km^3 beneath 52 m), yielding an effective replacement time of about 9 days. This value is applicable here for all the major intrusions.

There has been a variety of shorter duration studies of renewal of deep water in other fjord estuaries in western North America (Anderson and Devol, 1973; Muench and Heggie, 1977; Pickard, 1976; Stucchi and Farmer, 1976). None show the characteristics described here. Perhaps one difference is that the head of the main basin of Puget Sound is really a relatively narrow, shallow passage (The Narrows in Fig. 1) entering another basin. There are extremely high tidal currents through this passage, which appear to be capable of pumping deep water up from about twice the depth of the sill. Physically, the process would be explained by application of Bernoulli's principle, similar to that proposed by Stommel *et al.* (1973) for the Mediterranean deep water passing over its shallow sill into the Atlantic Ocean. Helseth *et al.* (1976) have shown this effect in Puget Sound using a hydraulic model. It is not yet known whether the process occurs continuously on all flood tides, or whether it occurs only on the largest, fortnightly tides. In either case, the effect of The Narrows is to greatly assist the movement of deep water through Puget Sound.

This information is of interest to city managers of wastewater discharges in the Puget Sound system. The renewal rates of 1-2 weeks indicated by these studies is significantly quicker than earlier estimates of 2-10 months based on water property observations. Additional field experiments are planned to provide better descriptions of the processes at the head of the main basin (The Narrows) and on the entrance sill (Admiralty Inlet).

SUMMARY

A year-long series of observations from moored instruments has provided significant new information concerning the replacement of bottom water below sill depth in Puget Sound, a fjord estuary. Deep water was shown to be replaced by intrusions at about fortnightly intervals (but not every fortnight) when flood-tide ranges exceeded 3.5 m at Seattle. At this range, Strait of Juan de Fuca water could completely transit the Admiralty Inlet entrance sill on one flood tide and thus be least mixed. An 8- to 10-day time lag was required for an intrusion to transit from the sill to the mooring.

Most dense water entered in late summer and early fall followed by an approximate steady decrease in density until early winter. Intrusions of colder and denser water were then observed lasting about 5 days each. The coldest was 1.6°C and followed flood tides concurrent with below-freezing air temperatures over Admiralty Inlet. However, following each intrusion, mixing resulted in the overall density of deep water continuing to decrease until February when the water column became least dense. Density then increased through a series of intrusions of higher salinity, the initial one being greater than 0.5 ‰. Some of the intrusions in late summer and in midwinter were not coincident with large flood-tide ranges, and these require additional study.

Mean daily near-bottom speeds at the mooring were up to 20 cm/sec during some of the intrusions. This speed implied a possible excursion exceeding the length of the Sound in 5 days. Computations of replacement time also were made using average daily currents, assumed uniform across the channel at the mooring. The landward transport was divided into the volume of water below the average depth of no-net-motion, yielding an effective replacement time of about 9 days. Thus, it seems possible that the deep water could move through Puget Sound in less than 2 weeks.

Water of greater density than the resident water in the Sound was always available in the Strait of Juan de Fuca at sill depth. Additional study is needed of the mixing processes across the finite-length sill in Admiralty Inlet to determine the conditions which will result in an intrusion when there are large flood tides. Further study also is required at The Narrows, the relatively shallow constriction at the head of the main basin which provides access to the southern basin. There is evidence that the large flood currents in The Narrows act as a hydraulic pump to remove deep water from the Puget Sound main basin.

ACKNOWLEDGMENTS

Supporting services by many in our Coastal Physics Group are gratefully acknowledged. Special thanks for the assistance in the field are due the crew and officers of the NOAA ship *McArthur*. One of us (GAC) is grateful to Professor Emeritus Clifford A. Barnes at the University of Washington for discussions concerning Puget Sound circulation. Financial support was provided by the Marine Ecosystems Analysis Program of the National Oceanic and Atmospheric Administration.

REFERENCES

Anderson, J. J. and Devol, A. H., 1973. Deep water renewal and Saanich Inlet, an intermittently anoxic basin. Estuarine and Coastal Marine Sci., 1:1-10.

Barnes, C. A. and Collias, E. E., 1958. Some consideration of oxygen utilization rates in Puget Sound. J. Marine Res., 17:68-80.

Barnes, C. A. and Ebbesmeyer, C. C., 1977. Some aspects of Puget Sound's circulation and water properties. In: B. Kjerfve (Editor), Transport Processes in Estuarine Environments. Proc. 7th Belle W. Baruch Symp., Univ. South Carolina Press, Columbia, South Carolina (in press).

Cannon, G. A., 1971. Statistical characteristics of velocity fluctuations at intermediate scales in a coastal plane estuary. J. Geophysical Res., 76:5852-5858.

Cannon, G. A., 1975. Observations of bottom-water flushing in a fjord-like estuary. Estuarine and Coastal Marine Sci., 3:95-102.

Cannon, G. A., and Ebbesmeyer, C. C., 1977. Winter replacement of bottom water in Puget Sound. In: B. Kjerfve (Editor), Transport Processes in Estuarine Environments. Proc. 7th Belle W. Baruch Symp., Univ. South Carolina Press, Columbia, South Carolina (in press).

Cannon, G. A. and Laird, N. P., 1976. Wind effects on tidally averaged current profiles in a fjord estuary. Trans. Am. Geophys. Union, 57:933 (Abstract).

Collias, E. E., McGary, N. and Barnes, C. A., 1974. Atlas of physical and chemical properties of Puget Sound and its approaches. Washington Sea Grant Publ., Univ. of Washington Press, Seattle, Washington, 235 pp.

Farmer, H. G. and Rattray, M., 1963. A model of the steady-state salinity distribution in Puget Sound. Technical Report No. 85, Dept. of Oceanography, Univ. of Washington, Seattle, Washington, 33 pp.

Helseth, J. M., Ebbesmeyer, C. C., Barnes, C. A. and Lincoln, J. H., 1976. Bathymetrically driven transport in a fjord: A simple demonstration using a physical model of Puget Sound. Trans. Am. Geophys. Union, 57:934 (Abstract).

Muench, R. C., and Heggie, D. T., 1977. Deep water exchange in Alaskan subarctic fjords. In: B. Kjerfve (Editor), Transport Processes in Estuarine Environments. Proc. 7th Belle W. Baruch Symp., Univ. South Carolina Press, Columbia, South Carolina (in press).

Pickard, G. L., 1975. Annual and longer term variations of deepwater properties in the coastal waters of southern British Columbia. J. Fish. Res. Bd. Canada, 32:1561-1587.

Rattray, M., 1967. Some aspects of the dynamics of circulation in fjords. In: G. H. Lauff (Editor), Estuaries. American Association Advancement Sci., Washington, D. C., pp. 52-62.

Stommel, H., Bryden, H. and Mangelsdorf, P., 1973. Does some of the Mediterranean outflow come from great depth? Basel, Birkhauser Verlag, Pageoph, 105:887-889.

Stucchi, D. and Farmer, D. M., 1976. Deep water exchange in Rupert-Holberg Inlet. Pacific Marine Science Report No. 76-10, Marine Sciences Directorate, Pacific Region, Victoria, B. C., Canada, 32 pp.

Winter, D. F., Banse, K. and Anderson, G. C., 1975. The dynamics of phytoplankton blooms in Puget Sound, a fjord in the northwestern United States. Marine Biology, 29:139-176.

SUBJECT INDEX

UNIVERSITY OF BRADFORD
WITHDRAWN AND DISCARDED
LIBRARY